Lecture Notes in Computer Science   9230

Commenced Publication in 1973
Founding and Former Series Editors:
Gerhard Goos, Juris Hartmanis, and Jan van Leeuwen

More information about this series at http://www.springer.com/series/7410

Kristin Lauter · Francisco Rodríguez-Henríquez (Eds.)

Progress in Cryptology – LATINCRYPT 2015

4th International Conference on Cryptology
and Information Security in Latin America
Guadalajara, Mexico, August 23–26, 2015
Proceedings

 Springer

Editors
Kristin Lauter
Microsoft Research
Redmond, WA
USA

Francisco Rodríguez-Henríquez
CINVESTAV-IPN
Mexico City, Distrito Federal
Mexico

ISSN 0302-9743   ISSN 1611-3349   (electronic)
Lecture Notes in Computer Science
ISBN 978-3-319-22173-1   ISBN 978-3-319-22174-8   (eBook)
DOI 10.1007/978-3-319-22174-8

Library of Congress Control Number: 2015944719

LNCS Sublibrary: SL4 – Security and Cryptology

Springer Cham Heidelberg New York Dordrecht London
© Springer International Publishing Switzerland 2015

Printed on acid-free paper

Springer International Publishing AG Switzerland is part of Springer Science+Business Media
(www.springer.com)

Preface

Latincrypt 2015, the 4th International Conference on Cryptology and Information Security in Latin America, took place August 23–26, 2015, in Guadalajara, Mexico. The main conference program was preceded by the Advanced School on Cryptology and Information Security in Latin America (ASCrypto) August 22–23, 2015. Latincrypt 2015 was organized by the Computer Science Department of CINVESTAV-IPN, in cooperation with The International Association for Cryptologic Research (IACR). The general chairs of the conference were Luis J. Dominguez Perez and Francisco Rodríguez-Henríquez.

The conference received 59 submissions, of which 10 were withdrawn under various circumstances, mainly at the authors' request. Each submission was assigned to at least three committee members. Submissions co-authored by members of the Program Committee were assigned to at least five committee members. The reviewing process was challenging owing to the large number of high-quality submissions, and we are deeply grateful to the committee members and external reviewers for their indefatigable work. Special thanks go out to the shepherds of this edition, who greatly helped us in making a better paper selection. Particularly, we would like to thank Paulo Barreto, Jérémie Detrey, Sorina Ionica, and Gregory Neven, for their outstanding reviewer activities.

After careful deliberation, the Program Committee, which was chaired by Kristin Lauter and Francisco Rodríguez-Henríquez, selected 20 submissions for presentation at the conference. In addition to these presentations, the program also included one session of talks by graduate students and four invited talks by Yuriy Bulygin and Andrew Furtak (Intel, USA), Jung Hee Cheon (Seoul National University, South Korea), Tal Rabin (Thomas J. Watson Research Center, USA), and Adi Shamir (Weizmann Institute, Israel).

The reviewing process was run using the WebSubRev software, written by Shai Halevi from IBM Research. We are grateful to him for releasing this software. Finally, we would like to thank our sponsors, namely, Intel Guadalajara, Microsoft Research, and Oracle for their financial support as well as all the people who contributed to the success of this conference. We are also indebted to the members of the Latincrypt Steering Committee and the general chairs for their diligent work and for making this conference possible. We would also like to thank Springer for accepting to publish the proceedings in the *Lecture Notes in Computer Science* series. It was a great honor to be Program Committee chairs for Latincrypt 2015 and we look forward to the next edition in the conference series.

August 2015   Kristin Lauter
                                                                                   Francisco Rodríguez-Henríquez

Latincrypt 2015

**The 4th International Conference
on Cryptology and Information Security in Latin America**

**Guadalajara, Mexico
August 23–26, 2015**

Organized by

Computer Science Department
CINVESTAV-IPN

In Cooperation with
The International Association for Cryptologic Research (IACR)

General Chairs

Luis J. Dominguez Perez   CIMAT, Mexico
Francisco Rodríguez-Henríquez   CINVESTAV-IPN, Mexico

Program Chairs

Kristin Lauter   Microsoft Research, USA
Francisco Rodríguez-Henríquez   CINVESTAV-IPN, Mexico

Steering Committee

Michel Abdalla   École Normale Supérieure, France
Paulo Barreto   Universidade de São Paulo, Brazil
Ricardo Dahab   Universidade Estadual de Campinas, Brazil
Alejandro Hevia   Universidad de Chile, Chile
Julio López   Universidade Estadual de Campinas, Brazil
Daniel Panario   Carleton University, Canada
Francisco Rodríguez-Henríquez   CINVESTAV-IPN, Mexico
Alfredo Viola   Universidad de la República, Uruguay

Local Organizing Committee

Nareli Cruz Cortés   CIC-IPN, Mexico
Luis J. Dominguez Perez   CIMAT, Mexico
Jorge E. González Díaz   Intel, Mexico

Program Committee

Additional Reviewers

Ali Akhavi
Elena Andreeva
Tomer Ashur
Christof Beierle
David Bernhard
Joppe Bos
Angelo De Caro
Wouter Castryck
Craig Costello
Thomas Dean
Leo Ducas
Keita Emura
Christian Franck
Felix Guenther
Hongjun Wu (Tao Huang)
Andreas Huelsing
Abhishek Jain
Antoine Joux

Philipp Jovanovic
Seny Kamara
Bhavana Kanukurthi
Koray Karabina
Thorsten Kranz
Stephan Krenn
Ravi Kumar
Gregor Leander
Hyung Tae Lee
Loick Lhote
Jean Martina
Ingo von Maurich
Filippo Melzani
Bart Mennink
Erdinc Ozturk
Louiza Papachristodoulou
Sylvain Pelissier
Geovandro Pereira

Luis J. Dominguez Perez
Srinivasan Raghuraman
Saeed Sadeghian
Amin Sakzad
Yunus Sarikaya
Tobias Schneider
Akshayaram Srinivasan
Pantelimon Stanica
Brett Stevens
Berk Sunar
Cihangir Tezcan
Qingju Wang
Keita Xagawa
Ye Zhang
Arash Afshar
Mehmet Sabir kiraz

Sponsoring Institutions

Centro de Investigación en Cómputo del Instituto Politécnico Nacional (CICIPN)
Centro de Investigación en Matemáticas (CIMAT)
Centro de Investigación y de Estudios Avanzados del Instituto Politécnico Nacional
    (CINVESTAV-IPN)
Consejo Nacional de Ciencia y Tecnología (Conacyt)
Intel Mexico
Microsoft Research
Oracle

Contents

Cryptographic Protocols

Efficient RKA-Secure KEM and IBE Schemes Against Invertible Functions

Eiichiro Fujisaki and Keita Xagawa[✉]

NTT Secure Platform Laboratories, Tokyo, Japan
{fujisaki.eiichiro,xagawa.keita}@lab.ntt.co.jp

Abstract. We propose efficient KEM and IBE schemes secure under the related-key attacks (RKAs) against almost all invertible related-key derivation (RKD) functions under the DBDH assumption. The class of RKD functions we consider is broader than the best known RKD function class: For example, the class contains polynomial functions of (bounded) polynomial degrees and the XOR functions simultaneously.

Keywords: Related-key attack security · Invertible related-key derivation functions

1 Introduction

1.1 Related-Key Secure Cryptography in the Read-Only Memory Model

This paper examines how to protect public-key cryptographic protocols efficiently from related-key attacks. Suppose Π is an IND-CCA secure public-key encryption (PKE) scheme, meaning that no adversary can correctly guess which of two messages is encrypted (with a significant probability more than one half) even if it is allowed to ask for decryption of ciphertext other than the challenge ciphertext. When the related-key attack (RKA) is mounted on the IND-CCA game, an adversary can change secret key sk to $\phi(sk)$ with oracle access to $D_{\phi(sk)}(\cdot)$. We can regard the RKA as a side-channel attack against a cryptographic hardware device on which an adversary physically provides noise in the stored secret and then learns some information from the input/output behavior of the device. Informally, a cryptographic protocol is Φ-RKA-secure if its (original) security level is not weakened when the RKA is mounted on the security game, where Φ is the class of the related-key derivation (RKD) functions that the adversary may choose when tampering with the secret, and a related-key derivation (RKD) function is an efficiently computable map defined over the secret key-space (of the target device). In this paper, we concentrate on the model where a stored secret (that the adversary may tamper with) is not overwritten (e.g., recorded in the read-only memory) by either an adversary or a device C. More precisely, an adversary can temporally provide noise in the secret and obtain the modified output of $C_{\phi(s)}(x)$, but s in the storage is unchanged. In addition, s is not overwritten by cryptographic device C. This setting is the same as many prior studies [3, 4, 6, 7, 18, 26, 34].

© Springer International Publishing Switzerland 2015
K. Lauter and F. Rodríguez-Henríquez (Eds.): LatinCrypt 2015, LNCS 9230, pp. 3–20, 2015.
DOI: 10.1007/978-3-319-22174-8_1

Table 1. Existing RKA-secure primitives in the read-only memory model. The table only considers the results achieving the strong, adaptive notions of security in [5]. The mark * means that [1,7] achieved a set of polynomials with non-zero linear coefficients. The mark + means that [26] achieved a set of polynomials with a similar restriction. The mark ✓ indicates our constructions.

Primitive	Linear	Affine	Polynomial	"Beyond Polynomial"
CCA-KEM	[34],[5] + [4]	[7], [21], [20]	[7]	✓,[19], [29]
IBE	[5] + [4]	[7]	[7]	✓,[19], [29]
Sig	[5] + [4]	[7]	[7]	✓,[19], [29]
CPA-SKE	[3], [5] + [4]	[18]	[18]	[19]
CCA-SKE	[5] + [4]	[7]	[7]*	[19]
PRF	[4]	[26], [1]	[26]+, [1]*	[19]

Prior work. To the best of our knowledge, Biham [8,9] and Knudsen [24] were the first to consider related-key attacks. They independently attacked symmetric-key encryption schemes, LOKI families and Lucifer, in the RKA scenario. In the context of public-key setting, Boneh, DeMillo, and Lipton [12] submitted the fault-injection attack against the RSA-CRT signature scheme, which can be considered a related-key attack. Bellare and Kohno [6] first treated the related-key attacks in the manner of provable security. A large body of research followed, e.g., [3,4,6,7,18,26,34]. In the literature, a stored secret is unchanged, as mentioned above. Another line of research exists for the model where the secret can be overwritten by device C, e.g., [2,14–16,22,27]. We concentrate on the read-only memory model where a stored secret is unchanged. In the read-only memory model, the broadest class of RKD functions to protect against attacks is $\Phi_{\mathsf{poly}(d)}$ [7] for IBE, (CCA-secure) PKE, and signature schemes, where $\Phi_{\mathsf{poly}(d)}$ is the class of polynomial functions of (at most) degree d (over field $\mathbb{F}_q$). We remark that degree d depends on the assumptions they adopt [7]. In the symmetric-key setting, the best class is $\Phi_{\mathsf{poly}(d)} \setminus \Phi_{\mathsf{const}}$ [1,7,18,26] for pseudo random functions (PRF) and pseudo random permutations (PRP), where Φ_{const} is the set of constant functions. (It is known that there exists no Φ_{const}-RKA-secure PRF or PRP in the definition of [6].) These schemes (in both public/secret-key settings) are proven secure based on slightly less standard assumptions such as d-dynamic assumptions: the d-extended Decision Bilinear Diffie-Hellman (d-eDBDH) assumption, d-DHI assumption, or d-multilinear Diffie-Hellman exponent assumption, where $d = \mathsf{poly}(\kappa)$.
   See the summary in Table 1.

1.2 Related-Key Derivation (RKD) Functions

Let κ denote the security parameter. A related-key deviation (RKD) function is an efficiently computable map on the secret key-space $\mathcal{S} \subseteq \{0,1\}^n$ (of the target device), where $n = \mathsf{poly}(\kappa)$. We write $\mathcal{F}[\mathcal{S}]$ to denote the set of all "efficiently computable" functions on $\mathcal{S}$. We will denote the identity map by $\mathfrak{id}$.

As we already indicate in Table 1, the main direction of research on RKA security is broadening classes of the RKD functions against which we can protect a system, for example, linear functions, affine functions, and polynomial functions.

Unfortunately, we know that in the read-only memory model there is a class of RKD functions such that there is no general method to protect device $\langle G, s \rangle$ against the RKD class. Let $\Phi_{\text{bit-fix}} = \{\phi_i \mid i = 1, \ldots, n\} \cup \{\text{id}\}$ be the RKD function class such that ϕ_i is the same as the identity map id except that it fixes the i-th bit of input to 0. A device $\langle G, s \rangle$ of a meaningful functionality is generally vulnerable against this RKD class if an adversary is allowed to have access to the device a-priori unbounded polynomial times. Consider that an adversary submits $(\text{id}, x), (\phi_1, x), \ldots, (\phi_n, x)$ for some (randomly chosen) x and checks whether $G_{\phi_i(s)}(x) = G_s(x)$ for every $i = 1, \ldots, n$. This attack appeared in [17] as the *testing-for-malfunctioning* attack, and we can expect that the adversary extracts the whole secret. Another class of RKD functions against which we cannot generally protect a system is $\Phi_{\text{bit-dependent}} = \{\phi_i^{a,b} \mid a \neq b \text{ and } i = 1, \ldots, n\} \cup \{\text{id}\}$, where $\phi_i^{a,b}$ maps s to constant a if the i-th bit of the secret s is 0, otherwise to constant b. When an adversary asks for query $(\phi_i^{a,b}, x)$ for $i = 1, \ldots, n$, if it obtains $G_a(x) = G_{\phi_i^{a,b}(s)}(x)$, then it means that $s_i = 0$, otherwise 1. In the end, the adversary will retrieve all bits of s.

Properties of RKD Function. For simplicity, we suppose that $\mathcal{S} = \{0, 1\}^n$, where $n = \text{poly}(\kappa)$. To analyze RKD functions, we define two classes of RKD functions for reals $\delta, \chi \in [0, 1]$ as follows:

- $\Phi_{\text{fp}}^\delta = \left\{\phi \in \mathcal{F}[\mathcal{S}] : \Pr_{x \leftarrow \mathcal{S}}[\phi(x) = x] = \delta\right\}$,
- $\Phi_{\text{pre}}^\chi = \left\{\phi \in \mathcal{F}[\mathcal{S}] : \max_{y \in \mathcal{S}} \Pr_{x \leftarrow \mathcal{S}}[\phi(x) = y] = \chi\right\}$.

Hereafter, we denote the fixed points of ϕ by S_{FP}. The former captures functions of which $\#S_{\text{FP}} = \delta \cdot \#\mathcal{S}$ and measures the distance from the identity function, since $\Phi_{\text{fp}}^1 = \{\text{id}\}$. The latter defines a class of functions whose min-entropy $H_\infty(\phi(x))$ is $-\lg \chi$ and measures the distance from the constant functions, since $\Phi_{\text{pre}}^1 = \Phi_{\text{const}}$. Using these two classes enables us to map the polynomial functions as in Fig. 1.

Impossible Classes. We note that $\Phi_{\text{fp}}^{1/\text{poly}(\kappa)}$ is a class of RKD functions on which the *testing-for-malfunctioning* attack works. For $i = 1, \ldots, n$ and $a \in \{0, 1\}^\ell$ with $\ell = c \log \kappa$, where c is a constant, we define $\phi_{i,a}(s)$ as s if $s_i \ldots s_{i+\ell-1} = a$ and $s \oplus 1^n$ otherwise. Such functions are included in $\Phi_{\text{fp}}^{1/\text{poly}(\kappa)}$, because the number of fixed points is $2^n/2^\ell = 2^n/\kappa^c$. An adversary can mount the testing-for-malfunctioning attack because it can determine $s_1 \ldots s_\ell$ by querying $\phi_{1,a}(s)$ for every $a \in \{0, 1\}^\ell$. Therefore, it is impossible to protect arbitrary $\langle G, s \rangle$ from this class of RKD functions (if an adversary may tamper with the secret a-priori unbounded polynomial times).

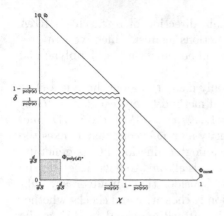

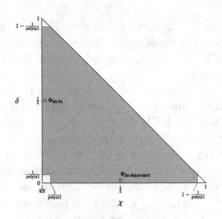

Fig. 1. Illustration of $\Phi_{\mathsf{poly}(d)}$ = $\Phi_{\mathsf{poly}(d)*} \cup \Phi_{\mathsf{const}} \cup \{\mathsf{id}\}$.

Fig. 2. Illustration of impossible RKD functions. Here, $\Phi_{\mathsf{bit\text{-}fix}}$ and $\Phi_{\mathsf{bit\text{-}dependent}}$ are part of the small purple rectangles at the left and the bottom, respectively. The green area indicates $\left(\Phi_{\mathsf{fp}}^{1/\mathrm{poly}(\kappa)} \cap \Phi_{\mathsf{fp}}^{1-1/\mathrm{poly}(\kappa)} \right) \cup \left(\Phi_{\mathsf{pre}}^{1/\mathrm{poly}(\kappa)} \cap \Phi_{\mathsf{pre}}^{1-1/\mathrm{poly}(\kappa)} \right)$.

Similarly, $\Phi_{\mathsf{fp}}^{1-1/\mathrm{poly}(\kappa)}$, $\Phi_{\mathsf{pre}}^{1/\mathrm{poly}(\kappa)}$, and $\Phi_{\mathsf{pre}}^{1-1/\mathrm{poly}(\kappa)}$ are also classes of RKD functions on which the *testing-for-malfunctioning attack* works. Taking the union of them, we can summarize the impossible class of RKD functions as the green area in Fig. 2. We remark that $\Phi_{\mathsf{bit\text{-}fix}} \subset \Phi_{\mathsf{fp}}^{1/2}$ and $\Phi_{\mathsf{bit\text{-}dependent}} \subset \Phi_{\mathsf{pre}}^{1/2}$.

Efficiently Invertible RKD Functions. We denote by Φ_{eInv} *the set of efficiently invertible functions* such that for every $\mathcal{S} \subseteq \{0,1\}^{\mathrm{poly}(\kappa)}$, every $\phi \in \Phi_{\mathsf{eInv}}$, and every $\hat{s} \in \mathcal{S}$, there is an efficient DPT algorithm that extracts *a description* of the set of *all* preimages of $\hat{s}$, denoted by $\phi^{-1}(\hat{s})$. For a technical reason, we additionally require that for every $\hat{s} \in \mathcal{S}$, $\#\phi^{-1}(\hat{s}) = \mathrm{poly}(\kappa)$ (i.e., the number of preimages of an element $\hat{s}$ is at most polynomial in κ), or that for some $\hat{s} \in \mathcal{S}$, $\#(\mathcal{S} \setminus \phi^{-1}(\hat{s})) = \mathrm{poly}(\kappa)$ (i.e., there is a unique $s_\phi \in \mathcal{S}$ such that the number of elements in $\mathcal{S} \setminus \phi^{-1}(s_\phi)$ is at most polynomial in κ). We define three subsets of Φ_{eInv} as follows:

- $\Phi_{\mathsf{A\text{-}const}} \subset \Phi_{\mathsf{eInv}}$ is said to be *a set of almost constant functions*, if for every $\phi \in \Phi_{\mathsf{A\text{-}const}}$, there is a unique $s_\phi \in \mathcal{S}$ such that $\#(\mathcal{S} \setminus \phi^{-1}(s_\phi)) = \mathrm{poly}(\kappa)$. We additionally require that there is an efficient algorithm that, given ϕ, outputs the unique point s_ϕ and the set $S_{\mathsf{NP}} := \mathcal{S} \setminus \phi^{-1}(s_\phi)$, whose cardinality is $\mathrm{poly}(\kappa)$.
- $\Phi_{\mathsf{RFP}} \subset \Phi_{\mathsf{eInv}}$ is said to be *a set of efficiently invertible functions with rare fixed-points*, if for every $\phi \in \Phi_{\mathsf{RFP}}$, $\#\phi^{-1}(\hat{s}) = \mathrm{poly}(\kappa)$ for all $\hat{s} \in \mathcal{S}$ and $\#S_{\mathsf{FP}} =$

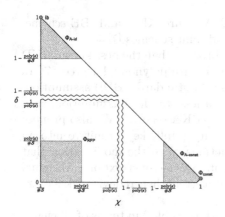 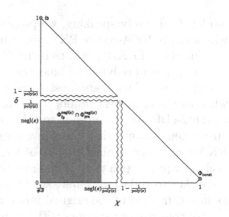

Fig. 3. Illustration of $\Phi_{\text{aInv}} = \Phi_{\text{RFP}} \cup \Phi_{\text{A-const}} \cup \Phi_{\text{A-id}}$. Φ_{RFP}, $\Phi_{\text{A-id}}$, and $\Phi_{\text{A-const}}$ consists of "efficiently invertible" RKD functions in the top-left, bottom-left, and bottom-right crosshatch areas, respectively.

Fig. 4. Illustration of $\Phi_{\text{h\&i}} = \left(\Phi_{\text{fp}}^{\text{negl}(\kappa)} \cap \Phi_{\text{pre}}^{\text{negl}(\kappa)} \right) \cup \Phi_{\text{const}} \cup \{\text{i}\partial\}$. $\Phi_{\text{h\&i}}$ consists of RKD functions covered by the blue square and circles.

$\text{poly}(\kappa)$ holds. We additionally require that there is an efficient algorithm that given ϕ only, outputs the set S_{FP}.

- $\Phi_{\text{A-id}} \subset \Phi_{\text{eInv}}$ is said to be *a set of almost identical functions*, if for every $\phi \in \Phi_{\text{A-id}}$, it holds that $\#(S \setminus S_{\text{FP}}) = \text{poly}(\kappa)$. We additionally require that there is an efficient algorithm that given ϕ only, outputs the set $S_{\text{NF}} := S \setminus S_{\text{FP}}$.

We then define $\Phi_{\text{aInv}} = \Phi_{\text{RFP}} \cup \Phi_{\text{A-id}} \cup \Phi_{\text{A-const}}$. See Fig. 3 for a summary of RKD functions.

Φ_{aInv} *contains* $\Phi_{\text{poly}(d)}$ *and more.* Suppose that S is a finite field. Then, it holds that $\Phi_{\text{poly}(d)} \subset \Phi_{\text{aInv}} = \Phi_{\text{RFP}} \cup \Phi_{\text{A-id}} \cup \Phi_{\text{A-const}}$ for every $d = \text{poly}(\kappa)$, because every $\phi \in \Phi_{\text{poly}(d)}$ is either an efficiently invertible function with at most d preimages for an element (see the textbooks, e.g., [31]) or a constant function. Actually, the class Φ_{aInv} is more convenient than $\Phi_{\text{poly}(d)}$. Let $S = \mathbb{F}_q$, where q is a power of an odd prime. Consider bit-flipping operations on S, regarding $s \in S$ as a string in $\{0,1\}^n$. Here letting $f_a(s) = s \oplus a$ for $a \in \{0,1\}^n$ ($a \neq 0^n$), we can represent it as a $\mathbb{F}_q$-coefficient univariate polynomial P_a, but the degree of P_a cannot be bounded by a polynomial in κ. (Indeed, we can present RKA with f_a against $\Phi_{\text{poly}(d)}$-RKA secure primitives. See the full paper.) On the other hand, it is obvious that f_a is efficiently invertible and has no fixed points; that is, $f_a \in \Phi_{\text{RFP}}$.

1.3 Our Results

We propose efficient RKA-secure public-key cryptographic primitives, against RKD function class $\Phi_{\text{aInv}} = \Phi_{\text{RFP}} \cup \Phi_{\text{A-id}} \cup \Phi_{\text{A-const}}$ in the read-only memory model. Concretely speaking, we proposed RKA-secure KEM and IBE scheme, which imply RKA-secure PKE, signatures, and joint schemes [7].

The class of RKD functions is significantly broader than the best known RKD function class of polynomial functions (on a field) with polynomial degrees [7]. In addition, our schemes are constructed based on the standard DBDH assumption, whereas the prior best results [7] are based on a less standard assumption, such as the d-eDBDH assumption, to achieve $\Phi_{\mathsf{poly}(d)}$-RKA security. We also propose an efficient system that can convert an arbitrary public-key primitive into an RKA-secure one against the set of RKD functions, with the modification that each user produces a tag τ of two group elements plus a ciphertext of the original secret-key to add to the public-key file and that a strengthened device takes (τ, x) as input, instead of the original input x.

We compare our schemes with the prior work in Table 2 in terms of efficiency and RKD function class.

CRS model. Our proposals are based on the common reference string model, where the common reference string is hard-wired into a device and an adversary cannot alter the CRS. We note that our setting is equivalent to prior works, i.e., [3,4,6,7,18,34].

1.4   Concurrent Related Works

Very recently, Jafargholi and Wichs [19] and Qin, Liu, Yuen, Deng, and Chen [29] reported new results on RKA security beyond polynomial functions.
Review of Qin et al. [29]. Let us review the class of the RKD functions, *high output entropy and input-output collision resistance (HOE&IOCR)*, defined in [29, Definition 1]. Adapting their definition to our notation, we can write the class $\Phi_{\mathsf{h\&i}}$ as $\Phi_{\mathsf{h\&i}} = \left(\Phi_{\mathsf{fp}}^{\mathsf{negl}(\kappa)} \cap \Phi_{\mathsf{pre}}^{\mathsf{negl}(\kappa)} \right) \cup \Phi_{\mathsf{const}} \cup \{\mathsf{id}\}.$ and can illustrate it as in Fig. 4.

Qin et al. proposed a conversion from an IND-FID-CPA-secure IBE scheme to IND-FID-CP-RKA secure IBE scheme with respect to $\Phi_{\mathsf{h\&i}}$ by using *continuous non-malleable* key-derivation functions (KDFs), which we call the QLY+15 conversion.

The QLY+15 conversion is summarized generally as follows: To set up the system, we run a sampling algorithm of a KDF scheme and obtain a secret $s \in \mathcal{S}_1$ and a corresponding public information $\tau \in \mathcal{S}_2$. We obtain a randomness σ by applying a KDF to s and τ and generate a key pair (MPK, MSK) by running the master-key-generation algorithm with the randomness σ. The new master public key is (MPK, τ), and the new master secret key is $(s, \tau) \in \mathcal{S}_1 \times \mathcal{S}_2$. The new extraction algorithm first reconstructs σ from its master secret key (s, τ), generates MSK by running the master-key-generation algorithm, runs the original extraction algorithm, and outputs a user secret key. The encryption and decryption algorithm is the same as the original.

We note that, strictly speaking, their construction achieves $\Phi_{\mathsf{QLY+15}}$-RKA security with respect to $\Phi_{\mathsf{h\&i}}[\mathcal{S}_1] \times \mathcal{F}[\mathcal{S}_2]$ instead of $\Phi_{\mathsf{h\&i}}[\mathcal{S}_1 \times \mathcal{S}_2]$; that is, they adopt the split-state model. In addition, we can give an RKA attack with respect

Table 2. Comparison of RKA-secure KEM and IBE schemes that are based on DBDH problems over the symmetric pairing setting. We apply two conversions in [19] and [29] to the basic KEM scheme in Sect. 4 and the Waters IBE scheme [32]. In [29], PP_{KDF} denotes the public parameters of a CNM-KDF scheme and $\mathcal{S}_1 \times \mathcal{S}_2$ is an output space of a sampler of the CNM-KDF scheme, a secret randomness and a corresponding public information. In [19], $\mathcal{S} = \{0,1\}^{|q|+v_1+v_2}$ and $h_1 : \{0,1\}^{v_1} \to \{0,1\}^{|q|}$ and $h_2 : \{0,1\}^{|q|+v_1} \to \{0,1\}^{v_2}$ is t-wise independent hash functions chosen randomly, where we can set $v_1 + v_2 = O(\kappa) + \log\log \#\Phi$ and $t = O(\log \#\Phi)$. In our schemes, Φ_{alnv} denotes $\Phi_{\text{A-id}} \cup \Phi_{\text{RFP}} \cup \Phi_{\text{A-const}} \subseteq \mathcal{F}[\mathbb{G}]$.

KEM schemes						
	class	pp	pk	sk	ct	assumption
[13]^a, [23]	−	$2|G|$	$2|G| + |G_T|$	$|G| + 2|q|$	$2|G|$	
[33]	−	$2|G|$	$2|G|$	$2|q|$	$2|G|$	
[34]^b	$\Phi^+[\mathbb{Z}_q]$	$3|G|$	$|G|$	$|q|$	$5|G| + 2|q|$	
[7]^a	$\Phi_{\text{poly}(d)}[\mathbb{Z}_q]$	$4|G|$	$|G|$	$|q|$	$2|G|$	$+d$-eDBDH
KEM$_B$ (Table 3)	−	$4|G|$	$|G|$	$|G|$	$2|G|$	
KEM$_B$+[29]	$\Phi_{\text{h\&i}}[\mathcal{S}_1] \times \mathcal{F}[\mathcal{S}_2]$	$4|G|+|\text{PP}_{\text{KDF}}|$	$|G|+|\mathcal{S}_2|$	$|\mathcal{S}_1|+|\mathcal{S}_2|$	$2|G|$	$+$CNM-KDF
KEM$_B$+[19]	$\Phi \subseteq \Phi_{\text{h\&i}}[\mathcal{S}]$	$4|G|+|h_1| + |h_2|$	$|G|$	$|\mathcal{S}|$	$2|G|$	
[Ours]	$\Phi_{\text{alnv}}[\mathbb{G}]$	$5|G|$	$2|G|$	$|G|$	$4|G|$	
IBE schemes						
	class	pp	mpk	msk	ct	assumption
[32]	−	$(n + 3)|G|$	$|G|$	$|q|$	$2|G|+|G_T|$	
[7]	$\Phi_{\text{poly}(d)}[\mathbb{Z}_q]$	$(n + 3)|G|$	$|G|$	$|q|$	$2|G|+|G_T|$	$+d$-eDBDH
[32]+[29]	$\Phi_{\text{h\&i}}[\mathcal{S}_1] \times \mathcal{F}[\mathcal{S}_2]$	$(n + 3)|G|+|\text{PP}_{\text{KDF}}|$	$|G|+|\mathcal{S}_2|$	$|\mathcal{S}_1|+|\mathcal{S}_2|$	$2|G|+|G_T|$	$+$CNM-KDF
[32]+[19]	$\Phi \subseteq \Phi_{\text{h\&i}}[\mathcal{S}]$	$(n + 3)|G|+|h_1|+|h_2|$	$|G|$	$|\mathcal{S}|$	$2|G|+|G_T|$	
[Ours]*	$\Phi_{\text{alnv}}[\mathbb{G}]$	$(n + 5)|G|$	$2|G|$	$|G|$	$2|G|+|G_T|$	

^{a}: We adopted the BMW05 and BPT12 KEM schemes into a symmetric paring setting. ^{b}: Wee proposed a PKE scheme. We translated it into KEM by adopting Wee's proof [34]. *: In our IBE scheme, an identity should be of the form (mpk, id).

to $\Phi_{\text{h\&i}}[\mathcal{S}_1 \times \mathcal{S}_2]$ against a concrete instantiation of their IBE scheme. For details, see the full paper.

Review of Jafargholi and Wichs [19]. Jafargholi and Wichs [19] define two properties of RKD functions, which are referred to as the numbered of fix points and the entropy of the outputs in [19, Definition 3.2] Jafargholi and Wichs considered the class $\Phi_{\text{JW15}} \subseteq (\Phi_{\text{fp}}^{\leq\delta} \cap \Phi_{\text{pre}}^{\leq 2^{-\mu}}) \cup \Phi_{\text{const}} \cup \{\text{id}\}$ in the RKA setting [19]

Jafargholi and Wichs gave a strengthening conversion of a cryptographic system in the context of non-malleable codes, which we call the JW15 conversion. Roughly speaking, the JW15 conversion with a code (E, D) converts an IBE scheme as follows; let MSK be an original master secret key. We convert it to a new master secret key $s = \mathsf{E}(\text{MSK})$. The new extraction algorithm first decodes $\text{MSK} = \mathsf{D}(s)$ and uses it to produce a user secret key.

We note that their parameter setting requires $\delta \leq 2^{-(\kappa+O(1))}$ and $\mu \geq \log\log \#\Phi + 2\kappa + O(1)$ assuming that $\#\Phi \geq 2^\kappa$. In addition, the JW15 conversion requires two t-wise independent hash functions h_1, h_2 as CRS with $t = \Omega(\log \#\Phi)$. Thus, for efficient construction, the number of RKD functions should be at most $2^{\text{poly}(\kappa)}$, and such a set of RKD functions is included with $\Phi_{\text{h\&i}}$.

Comparison among theirs and ours. For concreteness, we apply their conversions to the basic KEM scheme and the Waters IBE scheme. See Table 2 for a summary. We compare the classes of RKD functions, $\Phi_{\mathsf{alnv}} = \Phi_{\mathsf{RFP}} \cup \Phi_{\mathsf{A\text{-}id}} \cup \Phi_{\mathsf{A\text{-}const}}$, Φ_{JW15}, and $\Phi_{\mathsf{h\&i}}$. See Figs. 3 and 4. We first observe that $\Phi_{\mathsf{RFP}} \subseteq \Phi_{\mathsf{fp}}^{\mathsf{negl}(\kappa)} \cap \Phi_{\mathsf{pre}}^{\mathsf{negl}(\kappa)}$. Therefore, our class is narrower than $\Phi_{\mathsf{h\&i}}$ in the bottom-left area of RKD functions $\Phi_{\mathsf{fp}}^{1/\mathsf{poly}(\kappa)} \cap \Phi_{\mathsf{pre}}^{1/\mathsf{poly}(\kappa)}$. However, $\Phi_{\mathsf{h\&i}}$ only covers the identity function and constant functions, respectively, while our class covers broader areas by $\Phi_{\mathsf{A\text{-}id}}$ and $\Phi_{\mathsf{A\text{-}const}}$ in the top and right areas.

2   Preliminaries

Notations. A security parameter is denoted by κ. We use the standard O-notations, O, Ω, Θ, o, and ω. The abbreviations DPT and PPT stand for deterministic polynomial time and probabilistic polynomial time. A function $f(\kappa)$ is said to be negligible if $f(\kappa) = \kappa^{-\omega(1)}$. We denote a set of negligible functions by $\mathsf{negl}(\kappa)$. For a positive integer n, $[n]$ denotes $\{1, 2, \ldots, n\}$. For a finite set $\mathcal{S}$, we often write $x \leftarrow \mathcal{S}$, which indicates that we take a sample x from the uniform distribution over $\mathcal{S}$.

For two random variables X and Y, $\Delta(X, Y)$ denotes half of the total distance between X and Y. Two random variables X and Y are said to be statistically (and computationally) indistinguishable, denoted by $X \approx_s Y$ (and $X \approx_c Y$), if $\Delta(X, Y) = \mathsf{negl}(\kappa)$ (and for any PPT algorithm D, $\Delta(\mathsf{D}(1^\kappa, X), \mathsf{D}(1^\kappa, Y)) = \mathsf{negl}(\kappa)$), respectively.

Syntax and security definitions of basic cryptographic primitives are deferred to the full paper.

Assumption. Let GroupGen be a PPT algorithm that on inputting a security parameter 1^κ outputs a bilinear paring $\mathcal{G} = (\mathbb{G}, \mathbb{G}_T, e, q, g)$ such that; $\mathbb{G}$ and $\mathbb{G}_T$ are two cyclic groups of prime order q, g is a generator of $\mathbb{G}$, and a map $e : \mathbb{G} \times \mathbb{G} \to \mathbb{G}_T$ satisfies the following properties: $e(g^a, h^b) = e(g, h)^{ab}$ for any $g, h \in \mathbb{G}$ and any $a, b \in \mathbb{Z}_q$, $e(g, g)$ has order q in $\mathbb{G}_T$, and e is efficiently computable.

Definition 2.1 (DBDH Assumption). *We say the DBDH assumption holds for the bilinear group generator* GroupGen *if for any PPT adversary* A, *its advantage*

$$\left| \Pr\left[\mathsf{A}(\mathcal{G}, g^\alpha, g^\beta, g^\gamma, e(g, g)^{\alpha\beta\gamma}) = 1\right] - \Pr\left[\mathsf{A}(\mathcal{G}, g^\alpha, g^\beta, g^\gamma, e(g, g)^\delta) = 1\right] \right|$$

is negligible in κ, *where the probability is taken over the choices of* $\mathcal{G} = (\mathbb{G}, \mathbb{G}_T, q, e, g) \leftarrow$ GroupGen(1^κ), $\alpha, \beta, \gamma, \delta \leftarrow \mathbb{Z}_q$, *and the coin of* A.

3   RKA Security

We first define related-key derivation (RKD) functions and our class of RKD functions. We then review RKA security of key-encapsulation mechanisms.

3.1   RKD Functions

Let $\mathcal{S}$ denote the key space, which is defined as the public parameters of the specific scheme. We regard an RKD function $\phi \in \Phi$ simply as an *efficiently computable* function from $\mathcal{S}$ to $\mathcal{S}$, depending on the specification of the cryptographic protocol and the public parameters. We often write $\mathcal{F}[\mathcal{S}]$ to denote the set of all efficiently computable functions on $\mathcal{S}$. We denote an identity function on $\mathcal{S}$ by $\mathfrak{id}$. We often write $\Phi[\mathcal{S}]$ and $\Phi \subseteq \mathcal{F}[\mathcal{S}]$ when we want to stress the secret space $\mathcal{S}$.

We define the following function classes.

- Φ_{eInv} is *the class of efficiently invertible functions*; if given any $\kappa \in \mathbb{N}$, $\rho \in$ $\mathsf{Setup}(1^\kappa)$, $\phi \in \Phi_{\mathsf{eInv}}$ and $\hat{s} \in \mathcal{S}$, there is a DPT algorithm that computes the set of *all* the inverses $\phi^{-1}(\hat{s})$, denoted by S_{P}, where $\#S_{\mathsf{P}}$ is polynomial in κ and the subscript P indicates "preimages."
- $\Phi_{\mathsf{A\text{-}id}} \subset \Phi_{\mathsf{eInv}}$ is *the class of the almost identical functions*, which is the subclass of Φ_{eInv}; if given any $\kappa \in \mathbb{N}$, $\rho \in \mathsf{Setup}(1^\kappa)$, and $\phi \in \Phi_{\mathsf{A\text{-}id}}$, there is a DPT algorithm that computes S_{NF} such that $\#S_{\mathsf{NF}}$ is polynomial in κ and ϕ acts as an identity function over the restricted domain $\mathcal{S} \setminus S_{\mathsf{NF}}$; that is, $\phi(s) = s$ for any $s \in \mathcal{S} \setminus S_{\mathsf{NF}}$. The subscript NF indicates "not fixed."
- $\Phi_{\mathsf{RFP}} \subset \Phi_{\mathsf{eInv}}$ is *the class of invertible functions with rare fixed-points*, which is the subclass of Φ_{eInv}; if given any $\kappa \in \mathbb{N}$, $\rho \in \mathsf{Setup}(1^\kappa)$, and $\phi \in \Phi_{\mathsf{RFP}}$, there is a DPT algorithm that outputs the set of all fixed-points, S_{FP}, where $\#S_{\mathsf{FP}}$ is polynomial in κ and the subscript FP indicates "fixed points."
- $\Phi_{\mathsf{A\text{-}const}}$ is of *the class of the almost constant functions*; if given any $\kappa \in \mathbb{N}$, $\rho \in \mathsf{Setup}(1^\kappa)$, and $\phi \in \Phi_{\mathsf{A\text{-}const}}$, there is a DPT algorithm that computes unique $s_\phi \in \mathcal{S}$ and $S_{\mathsf{NP}} \subset \mathcal{S}$ such that $\#S_{\mathsf{NP}}$ is polynomial in κ and $\phi(s) = s_\phi$ for every $s \in \mathcal{S} \setminus S_{\mathsf{NP}}$. The subscript NP indicates "not preimage."

The function class $\Phi_{\mathsf{aInv}} = \Phi_{\mathsf{A\text{-}id}} \cup \Phi_{\mathsf{RFP}} \cup \Phi_{\mathsf{A\text{-}const}}$ is the class of RKD functions against which we achieve RKA-security with respect to a general circuit.

Remark 3.1. We will exploit the fact that any function ϕ in $\Phi_{\mathsf{A\text{-}const}}$ has few fixed-points, and such points can be computed efficiently; almost all points in $\mathcal{S}$ are mapped to a point s_ϕ by the definition of $\Phi_{\mathsf{A\text{-}const}}$. Therefore, at most $\#S_{\mathsf{NP}} + 1$ points can be fixed points of ϕ. Formally, we can compute a set of all fixed points S_{FP} as follows: (1) Compute $s_\phi \in \mathcal{S}$ and $S_{\mathsf{NP}} = \{s \in \mathcal{S} \mid \phi(s) \neq s_\phi\}$; (2) Compute $S^*_{\mathsf{NP}} = \{s \in S_{\mathsf{NP}} \mid \phi(s) = s\}$; (3) Check if $\phi(s_\phi) = s_\phi$ or not; if so, output $S_{\mathsf{FP}} \leftarrow S^*_{\mathsf{NP}} \cup \{s_\phi\}$; otherwise, output $S_{\mathsf{FP}} \leftarrow S^*_{\mathsf{NP}}$.

Remark 3.2. In the proofs, we require two classifying algorithms for a technical reason; one classifies a function $\phi \in \Phi$ into $\Phi_{\mathsf{A\text{-}id}}$ or $\Phi_{\mathsf{RFP}} \cup \Phi_{\mathsf{A\text{-}const}}$. It is easy to classify ϕ into $\Phi_{\mathsf{A\text{-}id}}$ or $\Phi_{\mathsf{RFP}} \cup \Phi_{\mathsf{A\text{-}const}}$ by checking the equalities of random inputs and their outputs many times. The other classifies a function ϕ into $\Phi_{\mathsf{RFP}} \cup \Phi_{\mathsf{A\text{-}id}}$ or $\Phi_{\mathsf{A\text{-}const}}$. Note that we can easily implement the algorithm by checking if the outputs of sufficiently many random inputs converge.

Examples. Suppose that $\mathcal{S}$ is a finite field and the additive and multiplicative operations on $\mathcal{S}$ are efficiently computable. Define $\Phi_{\text{const}} = \{\phi_c : s \mapsto c \mid c \in \mathcal{S}\}$, $\Phi_{\text{aff}} = \{\phi_{a,b} : s \mapsto as + b \mid a, b \in \mathcal{S}\}$, and $\Phi_{\text{poly}(d)} = \{\phi_f : s \mapsto f(s) \mid f(X) \in \mathcal{S}[X], \deg(f) \le d\}$ as classes of algebraic functions. We have that $\Phi_{\text{const}} \subset \Phi_{\text{A-const}}$, $(\Phi_{\text{aff}} \setminus \Phi_{\text{const}}) \subset \Phi_{\text{RFP}}$, and $(\Phi_{\text{poly}(d)} \setminus \Phi_{\text{const}}) \subset \Phi_{\text{RFP}}$. Therefore, the class $\Phi_{\text{A-id}} \cup \Phi_{\text{RFP}} \cup \Phi_{\text{A-const}}$ over a finite field $\mathcal{S}$ includes $\Phi_{\text{poly}(d)}$.

Moreover, the definition captures interesting and practical RKAs such as bit-flipping functions. Suppose that a secret is written in $\mathbb{F}_p^k$; e.g., a vector $(a_1, \ldots, a_k) \in \mathbb{F}_p^k$ or a point over the elliptic curve, say, $(x, y) \in E(\mathbb{F}_p) \subseteq \mathbb{F}_p^2$. In the real world, we have a (partially invertible) mapping M form $\mathbb{F}_p^k$ to a field $\mathbb{F}_Q$; e.g., it is very natural to map a vector or a point to an element in $\mathbb{F}_{p^k}$ or a "bit string" in $\mathbb{F}_{2^\kappa}$. By the invertibility of M, "polynomials" $\phi_P : s \mapsto (M \circ P \circ M^{-1})(s)$ for a $\mathbb{F}_Q$ coefficient polynomial $P \in \mathbb{F}_Q[X]$ is invertible with rare fixed points. When $Q = 2^\kappa$, "bit-wise xor" functions, $\phi_\delta : s \mapsto M^{-1}(M(s) \oplus \delta)$ for $\delta \in \mathbb{F}_Q$, are invertible with rare fixed points.

3.2   RKA Security of Key-Encapsulation Mechanism

Syntax. A KEM scheme consists of four algorithms: the setup algorithm Setup that, on input 1^κ, outputs public parameter pp; the key-generation algorithm Gen that, on input pp, outputs a key pair (ek, dk); the encapsulation algorithm Encaps that, on inputs pp and ek, outputs ciphertext ct and key $\xi \in \mathcal{K}$; and the decapsulation algorithm Decaps that, on inputs pp, dk, and ct, outputs key ξ or rejection symbol $\perp \notin \mathcal{K}$.

We say that the scheme is correct if for any pp and (ek, dk) generated by Setup and Gen, we have $\Pr[\xi' = \xi : (ct, \xi) \leftarrow \mathsf{Encaps}(pp, ek, ct); \xi' \leftarrow \mathsf{Decaps}(pp, dk, ct)] = 1$.

Security. Let us define IND-CC-RKA security (indistinguishability against chosen-ciphertext and related-key attacks) of KEM schemes. We follow the definition in [5].

Definition 3.1 (IND-CC-RKA Security). *Define experiment* $\mathsf{Expt}_{\mathsf{A},\mathsf{KEM},\Phi}^{\text{ind-cc-rka}}$ (κ, b) *between adversary* A *and the challenger as follows:*

Initialization: $pp \leftarrow \mathsf{Setup}(1^\kappa)$, $(ek, dk) \leftarrow \mathsf{Gen}(1^\kappa, pp)$, $ct^* \leftarrow \perp$. *Run the adversary with* (pp, ek).

Learning: The adversary can query the challenge oracle and the related-key decapsulation oracle.

   CHALLENGE: *This oracle is queried only once. It generates* $(ct^*, \xi_0) \leftarrow \mathsf{Encaps}(pp, ek)$. *It additionally generates* $\xi_1 \leftarrow \mathcal{K}$. *It returns* (ct^*, ξ_b).

   RK-DECAPSULE: *It receives* (ϕ, ct). *If* $\phi \notin \Phi$, *then return* $\perp$. $dk' \leftarrow \phi(dk)$. *If* $(dk', ct) = (dk, ct^*)$, *then return* $\perp$. *Else, return* $\xi \leftarrow \mathsf{Decaps}(pp, dk', ct)$.

Finalization: The adversary stops with output b'. *Output* b'.

We define the advantage of A *as*

$$\mathsf{Adv}^{\mathsf{ind\text{-}cc\text{-}rka}}_{\mathsf{A,KEM},\Phi}(\kappa) = \left| \Pr\left[\mathsf{Expt}^{\mathsf{ind\text{-}cc\text{-}rka}}_{\mathsf{A,KEM},\Phi}(\kappa,0) = 1\right] - \Pr\left[\mathsf{Expt}^{\mathsf{ind\text{-}cc\text{-}rka}}_{\mathsf{A,KEM},\Phi}(\kappa,1) = 1\right] \right|.$$

We say that the scheme KEM *is* indistinguishable against chosen-ciphertext and Φ-related key attacks *(*Φ-IND-CC-RKA secure *or* IND-CC-RKA secure with respect to Φ*) if for any PPT adversary* A, *its advantage is negligible in* κ.

If $\Phi = \{\mathsf{id}\}$, the definition is equivalent to that of IND-CCA2 security.

4   RKA-Secure KEM Scheme

We first review a DBDH-based KEM scheme and Bellare et al.'s KEM scheme [7]. Then, we explain our idea and construct an RKA-secure KEM scheme with respect to $\Phi_{\mathsf{alnv}}[\mathbb{G}]$.

4.1   Reviews

We quickly review the DBDH-based CCA-secure KEM scheme $\mathsf{KEM_B}$ in the left column of Table 3, which is a variant of the Boneh-Boyen IBE scheme [10] and similar to KEM schemes proposed in [13,23,33]. We call it the basic KEM scheme. The CCA security of $\mathsf{KEM_B}$ is proven by adapting the proofs in [10,13, 23,33].

Theorem 4.1. *The basic KEM scheme* $\mathsf{KEM_B}$ *is* IND-CCA *secure if the DBDH assumption holds and* $\mathcal{H}_\kappa$ *is a family of target-collision resistant hash functions.*

This basic KEM scheme is vulnerable to a simple RKA; let $ct = (u, \psi)$ be a target ciphertext. Define $\phi : s \mapsto s \cdot g$. Upon sending a query (ϕ, ct) to the RK-decapsulation oracle, we receive $\xi' = e(u, s \cdot g) = \xi \cdot e(u, g)$ and are able to compute $\xi = \xi'/e(u, g)$.

Review of Bellare et al.'s KEM Scheme. To avoid this attack, Bellare et al. employed key-fingerprinting [7]. Very roughly speaking, $x = g^\sigma$ is a key-fingerprinting of the secret σ. They modify the encapsulation and decapsulation algorithms to check the validity of ψ with respect to u and x. See the center column of Table 3 for the details of $\mathsf{KEM_{BPT12}}$. The previous simple RKA is prevented because we cannot compute a valid ciphertext (u, ψ') corresponding to the tampered secret key $\sigma + 1$; that is, it is hard to compute $\psi' = (yw^{\mathsf{H}(x \cdot g, u)})^\gamma$. By following the proof for IBE schemes in Bellare et al., we can show $\Phi_{\mathsf{poly}(d)}$-RKA security of KEM by Bellare et al. under the d-eDBDH assumption. (instead of showing $\Phi_{\mathsf{poly}(1)}$-RKA security of the KEM scheme under the DBDH assumption).

Theorem 4.2 (Adapted version of [7]). *The KEM scheme* $\mathsf{KEM_{BPT12}}$ *is* $\Phi_{\mathsf{poly}(d)}[\mathbb{Z}_q]$-IND-CC-RKA *secure if the d-eDBDH assumption holds and* $\mathcal{H}_\kappa$ *is a family of target-collision resistant hash functions.*

However, this KEM scheme is still vulnerable to related-key attacks employing bit-flipping functions, which lie outside $\Phi_{\mathsf{poly}(d)}[\mathbb{Z}_q]$. See the full paper for details of the attack.

Table 3. The KEM schemes based on the DBDH assumption in the symmetric pairing group setting. $\mathcal{G}$ denotes the description of the symmetric paring group, $(\mathbb{G}, \mathbb{G}_T, q, e, g)$. In all schemes, H is a target-collision resistant hash function chosen from $\mathcal{H}_\kappa = \{\mathsf{H} : \{0,1\}^* \to \mathbb{Z}_q\}$. In the proof of all schemes, we will be given $(g, g^\alpha, g^\beta, g^\gamma, T)$ as an instance of the DBDH problem and set $x = g^\sigma = g^\alpha$, $h = g^\beta$, $u = g^\gamma$, and $\xi = T$.

Scheme KEM$_\mathsf{B}$	Scheme KEM$_{\mathsf{BPT12}}$ [7]	Scheme KEM$_{\mathsf{Ours}}$
Algorithm Setup(1^κ):	Algorithm Setup(1^κ):	Algorithm Setup(1^κ):
$\mathcal{G} = (\mathbb{G}, \mathbb{G}_T, e, q, g)$	$\mathcal{G} = (\mathbb{G}, \mathbb{G}_T, e, q, g)$	$\mathcal{G} = (\mathbb{G}, \mathbb{G}_T, e, q, g)$
$\quad \leftarrow \mathsf{GroupG}(1^\kappa)$	$\quad \leftarrow \mathsf{GroupG}(1^\kappa)$	$\quad \leftarrow \mathsf{GroupG}(1^\kappa)$
$h, y, w \leftarrow \mathbb{G}$	$h, y, w \leftarrow \mathbb{G}$	$h, y, v, w \leftarrow \mathbb{G}$
$\mathsf{H} \leftarrow \mathcal{H}_\kappa$	$\mathsf{H} \leftarrow \mathcal{H}_\kappa$	$\mathsf{H} \leftarrow \mathcal{H}_\kappa$
Return $pp = (\mathcal{G}, h, y, w, \mathsf{H})$	Return $pp = (\mathcal{G}, h, y, w, \mathsf{H})$	Return $pp = (\mathcal{G}, h, y, v, w, \mathsf{H})$
Algorithm Gen(pp):	Algorithm Gen(pp):	Algorithm Gen(pp):
$\sigma \leftarrow \mathbb{Z}_q$	$\sigma \leftarrow \mathbb{Z}_q$	$\sigma \leftarrow \mathbb{Z}_q$
$(x, s) \leftarrow (g^\sigma, h^\sigma)$	$x \leftarrow g^\sigma$	$(x, \pi, s) \leftarrow (g^\sigma, (vw^{\mathsf{H}(x)})^\sigma, h^\sigma)$
Return $ek = x$ and $dk = s$	Return $ek = x$ and $dk = \sigma$	Return $ek = (x, \pi)$ and $dk = s$
Algorithm Enc(ek, m):	Algorithm Enc(ek, m):	Algorithm Enc(ek, m):
$\gamma \leftarrow \mathbb{Z}_q$	$\gamma \leftarrow \mathbb{Z}_q$	$\gamma \leftarrow \mathbb{Z}_q$
$(u, \psi) \leftarrow (g^\gamma, (yw^{\mathsf{H}(u)})^\gamma)$	$(u, \psi) \leftarrow (g^\gamma, (yw^{\mathsf{H}(x,u)})^\gamma)$	$(u, \psi) \leftarrow (g^\gamma, (yw^{\mathsf{H}(u)})^\gamma)$
$\xi \leftarrow e(x, h)^\gamma$	$\xi \leftarrow e(x, h)^\gamma$	$\xi \leftarrow e(x, h)^\gamma$
Return $ct = (u, \psi)$ and ξ	Return $ct = (u, \psi)$ and ξ	Return $ct = (x, \pi, u, \psi)$ and ξ
Algorithm Dec(dk, ct):	Algorithm Dec(dk, ct):	Algorithm Dec(dk, ct):
		If $e(x, h) \neq e(g, s)$, return $\bot$
		If $e(u, vw^{\mathsf{H}(x)}) \neq e(g, \pi)$,
		$\quad$ return $\bot$
If $e(u, yw^{\mathsf{H}(u)}) \neq e(g, \psi)$	If $e(u, yw^{\mathsf{H}(g^\sigma, u)}) \neq e(g, \psi)$,	If $e(u, yw^{\mathsf{H}(u)}) \neq e(g, \psi)$,
$\quad$ return $\bot$	$\quad$ return $\bot$	$\quad$ return $\bot$
$\xi \leftarrow e(u, s)$	$\xi \leftarrow e(u, h)^\sigma$	$\xi \leftarrow e(u, s)$
Return ξ	Return ξ	Return ξ

4.2 Our Idea

We briefly explain our idea to construct $\mathsf{KEM}_{\mathsf{Ours}} = (\overline{\mathsf{Setup}}, \overline{\mathsf{Gen}}, \overline{\mathsf{Encaps}}, \overline{\mathsf{Decaps}})$ upon KEM_B. In the RKA setting, an adversary tampers with a decapsulation key in a decapsulation circuit, $\overline{\mathsf{Decaps}}$, by querying a ciphertext $\overline{ct}$ and an RKD function $\phi \in \Phi$ (see Sect. 3.2). Our main strategy is to neutralize the RK-decapsulation oracle and simulate it using the ordinal decapsulation oracle and the trapdoor embedded in the public parameter $\overline{pp}$. Intuitively speaking, if we can force the adversary to submit a "proof" of a simulation-sound non-interactive proof of knowledge (SS-NIPoK) that it knows $\phi(dk)$ to the RK-decapsulation oracle, then the oracle can extract $\phi(dk)$ by using the trapdoor of SS-NIPoK and can return $\xi/\bot$ computed by $\overline{\mathsf{Decaps}}$ with the extracted secret $\phi(dk)$. In order to do so, we define

$$ek = (x, \pi), \ dk = s, \text{ and } \overline{ct} = (x, \pi, ct),$$

where π is a SS-NIPoK by which one can show the possession of $s = h^\sigma$ corresponding to $x = g^\sigma$. In our scheme, the decapsulation algorithm checks keys' validity by $e(x, h) = e(g, s)$ and by checking the validity of π.

We explain how our construction allows us to prove Φ-RKA security, where $\Phi_{\text{aInv}} = \Phi_{\text{RFP}} \cup \Phi_{\text{A-id}} \cup \Phi_{\text{A-const}}$. For brevity, we ignore the target ciphertext in the security definition. Let $\overline{ct} = (x, \pi, ct)$ and $\phi \in \Phi_{\text{aInv}}$ be a query from the adversary. Abstractly speaking, our simulator simulates the RK-decapsulation oracle as follows:

- $x = x^*$ and $\phi \in \Phi_{\text{A-id}}$: We expect that $\phi(s^*) = s^*$ because ϕ acts as an identity function for almost all elements in $\mathbb{G}$. If so, the RK-decapsulation oracle returns $\xi = \mathsf{Decaps}(pp, s^*, ct)$. The simulator returns it also by using its decapsulation oracle.
- $x = x^*$ and $\phi \in \Phi_{\text{RFP}} \cup \Phi_{\text{A-const}}$: We expect that $\phi(s^*) \neq s^*$. If so, the RK-decapsulation oracle returns $\perp$ since it checks if $e(g, \phi(s^*)) = e(x^*, h)$. Therefore, the simulator returns $\perp$.
- $x \neq x^*$ and $\phi \in \Phi_{\text{RFP}} \cup \Phi_{\text{A-id}}$: The simulator extracts $\hat{s}$ corresponding to x. We expect that $\phi(s^*) \neq \hat{s}$ by the property of ϕ. If so, the RK-decapsulation oracle returns $\perp$ since $e(g, \phi(s^*)) \neq e(g, \hat{s}) = e(x, h)$. Then, the simulator returns $\perp$.
- $x \neq x^*$ and $\phi \in \Phi_{\text{A-const}}$: In the case, the simulator extracts $\hat{s}$ corresponding to x. By the property of ϕ, all but negligible secrets are mapped to s_ϕ. We expect that either $\phi(s^*) = s_\phi \neq \hat{s}$ or $\phi(s^*) = s_\phi = \hat{s}$. In the former case, the simulator returns $\perp$. In the latter case, the simulator computes $\xi = \mathsf{Decaps}(pp, \hat{s}, ct)$ by using $\hat{s}$.

By using the above simulation, we can reduce the RKA security to the ordinal CCA security of the basic KEM scheme.

We note that our simulation technique is unlike the simulators in the previous papers on the RKA security: Most of the previous papers employed the framework of Bellare and Cash [4], which builds RKA-secure PRFs based on *key-homomorphic* PRFs and *fingerprinting of the secret key*. In the RKA-secure PKE/KEM context, Wee [34] and Bellare, Paterson, and Thomson [7] employed this framework; in their proofs, the simulator generates a message m under a tampered secret key $\phi(dk)$ by querying its decryption oracle $\mathsf{Decaps}(pp, dk, \cdot)$ on $ct' = T(pp, ek, ct, \phi)$, where T is a deterministic efficient function satisfying $\mathsf{Decaps}(pp, dk, ct') = \overline{\mathsf{Decaps}}(pp, \phi(dk), ct)$. Intuitively speaking, their schemes achieve RKA security, which is similar to non-malleability, by using malleable basic primitives. In contrast, our schemes directly employ non-malleable primitives, SS-NIPoK of the secret key $s = h^\sigma$.

Finally, we explain the details of SS-NIPoK. We set $\pi = (vw^{\mathsf{H}(x)})^\sigma$, and the verification checks whether $e(g, \pi) = e(x, vw^{\mathsf{H}(x)})$ or not. We note that (x, π) is of the form of a *ciphertext* of the basic KEM scheme, whose public parameter is (g, h, v, w) (instead of (g, h, y, z)). For given $x^* = g^{\sigma^*}$, we define the simulated CRS of NIPoK as $w = h^\theta$ and $v = w^{-\mathsf{H}(x^*)} \cdot g^\eta$, where $\theta, \eta \leftarrow \mathbb{Z}_q$. This construction has two well-known important properties:

- (simulation:) one can generate $\pi^* = (vw^{H(x^*)})^{\sigma^*} = (x^*)^\eta$ without knowing σ^* and
- (knowledge extractor:) for any (x, π) satisfying $H(x^*) \neq H(x)$ and $e(g, \pi) = e(x, vw^{H(x)})$, one can extract $s = (\pi \cdot x^{-\eta})^{1/(\theta(H(x) - H(x^*)))} = h^\sigma$, where $x = g^\sigma$, by using trapdoors η and θ.[1]

We will exploit these properties in the proof.

4.3   Construction

Below, we propose an optimized version of the KEM scheme described in the previous subsection. We set $z = w$ below in order to reduce the number of group elements, which results in a slightly complex proof.

Our proposal is in the right column of Table 3.

Remark 4.1. We stress that we did not change anything about the security model. We just attached an encapsulation key (x, π) to an original ciphertext (u, ψ).

Theorem 4.3 (Security of KEM$_{\mathsf{Ours}}$). *Let* $\Phi_{\mathsf{aInv}} = \Phi_{\mathsf{A\text{-}id}} \cup \Phi_{\mathsf{RFP}} \cup \Phi_{\mathsf{A\text{-}const}} \subseteq \mathcal{F}[\mathbb{G}]$. *The KEM scheme* KEM$_{\mathsf{Ours}}$ *is* Φ-IND-CC-RKA *secure if the DBDH assumption holds and* H *is target collision-resistant.*

We defer the formal proof to the full paper, and here, we give a proof sketch.

Proof sketch. We define 14 games $\mathsf{Game}_i(b)$ with $i \in \{0, \ldots, 6\}$ and $b \in \{0, 1\}$ and show that the differences between them are negligible under our assumptions.

$\mathsf{Game}_0(b)$: This is the original game $\mathsf{Expt}^{\mathsf{ind\text{-}cc\text{-}rka}}_{\mathsf{A, KEM_{Ours}, \Phi_{aInv}}}(\kappa, b)$. We denote the target master public key by $ek^* = (x^*, \pi^*) = (g^{\sigma^*}(vw^{H(x^*)})^{\sigma^*})$ and the target ciphertext by $ct^* = (x^*, \pi^*, u^*, \psi^*) = (x^*, \pi^*, g^{\gamma^*}(yw^{H(u^*)})^{\gamma^*})$. We will show that the adversary's advantage $\left| \Pr[\mathsf{Game}_0(0) = 1] - \Pr[\mathsf{Game}_0(1) = 1] \right|$ is negligible.

$\mathsf{Game}_1(b)$: We change the order of the generation of the target ciphertext. Since this is just a conceptual change, this introduces only negligible differences.

$\mathsf{Game}_2(b)$: This game is the same as $\mathsf{Game}_1(b)$ except that the RK-extract oracle returns $\bot$ once it finds the collision of H. Correctly speaking, the oracle returns $\bot$ on i-th query $ct_i = (x_i, \pi_i, u_i, \psi_i)$ if $(H(x_i) = H(x^*) \wedge x_i \neq x^*)$ or $(H(u_i) = H(u^*) \wedge u_i \neq u^*)$. It is obvious that if H is target collision-resistant, then $\mathsf{Game}_2(b)$ is computationally indistinguishable from $\mathsf{Game}_1(b)$.

$\mathsf{Game}_3(b)$: This game is the same as $\mathsf{Game}_2(b)$ except that the challenger generates v, w, y, π^*, and ψ^* after generating $x^* = g^{\sigma^*}$ and $u^* = g^{\gamma^*}$ as in [10]. Formally, the challenger sets $w = h^\theta$, $v = w^{-H(x^*)}g^\eta$, $y = w^{H(u^*)}g^\zeta$, $\pi^* = (x^*)^\eta = (vw^{H(x^*)})^{\sigma^*}$, $\psi^* = (u^*)^\zeta = (yw^{H(u^*)})^{\gamma^*}$, where $\theta, \eta, \zeta \leftarrow \mathbb{Z}_q$. This modification changes nothing as in the proof in Boneh and Boyen [10] and, thus, $\mathsf{Game}_2(b)$ is equivalent to $\mathsf{Game}_3(b)$.

---

[1] To the best of our knowledge, this knowledge extractor is implicitly due to Kiltz [23, Sect. 5.2] and is formalized as an all-but-one extractable hash proof for the Diffie-Hellman relation by Wee [33, Sect. 5.1].

$\mathsf{Game}_4(b)$: We then modify the RK-decapsulation oracle to modify the generation of session keys in the case $x = x^*$ and reverse the check in the case $x \neq x^*$. The oracle first rejects the query if $e(g, \pi) \neq e(x, vw^{\mathsf{H}(x)})$ or $e(g, \psi) \neq e(u, yw^{\mathsf{H}(u)})$.

- If $(x, u) = (x^*, u^*)$, then the oracle rejects the query because the query is either prohibited or invalid.
- If $x = x^*$ but $u \neq u^*$, then the oracle extracts $\hat{a}$ satisfying $e(u, h) = e(g, \hat{a})$ from ψ by using the trapdoors ζ and θ. The oracle rejects the query if $e(g, \phi(s^*)) \neq e(x^*, h)$. Otherwise, the oracle uses $\hat{a}$ to produce $\xi = e(u, s^*) = e(x^*, \hat{a})$.
- If $x \neq x^*$, then the oracle extracts $\hat{s}$ satisfying $e(g, \hat{s}) = e(x, h)$ from π by using the trapdoors η and θ. The oracle checks whether $\phi(s^*) = \hat{s}$ or not. (In $\mathsf{Game}_3(b)$, the oracle checks whether $e(g, \phi(s^*)) = e(x, h)$ or not.) If so, the oracle returns $\xi = e(u, \hat{s}) = e(u, \phi(s^*))$.

Because of the property of the pairing groups, $\mathsf{Game}_4(b)$ is equivalent to $\mathsf{Game}_3(b)$.

$\mathsf{Game}_5(b)$: We again modify the RK-decapsulation oracle to forget s^*. In order to forget s^*, we change how the challenger computes $s' = \phi(s^*)$. The challenger now computes s' from ϕ and $ct = (x, \pi, u, \psi)$ by exploiting the properties of ϕ and the pairing. (For the details, see the full paper. This part is the core of our proof.) We show that $\mathsf{Game}_5(b)$ is equal to $\mathsf{Game}_4(b)$.

$\mathsf{Game}_6(b)$: Finally, we simplify the RK-extract oracle. The oracle is defined as follows: The oracle rejects the query if it finds the collision of H. If $e(g, \pi) \neq e(x, vw^{\mathsf{H}(x)})$ or $e(g, \psi) \neq e(u, yw^{\mathsf{H}(u)})$, then return $\perp$.

- If $(x, u) = (x^*, u^*)$, then return $\perp$.
- If $x = x^*$, $u \neq u^*$, and $\phi \in \Phi_{\mathsf{A-id}}$: Compute $\hat{a}$ from (u, ψ) by using the trapdoors ζ and θ. Return $\xi = e(x^*, \hat{a})$.
- If $x = x^*$, $u \neq u^*$, and $\phi \in \Phi_{\mathsf{RFP}} \cup \Phi_{\mathsf{A-const}}$: Return $\perp$.
- If $x \neq x^*$ and $\phi \in \Phi_{\mathsf{RFP}} \cup \Phi_{\mathsf{A-id}}$: Return $\perp$.
- If $x \neq x^*$ and $\phi \in \Phi_{\mathsf{A-const}}$: Compute s_ϕ. Extract $\hat{s}$ from (x, π) by using the trapdoors η and θ. If $\hat{s} \neq s_\phi$, then return $\perp$. Otherwise, return $\xi = e(u, \hat{s})$.

We show that if the adversary distinguishes $\mathsf{Game}_6(b)$ from $\mathsf{Game}_5(b)$, then we can solve the CDH problem. We finally show that $\mathsf{Game}_6(0)$ is indistinguishable from $\mathsf{Game}_6(1)$ if the DBDH assumption holds.

5   RKA-Secure IBE Scheme

Adapting our technique to the Waters IBE scheme [32], we can obtain Φ_{aInv}-IND-FID-CP-RKA secure IBE scheme. We defer the details to the full version of this paper.

Bellare et al. [5] showed that a Φ-IND-FID-CP-RKA secure IBE scheme can be converted into a Φ-EUF-CM-RKA secure signature scheme (by the Naor transformation) and a Φ-IND-CC-RKA secure PKE scheme (by the CHK or BK conversion [11]). Bellare et al. [7] showed that the conversion from an IND-FID-CPA-secure IBE scheme and a strongly-secure one-time signature scheme to a combined signature and encryption scheme [28] can be translated in the RKA setting. Therefore, a combined scheme obtained from a Φ-IND-FID-CP-RKA secure IBE scheme and a strongly secure one-time signature scheme enjoys joint security.

Because we constructed a Φ_{alnv}-IND-FID-CP-RKA secure IBE scheme with $\Phi_{\mathsf{alnv}} = \Phi_{\mathsf{RFP}} \cup \Phi_{\mathsf{A\text{-}const}} \cup \Phi_{\mathsf{A\text{-}id}}$, we can convert it to a Φ_{alnv}-EUF-CM-RKA secure signature scheme, a Φ_{alnv}-IND-CC-RKA secure PKE scheme, and a Φ_{alnv}-RKA secure combined signature and encryption scheme.

References

1. Abdalla, M., Benhamouda, F., Passelègue, A., Paterson, K.G.: Related-key security for pseudorandom functions beyond the linear barrier. In: Garay, J.A., Gennaro, R. (eds.) CRYPTO 2014, Part I. LNCS, vol. 8616, pp. 77–94. Springer, Heidelberg (2014). https://eprint.iacr.org/2014/488
2. Aggarwal, D., Dodis, Y., Lovett, S.: Non-malleable codes from additive combinatorics. In: Shmoys, D.B. (ed.) STOC 2013, pp. 774–783. ACM (2014). https://eprint.iacr.org/2013/201
3. Applebaum, B., Harnik, D., Ishai, Y.: Semantic security under related-key attacks and applications. In: Chazelle, B. (ed.) ICS 2011, pp. 45–60. Tsinghua University Press (2011). https://eprint.iacr.org/2010/544
4. Bellare, M., Cash, D.: Pseudorandom functions and permutations provably secure against related-key attacks. In: Rabin [30], pp. 666–684. https://eprint.iacr.org/2010/397
5. Bellare, M., Cash, D., Miller, R.: Cryptography secure against related-key attacks and tampering. In: Lee and Wang [25], pp. 486–503. https://eprint.iacr.org/2011/252
6. Bellare, M., Kohno, T.: A theoretical treatment of related-key attacks: RKA-PRPs, RKA-PRFs, and applications. In: Biham, E. (ed.) EUROCRYPT 2003. LNCS, vol. 2656, pp. 491–506. Springer, Heidelberg (2003)
7. Bellare, M., Paterson, K.G., Thomson, S.: RKA Security beyond the linear barrier: IBE, encryption and signatures. In: Wang, X., Sako, K. (eds.) ASIACRYPT 2012. LNCS, vol. 7658, pp. 331–348. Springer, Heidelberg (2012). https://eprint.iacr.org/2012/514
8. Biham, E.: New types of cryptanalytic attacks using related keys. In: Helleseth, T. (ed.) EUROCRYPT 1993. LNCS, vol. 765, pp. 398–409. Springer, Heidelberg (1994)
9. Biham, E.: New types of cryptanalytic attacks using related keys. J. Cryptol. **7**(4), 229–246 (1994). A preliminary version appeared in EUROCRYPT 1993 (1993)
10. Boneh, D., Boyen, X.: Efficient selective identity-based encryption without random oracles. J. Cryptol. **24**(4), 659–693 (2011). A preliminary version appeared in EUROCRYPT 2004, 2004
11. Boneh, D., Canetti, R., Halevi, S., Katz, J.: Chosen-ciphertext security from identity-based encryption. SIAM J. Comput. **36**(5), 1301–1328 (2006)
12. Boneh, D., DeMillo, R.A., Lipton, R.J.: On the importance of eliminating errors in cryptographic computations. J. Cryptol. **14**(2), 101–119 (2001). A preliminary version appeared in EUROCRYPT 1997 (1997)

13. Boyen, X., Mei, Q., Waters, B.: Direct chosen ciphertext security from identity-based techniques. In: Atluri, V., Meadows, C., Juels, A. (eds.) CCS 2005, pp. 320–329. ACM (2005). https://eprint.iacr.org/2005/288

14. Choi, S.G., Kiayias, A., Malkin, T.: BiTR: built-in tamper resilience. In: Lee and Wang [25], pp. 740–758. https://eprint.iacr.org/2010/503

15. Dziembowski, S., Pietrzak, K., Wichs, D.: Non-malleable codes. In: Yao, A.C.-C. (ed.) ICS 2010, pp. 434–452. Tsinghua University Press (2010). https://eprint.iacr.org/2009/608

16. Faust, S., Mukherjee, P., Venturi, D., Wichs, D.: Efficient non-malleable codes and key-derivation for poly-size tampering circuits. In: Nguyen, P.Q., Oswald, E. (eds.) EUROCRYPT 2014. LNCS, vol. 8441, pp. 111–128. Springer, Heidelberg (2014). https://eprint.iacr.org/2013/702

17. Gennaro, R., Lysyanskaya, A., Malkin, T., Micali, S., Rabin, T.: Algorithmic tamper-proof (ATP) security: theoretical foundations for security against hardware tampering. In: Naor, M. (ed.) TCC 2004. LNCS, vol. 2951, pp. 258–277. Springer, Heidelberg (2004)

18. Goyal, V., O'Neill, A., Rao, V.: Correlated-input secure hash functions. In: Ishai, Y. (ed.) TCC 2011. LNCS, vol. 6597, pp. 182–200. Springer, Heidelberg (2011). https://eprint.iacr.org/2011/233

19. Jafargholi, Z., Wichs, D.: Tamper detection and continuous non-malleable codes. In: Dodis, Y., Nielsen, J.B. (eds.) TCC 2015, Part I. LNCS, vol. 9014, pp. 451–480. Springer, Heidelberg (2015). https://eprint.iacr.org/2014/956

20. Jia, D., Li, B., Lu, X., Mei, Q.: Related key secure PKE from hash proof systems. In: Yoshida, M., Mouri, K. (eds.) IWSEC 2014. LNCS, vol. 8639, pp. 250–265. Springer, Heidelberg (2014)

21. Jia, D., Lu, X., Li, B., Mei, Q.: RKA secure PKE based on the DDH and HR assumptions. In: Susilo, W., Reyhanitabar, R. (eds.) ProvSec 2013. LNCS, vol. 8209, pp. 271–287. Springer, Heidelberg (2013)

22. Kalai, Y.T., Kanukurthi, B., Sahai, A.: Cryptography with tamperable and leaky memory. In: Rogaway, P. (ed.) CRYPTO 2011. LNCS, vol. 6841, pp. 373–390. Springer, Heidelberg (2011)

23. Kiltz, E.: Chosen-ciphertext security from tag-based encryption. In: Halevi, S., Rabin, T. (eds.) TCC 2006. LNCS, vol. 3876, pp. 581–600. Springer, Heidelberg (2006)

24. Knudsen, L.R.: Cryptanalysis of LOKI91. In: Seberry, J., Zheng, Y. (eds.) AUSCRYPT '92. LNCS, vol. 718, pp. 196–208. Springer, Heidelberg (1993)

25. Lee, D.H., Wang, X. (eds.): ASIACRYPT 2011. LNCS, vol. 7073. Springer, Heidelberg (2011)

26. Lewi, K., Montgomery, H., Raghunathan, A.: Improved constructions of PRFs secure against related-key attacks. In: Boureanu, I., Owesarski, P., Vaudenay, S. (eds.) ACNS 2014. LNCS, vol. 8479, pp. 44–61. Springer, Heidelberg (2014)

27. Liu, F.-H., Lysyanskaya, A.: Tamper and leakage resilience in the split-state model. Manuscript, February 2012. Available at the authors' cite

28. Paterson, K.G., Schuldt, J.C.N., Stam, M., Thomson, S.: On the joint security of encryption and signature, revisited. In: Lee and Wang [25], pp. 161–178. https://eprint.iacr.org/2011/486

29. Qin, B., Liu, S., Yuen, T.H., Deng, R.H., Chen, K.: Continuous non-malleable key derivation and its application to related-key security. In: Katz, J. (ed.) PKC 2015. LNCS, vol. 9020, pp. 557–578. Springer, Heidelberg (2015). https://eprint.iacr.org/2015/003

30. Rabin, T. (ed.): CRYPTO 2010. LNCS, vol. 6223. Springer, Heidelberg (2010)
31. von zur Gathen, J., Gerhard, J.: Modern Computer Algebra, 3rd edn. Cambridge University Press, Cambridge (2013)
32. Waters, B.: Efficient Identity-based encryption without random oracles. In: Cramer, R. (ed.) EUROCRYPT 2005. LNCS, vol. 3494, pp. 114–127. Springer, Heidelberg (2005). https://eprint.iacr.org/2004/180
33. Wee, H.: Efficient chosen-ciphertext security via extractable hash proofs. In: Rabin [30], pp. 314–332
34. Wee, H.: Public key encryption against related key attacks. In: Fischlin, M., Buchmann, J., Manulis, M. (eds.) PKC 2012. LNCS, vol. 7293, pp. 262–279. Springer, Heidelberg (2012)

Simulation-Based Secure Functional Encryption in the Random Oracle Model

Vincenzo Iovino[1]([✉]) and Karol Żebroski[2]

[1] University of Luxembourg, Luxembourg, Luxembourg
vincenzo.iovino@uni.lu
[2] University of Warsaw, Warsaw, Poland
kz277580@students.mimuw.edu.pl

Abstract. One of the main lines of research in functional encryption (FE) has consisted in studying the security notions for FE and their achievability. This study was initiated by [Boneh et al. – TCC'11, O'Neill – ePrint'10] where it was first shown that for FE the indistinguishability-based (IND) security notion is not sufficient in the sense that there are FE schemes that are provably IND-Secure but concretely insecure. For this reason, researchers investigated the achievability of Simulation-based (SIM) security, a stronger notion of security. Unfortunately, the above-mentioned works and others [e.g., Agrawal et al. – CRYPTO'13] have shown strong impossibility results for SIM-Security. One way to overcome these impossibility results was first suggested in the work of Boneh et al. where it was shown how to construct, in the Random Oracle (RO) model, SIM-Secure FE for restricted functionalities and was asked the generalization to more complex functionalities as a challenging problem in the area. Subsequently, [De Caro et al. – CRYPTO'13] proposed a candidate construction of SIM-Secure FE for all circuits in the RO model assuming the existence of an IND-Secure FE scheme for circuits *with RO gates*. To our knowledge there are no proposed candidate IND-Secure FE schemes for circuits with RO gates and they seem unlikely to exist. We propose the first constructions of SIM-Secure FE schemes in the RO model that overcome the current impossibility results in different settings. We can do that because we resort to the two following models:

- In the public-key setting we assume a bound on the number of queries but this bound only affects the *running-times* of our encryption and decryption procedures. We stress that our FE schemes in this model are SIM-Secure and have ciphertexts and tokens of *constant-size*, whereas in the standard model, the current SIM-Secure FE schemes for general functionalities [De Caro et al., Gorbunov et al. – CRYPTO'12] have ciphertexts and tokens of size growing as the number of queries.
- In the symmetric-key setting we assume a timestamp on both ciphertexts and tokens. In this model, we provide FE schemes with short ciphertexts and tokens that are SIM-Secure against adversaries asking an *unbounded* number of queries.

Both results also assume the RO model, but not functionalities with RO gates and rely on extractability obfuscation [Boyle et al. – TCC'14] (and other standard primitives) secure only in the *standard* model.

© Springer International Publishing Switzerland 2015
K. Lauter and F. Rodríguez-Henríquez (Eds.): LatinCrypt 2015, LNCS 9230, pp. 21–39, 2015.
DOI: 10.1007/978-3-319-22174-8_2

Keywords: Functional encryption · Random oracle model · Simulation-based security · Obfuscation

1 Introduction

The study of simulation-based (SIM) notions of security for functional encryption were initiated only recently by Boneh, Sahai, and Waters [1] and O'Neill [2]. Quite interestingly, they show there exists clearly insecure FE schemes for certain functionalities that are nonetheless deemed secure by IND-Security, whereas these schemes do not meet the stronger notion of SIM-Security.

For this reason, researchers have started a further theoretical study of FE that includes either negative results showing SIM-Security is not always achievable [1,3,4] or alternative models overcoming the impossibility results [5,6]. On the positive direction, Boneh et al. [1] showed the existence of SIM-Secure FE schemes in the Random Oracle (RO, in short) [7] model for restricted functionalities (i.e., Attribute-based Encryption), and at the same time they left as a challenging problem the construction of FE for more sophisticated functionalities that satisfy SIM-Security in the RO model. More recently, De Caro et al. [8] showed how to overcome all known impossibility results assuming the existence of IND-Secure schemes for circuits with *random oracle gates*. This is a very strong assumption for which we do *not* know any candidate scheme and their existence seems unlikely[1].

Furthermore, their scheme incurs the following theoretical problem. First of all, recall that in a FE system for functionality $F : K \times X \to \Sigma$, defined over *key space* K, *message space* X and *output space* Σ, for every *key* $k \in K$, the owner of the master secret key Msk associated with master public key Pk can generate a secret key Tok_k that allows the computation of $F(k,x)$ from a ciphertext of x computed under master public key Pk. Thus, in a standard FE scheme the functionality is fixed in advance, and the scheme should allow to compute over encrypted data accordingly to this functionality. Instead, in the scheme for the RO model of De Caro et al. [8], the functionality does *depend* on the RO, and thus, even their implicit definition of functionality and FE scheme is not standard. Therefore, their scheme is not satisfactory.

This leads to the main question that we study in this work:

Can we achieve standard FE schemes in the (conventional) Programmable RO model from reasonable assumptions?

Our results answer affirmatively to this question demonstrating the existence of SIM-Secure schemes in the RO model with short parameters. The impossibility result of [4] shows that in a SIM-Secure FE scheme, the size of the ciphertexts has to grow as the number of token queries (see also [10] and [5]). On the other hand, our results also provide schemes for the RO model that are SIM-Secure but

---

[1] This issue was first noticed by several researchers [9] and personally communicated by Jonathan Katz to the authors of the work [8].

in which the size of the tokens and the ciphertexts is constant[2]. Before presenting our positive results in more detail, we prefer to first sketch our techniques so to highlight the technical problems that led to our constructions and models.

Our Techniques. We recall (a simplified version of) the transformation of De Caro *et al.* [8] to bootstrap an IND-Secure scheme for Boolean circuits to a SIM-Secure scheme for the same functionality. For sake of simplicity we focus on the non-adaptive setting, specifically on SIM-Security against adversaries asking q non-adaptive queries[3]. The idea of their transformation is to replace the original circuit with a "trapdoor" one that the simulator can use to program the output in some way. In the transformed scheme, they put additional "slots" in the plaintexts and secret keys that will only be used by the simulator. A plaintext has $2 + 2q$ slots and a secret key will have one. In the plaintext, the first slot is the actual message m and the second slot will be a bit flag indicating whether the ciphertext is in trapdoor mode. and the last $2q$ slots will be q pairs (r_i, z_i), where r_i is a random string and z_i is a programmed string. These $2q$ slots are used to handle q non-adaptive queries. On the other hand, in a secret key for circuit C, the slot is a random string r, that will be equal to one of the r_i in the challenge ciphertext. For evaluation, if the ciphertext is not in trapdoor mode (flag $= 0$) then the new circuit simply evaluates the original circuit C of the message m. If the ciphertext is in trapdoor mode, if $r = r_i$ for some $i \in [q]$ then the transformed circuit outputs z_i.

A natural approach to shorten the size of the ciphertexts in the RO model would be the following. Recall that a Multi-Input FE (MI-FE) scheme [11–13] is a FE over multiple ciphertexts. Let our starting scheme be a MI-FE over 2-inputs. Then, instead of encoding the slots (r_i, z_i)'s in the ciphertext, we could add to the ciphertext a tag $\mathsf{tag_c}$ (that is, the encryption would consist of a ciphertext plus the tag in clear) such that the simulator can program the RO on this point to output the values (r_i, z_i)'s. Now the ciphertext would consist of only a short ciphertext ct and the short tag $\mathsf{tag_c}$. At decryption time, we could "decompress" $\mathsf{tag_c}$ to get some string y, encrypt it with the public-key of the of the MI-FE scheme to produce $\mathsf{ct_2}$ and finally feed $\mathsf{ct_1}$ and $\mathsf{ct_2}$ to the multi-input token. Then, the simulator could program the RO so to output the values (r_i, z_i)'s on the point $\mathsf{tag_c}$. The functionality would be thus modified so that, in trapdoor

---

[2] Specifically, in our main transformation the size of the tokens is constant if we employ a collision-resistant hash function of variable-length, otherwise their size only depends on the encoding of the value and thus can be sub-logarithmic. Similarly, for the timestamp model of Sect. 3.3, both tokens and ciphertexts need to encode a temporal index that being a number at most equal to the number of queries issued by any PPT adversary, will be at most super-logarithmic, and thus can be encoded with a string of poly-logarithmic size. For simplicity, henceforth, we will claim that our constructions have tokens of constant size omitting to specify this detail.

[3] Henceforth, we mean by non-adaptive queries the queries that the adversary asks before seeing the challenge ciphertext and adaptive queries the queries the adversary asks after seeing it.

mode, the output would be taken by the values (r_i, z_i)'s (specifically, choosing the values z_i corresponding to the string r_i in the token). Therefore, any token applied to the simulated ciphertext (that is in trapdoor mode) should decrypt the same output as the real ciphertext would do.

This simple approach incurs more problems, the most serious being that the adversary could feed the multi-input token with a *different* ciphertext that does not correspond to $\mathcal{RO}(\mathsf{tag_c})$, thus detecting whether the first ciphertext is in normal or trapdoor mode. In fact, notice that in normal mode the second ciphertext does not affect the final output, whereas in trapdoor mode, the output only depends on the second ciphertext fed to the multi-input token. Another problem here is that $\mathcal{RO}(\mathsf{tag_c})$ should not contain the values (r_i, z_i)'s in clear, but this is easily solved by letting $\mathcal{RO}(\mathsf{tag_c})$ be an encryption of them and putting the corresponding secret-key in the first ciphertext. So, the main technical problem is:

> How can we force the adversary to feed the 2-inputs token with a second ciphertext that encrypts $\mathcal{RO}(\mathsf{tag_c})$?

Note that in the case of FE schemes that support functionalities with RO gates, this can be easily done by defining a new functionality that first tests whether the second input equals $\mathcal{RO}(\mathsf{tag_c})$, but in the "pure" RO model this solution can not be applied. Our patch is to add a new slot h of *short* size in the first ciphertext. Such value h is set to the hash of $\mathcal{RO}(\mathsf{tag_c})$ with respect to a Collision-Resistant Hash Function (CRHF) Hash, i.e., $h = \mathsf{Hash}(\mathcal{RO}(\mathsf{tag_c}))$. Furthermore, we modify the transformed functionality so that it first checks whether $\mathsf{Hash}(\mathcal{RO}(\mathsf{tag_c})) = h$. If this test fails, the functionality outputs an error $\bot$. The intuition is that with this modification, the adversary is now forced to use a second ciphertext that encrypts $\mathcal{RO}(\mathsf{tag_c})$ since otherwise it gets $\bot$ on both real or simulated ciphertext, and so, under the IND-Security of the MI-FE scheme, it seems that the adversary can not tell apart a real ciphertext from a simulated ciphertext. Unfortunately, we are not able to prove the security of this transformation assuming only the standard notion of IND-Security for MI-FE. In fact, notice that there *exist* second inputs for the modified functionality that distinguish whether the first input has the flag set to normal or trapdoor mode, namely inputs that correspond to collisions of Hash with respect to h and $\mathcal{RO}(\mathsf{tag_c})$. That is, any another second ciphertext that encrypt a value $y \neq \mathcal{RO}(\mathsf{tag_c})$ such that $\mathsf{Hash}(y) = h$ allows to distinguish whether the first ciphertext is in normal or trapdoor mode. Furthermore, it is not possible to make direct use of the security of the CRHF. The problem is that the definition of (2-inputs) MI-FE is too strong in that it requests the adversary to output two challenge message pairs (x_0, y) and (x_1, y) such that for any function f for which the adversary asked a query $f(x_0, \cdot) = f(x_1, \cdot)$. In our case, this does not hold: there exists a set of collisions C such that for any $z \in C$, $\mathsf{Hash}(z) = h$ and $f(x_0, z) \neq f(x_1, z)$. However, notice that it seems difficult for the adversary to find such collisions.

Our Assumptions and CRIND-Security. For these reasons, we need to extend the notion of MI-FE to what we call *collision-resistant indistinguishability* (CRIND, in short)[4]. In Sect. 4 we provide an instantiation of this primitive from extractability obfuscation w.r.t. distributional auxiliary input [14] (cf. Remark 3). We think that this definition can be of independent interest since it is more tailored for the applicability of MI-FE to other primitives. The reader may have noticed that the security of the second ciphertext guaranteed by the underlying MI-FE is not *necessary*. That is, our transformation would work even assuming 2-inputs MI-FE systems that take the second input *in clear*. In fact, CRIND-Security does *not* imply IND-Security for MI-FE schemes but this suffices for our scopes.

Roughly speaking, in CRIND-Security the security is quantified only with respect to *valid* adversaries[5], where an adversary is considered valid if it only submits challenges m_0, m_1 and asks a set of queries K that satisfy some "hardness" property called *collision-resistance compatibility*, namely that it is difficult to find a second input m_2 such that $F(k, m_0, m_2) \neq F(k, m_1, m_2)$ for some $k \in K$. Since in the reductions to CRIND-Security it is not *generally* possible to directly check the hardness of (K, m_0, m_1), the definition dictates (1) the existence of an efficient *checker* algorithm that approves *only* (but possibly not all) valid triples (K, m_0, m_1) (i.e., the checker can detect efficiently if a triple is collision-resistant compatible) and (2) that an adversary is valid if it only asks triples approved by the checker. We defer the details of the definition to Sect. 2.1.

Next, in a security reduction to CRIND-Security (i.e., when we need to prove the indistinguishability of two hybrid experiments assuming the CRIND-Security), the main task is to define an appropriate checker and prove that triples that are not collision-resistant compatible are rejected by it. This is usually done by checking that messages and keys satisfy an appropriate format. For instance, in the above case, the checker will check whether the machine (corresponding to the token) uses as sub-routine the specified CRHF and that the challenge messages and such machine have the right format. The construction of CRIND-Secure schemes follows the lines of the construction of fully IND-Secure FE schemes of Boyle *et al.* Namely, the encryption of the first input m_1 will be an obfuscation of a machine that has embedded m_1 and a verification key for a signature scheme and takes as input a signature of a machine M and a second input m_2 and (1) checks the validity of the signature and (2) if such test passes outputs $M(m_1, m_2)$. For the same resonas of Boyle *et al.* we need to resort to functional signatures. Details along with a broader overview can be found in Sect. 4.

The above presentation is an oversimplification and we defer the reader to the Sects. 3.1 and 3.2 for more details.

---

[4] Maybe, a better name would have been "differing-inputs indistinguishability" but we do not adopt this name to not overlap with differing-inputs obfuscation and because it recalls the reason to advocate this stronger notion for our transformations.

[5] We can recast IND-Security in a similar way by defining valid adversaries that only ask queries and challenges satisfying the compatibility property.

Our Models and Results. The reader may have noticed that the output of $\mathcal{RO}$ has to be "big", i.e., its size depends on the number of queries. Of course, we could assume that its range has constant size and replace a single invocation of the RO with range of size $> q$ with many invocation of a RO with range of constant size, but also in this case the *running-time* of the encryption and decryption procedures would have to depend on the number of queries. Here, it is the novelty of our approach. All the parameters of our SIM-Secure public-key FE scheme (including ciphertexts and tokens) have *constant* size but the cost of the "expansion" is moved from the length of ciphertexts and tokens to the running-time of the encryption and decryption procedures. That is, our SIM-Secure public-key FE scheme stills depends on q in the setup and running-time, but the *size* of the ciphertexts and tokens is *constant*. The results we achieve can be summarized as follows:

- $(q_1, q_c, \mathsf{poly})$-SIM-Security with ciphertext of constant size and tokens of size q_c. That is, SIM-Security against adversaries asking bounded non-adaptive token queries, bounded ciphertext queries, and unbounded adaptive token queries. In this case the size of the ciphertexts is constant but the size of the tokens grows as the number of ciphertext queries (and thus is constant in the case of 1 ciphertext query). This is known to be impossible in the standard model due to the impossibility of Agrawal *et al.* [4] (precisely this impossibility does not rule out the existence of schemes that satisfy this notion of security but it rules out the existence of schemes that satisfy both this notion of security *and* have short ciphertexts). Moreover, in this case the encryption and decryption procedures have running-times depending on q_1.
- (q_1, q_c, q_2)-SIM-Security with both ciphertexts and tokens of constant size. That is, SIM-Security against adversaries asking bounded token (both non-adaptive and adaptive) and ciphertext queries but with both ciphertext and token of constant size. In the standard model this is known to be impossible due to the impossibility result of De Caro and Iovino [5] for SIM-Security against adversaries asking unbounded ciphertext queries and bounded adaptive token queries (this impossibility is essentially an adaptation of the impossibility of Agrawal *et al.* [4]). In this case, the encryption and decryption procedures have running-times depending on $\max\{q_1, q_c, q_2\}$.
- We show how to remove the afore-mentioned limitation in a variant of the symmetric-key model where ciphertexts and tokens are tagged with a timestamp that imposes an order on their generation (i.e., the i-th token/ciphertext generated in the system is tagged with the value i). This model is reasonable because in the symmetric-key setting, the user that set-up the system is the same entity that generates tokens and ciphertexts as well[6]. We defer the reader to Sect. 3.3 for more details.

---

[6] The same considerations also hold in applications where many users share the same secret-key. Indeed, in this case one needs to assume that such users trust each other, and thus they will tag the ciphertexts and tokens with the correct timestamp.

For the sake of providing constructions with optimal parameters we work in the *Turing Machine* model of computation[7].

The Optimality and the Soundness of Our Results. It is easy to see that SIM-Security in the standard model but for schemes with procedures of running-time dependent on the number of queries is impossible as well. Moreover, we think that SIM-Security in the RO model with a constant number of RO calls is impossible to achieve as well, though we were not able to prove an impossibility result for it and leave to future work to set positive or negative results. Anyway, one could object that if we instantiate the RO with any concrete hash function, the resulting scheme is *not* "SIM-Secure" due to the impossibility results. This problem is also shared with the constructions for the RO model of De Caro *et al.* [8] and Boneh *et al.* [1]. What does it mean?

What the impossibility results say is that there are adversaries for which there exists no simulator though we do not know any concrete attacks on these schemes. This is different from the counter-examples of Canetti *et al.* [15] where they were presented signature schemes provably secure in the RO model but *concretely* insecure when instantiated with *any* hash function. In our view, this could merely mean that general definitions of SIM-Security are too strong. Along this direction, the works of De Caro and Iovino [5] and Agrawal *et al.* [6] provide another way to overcome this limitation.

2   Definitions

Due to space constraints we defer to the full version [16] the definitions of functional signature scheme, collision-resistant hash function, pseudo-random function family, single and multi-inputs functional encryption scheme.

2.1   Collision-Resistant Indistinguishability Security for MI-FE

As we mentioned in the construction overview sketched in Sect. 1, we need a different notion of MI-FE security. Here, we consider only the 2-inputs case, since this is suited for our main transformation but it is straightforward to extend it to the *n*-ary case.

Furthermore, in Sect. 4 we will show how to construct a CRIND-Secure MI-FE scheme from extractability obfuscation w.r.t. distributional auxiliary input [14] (cf. Remark 3). We presented an informal discussion of the definition in Sect. 1. We now present the formal definition.

The collision-resistant indistinguishability-based notion of security for a multi-input functional encryption scheme MI-FE = (Setup, KeyGen, Enc, Eval) for

---

[7] The main focus of the work of Goldwasser *et al.* [13] is for the circuit model but they sketch how to extend it to the Turing Machine model. Similar considerations hold for the schemes of Gordon *et al.* [12]. Further details will be given in the Master's Thesis of the second author.

functionality F defined over (K, M) is formalized by means of the following experiment $\text{CRIND}_{\mathcal{A}}^{\text{MI-FE}}$ with an adversary $\mathcal{A} = (\mathcal{A}_0, \mathcal{A}_1)$. Below, we present the definition for only one message; it is easy to see the definition extends naturally for multiple messages.

$\text{CRIND}_{\mathcal{A}}^{\text{MI-FE}}(1^\lambda)$

1. $(\text{Mpk}, \text{Msk}) \leftarrow \text{MI-FE.Setup}(1^\lambda)$;
2. $r \xleftarrow{R} \{0,1\}^\lambda$;
3. $(x_0, x_1, \text{st}) \leftarrow \mathcal{A}_0^{\text{MI-FE.KeyGen}(\text{Msk}, \cdot)}(\text{Mpk}, r)$ where we require that $|x_0| = |x_1|$;
4. $b \xleftarrow{R} \{0,1\}$;
5. $\text{Ct}_1 \leftarrow \text{MI-FE.Encrypt}(\text{Mpk}, (x_b \| r))$;
6. $b' \leftarrow \mathcal{A}_1^{\text{MI-FE.KeyGen}(\text{Msk}, \cdot)}(\text{Mpk}, \text{Ct}_1, \text{st})$;
7. **Output:** $(b = b')$.

We make the following additional requirements:

- Collision-resistance compatibility. Let K denote a set of keys. We say that a pair of messages x_0 and x_1 is *collision-resistant compatible* with K if it holds that: any (possibly, non-uniform) PPT algorithm $\mathcal{B}$ given the security parameter 1^λ and (r, x_0, x_1) for r uniformly distributed in $\{0,1\}^\lambda$, can find y satisfying inequality $F(k, (x_0 \| r), y) \neq F(k, (x_1 \| r), y)$ for some $k \in K$ with at most negligible (in λ) probability, where probability is taken over $r \xleftarrow{R} \{0,1\}^\lambda$ and the random coins of $\mathcal{B}$.[8]
- Efficient checkability. We assume that there exist efficient checker algorithm Checker, which takes as input (k, x_0, x_1) and outputs false if x_0 and x_1 are *not* collision-resistant compatible with $\{k\}$ (i.e., the singleton set containing the key k).
- Validity. We say that an adversary $\mathcal{A}$ in the above game is *valid* with respect to a checker Checker with efficient checkability if during the execution of the above game, $\mathcal{A}$ outputs challenge messages x_0 and x_1 of the same length and asks a set of queries K in the key space of the functionality such that for any $k \in K$, $\text{Checker}(k, x_0, x_1) = \text{true}$ (i.e., the adversary only asks queries and challenges approved by the checker).

The advantage of a valid adversary $\mathcal{A}$ in the above game is defined as

$$\text{Adv}_{\mathcal{A}}^{\text{MI-FE,IND}}(1^\lambda) = |\text{Prob}[\text{CRIND}_{\mathcal{A}}^{\text{MI-FE}}(1^\lambda) = 1] - 1/2|.$$

---

[8] It can appear that the definition be not well-defined because we do not specify how the key k is related to the security parameter. To understand this, you may imagine that k be the code of some algorithm P (ant thus of constant-size) to compute a keyed hash function $\text{Hash}(\cdot, \cdot)$. The program P takes an hashing key s computed with respect to an arbitrarily long security parameter λ and an input x and computes $\text{Hash}(s, x)$. Therefore in the above definition, k (along with the functionality F) plays the role of P and thus can have constant size whereas r plays the role of the hashing key s that depends instead on the security parameter.

Definition 1. We say that a 2-inputs MI-FE scheme is *collision-resistant indistinguishably secure* (CRIND-Secure, in short) if for any checker Checkersatisfying efficient checkability, it holds that all PPT adversaries $\mathcal{A}$ *valid* with respect to Checkerhave at most negligible advantage in the above game.

2.2 Extractability Obfuscation w.r.t. Distributional Auxiliary Input

Boyl *et al.* [14] defined obfuscators secure against general distributional auxiliary input. We recall their definition (cf. Remark 3).

Definition 2. A uniform PPT machine $e\mathcal{O}$ is an *extractability obfuscator w.r.t. (general) distributional auiliary input* for the class of Turing Machines $\{\mathcal{M}_\lambda\}_{\lambda \in \mathbb{N}}$ if it satisfies the following properties:

- (Correctness): For all security parameter λ, all $M \in \mathcal{M}$, all inputs x, we have
$$\Pr\left[M' \leftarrow e\mathcal{O}(1^\lambda, M) \colon M'(x) = M(x) \right] = 1.$$

- (Polynomial slowdown): There exist a universal polynomial p such that for any machine M, we have $|M'| \leq p(|M|)$ for all $M' = e\mathcal{O}(1^\lambda, M)$ under all random coins.

- (Security): For every non-uniform PPT adversary $\mathcal{A}$, any polynomial $p(\lambda)$, and efficiently sampleable distribution $\mathcal{D}$ over $\mathcal{M}_\lambda \times \mathcal{M}_\lambda \times \{0,1\}^*$, there exists a non-uniform PPT extractor E and polynomial $q(\lambda)$ and negligible function $\mathsf{negl}(\lambda)$ such that, for every $\lambda \in \mathbb{N}$, with probability $\geq 1 - \mathsf{negl}(\lambda)$ over $(M_0, M_1, z) \leftarrow \mathcal{D}(1^\lambda)$, it holds that:
If $\Pr\left[b \xleftarrow{R} \{0,1\}; M' \leftarrow e\mathcal{O}(1^\lambda, M_b) : \mathcal{A}(1^\lambda, M', M_0, M_1, z) = b \right] \geq \frac{1}{2} + \frac{1}{p(\lambda)}$,
then $\Pr\left[w \leftarrow E(1^\lambda, M_0, M_1, z) : M_0(w) \neq M_1(w) \right] \geq \frac{1}{q(\lambda)}$

Remark 3. In light of the recent "implausibility" results on extractability obfuscation with auxiliary input [17,18], we would like to point out that our results are based on *specific* distributions for which no implausibility result is known. Same considerations were also made in [14].

3 Our Transformations

3.1 $(q_1, q_c, \mathsf{poly})$-SIM-Security

In this section, we show that assuming a CRIND-Secure (in the standard model) MI-FE scheme for p-TM_2 (2-inputs Turing machines with run time equal to a polynomial p) for any polynomial p, it is possible to construct a SIM-Secure functional encryption scheme in the RO model for functionality p-TM for any polynomial p. Moreover, this is possible also for FE schemes with input-specific run time. The resulting scheme is secure for a bounded number of messages and non-adaptive token queries, and unbounded number of adaptive key queries. Moreover, it enjoys ciphertexts and tokens of size not growing with the number of non-adaptive queries, overcoming the impossibility result of Agrawal *et al.* [4] for the standard model. In Sect. 4 we will show how to construct a CRIND-Secure MI-FE from extractability obfuscation w.r.t. distributional auxiliary input [14] (cf. Remark 3).

Trapdoor Machines. The idea of our transformations is to replace the original machine with a "trapdoor" one that the simulator can use to program the output in some way. This approach is inspired by the FLS paradigm introduced by Feige, Lapidot and Shamir [19] to obtain zero-knowledge proof systems from witness indistinguishable proof systems. Below we present the construction of trapdoor machine, which works in standard model.

Definition 4. [Trapdoor Machine] Fix $q > 0$. Let M be a Turing machine with one input. Let $\mathsf{SE} = (\mathsf{SE.Enc}, \mathsf{SE.Dec})$ be a symmetric-key encryption scheme with key-space $\{0,1\}^\lambda$, message-space $\{0,1\}^{\lambda+1}$, and ciphertext-space $\{0,1\}^\nu$. We require for simplicity that SE has pseudo-random ciphertexts (see the full version [16]) and can encrypt messages of variable length (at most $\lambda + 1$). Let $\mathsf{Hash}\colon \{0,1\}^\lambda \times \{0,1\}^{q \cdot \nu} \to \{0,1\}^\lambda$ be a collision resistant hash function[9]. For $\mathsf{tag}_k = (id_k, c) \in \{0,1\}^{\lambda+\nu}$ define the corresponding *trapdoor machine* $\mathsf{Trap}[M, \mathsf{Hash}, \mathsf{SE}]^{\mathsf{tag}_k}$ on two inputs as follows:

$$
\begin{aligned}
&\textbf{Machine } \mathsf{Trap}[M, \mathsf{Hash}, \mathsf{SE}]^{\mathsf{tag}_k}(m', R) \\
&\quad (m, \mathsf{flag}, sk, h, k_\mathsf{H}) \leftarrow m' \\
&\quad \text{If } \mathsf{Hash}(k_\mathsf{H}, R) \neq h \text{ then return } \bot \\
&\quad \text{If } \mathsf{flag} = 0 \text{ then return } M(m) \\
&\quad (id_k, c) \leftarrow \mathsf{tag}_k \\
&\quad (R_1, \ldots, R_q) \leftarrow R \\
&\quad \text{For } i = 1, \ldots, q \text{ do} \\
&\quad\quad (id_k{}', v) \leftarrow \mathsf{SE.Dec}(sk, R_i) \\
&\quad\quad \text{If } id_k{}' = id_k \text{ then return } v \\
&\quad \text{return } \mathsf{SE.Dec}(sk, c)
\end{aligned}
$$

RO-based Transformation

Overview. In Sect. 1 we sketched a simplified version of our transformation. Here, we present an overview with more details. The idea is to put additional "slots" in the plaintexts and secret keys that will only be used by the simulator. A plaintext contains five slots and a secret key contains two slots. In the plaintext, the first slot is the actual message m. The second slot is a bit flag indicating whether the ciphertext is in trapdoor mode. The third slot is a random key sk used by SE scheme, the fourth slot is a hash h of $\mathcal{RO}(\mathsf{tag}_c)$ (computed with respect to a CRHF) attached to the ciphertext and finally the fifth slot contains a hash function key k_H.

In the secret key, the first slot encodes the actual machine M and the second slot is a random tag $\mathsf{tag}_k = (id_k, c)$. Slot id_k is used by simulator to identify pre-challenge tokens and c is used to convey programmed output value in post-challenge tokens. For evaluation, if the ciphertext is not in trapdoor mode

---

[9] For sake of simplicity as Hash key we will use a random string of length λ, instead of key generated by Gen. Alternatively, we could feed the Gen algorithm with this randomness.

(i.e., flag $= 0$) then the functionality simply evaluates the original machine M of the message m. If the ciphertext is in trapdoor mode, depending on the nature of the secret key (non-adaptive or adaptive), for $\mathsf{tag}_k = (id_k, c)$, if for some $i \in [q]$, $(id_k, v) = \mathsf{SE.Dec}(sk, R_i)$, then the functionality outputs v, otherwise it outputs $\mathsf{SE.Dec}(sk, c)$. Here R_i is the i-th element in the string R that the machine takes as second input, and is set by the (honest) evaluation procedure to $\mathcal{RO}(\mathsf{tag}_c)$.

For sake of simplicity we assume that TM functionality, for which our scheme is constructed, has output space $\{0, 1\}$. The construction can be easily extended to work for any TM functionality with bounded output length.

Definition 5. [RO-Based Transformation] Let $p(\cdot)$ be any polynomial. Let $\mathsf{SE} = (\mathsf{SE.Enc}, \mathsf{SE.Dec})$ be a symmetric-key encryption scheme with key-space $\{0, 1\}^\lambda$, message-space $\{0, 1\}^{\lambda+1}$, and ciphertext-space $\{0, 1\}^\nu$. We require for simplicity that SE has pseudo-random ciphertexts (see the full version [16]) and can encrypt messages of variable length (at most $\lambda+1$). Let $\mathsf{Hash} \colon \{0, 1\}^\lambda \times \{0, 1\}^{q \cdot \nu} \to \{0, 1\}^\lambda$ be a collision-resistant hash function (not modeled as a random oracle). Assuming that the running time of machine M equals exactly $p(|m|)$ on input m, the running time of trapdoor machine $\mathsf{Trap}[M, \mathsf{Hash}, \mathsf{SE}]^{\mathsf{tag}_k}$ is bounded by some polynomial $p'(\cdot)$. Let $\mathsf{MI\text{-}FE} = (\mathsf{MI\text{-}FE.Setup}, \mathsf{MI\text{-}FE.Enc}, \mathsf{MI\text{-}FE.KeyGen}, \mathsf{MI\text{-}FE.Eval})$ be a multi-input functional encryption scheme for the functionality $p'\text{-}\mathsf{TM}_2$.

In our construction we assume that the output length of the programmable random oracle $\mathcal{RO}$ equals $q \cdot \nu$.

We define a new (single-input) functional encryption scheme $\mathsf{SimFE}[\mathsf{Hash}, \mathsf{SE}] = (\mathsf{Setup}, \mathsf{KeyGen}, \mathsf{Enc}, \mathsf{Eval})$ for functionality $p\text{-}\mathsf{TM}$ as follows.

- $\mathsf{Setup}(1^\lambda)$: runs $(\mathsf{Mpk}, \mathsf{Msk}) \leftarrow \mathsf{MI\text{-}FE.Setup}(1^\lambda)$ and chooses random $r \xleftarrow{R} \{0, 1\}^\lambda$ and returns a pair (Mpk, r) as public key Pk and Msk as master secret key.
- $\mathsf{Enc}(\mathsf{Pk}, m)$: on input $\mathsf{Pk} = (\mathsf{Ek}_1, \mathsf{Ek}_2, r)$ and $m \in \{0, 1\}^*$, the algorithm chooses random $sk \xleftarrow{R} \{0, 1\}^\lambda$, $\mathsf{tag}_c \xleftarrow{R} \{0, 1\}^\lambda$ and sets $k_\mathsf{H} = r$, then it takes $m' = (m, 0, sk, \mathsf{Hash}(k_\mathsf{H}, \mathcal{RO}(\mathsf{tag}_c), k_\mathsf{H})$, computes $c \leftarrow \mathsf{MI\text{-}FE.Enc}(\mathsf{Ek}_1, m')$ and returns a pair (c, tag_c) as its own output.
- $\mathsf{KeyGen}(\mathsf{Msk}, M)$: on input Msk and a machine M, the algorithm chooses random $id_k \xleftarrow{R} \{0, 1\}^\lambda$, $c \xleftarrow{R} \{0, 1\}^\nu$ and returns $(\mathsf{Tok}, \mathsf{tag}_k)$ where $\mathsf{Tok} \leftarrow \mathsf{MI\text{-}FE.KeyGen}(\mathsf{Msk}, \mathsf{Trap}[M, \mathsf{Hash}, \mathsf{SE}]^{\mathsf{tag}_k})$ and $\mathsf{tag}_k = (id_k, c)$.
- $\mathsf{Eval}(\mathsf{Pk}, \mathsf{Ct}, \mathsf{Tok})$: on input $\mathsf{Pk} = (\mathsf{Ek}_1, \mathsf{Ek}_2, r)$, $\mathsf{Ct} = (c, \mathsf{tag}_c)$ and $\mathsf{Tok} = (\mathsf{Tok}', \mathsf{tag}_k)$, computes $c' = \mathsf{MI\text{-}FE.Enc}(\mathsf{Ek}_2, \mathcal{RO}(\mathsf{tag}_c))$ and returns the output $\mathsf{MI\text{-}FE.Eval}(\mathsf{Tok}', c, c')$.

Theorem 6. *Suppose* MI-FE *is* CRIND-*Secure in the standard model. Then* SimFE *is* $(q, 1, \mathsf{poly})$-SIM-*Secure in the random oracle model. Furthermore, this can be extended to* $(q_1, q_c, \mathsf{poly})$-SIM-*Security (see the full version [16]) and if* MI-FE *satisfies the properties of succinctness and input-specific time, so* SimFE *does.*

Security proof overview. We conduct the security proof of our construction by a standard hybrid argument. To move from real world experiment to ideal one we use the following hybrid experiments:

- The first hybrid experiment corresponds to the real experiment.
- The second hybrid experiment is identical to the previous one except that the random oracle is programmed at point $\mathsf{tag_c}$ so to output the encryption of the desired output values on pre-challenge queries, and post-challenge queries are answered with tokens that have embedded appropriate encrypted output values. Moreover, *all* these values are encrypted using the underlying SE scheme with a *randomly chosen* secret-key sk'. Notice that in this experiment the secret-key sk' is *uncorrelated* to the secret-key sk embedded in ciphertext.
- The third hybrid experiment is identical to the previous one except that the ciphertext contains the same secret-key sk' used to program the RO.
- In last step we switch the flag slot in the ciphertext to 1 indicating the trapdoor mode. At the same time we change the content of message slot m to $0^{|m|}$. This is necessary due to the fact that simulator only knows the challenge message length, but not the message itself.

One can reduce the security of first two transitions to the ciphertext pseudo-randomness of SE scheme and to the CRIND-Security of underlying MI-FE scheme. The proof in these cases is pretty straightforward.

One could be tempted to reduce the indistinguishability security of last two hybrids to both collision resistance of used hash function and IND-Security on MI-FE. However, the security reduction is not obvious. The adversary could recognize the simulation by finding a string R different from $\mathcal{RO}(\mathsf{tag_c})$ for which $\mathsf{Hash}(k_\mathsf{H}, R) = \mathsf{Hash}(k_\mathsf{H}, \mathcal{RO}(\mathsf{tag_c}))$, and applying the evaluation algorithm to this value as second input. The output of evaluation algorithm in this case would be different than expected. Although the adversary would contradict the collision resistance of Hash, we are not able to construct algorithm based on that adversary, which breaks the hash function security.

Therefore we need to rely on the CRIND-Security of MI-FE. Moreover, for completeness we will only assume CRIND-Security and never IND-Security.

We defer the proof of Theorem 6 to the full version [16].

3.2 (q_1, q_c, q_2)-SIM-Security with Short Tokens

The previous transformation suffers from the problem that the size of the tokens grows as the number of ciphertext queries q_c. In this Section we show how to achieve (q_1, q_c, q_2)-SIM-Security. Notice that in the standard model constructions satisfying this level of security like the scheme of Gorbunov *et al.* [10] have short ciphertexts and tokens. Moreover, De Caro and Iovino [5] showed an impossibility result for this setting. Our transformation assumes a 3-inputs MI-FE scheme (CRIND-Secure in the standard model). The resulting scheme is (q_1, q_c, q_2)-SIM-Secure according to the definition with simulated setup (see the full version [16]). The idea is very similar to the transformation presented in Sect. 3.1, but due to space constraints, we defer the sketch of the transformation to the full version [16].

3.3   (poly, poly, poly)-SIM-Security in the Timestamp Model

We recall that any known impossibility results in the standard model also apply
to the symmetric-key setting. In this Section we show how to achieve unbounded
SIM-Security in the RO model in a variant of the symmetric-key setting that we
call the *timestamp* model. Moreover, our scheme enjoys ciphertexts and tokens
of constant size. The timestamp model is identical to the symmetric-key mode
except for the following changes:

- The encryption and key generation procedures also take as input a *temporal
  index* or *timestamp*. The security of this index is not required. The security
  experiments are identical to those of the symmetric-key model except that
  the queries are answered by providing tokens and ciphertexts with *increasing*
  temporal index (the exact value does not matter as long as they are ordered
  in order of invocation). Roughly speaking, this is equivalent to saying that the
  procedures are stateful. Notice that in the symmetric-key model, this change
  has no cost since the user who set-up the system can keep the value of the
  current timestamp and guarantee that ciphertexts and tokens are generated
  with timestamps of increasing order.
- For simplicity, we also assume that there is a decryption key. Precisely, the
  evaluation algorithm takes as input a token, a ciphertext and a decryption
  key. It is easy to see that this decryption key can be removed at the cost of
  including it in any token or ciphertext.

Sketch of the Transformation. With these changes in mind it is easy to
modify the scheme of Sect. 3.2 to satisfy (poly, poly, poly)-SIM-Security in the
RO model. Precisely, the slots for the identifiers will contain the temporal index
in *clear*[10]. In the scheme, the tags will be such that the (both non-programmed
and programmed) RO on these input will output a string of size proportional
to i, where i is the temporal index. This can be done by assuming that the RO
outputs a single bit and invoking it many times. That is, instead of computing
$\mathcal{RO}(\text{tag}_c)$, the procedures will compute $\mathcal{RO}(\text{tag}_c \| j)$ for $j = 1, \ldots, m$ where m
is the needed size. For simplicity, henceforth, we assume that the RO outputs
strings of variable-length. As byproduct, we need to program the RO on the tag
tag_c (resp. tag_k) of a ciphertext (resp. token) with timestamp i to only output
i ciphertexts. Thus, we do not need to fix in *advance* any bound, which was
the only limitation of the previous transformations. Notice that the evaluation
algorithm needs to encrypt the output of $\mathcal{RO}(\text{tag}_c)$ and $\mathcal{RO}(\text{tag}_k)$ and this is
done by using the encryption keys Ek_2 and Ek_3 for the second and third input.
This is the reason why we assume that there is a decryption key that in this
scheme consists of the pair $(\text{Ek}_2, \text{Ek}_3)$.

---

[10] We stress that we could also assume that the temporal index is appended in clear
to the final ciphertext.

Security. The security proof is identical to that of the transformation of Sect. 3.2 with further simplifications due to the fact that we do not need to have the temporal index encrypted. We first need to switch the slot of the identifiers in both ciphertext and tokens to be encryption of the the temporal index. Then the proof proceeds as before.

4   Constructions of CRIND-Secure MI-FE from e$\mathcal{O}$

Overview of the construction and security reduction. In order to achieve CRIND-security of MI-FE (as defined in Sect. 2.1), we make use of the following ideas inspired by the construction of fully IND-Secure FE of Boyle *et al.* [14]. We assume a functional signature scheme FS [20]. Namely, our encryption procedure takes as input the first input m_1 and produces an obfuscation of a machine that has embedded m_1 and takes as input a second message m_2 and a functional signature for some function f and (1) verifies the signature and (2) outputs $f(m_1, m_2)$. Roughly speaking, we want prevent the adversary to be able to find distinguishing inputs. To this scope, we need to forbid the adversary from evaluating the machine on functions for which it did not see a signature. For the same reasons as in Boyle *et al.* it is not possible to use a standard signature scheme. This is because, the adversary $\mathcal{A}$ against e$\mathcal{O}$ needs to produce a view to the adversary $\mathcal{B}$ against CRIND-Security, and in particular to simulate the *post-challenge* tokens. To that aim, $\mathcal{A}$ would need to receive an auxiliary input z containing the signing key of the traditional signature scheme but in this case an extractor with access to z could easily find a distinguishing input. As in Boyle *et al.* we resort to functional signatures. We recall their ideas. In the scheme, they put a functional signing key that allows to sign any function. In the security proof, they use the property of function privacy to show that the original experiment is computationally indistinguishable to an experiment where the post-challenge queries are answered with respect to a *restricted* signing key for the Boolean predicate that is verified on all and only the machines T for which $T(m_0) = T(m_1)$, where m_0 and m_1 are the challenges chosen by the adversary against IND-Security. Thus, putting this restricted signing key in the auxiliary distribution does *not* hurt of the security since an extractor can not make use of it to find a distinguishing input. In the case of CRIND-Security, it is not longer true that $T(m_0) = T(m_1)$ but we will invoke the properties of the checker and we set the Boolean predicate to one that is verified for all machines T approved by the checker with respect to the challenges, i.e., such that $\mathsf{Checker}(T, m_0, m_1) = 1$. Then, by the property of the checker and valid adversaries, it follows that for any machine T for which the valid adversary asked a query, it is difficult to find a second input m_2 such that $T(m_0, m_2) \neq T(m_1, m_2)$. Thus, the existence of an extractor for our distribution would contradict the hypothesis that the adversary is valid and only asked queries for machines and challenges satisfying the collision-resistance compatibility. Precisely, we have the following transformation.

Definition 7. [$e\mathcal{O}$-Based Transformation] Let $e\mathcal{O}$ be an extractability obfuscator w.r.t. distributional auxiliary input (cf. Remark 3). Let FS = (FS.Setup, FS.KeyGen, FS.Sign, FS.Verify) be a signature scheme.

For any message m_1 and verification key vk of FS, let us define a machine M_{m_1} (where for simplicity we omit the other parameters in subscript) that takes two inputs, a signature σ_T of machine T and a message m_2, and (1) the machine verifies the signature σ_T according to vk, and if it is an invalid signature, it returns $\perp$; and (2) the machine returns $T(m_1, m_2)$.

We define a new 2-inputs functional encryption scheme CRFE[$e\mathcal{O}$, FS] = (Setup, KeyGen, Enc, Eval) for functionality TM_2 as follows[11]

- Setup(1^λ): chooses a pair (msk, vk) $\leftarrow$ FS.Setup(1^λ) and generates a key $\mathsf{sk}_1 \leftarrow$ FS.KeyGen(msk, 1) that allows signing all messages (i.e., for the always-accepting predicate $1(T) = T \ \forall \ T$). It sets $\mathsf{Ek}_1 = \mathsf{Ek}_2 = \mathsf{vk}$ and outputs Mpk = Ek_1 and Msk = (sk_1, vk).
- Enc(Ek, m): depending on whether Ek is an encryption key for first or second input:
  - if Ek = Ek_1 then outputs $e\mathcal{O}(M_m)$ where M_m is defined as above with respect to m and vk (recall that for simiplicity we omit the subscript for vk).
  - if Ek = Ek_2 then outputs the message m in clear (recall that we are not interested in the security of the second input and we adopted the formalism of multi-input FE to avoid the need of a new syntax and for sake of generality, e.g., providing in future constructions that satisfy *both* IND-Security and CRIND-Security).
- KeyGen(Msk, T): on input Msk = (sk_1, vk) and a machine T, the algorithm generates a signature on T via $\sigma_T \leftarrow$ Signature.Sign(sk_1, T) and outputs token σ_T.
- Eval(Mpk, Ct_1, Ct_2, Tok): on input Mpk = vk, Ct_1 which is an obfuscated machine M of machine M_{m_1}, $\mathsf{Ct}_2 = m_2$, Tok = σ_T, runs machine M with the other inputs as arguments, and returns the machine output as its own.

Correctness. It is easy to see that the scheme satisfies correctness assuming the correctness of $e\mathcal{O}$ and FS.

We now state the security of the constructed scheme.

Theorem 8. *If $e\mathcal{O}$ is an extractable obfuscator w.r.t. distributional auxiliary input, FS is an unforgeable functional signature scheme with function privacy, then, for any* Checker *satisfying the requirement of the* CRIND-*Security, it holds that* CRFE[$e\mathcal{O}$, FS] *is* CRIND-*Secure. Furthermore such scheme satisfies input-specific run time and assuming that* FS *is also succinct, so* CRFE *does.*

Due to space constraints, we defer the formal proof of the theorem to Appendix A.

---

[11] For simplicity, henceforth we omit to specify whether the functionality is with respect to machine of fixed time or input-specific. Both cases can be taken in account with small changes.

Acknowledgments. We thank Abhishek Jain, Adam O'Neill, Anna Sorrentino and the anonymous reviewers for useful comments. Part of this work was done while Vincenzo Iovino was at the University of Warsaw. This work was supported by the WELCOME/2010-4/2 grant founded within the framework of the EU Innovative Economy Operational Programme and by the National Research Fund of Luxembourg.

A   Proof of Theorem 8

Proof. We define the following hybrids. Let $q(\lambda)$ be a bound on the number of *post-challenge* token queries asked by $\mathcal{B}$ in any execution with security parameter 1^λ. Such bound exists because $\mathcal{B}$ is a PPT algorithm.

- Hybrid $H_0^{\mathcal{B}}$: This is the real experiment $\mathsf{CRIND}_{\mathcal{B}}^{\mathsf{CRFE[e\mathcal{O},FS]}}$.
- Hybrid $H_i^{\mathcal{B}}, i = 0, \ldots, q$: Same as the previous hybrid, except that the first i post-challenge token queries are answered with respect to a *restricted* signing key sk_C for the Boolean predicate C that allows one to sign exactly Turing machines T for which $\mathsf{Checker}(T, m_0, m_1) = 1$. (This is one of the differences with the proof of Boyle *et al.* wherein, being the scope to prove IND-Security, the signing key is for a predicate that allows one to sign exactly the machines T for which $T(m_0) = T(m_1)$. This is not possible in our case, but we make use of the definition of valid adversary that dictates that such adversary will only make queries for machines approved by the checker.). Specifically, at the beginning of the game the challenger generates a restricted signing key $\mathsf{sk}_C \leftarrow \mathsf{FS.KeyGen}(\mathsf{Msk}, C)$. The pre-challenge queries are answered using the standard signing key sk_1 as in hybrid H_0. The first i post-challenge token queries are answered using the restricted key sk_C, that is a token query for machine T is answered with $\sigma_T \leftarrow \mathsf{FS.Sign}(\mathsf{sk}_C, T)$. All remaining token queries are anwered using the standard key sk_1.

Claim 9. For $i = 1, \ldots, q$, the advantage of $\mathcal{B}$ in guessing the bit b in hybrid $H_i^{\mathcal{B}}$ is equal to the advantage of $\mathcal{B}$ in guessing the bit b in hybrid $H_{i-1}^{\mathcal{B}}$ up to a negligible factor.

We prove the claim by using the function privacy property of FS. Namely, for any $i \in [q]$, consider the following adversary $\mathcal{A}_{\mathsf{priv}}(1^\lambda)$ against function privacy of FS.

- $\mathcal{A}_{\mathsf{priv}}^i$ is given keys $(\mathsf{vk}, \mathsf{msk}) \leftarrow \mathsf{FS.Setup}(1^\lambda)$ from the function privacy challenger.
- $\mathcal{A}_{\mathsf{priv}}^i$ submits the all-accepting function 1 as the first of its two challenge functions, and receives a corresponding signing key $\mathsf{sk}_1 \leftarrow \mathsf{FS.KeyGen}(\mathsf{msk}, 1)$.
- $\mathcal{A}_{\mathsf{priv}}^i$ simulates interaction with $\mathcal{B}$. First, it forwards vk to $\mathcal{B}$ as the public-key and chooses a random string $r \in \{0, 1\}^\lambda$. For each token query T made by $\mathcal{B}$, it generates a signature on T using key sk_1.
- $\mathcal{A}_{\mathsf{priv}}^i$ At some point $\mathcal{B}$ outputs a pair of messages m_0, m_1. $\mathcal{A}_{\mathsf{priv}}^i$ generates a challenge ciphertext in the CRIND-Security game by sampling a random bit b and encrypting $(m_b||r)$ and sending it to $\mathcal{B}$.

- $\mathcal{A}^i_{\mathsf{priv}}$ submits as its second challenge function C (as defined above). It receives a corresponding signing key $\mathsf{sk}_C \leftarrow \mathsf{FS.KeyGen}(\mathsf{msk}, \mathsf{P}_C)$.
- $\mathcal{A}^i_{\mathsf{priv}}$ now simulates interaction with $\mathcal{B}$ as follows.
  For the first $i-1$ post-challenge token queries T made by $\mathcal{B}$, $\mathcal{A}^i_{\mathsf{priv}}$ generates a signature using key sk_C, i.e., $\sigma_T \leftarrow \mathsf{FS.Sign}(\mathsf{sk}_C, T)$. For $\mathcal{B}$'s i-th post-challenge query, $\mathcal{A}^i_{\mathsf{priv}}$ submits the pair of preimages (T, T) to the function privacy challenger (note that $1(T) = C(T) = T$) since, being $\mathcal{B}$ a valid adversary, it only asks queries T such that $\mathsf{Checker}(T, m_0, m_1) = 1$), and receives a signature σ_T generated either using key sk_1 or key sk_C. $\mathcal{A}^i_{\mathsf{priv}}$ generates the remaining post-challenge queries of $\mathcal{B}$ using key sk_1.
- Eventually $\mathcal{B}$ outputs a bit b'. If $b' = b$ is a correct guess, then $\mathcal{A}^i_{\mathsf{priv}}$ outputs function 1; otherwise, it outputs function C.

Note that if the function privacy challenger selected the function 1, then $\mathcal{A}^i_{\mathsf{priv}}$ perfectly simulates hybrid $H^{\mathcal{B}}_{i-1}$, otherwise it perfectly simulates hybrid $H^{\mathcal{B}}_i$. Thus, the advantage of $\mathcal{A}^i_{\mathsf{priv}}$ is exactly the difference in guessing the bit b in the two hybrids, $H^{\mathcal{B}}_i$ and $H^{\mathcal{B}}_{i-1}$ and the claim follows from the function privacy property.

Next, we define the following distribution $\mathcal{D}$ depending on $\mathcal{B}$.

- $\mathcal{D}(1^\lambda)$ gets $r \xleftarrow{R} \{0, 1\}^\lambda$, samples a key pair $(\mathsf{vk}, \mathsf{msk} \leftarrow \mathsf{FS.Setup}(1^\lambda))$ and generates the signing key for the all-accepting function 1 by $\mathsf{sk}_1 \leftarrow \mathsf{FS.KeyGen}(\mathsf{msk}, 1)$.
- Using sk_1 and vk, $\mathcal{D}$ simulates the action of $\mathcal{B}$ in experiment $H^{\mathcal{B}}_q$ up to the point in which $\mathcal{B}$ outputs a pair of challenge messages m_1, m_1. Denote by $\mathsf{view}_{\mathcal{B}}$ the current view of $\mathcal{B}$ up to this point of the simulation.
- $\mathcal{D}$ generates a signing key sk_C for the function C as defined above and machines $M_{m_0||r}$ and $M_{m_1||r}$ as defined above (recall that as usual we omit to specify the subscript relative to vk).
- $\mathcal{D}$ outputs the tuple $(M_{m_0||r}, M_{m_1||r}, z = (\mathsf{view}_{\mathcal{B}}, \mathsf{sk}_C))$.

We now can construct an adversary $\mathcal{A}(1^\lambda, M', M_0, M_1, z)$ against the security of $e\mathcal{O}$.

- $\mathcal{A}$ takes as input the security parameter 1^λ, an obfuscation M' of machine M_b for randomly chosen bit b, two machines M_0 and M_1, and auxiliary input $z = (\mathsf{view}_{\mathcal{B}}, \mathsf{sk}_C)$.
- Using $\mathsf{view}_{\mathcal{B}}$, $\mathcal{A}$ returns $\mathcal{B}$ to the state of execution as in the corresponding earlier simulation during the $\mathcal{D}$ sampling process.
- Simulate the challenge ciphertex to $\mathcal{B}$ as M'. For each subsequent token query M made by $\mathcal{B}$, $\mathcal{A}$ answers it by producing a signature on M using sk_C.
- Eventually, $\mathcal{B}$ outputs a bit b' for the challenge ciphertext that $\mathcal{A}$ returns as its own guess.

Note that the interaction with the adversary $\mathcal{B}$ in sampling from $\mathcal{D}$ is precisely a simulation in hybrid $H^{\mathcal{B}}_q$ up to the point in which $\mathcal{B}$ outputs the challenge messages, and the interaction with $\mathcal{B}$ made by $\mathcal{A}$ is precisely a simulation of the remaining steps in hybrid $H^{\mathcal{B}}_q$. We are assuming that the advantage of $\mathcal{B}$ in hybrid H^i_q is

$\geq 2a(\lambda)$ for some non-negligible function $a(\lambda)$. This implies that there is a polynomial $p(\lambda)$ such that for an infinite set S of values λ, it holds that the advantage of $\mathcal{B}$ in hybrid H_q^i for parameter λ is greater than $1/2p(\lambda)$. Thus, by an averaging argument, for all $\lambda \in S$, $\mathcal{A}$'s advantage (with respect to λ) in guessing the bit b on which it is challenged upon is greater than $1/p$ with probability greater than $1/p$ over the output of $\mathcal{D}$. By the security of $\mathsf{e}\mathcal{O}$ this implies a corresponding PPT extractor E and polynomial $q(\lambda)$ and negligible function $\mathsf{negl}(\lambda)$ such for all $\lambda \in S$, with probability $1 - \mathsf{negl}(\lambda)$ over the output (M_0, M_1, z) of $\mathcal{D}$, it holds that:

if $\Pr\left[b \overset{R}{\leftarrow} \{0,1\}; M' \leftarrow \mathsf{e}\mathcal{O}(1^\lambda, M_b) : \mathcal{A}(1^\lambda, M', M_0, M_1, z) = b \right] \geq \frac{1}{2} + \frac{1}{p(\lambda)}$,

then $\Pr\left[w \leftarrow \mathsf{E}(1^\lambda, M_0, M_1, z) : M_0(w) \neq M_1(w) \right] \geq 1/q(\lambda)$. This implies that for an infinite number of values λ, with probability $\geq 1/p(\lambda) - \mathsf{negl}(\lambda)$ over the output (M_0, M_1, z) of $\mathcal{D}$, it holds that

$\quad \Pr\left[w \leftarrow \mathsf{E}(1^\lambda, M_0, M_1, z) : M_0(w) \neq M_1(w) \right] \geq 1/q(\lambda)$.

We now show that such PPT extractor can not exist.

Claim 10. There can not exist a PPT extractor as above.

Suppose toward a contradiction that there exists such extractor that outputs a signature σ_A for some machine A , and a second input m_2 that distinguishes $M_{m_0\|r}$ from $M_{m_1\|r}$. We note that any signature output by the extractor must be a valid signature for a machine A for which the adversary asked a query. This follows from the unforgeability of FS. From this fact, and from the fact that the checker approved the triple (A, m_0, m_1), it follows that m_0 and m_1 are collision-resistant compatible with $\{A\}$. Therefore, this adversary can be used to break the collision-resistance compatibility with respect to m_0 and m_1 and $\{A\}$, contradicting the hypothesis.

It is trivial to see that the claim on input-specific run time holds if the scheme is used with Turing machines of input-specific run time and that the claim on the succinctness follows easily from our construction and the succinctness of FS. This concludes the proof.

References

1. Boneh, D., Sahai, A., Waters, B.: Functional encryption: definitions and challenges. In: Ishai, Y. (ed.) TCC 2011. LNCS, vol. 6597, pp. 253–273. Springer, Heidelberg (2011)
2. O'Neill, A.: Definitional issues in functional encryption. Cryptology ePrint Archive, Report 2010/556 (2010). http://eprint.iacr.org/
3. Bellare, M., O'Neill, A.: Semantically-secure functional encryption: possibility results, impossibility results and the quest for a general definition. In: Abdalla, M., Nita-Rotaru, C., Dahab, R. (eds.) CANS 2013. LNCS, vol. 8257, pp. 218–234. Springer, Heidelberg (2013)
4. Agrawal, S., Gorbunov, S., Vaikuntanathan, V., Wee, H.: Functional encryption: new perspectives and lower bounds. In: Canetti, R., Garay, J.A. (eds.) CRYPTO 2013, Part II. LNCS, vol. 8043, pp. 500–518. Springer, Heidelberg (2013)
5. De Caro, A., Iovino, V.: On the power of rewinding simulators in functional encryption. IACR Cryptology ePrint Archive, 2013:752 (2013)

6. Agrawal, S., Agrawal, S., Badrinarayanan, S., Kumarasubramanian, A., Prabhakaran, M., Sahai, A.: Function private functional encryption and property preserving encryption : new definitions and positive results. Cryptology ePrint Archive, Report 2013/744 (2013). http://eprint.iacr.org/
7. Bellare, M., Rogaway, P.: Random oracles are practical: a paradigm for designing efficient protocols. In: Ashby, V. (ed.) ACM CCS 93: 1st Conference on Computer and Communications Security, Fairfax, Virginia, USA, pp. 62–73. ACM Press, 3–5 November 1993
8. De Caro, A., Iovino, V., Jain, A., O'Neill, A., Paneth, O., Persiano, G.: On the achievability of simulation-based security for functional encryption. In: Canetti and Garay [22], pp. 519–535
9. Apon, D., Gordon, D., Katz, J., Liu, F.-H., Zhou, H.-S., Shi, E.: Personal Communication, July 2013
10. Gorbunov, S., Vaikuntanathan, V., Wee, H.: Functional encryption with bounded collusions via multi-party computation. In: Safavi-Naini, R., Canetti, R. (eds.) CRYPTO 2012. LNCS, vol. 7417, pp. 162–179. Springer, Heidelberg (2012)
11. Goldwasser, S., Gordon, S.D., Goyal, V., Jain, A., Katz, J., Liu, F.-H., Sahai, A., Shi, E., Zhou, H.-S.: Multi-input functional encryption. In: Nguyen, P.Q., Oswald, E. (eds.) EUROCRYPT 2014. LNCS, vol. 8441, pp. 578–602. Springer, Heidelberg (2014)
12. Dov Gordon, S., Katz, J., Liu, F.-H., Shi, E., Zhou, H.-S.: Multi-input functional encryption. IACR Cryptology ePrint Archive, 2013:774 (2013)
13. Goldwasser, S., Goyal, V., Jain, A., Sahai, A.: Multi-input functional encryption. Cryptology ePrint Archive, Report 2013/727 (2013). http://eprint.iacr.org/
14. Boyle, E., Chung, K.-M., Pass, R.: On extractability obfuscation. In: Lindell, Y. (ed.) TCC 2014. LNCS, vol. 8349, pp. 52–73. Springer, Heidelberg (2014)
15. Canetti, R., Goldreich, O., Halevi, S.: The random oracle methodology, revisited (preliminary version). In: 30th ACM STOCAnnual ACM Symposium on Theory of Computing, Dallas, Texas, USA, pp. 209–218. ACM Press, 23–26 May 1998
16. Iovino, V., Żebrowksi, K.: Simulation-based secure functional encryption in the random oracle model. Cryptology ePrint Archive, Report 2014/810 (2014). http://eprint.iacr.org/
17. Garg, S., Gentry, C., Halevi, S., Wichs, D.: On the implausibility of differing-inputs obfuscation and extractable witness encryption with auxiliary input. Cryptology ePrint Archive, Report 2013/860 (2013). http://eprint.iacr.org/
18. Boyle, E., Pass, R.: Limits of extractability assumptions with distributional auxiliary input. Cryptology ePrint Archive, Report 2013/703 (2013). http://eprint.iacr.org/
19. Feige, U., Lapidot, D., Shamir, A.: Multiple non-interactive zero knowledge proofs based on a single random string (extended abstract). In: 31st Annual Symposium on Foundations of Computer Science, St. Louis, Missouri, USA, vol. I, pp, 308–317. IEEE Computer Society, 22–24 October 1990
20. Boyle, E., Goldwasser, S., Ivan, I.: Functional signatures and pseudorandom functions. IACR Cryptology ePrint Archive, 2013:401 (2013)
21. Canetti, R., Garay, J.A. (eds.): CRYPTO 2013, Part II. LNCS, vol. 8043. Springer, Heidelberg (2013)

The Simplest Protocol for Oblivious Transfer

Tung Chou[1] and Claudio Orlandi[2(✉)]

[1] Technische Universieit Eindhoven, Eindhoven, The Netherlands
[2] Aarhus University, Aarhus, Denmark
orlandi@cs.au.dk

Abstract. Oblivious Transfer (OT) is the fundamental building block of cryptographic protocols. In this paper we describe the simplest and most efficient protocol for 1-out-of-n OT to date, which is obtained by tweaking the Diffie-Hellman key-exchange protocol. The protocol achieves UC-security against active and adaptive corruptions in the random oracle model. Due to its simplicity, the protocol is extremely efficient and it allows to perform m 1-out-of-n OTs using only:
- **Computation:** $(n+1)m + 2$ exponentiations (mn for the receiver, $mn + 2$ for the sender) and
- **Communication:** $32(m+1)$ bytes (for the group elements), and $2mn$ ciphertexts.

We also report on an implementation of the protocol using elliptic curves, and on a number of mechanisms we employ to ensure that our software is secure against active attacks too. Experimental results show that our protocol (thanks to both algorithmic and implementation optimizations) is at least one order of magnitude faster than previous work.

1 Introduction

Oblivious Transfer (OT) is a cryptographic primitive defined as follows: in its simplest flavour, 1-out-of-2 OT, a sender has two input messages M_0 and M_1 and a receiver has a choice bit c. At the end of the protocol the receiver is supposed to learn the message M_c and nothing else, while the sender is supposed to learn nothing. Perhaps surprisingly, this extremely simple primitive is sufficient to implement any cryptographic task [Kil88]. OT can also be used to implement most advanced cryptographic tasks, such as secure two- and multi-party computation (e.g., the millionaire's problem) in an efficient way [NNOB12,BLN+15].

Sender	Receiver
Input: (M_0, M_1)	Input: c
Output: none	Output: M_c

$$a \leftarrow \mathbb{Z}_p \qquad\qquad b \leftarrow \mathbb{Z}_p$$
$$\xrightarrow{\quad A = g^a \quad}$$

if $c = 0$: $B = g^b$
if $c = 1$: $B = Ag^b$

$$\xleftarrow{\quad B \quad}$$

$$k_0 = H\left(B^a\right) \qquad\qquad k_R = H(A^b)$$
$$k_1 = H\left(\left(\tfrac{B}{A}\right)^a\right)$$

$$e_0 \leftarrow E_{k_0}(M_0)$$
$$e_1 \leftarrow E_{k_1}(M_1)$$
$$\xrightarrow{\hspace{3cm}}$$

$$M_c = D_{k_R}(e_c)$$

Fig. 1. Our protocol in a nutshell

© Springer International Publishing Switzerland 2015
K. Lauter and F. Rodríguez-Henríquez (Eds.): LatinCrypt 2015, LNCS 9230, pp. 40–58, 2015.
DOI: 10.1007/978-3-319-22174-8_3

Given the importance of OT, and the fact that most OT applications require a very large number of OTs, it is crucial to construct OT protocols which are at the same time efficient and secure against realistic adversaries.

A Novel OT Protocol. In this paper we present a novel and extremely *simple, efficient* and *secure* OT protocol. The protocol is a simple tweak of the celebrated Diffie-Hellman (DH) key exchange protocol. Given a group $\mathbb{G}$ and a generator g, the DH protocol allows two players Alice and Bob to agree on a key as follows: Alice samples a random a, computes $A = g^a$ and sends A to Bob. Symmetrically Bob samples a random b, computes $B = g^b$ and sends B to Alice. Now both parties can compute $g^{ab} = A^b = B^a$ from which they can derive a key k. The key observation is now that Alice can also derive a different key from the value $(B/A)^a = g^{ab-a^2}$, and that Bob cannot compute this group element (assuming that the computational DH problem is hard).

We can now turn this into an OT protocol by letting Alice play the role of the sender and Bob the role of the receiver (with choice bit c) as shown in Fig. 1. The first message (from Alice to Bob) is left unchanged (and can be reused over multiple instances of the protocol) but now Bob computes B as a function of his choice bit c: if $c = 0$ Bob computes $B = g^b$ and if $c = 1$ Bob computes $B = Ag^b$. At this point Alice derives two keys k_0, k_1 from $(B)^a$ and $(B/A)^a$ respectively. It is easy to check that Bob can derive the key k_c corresponding to his choice bit from A^b, but cannot compute the other one. This can be seen as a *random OT* i.e., an OT where the sender has no input but instead receives two random messages from the protocol, which can be used later to encrypt his inputs.

We show that combining the above random OT protocol with the right symmetric encryption scheme (e.g., a *robust encryption scheme* [ABN10, FLPQ13]) achieves security in a strong, simulation based sense and in particular we prove UC-security against active and adaptive corruptions in the random oracle model.

A Secure and Efficient Implementation. We report on an efficient and secure implementation of the 1-out-of-2 random OT protocol: Our choice for the group is a twisted Edwards curve that has been used by Bernstein, Duif, Lange, Schwabe and Yang for building the Ed25519 signature scheme [BDL+11]. The security of the curve comes from the fact that it is birationally equivalent to Bernstein's Montgomery curve Curve25519 [Ber06] where ECDLP is believed to be hard: Bernstein and Lange's SafeCurves website [BL14] reports cost of $2^{125.8}$ for solving ECDLP on Curve25519 using the *rho method*. The speed comes from the complete formulas for twisted Edwards curves proposed by Hisil, Wong, Carter, and Dawson in [HWCD08].

We first modify the code in [BDL+11] and build a fast implementation for a single OT. Later we build a vectorized implementation that runs OTs in batches. A comparison with the state of the art shows that our vectorized implementation is at least an order of magnitude faster than previous work (we compare in particular with the implementation reported by Asharov, Lindell, Schneider and Zohner in [ALSZ13]) on recent Intel microarchitectures. Furthermore, we take great care to make sure that our implementation is secure against both passive attacks (our software is *immune to timing attacks*, since the implementation is *constant-time*) and active attacks (by designing an appropriate encoding of group elements, which can be efficiently verified and computed on). Our code can be downloaded from http://orlandi.dk/simpleOT.

Related Work. OT owes its name to Rabin [Rab81] (a similar concept was introduced earlier by Wiesner [Wie83] under the name of "conjugate coding"). There are different flavours of OT, and in this paper we focus on the most common and useful flavour, namely $\binom{n}{1}$-OT, which was first introduced in [EGL85]. Many efficient protocols for OT have been proposed over the years. Some of the protocols which are most similar to ours are those of Bellare-Micali [BM89] and Naor-Pinkas [NP01]: those protocols are (slightly) less efficient than ours and, most importantly, are not known to achieve full simulation based security. More recent OT protocols such as [HL10, DNO08, PVW08] focus on achieving a strong level of security in concurrent settings[1] without relying on the random oracle model. Unfortunately this makes these protocols more cumbersome for practical applications: even the most efficient of these protocols i.e., the protocol of Peikert, Vaikuntanathan, and Waters [PVW08] requires 11 exponentiations for a single $\binom{2}{1}$-OT and a common random string (which must be generated by some trusted source of randomness at the beginning of the protocol). In comparison our protocol uses fewer exponentiations (e.g., 5 for $\binom{2}{1}$-OT), generalizes to $\binom{n}{1}$-OT and does not require any (hard to implement in practice) setup assumptions.

OT Extension. While OT provably requires "public-key" type of assumptions [IR89] (such as factoring, discrete log, etc.), OT can be "extended" [Bea96] in the sense that it is enough to generate few "seed" OTs based on public-key cryptography which can then be extended to any number of OTs using symmetric-key primitives only (PRG, hash functions, etc.). This can be seen as the OT equivalent of *hybrid encryption* (where one encrypts a large amount of data using symmetric-key cryptography, and then encapsulates the symmetric-key using a public-key cryptosystem). OT extension can be performed very efficiently both against passive [IKNP03, ALSZ13] and active [Nie07, NNOB12, Lar14, ALSZ15, KOS15] adversaries. Still, to bootstrap OT extension we need a secure and efficient OT protocol for the seed OTs (as much as we need secure and efficient public-key encryption schemes to bootstrap hybrid encryption): The OT extension of [ALSZ15] reports that it takes time $(7 \cdot 10^5 + 1.3m)\mu s$ to perform m OTs, where the fixed term comes from running 190 base OTs. Using our protocol as the base OT in [ALSZ15] would reduce the initial cost to approximately $190 \cdot 114 \approx 2 \cdot 10^4 \mu s$ [Sch15], which leads to a significant overall improvement (e.g., a factor 10 for up to $4 \cdot 10^4$ OTs and a factor 2 for up to $5 \cdot 10^5$ OTs).

2   The Protocol

Notation. If S is a set $s \leftarrow S$ is a random element sampled from S. We work over an additive group $(\mathbb{G}, B, p, +)$ of prime order p (with $\log(p) > \kappa$) generated

---

[1] I.e., UC security [Can01], which is impossible to achieve without some kind of trusted setup assumptions [CF01].

by B (the base point), and we use the additive notation for the group since we later implement our protocol using elliptic curves. Given the representation of some group element P we assume it is possible to efficiently verify if $P \in \mathbb{G}$. We use $[n]$ as a shortcut for $\{0, 1, \ldots, n-1\}$.

Building Blocks. We use a hash-function $H : (\mathbb{G} \times \mathbb{G}) \times \mathbb{G} \to \{0, 1\}^\kappa$ as a key-derivation function to extract a κ bit key from a group element, and the first two inputs are used to seed the function.[2] We model H as a random oracle when arguing about the security of our protocol.

The Ideal Functionality. We want to implement m $\binom{n}{1}$-OT's for ℓ-bit messages with κ-bit security between a sender $\mathcal{S}$ and a receiver $\mathcal{R}$. We define a functionality $\mathcal{F}_{OT}^-(n, m, \ell)$ as follows:

Honest Use: the functionality receives a vector of indices $(c^1, \ldots, c^m) \in [n]^m$ from the receiver $\mathcal{R}$ and m vectors of message $\{(M_0^i, \ldots, M_{n-1}^i)\}_{i \in [m]}$ from the sender $\mathcal{S}$ where for all $i, j : M_i^j \in \{0, 1\}^\ell$. The functionality outputs a vector of ℓ-bit strings $(z^1, \ldots, z^n)$ to the receiver $\mathcal{R}$, such that for all $i \in [m]$, $z^i = M_{c^i}^i$.

Dishonest Use: We weaken the functionality (hence the minus in the name) in the following way: a corrupted receiver $\mathcal{R}^*$ can input the choice values in an adaptive fashion i.e., the ideal adversary can input the choice indices c^i one by one and learn the message z^i before choosing the next index.

Note that when $m = 1$ the weakening has no effect. We choose to describe the protocol for m OTs in parallel since we can do this more efficiently than simply repeating m times the protocol for a single OT.

2.1 Random OT

We split the presentation in two parts: first, we describe and analyze a protocol for *random OT* where the sender outputs n random keys and the receiver only learns one of them; then, we describe how to combine this protocol with an appropriate encryption scheme to complete the OT. We are now ready to describe our novel *random OT* protocol:

Setup: (only once, independent of m):
  1. $\mathcal{S}$ samples $y \leftarrow \mathbb{Z}_p$ and computes $S = yB$ and $T = yS$;
  2. $\mathcal{S}$ sends S to $\mathcal{R}$, who aborts if $S \notin \mathbb{G}$;
Choose: (in parallel for all $i \in [m]$)
  1. $\mathcal{R}$ (with input $c^i \in [n]$) samples $x^i \leftarrow \mathbb{Z}_p$ and computes

$$R^i = c^i S + x^i B$$

  2. $\mathcal{R}$ sends R^i to $\mathcal{S}$, who aborts if $R^i \notin \mathbb{G}$;

---

[2] Standard hash functions do not take group elements as inputs, and in later sections we will give explicit encodings of group elements into bitstrings.

Key Derivation: (in parallel for all $i \in [m]$)
1. For all $j \in [n]$, $\mathcal{S}$ computes

$$k_j^i = H_{(S,R^i)}(yR^i - jT)$$

2. $\mathcal{R}$ computes

$$k_R^i = H_{(S,R^i)}(x^i S)$$

Basic Properties. The key k_j^i is computed by hashing $x^i y B + (c^i - j)T$ and therefore at the end of the protocol $k_R^i = k_{c^i}^i$ if both parties are honest. It is also easy to see that:

Lemma 1. *No (computationally unbounded) $\mathcal{S}^*$ on input R^i can guess c^i with probability greater than $1/n$.*

Proof. Since B generates $\mathbb{G}$, fixed any $P = x_0 B$ the probability that $R^i = P$ when $c^i = j$ is the probability that $x^i = (x_0 - c^i y)$, therefore $\forall S, P \in \mathbb{G}, j \in [n]$, $\Pr[R^i = P | c^i = j] = 1/p$, which is independent of j.

Lemma 2. *No (computationally bounded) $\mathcal{R}^*$ can output any two keys $k_{j_0}^i$ and $k_{j_1}^i$ with $j_0 \neq j_1 \in [n]$ if the computational Diffie-Hellman problem is hard in $\mathbb{G}$.*

Proof. In the random oracle model $\mathcal{R}^*$ can only (except with negligible probability) compute $k_{j_0}^i, k_{j_1}^i$ by querying the oracle on points of the form $U_0^i = (yR^i - j_0T)$ and $U_1^i = (yR^i - j_1T)$. Assume for the sake of contradiction that there exist a PPT $\mathcal{R}^*$ who outputs $(R, j_0, j_1, U_0, U_1) \leftarrow \mathcal{R}^*(B, S)$ such that $(j_1 - j_0)^{-1}(U_0 - U_1) = T = \log_B(S)^2 B$ with probability at least ϵ. We show an algorithm $\mathcal{A}$ which on input $(B, X = xB, Y = yB)$ outputs $Z = xyB$ with probability greater than ϵ^3. Run $(R^X, U_0^X, U_1^X) \leftarrow \mathcal{R}^*(B, X)$, $(R^Y, U_0^Y, U_1^Y) \leftarrow \mathcal{R}^*(B, Y)$, then run $(R^+, U_0^+, U_1^+) \leftarrow \mathcal{R}^*(B, X+Y)$ and finally output

$$Z = \frac{(p+1)}{2}\left((U_0^+ + U_1^+) - (U_0^X + U_1^X) - (U_0^Y + U_1^Y)\right)$$

Now $Z = xyB$ with probability at least ϵ^3, since when all three executions of $\mathcal{R}^*$ are successful, then $U_0^X + U_1^X = (x^2)B$, $U_0^Y + U_1^Y = (y^2)B$ and $U_0^+, U_1^+ = (x+y)^2 B$ and therefore $Z = \frac{p+1}{2}2xyB = xyB$. $\square$

Note that the above proof loses a cubic factor. A better proof for this lemma, which only loses a quadratic factor, can be found in [BCP04].

2.2 How to Use the Protocol and UC Security

We now show how to combine our random OT protocol with an appropriate encryption scheme to achieve UC security.

Motivation. Lemmas 1 and 2 only state that some form of "privacy" holds for both the sender and the receiver. However, since OT is mostly used as a building block into more complex protocols, it is important to understand to which extent our protocol offers security when composed arbitrarily with itself or other protocols: Simulation based security is the minimal requirement which enables to argue that a given protocol is secure when composed with other protocols. Without simulation based security, it is not even possible to argue that a protocol is secure if it is executed twice in a sequential way! (See e.g., [DNO08] for a concrete counterexample for OT). The UC theorem [Can01] allows us to say that if a protocol satisfies the UC definition of security, then that protocol will be secure even when arbitrarily composed with other protocols. Among other things, to show that a protocol is UC secure one needs to show that a simulator can *extract* the input of a corrupted party: intuitively, this is a guarantee that the party *knows* its input, and its not reusing/modifying messages received in other protocols (aka malleability attack).

From Random OT to *standard* OT. We start by adding a transfer phase to the protocol, where the sender sends the encryption of his messages to the receiver:

Transfer: (in parallel for all $i \in [m]$)
    1. For all $j \in [n]$, S computes $e_j^i \leftarrow E(k_j^i, M_j^i)$
    2. S sends $(e_0^i, \ldots, e_{n-1}^i)$ to $\mathcal{R}$;
Retrieve: (in parallel for all $i \in [m]$)
    1. $\mathcal{R}$ computes and outputs $z^i = D(k^i, e_{c^i}^i)$.

The Encryption Scheme. The protocol uses a symmetric encryption scheme (E, D). We call $\mathcal{K}, \mathcal{M}, \mathcal{C}$ the key space, message space and ciphertext space respectively and κ the security parameter. We allow the decryption algorithm to output a special symbol $\perp$ to indicate an invalid ciphertext. We need the encryption scheme to satisfy the following properties:

Definition 1. *We say a symmetric encryption scheme (E, D) is non-committing if there exist PPT algorithms S_1, S_2 such that $\forall M \in \mathcal{M}$ (e', k') and (e, k) are computationally indistinguishable where $e' \leftarrow S_1(1^\kappa)$, $k' \leftarrow S_2(e', M)$, $k \leftarrow \mathcal{K}$ and $e \leftarrow E(k, M)$ (S_1, S_2 are allowed to share a state).*

The definition says that it is possible for a simulator to come up with a ciphertext e which can later be "explained" as an encryption of any message, in such a way that the joint distribution of the ciphertext and the key in this simulated experiment is indistinguishable from the normal use of the encryption scheme, where a key is first sampled and then an encryption of M is generated.

Definition 2. *Let S be a set of random keys from $\mathcal{K}$ and $V_{S,e} \subseteq S$ the subset of valid keys for a given ciphertext e i.e., the keys in S such that $D(k, e) \neq \perp$.*
    We say (E, D) satisfies robustness if for all ciphertexts $e \leftarrow \mathcal{A}(1^\kappa, S)$ adversarially generated by a PPT $\mathcal{A}$, $|V_{S,e}| \leq 1$ except with negligible probability.

The definition says that it should be hard for an adversary to generate a ciphertext which can be decrypted to more than one valid ciphertext using any polynomial number of randomly generated keys (even for adversaries who see those keys before generating the ciphertext).

A Concrete Example. We give a concrete example of a very simple scheme which satisfies Definitions 1 and 2: let $\mathcal{M} = \{0,1\}^\ell$ and $\mathcal{K} = \mathcal{C} = \{0,1\}^{\ell+\kappa}$. The encryption algorithm $E(k,m)$ parses k as (α, β) and $e = (m \oplus \alpha, \beta)$. The decryption algorithm $D(k,e)$ parses $k = (\alpha, \beta)$ and $e = (e_1, e_2)$ and outputs $\bot$ if $e_2 \neq \beta$ or outputs $m = e_1 \oplus \alpha$ otherwise. It can be shown that:

Lemma 3. *The scheme (E, D) defined above satisfies Definitions 1 and 2.*

Proof. We show that the scheme satisfies Definitions 1 and 2 in a strong, information theoretic sense. For Definition 1: $\mathcal{S}_1$ outputs a random $e \leftarrow \{0,1\}^{\ell+\kappa}$; $\mathcal{S}_2(e, M)$ parses $e = (e_1, e_2)$ and outputs $k = (e_1 \oplus M, e_2)$. The simulated distribution is trivially identical to the real one. For Definition 2: given any ciphertext $e = (e_1, e_2)$, $D((\alpha, \beta), e) \neq \bot$ implies that $\beta = e_2$. Thus even an unbounded adversary can break robustness of the scheme only if there are two keys $k_i, k_j \in S$ such that $\beta_i = \beta_j$ which only happens with probability negligible in κ.

2.3   Simulation Based Security (UC)

We can finally argue UC security of our protocol.[3] The main ideas behind the proof are: it is possible to extract the choice value by checking whether a corrupted receiver queries the random oracle on points of the form $yR^i - cT$ for some c, since no adversary can query on points of this form for more than one c (without breaking the CDH assumption) and the *non-committing* property of (E, D) allows us to complete a successful simulation even if the corrupted receiver queries the oracle *after* he receives the ciphertexts; it is also possible to extract the sender messages by decrypting the ciphertexts with every key which the receiver got from the random oracle and Definition 2) allows us to conclude that except with negligible probability D returns $\bot$ for all keys different from the correct one.

Theorem 1. *The above protocol securely implements the functionality $\mathcal{F}_{OT}^-(n, m, \ell)$ under the following conditions:*

Corruption Model: *any active, adaptive corruption;*
Hybrid Functionalities: *we model H as a random oracle and we assume an authenticated channel (but not confidential) between the parties;*
Computational Assumptions: *we assume that the symmetric encryption scheme (E, D) satisfies Definitions 1 and 2 and the computational Diffie-Hellman problem is hard in $\mathbb{G}$.*

Proof. We prove our theorem in steps: first, we show that the protocol is secure if the adversary corrupts the sender or the receiver at the beginning of the

---

[3] This subsection assumes that the reader is familiar with standard security definitions and proofs for two-party computation protocols such as those presented in [HL10].

protocol (i..e, *static* corruptions). Then we show that the protocol is secure if the adversary corrupts both parties at the end of the protocol (*post-execution corruption*).

(*Corrupted Sender*) First we argue that our protocol securely implements the functionality against a corrupted sender in the random oracle model (we will in particular use the property that the simulator can learn on which points the oracle was queried on), by constructing a simulator for a corrupted S^* in the following way:[4] (1) in the first phase, the simulator answers random oracle queries $H_{(\cdot,\cdot)}(\cdot)$ at random; (2) at some point S^* outputs S and the simulator checks that $S \in \mathbb{G}$ or aborts otherwise; (3) the simulator now chooses a random x^i for all $i \in [m]$ and sends $R^i = x^i B$ to S^*. Note that since x^i is chosen at random the probability that S^* had queried any oracle $H_{(S,R^i)}(\cdot)$ before is negligible. At this point, any time S^* makes a query of the form $H_{(S,R^i)}(P^q)$, the simulator stores its random answers in $k^{i,q}$; (4) Now S^* outputs $(e_0^i, \ldots, e_{n-1}^i)$ and the simulator computes for all i, j the value M_j^i in the following way: for all q compute $D(k^{i,q}, e_j^i)$ and set M_j^i to be the first such value which is $\neq \bot$ (if any), or $\bot$ otherwise; (5) finally it inputs all the vectors $(M_0^i, \ldots, M_{n-1}^i)$ to the ideal functionality. We now argue that no distinguisher can tell a real-world view apart from a simulated view. This follows from Lemma 1 (the distribution of R^i does not depend on c^i), and that the output of the honest receiver can only be different if there exists a pair (i, j) such that the adversary queried the random oracle on a point $P' \neq yR^i - jT$ and $M' = D(k', e_i^j) \neq \bot$, where $k' = H_{(S,R^i)}(P')$. In this case the simulator will input $M_j^i = M'$ to the ideal functionality which could cause the honest receiver in the ideal world to output a different value than it would in the real world (if $c^i = j$). But this happens only with negligible probability thanks to the property of the encryption scheme (Definition 2).

(*Corrupted Receiver*) We now construct a simulator for a corrupted receiver[5]: (1) In the first phase, the simulator answers random oracle queries $H_{(\cdot,\cdot)}(\cdot)$ truly at random; (2) at some point the simulator samples a random y and outputs $S = yB$. Afterwards it keeps answering oracle queries at random, but for each query of the form $k^q = H_{(S,P^q)}(Q^q)$ it saves the triple (k^q, P^q, Q^q) (since y is random the probability that any query of the form $H_{(S,\cdot)}(\cdot)$ was performed before is negligible); (3) at some point the simulator receives a vector of elements R^i and aborts if $\exists i : R^i \notin \mathbb{G}$; (4) the simulator now initializes all $c^i = \bot$; for each tuple q in memory such that for some i it holds that $P^q = R^i$ the simulator checks if $Q^q = y(R^i - dS)$ for some $d \in [n]$. Now the simulator saves this value d in c^i if c^i had not been defined before or aborts otherwise. In other words, when the simulator finds a candidate choice value d for some i it checks if it had already found a choice value for that i (i.e., $c^i \neq \bot$) and if so it aborts and

---

[4] The main goal of this argument is to show that a corrupted sender *knows* the message vectors.

[5] The main goal of this argument is to show that a corrupted receiver *knows* the choice value.

outputs `fail`, otherwise if it had not found a candidate choice bit for i before (i.e., $c^i = \bot$) it sets $c^i = d$; (5) When the adversary is done querying the random oracle, the simulator has to send all ciphertexts vectors $\{(e_0^i, \ldots, e_{n-1}^i)\}_{i \in [m]}$: $\forall i \in [m], j \in [n]$ the simulator sets a) if $c^i = \bot$: $e_j^i = \mathcal{S}_1(1^\kappa)$ b) if $j \neq c^i$: $e_j^i = \mathcal{S}_1(1^\kappa)$ and c) if $j = c^i$: $e_j^i = E(k_{c^i}^i, z^i)$; (6) at this point the protocol is almost over but the simulator can still receive random oracle queries. As before, the simulator answers them at random except if the adversary queries on some point $H_{(S,R^i)}(Q^q)$ with $Q^q = y(R^i - dS)$. If this happens for any i such that $c^i \neq \bot$ the simulator aborts and outputs `fail`. Otherwise the simulator sets $c^i = d$, inputs c^i to the ideal functionality, receives z^i and programs the random oracle to output $k' \leftarrow \mathcal{S}_2(e_{c^i}^i, z^i)$.

Now to conclude our proof, we must argue that a simulated view is indistinguishable from the view of a corrupted party in an execution of the protocol. When the simulator does not output `fail` indistinguishability follows immediately from Definition 1. Finally the simulator only outputs `fail` if $\mathcal{R}^*$ queries the oracle on two points U_0, U_1 such that $U_1 - U_0$ is a known multiple of $y^2 B$, and as argued in Lemma 2 such an adversary can be used to break the CDH assumption.

(Post-Execution Corruptions) We now construct a simulator for an adversary that corrupts adaptively either/both of the two parties after the protocol is over. This is the hardest case and it is easy to see how our simulator can be adapted for an adversary who corrupts either party during the protocol execution. Since we are not assuming confidential channels the simulator needs to produce a view even while both parties are honest: the simulator (1) samples random $y, x_0^i \leftarrow \mathbb{Z}_p$ and computes $S = yB$ and $R^i = x_0^i B$ for all $i \in [m]$; computes and stores the values x_j^i for all $j \in [n]$ as $x_j^i = x_0^i - jy$ (those are the values which are consistent with the view of the protocol for a receiver with input $c^i = j$, since $R^i = x_0^i B = jS + x_j^i B$); (3) computes and stores the values $Q_j^i = y(R^i - jS)$ for all $i \in [m], j \in [n]$; (4) the simulator computes $e_j^i \leftarrow \mathcal{S}_1(1^\kappa)$ and outputs S, R^i, e_j^i for all $i \in [m], j \in [n]$; (5) The simulator starts answering all random oracle queries at random except for queries of the form $H_{(S,R^i)}(Q_j^i)$, in which case it aborts. When/if the adversary corrupts the sender, the simulator learns M_i^j for all $i \in [m], j \in [n]$, runs $k_j^i \leftarrow \mathcal{S}_2(e_j^i, M_j^i)$ and programs the random oracle to answer $k_j^i = H_{(S,R^i)}(Q_j^i)$. When/if the adversary corrupts the receiver, the simulator learns $c^i, z^i = M_{c^i}^i$ for all $i \in [m]$, runs $k_{c^i}^i \leftarrow \mathcal{S}_2(e_{c^i}^i, z^i)$ and programs the random oracle to answer $k_{c^i}^i = H_{(S,R^i)}(Q_j^i)$. If the simulator does not abort the simulated view is indistinguishable from a real view of the protocol, as the distribution of S, R^i is identical in both cases (see Lemma 1) and thanks to non-committing property of (E, D) (Definition 1). Using the same argument as above, we can show that an adversary that makes the simulator abort (i.e., queries the oracle on a point of the form (S, R^i, Q_j^i)) can be used to break the CDH assumption.

Non-Malleability in Practice. Clearly, a proof that $a \to b$ only says that b is true when a is true, and since cryptographic security models (a) are not always a good approximation of the real world, we discuss some of these discrepancies here and therefore to which extent our protocol achieves security in practice (b), with particular focus on malleability attacks.

When instantiating our protocol we must replace the random oracle with a hash function: UC proofs crucially rely on the fact that the oracle is *local* to the protocol i.e., it can be only queried by the protocol participants, and different instances of the protocol run with different random oracles: clearly, there is no such a thing in the real world. To approximate the model, one can "localize" the random oracle by prepending the parties id's and the session id to the hash function. We argue here that our choice of using the transcript of the protocol (S, R^i) as salt for the hash function helps in making sure that the oracle is *local* to the protocol, and helps against malleability attacks in cases where the parties' and session id's are unavailable. Consider the following man-in-the middle attack, where an adversary $\mathcal{A}$ plays two copies of the $\binom{n}{1}$-OT, one as the sender with $\mathcal{R}$ and one as the receiver with $\mathcal{S}$. Here is how the attack works: (1) $\mathcal{A}$ receives S from $\mathcal{S}$ and forwards it to $\mathcal{R}$; (2) Then the adversary receives R from $\mathcal{R}$ and sends $R' = S + R$ to $\mathcal{S}$; (3) Finally $\mathcal{A}$ receives the $\{e_i\}_{i \in [n]}$ from $\mathcal{S}$ and sets $e'_i = e_{(i-1 \mod n)}$ to $\mathcal{R}$. It is easy to see that if the same hash function is used to instantiate the random oracle in the two protocols (and if $c \neq 0$), then the honest receiver outputs $z = M_{c+1}$, which is clearly a breach of security (i.e., this attack could not be run if the protocols are replaced with OT functionalities).

The previous can be seen as a malleability attack on the choice bit. An adversary can also try a malleability attack on the sender messages by forwarding $(S', R') = (S, R)$ but then manipulating the e_i's into ciphertexts e'_i which decrypt to related messages. In the $\binom{2}{1}$-OT, these attacks can be mitigated by using *authenticated encryption* for (E, D) (which also satisfies *robustness* as in Definition 2). Now an adversary who changes both ciphertext is equivalent to an ideal adversary using input $(\bot, \bot)$, while an adversary who only changes one ciphertext, say e_c, is equivalent to an adversary which uses input bit $1 - c$ on the left and inputs $(m_{1-c}, \bot)$ on the right. Unfortunately for $\binom{n}{1}$-OT (with $n > 2$) this does not work. For instance, an adversary who corrupts only 1 out of m ciphertext cannot be simulated having access to ideal functionalities.

Finally we note that no practical instantiation of the encryption scheme leads to a *non-committing* encryption scheme (as required in Definition 1), but we conjecture that this an artificial requirement and does not lead to any concrete vulnerabilities.

3 The Random OT Protocol in Practice

This section describes how the random OT protocol can be realized in practice. In particular, this section focuses on describing how group elements are represented as bitstrings, i.e., the *encodings*. In the abstract description of the random OT protocol, the sender and the receiver transmit and compute on "group elements", but clearly any implementation of the protocol transmits and computes on bitstrings. We describe how the encodings are designed to achieve efficiency (both for communication and computation) and security (particularly against a malicious party who might try to send malformed encodings).

The Group. The group $\mathbb{G}$ we choose for the protocol is a subset of $\bar{\mathbb{G}}$; $\bar{\mathbb{G}}$ is defined by the set of points on the twisted Edwards curve

$$\{(x, y) \in \mathbb{F}_{2^{255}-19} \times \mathbb{F}_{2^{255}-19} : -x^2 + y^2 = 1 + dx^2 y^2\}$$

and the twisted Edwards addition law

$$(x_1, y_1) + (x_2, y_2) = \left(\frac{x_1 y_2 + x_2 y_1}{1 + dx_1 x_2 y_1 y_2}, \frac{y_1 y_2 + x_1 x_2}{1 - dx_1 x_2 y_1 y_2} \right)$$

introduced by Bernstein, Birkner, Joye, Lange, and Peters in [BBJ+08]. The constant d and the generator B can be found in [BDL+11]. The two groups $\bar{\mathbb{G}}$ and $\mathbb{G}$ are isomorphic respectively to $\mathbb{Z}_p \times \mathbb{Z}_8$ and $\mathbb{Z}_p$ with $p = 2^{252} + 27742317777372353535851937790883648493$.

Encoding of Group Element. An *encoding* $\mathcal{E}$ for a group $\mathbb{G}_0$ is a way of representing group elements as fixed-length bitstrings. We write $\mathcal{E}(P)$ for a bitstring which represents $P \in \mathbb{G}_0$. Note that there can be multiple bitstrings that represent P; if there is only one bitstring for each group element, $\mathcal{E}$ is said to be *deterministic* ($\mathcal{E}$ is said to be *non-deterministic* otherwise[6]). Also note that some bitstrings (of the fixed length) might not represent any group element; we write $\mathcal{E}(\mathbb{G}_1)$ for the set of bitstrings which represent some element in $\mathbb{G}_1 \subseteq \mathbb{G}_0$. $\mathcal{E}$ is said to be *verifiable* if there exists an efficient algorithm that, given a bitstring as input, outputs whether it is in $\mathcal{E}(\mathbb{G}_0)$ or not.

The Encoding $\mathcal{E}_X$ for Group Operations. The non-deterministic encoding $\mathcal{E}_X$ for $\bar{\mathbb{G}}$, which is based on the *extended coordinates* in [HWCD08], represents each point using the tuple $(X : Y : Z : T)$ with $XY = ZT$, representing $x = X/Z$ and $y = Y/Z$. We use $\mathcal{E}_X$ whenever we need to perform group operations since given $\mathcal{E}_X(P), \mathcal{E}_X(Q)$ where $P, Q \in \bar{\mathbb{G}}$, it is efficient to compute $\mathcal{E}_X(P+P), \mathcal{E}_X(P+Q)$, and $\mathcal{E}_X(P-Q)$. In particular, given an integer scalar $r \in \mathbb{Z}_p$ it is efficient to compute $\mathcal{E}_X(rB)$, and given r and $\mathcal{E}_X(P)$ it is efficient to compute $\mathcal{E}_X(rP)$.

The Encoding $\mathcal{E}_0$ and Related Encodings. The deterministic encoding $\mathcal{E}_0$ for $\bar{\mathbb{G}}$ represents each group element as a 256-bit bitstring: the natural 255-bit encoding of y followed by a sign bit which depends only on x. The way to recover the full value x is described in [BDL+11, Sect. 5], and group membership can be verified efficiently by checking whether $x^2(y^2 - 1) = dy^2 + 1$ holds; therefore $\mathcal{E}_0$ is verifiable. See [BDL+11] for more details of $\mathcal{E}_0$.

For the following discussions, we define deterministic encodings $\mathcal{E}_1$ and $\mathcal{E}_2$ for $\mathbb{G}$ as

$$\mathcal{E}_1(P) = \mathcal{E}_0(8P), \mathcal{E}_2(P) = \mathcal{E}_0(64P), P \in \mathbb{G}.$$

---

[6] We stress that non-deterministic in this context does not mean that the encoding involves any randomness.

We also define non-deterministic encodings $\mathcal{E}^{(0)}$ and $\mathcal{E}^{(1)}$ for $\mathbb{G}$ as

$$\mathcal{E}^{(0)}(P) = \mathcal{E}_0(P + t), \mathcal{E}^{(1)}(P) = \mathcal{E}_0(8P + t'), P \in \mathbb{G},$$

where t, t' can be any 8-torsion point. Note that each element in $\mathbb{G}$ has exactly 8 representations under $\mathcal{E}^{(0)}$ and $\mathcal{E}^{(1)}$.

Point Compression/Decompression. It is efficient to convert from $\mathcal{E}_X(P)$ to $\mathcal{E}_0(P)$ and back; since $\mathcal{E}_0$ represents points as much shorter bitstrings, these operations are called *point compression* and *point decompression*, respectively. Roughly speaking, point compression outputs $y = Y/Z$ along with the sign bit of $x = X/Z$, and point decompression first recovers x and then outputs $X = x, Y = y, Z = 1, T = xy$. We always check for group membership during point decompression.

We use $\mathcal{E}_0$ for data transmission: the parties send bitstrings in $\mathcal{E}_0(\bar{\mathbb{G}})$ and expect to receive bitstrings in $\mathcal{E}_0(\bar{\mathbb{G}})$. This means a computed point encoded by $\mathcal{E}_X$ has to be compressed before it is sent, and a received bitstring has to be decompressed for subsequent group operations. Sending compressed points helps to reduce the communication complexity: the parties only need to transfer $32 + 32m$ bytes in total.

Secure Data Transmission. At the beginning of the protocol S computes and sends $\mathcal{E}_0(S)$. In the ideal case, R should receive a bitstring in $\mathcal{E}_0(\mathbb{G})$ which he interprets as $\mathcal{E}_0(S)$. However, an attacker (a corrupted S^* or a man-in-the-middle) can send R 1) a bitstring that is not in $\mathcal{E}_0(\bar{\mathbb{G}})$ or 2) a bitstring in $\mathcal{E}_0(\bar{\mathbb{G}} \setminus \mathbb{G})$. In the first case, R detects that the received bitstring is not valid during point decompression and ignores it. In the second case, R can check group membership by computing the pth multiple of the point, but a more efficient way is to use a new encoding $\mathcal{E}'$ such that each bitstrings in $\mathcal{E}_0(\bar{\mathbb{G}})$ represents a point in $\mathbb{G}$ under $\mathcal{E}'$. Therefore R considers the received bitstring as $\mathcal{E}^{(0)}(S) = \mathcal{E}_0(S+t)$, where t can be any 8-torsion point.

The encoding $\mathcal{E}^{(0)}$ (along with point decompression) makes sure that R receives bitstrings representing elements in $\mathbb{G}$. However, an attacker can derive c^i by exploiting the extra information given by a nonzero t: a naive R would compute and send $\mathcal{E}_0(c^i(S + t) + x^i B) = \mathcal{E}_0(c^i t + R^i)$; now by testing whether the result is $\mathcal{E}_0(\mathbb{G})$ the attacker learns whether $c^i = 0$.

To get rid of the 8-torsion point, R can multiply received point by $8 \cdot (8^{-1} \mod p)$, but a more efficient way is to just multiply by 8 and then operate on $\mathcal{E}_X(8S)$ and $\mathcal{E}_X(8x^i B)$ to obtain and send $\mathcal{E}_1(R^i) = \mathcal{E}_0(8R^i)$, i.e., the encoding switches to $\mathcal{E}_1$ for R^i. After this S works similarly as R: to ensure that the received bitstring represents an element in $\mathbb{G}$, S interprets the bitstring as $\mathcal{E}^{(1)}(R^i) = \mathcal{E}_0(8R^i + t)$; to get rid of the 8-torsion point S also multiplies the received point by 8, and then S operates on $\mathcal{E}_X(64R^i)$ and $\mathcal{E}_X(64T)$ to obtain $\mathcal{E}_X(64(yR^i - jT))$.

Key Derivation. The protocol computes $H_{S,R^i}(P)$ where P can be $x^i S, yR^i$, or $yR^i - jT$ for $j \in [n]$. This is implemented by hashing $\mathcal{E}_1(S) \parallel \mathcal{E}_2(R^i) \parallel \mathcal{E}_2(P)$

with Keccak [BDPVA09] with 256-bit output. The choice of encodings is natural: $\mathcal{S}$ computes $\mathcal{E}_X(S)$, and $\mathcal{R}$ computes $\mathcal{E}_X(8S)$; since multiplication by 8 is much cheaper than multiplication by $(8^{-1} \mod p)$, we use $\mathcal{E}_1(S) = \mathcal{E}_0(8S)$ for hashing. For similar reasons we use $\mathcal{E}_2$ for R^i and P.

Table 1. How the parties compute encodings of group elements: each row shows that the "Output" is computed given "Input" using the operations "Operations". The input might come from the output of a previous row, a received string (e.g., $\mathcal{E}^{(1)}(R^i)$), or a random scalar that the party generates (e.g., $8x^i$). The upper half of the table are the operations that does not depend on i, which means the operations are performed only once for the whole protocol. $\mathcal{E}_X$ is suppressed: group elements written without encoding are actually encoded by $\mathcal{E}_X$. $\mathcal{C}$ and $\mathcal{D}$ stand for point compression and point decompression respectively. Computation of the rth multiple of P is denoted as "$r \cdot P$". In particular, $8 \cdot P$ can be carried out with only 3 point doublings.

$\mathcal{S}$			$\mathcal{R}$		
Output	Input	Operations	Output	Input	Operations
S	y	$y \cdot B$	$8S$	$\mathcal{E}^{(0)}(S)$	$8 \cdot \mathcal{D}(\mathcal{E}^{(0)}(S))$
$\mathcal{E}^{(0)}(S)$	S	$\mathcal{C}(S)$	$\mathcal{E}_1(S)$	$8S$	$\mathcal{C}(8S)$
$8S$	S	$8 \cdot S$			
$\mathcal{E}_1(S)$	$8S$	$\mathcal{C}(8S)$			
$64T$	$8y, 8S$	$8 \cdot (y \cdot 8S)$			
$64R^i$	$\mathcal{E}^{(1)}(R^i)$	$8 \cdot \mathcal{D}(\mathcal{E}^{(1)}(R^i))$	$8x^i B$	$8x^i$	$8x^i \cdot B$
$\mathcal{E}_2(R^i)$	$64R^i$	$\mathcal{C}(64R^i)$	$8x^i B + 8S$	$8S, 8x^i B$	$8x^i B + 8S$
$64yR^i$	$y, 64R^i$	$y \cdot 64R^i$	$\mathcal{E}^{(1)}(R^i)$	$8R^i$	$\mathcal{C}(8R^i)$
$\mathcal{E}_2(yR^i)$	$64yR^i$	$\mathcal{C}(64yR^i)$	$\mathcal{E}_2(R^i)$	$8R^i$	$\mathcal{C}(8 \cdot 8R^i)$
$64(yR^i - T)$	$64T, 64yR^i$	$64yR^i - 64T$	$64x^i S$	$8x^i, 8S$	$8x^i \cdot 8S$
$\mathcal{E}_2(yR^i - T)$	$64(yR^i - T)$	$\mathcal{C}(64(yR^i - T))$	$\mathcal{E}_2(x^i S)$	$64x^i S$	$\mathcal{C}(64x^i S)$

Actual Operations. For completeness, we present in Table 1 a full overview of operations performed during the protocol for the case of 1 out of 2 OT (i.e., $n = 2$).

4   Field Arithmetic

This section describes our implementation strategy for arithmetic operations in $\mathbb{F}_{2^{255}-19}$, which serve as low-level building blocks for operations on the curve. Field operations are decomposed into double-precision floating-point operations using our strategy. A straightforward way for implementation is then using double-precision floating-point instructions. However, a better way to utilize the $64 \times 64 \to 128$-bit serial multiplier is to decompose field operations into integer instructions as [BDL+11] does. The real reason we decide to use floating-point operations is that it allows us to use 256-bit vector instructions on the target microarchitectures, which are functionally equivalent to 4 double-precision floating-point instructions. The technique, which is called *vectorization*, makes our vectorized implementation achieve much higher throughtput than our non-vectorized implementation based on [BDL+11].

Representation of Field Elements. Each field element $x \in \mathbb{F}_{2^{255}-19}$ is represented as 12 *limbs* $(x_0, x_1, \ldots, x_{11})$ such that $x = \sum x_i$ and $x_i/2^{\lceil 21.25i \rceil} \in \mathbb{Z}$. Each x_i is stored as a double-precision floating-point number. Field operations are then carried out by limb operations such as floating-point additions and multiplications.

When a field element gets initialized (e.g., when obtained from a table lookup), each x_i uses no more than 21 bits of the 53-bit mantissa. However, after a series of limb operations, the number of bits x_i takes can grow. It is thus necessary to reduce the number of bits (in the mantissa) with carries before any precision is lost; see below for more discussions.

Field Arithmetic. Additions and subtractions of field elements are implemented in a straightforward way: simply adding/subtracting the corresponding limbs. This does increase the number of bits in the mantissa, but in our application it suffices to reduce bits only at the end of the multiplication function.

A field multiplication is divided into two steps. The first step is a schoolbook multiplication on the $2 \cdot 12$ input limbs, with reduction modulo $2^{255} - 19$ to bring the result back to 12 limbs. The schoolbook multiplication takes 132 floating-point additions, 144 floating-point multiplications, and a few more multiplications by constants to handle the reduction.

Let $(c_0, c_1, \ldots, c_{11})$ be the result after schoolbook multiplication. The second step is to perform carries to reduce number of bits in c_i. Carry from c_i to c_{i+1} (indices work modulo 12), which we denote as $c_i \to c_{i+1}$, is performed with 4 floating-point operations: $c \leftarrow c_i + \alpha_i$; $c \leftarrow c - \alpha_i$; $c_i \leftarrow c_i - c$; $c_{i+1} \leftarrow c_{i+1} + c$. The idea is to use $\alpha_i = 3 \cdot 2^{k_i}$ where k_i is big enough so that the less significant part of c_i are discarded in $c_i + \alpha_i$, forcing c to contain only the more significant part of c_i. For $i = 11$, one extra multiplication is required to scale c by $19 \cdot 2^{-255}$ before it is added to c_0.

A straightforward way to reduce number of bits in all limbs is to use the carry chain $c_0 \to c_1 \to c_2 \to \cdots \to c_{11} \to c_0 \to c_1$. The problem with the straightforward carry chain is that there is not enough instruction level parallelism to hide the 3-cycle latencies (see discussion below). To hide the latencies we thus interleave the following 3 carry chains:

$$c_0 \to c_1 \to c_2 \to c_3 \to c_4 \to c_5,$$
$$c_4 \to c_5 \to c_6 \to c_7 \to c_8 \to c_9,$$
$$c_8 \to c_9 \to c_{10} \to c_{11} \to c_0 \to c_1.$$

In total the multiplication function takes 192 floating-point additions/subtractions and 156 floating-point multiplications.

When the input operands are the same, many limb products will repeat in the schoolbook multiplication; a field squaring is therefore cheaper than a

Table 2. 256-bit vector instructions used in our implementation. Note that `vxorpd` has throughput of 4 when it has only one source operand.

instruction	latency	throughput	description
vandpd	1	1	bitwise and
vorpd	1	1	bitwise or
vxorpd	1	1 (4)	bitwise xor
vaddpd	3	1	4-way parallel double-precision floating-point additions
vsubpd	3	1	4-way parallel double-precision floating-point subtractions
vmulpd	5	1	4-way parallel double-precision floating-point multiplications

field multiplication. In total the squaring function takes 126 floating-point additions/subtractions and 101 floating-point multiplications.

Field inversion is implemented as a fix sequence of field squarings and multiplications.

Vectorization. We decompose field operations into 64-bit floating-point and logical operations. The Intel Sandy Bridge and Ivy Bridge microarchitectures, as well as many recent microarchitectures, offer instructions that operate on 256-bit registers. Some of these instructions treat the registers as vectors of 4 double-precision floating-point numbers and perform 4 floating-point operations in parallel; there are also 256-bit logical instructions that can be viewed as 4 64-bit logical instructions. We thus use these instructions to run 4 scalar multiplications in parallel. Table 2 shows the instructions we use, along with their latencies and throughputs on the Sandy Bridge and Ivy Bridge given in Fog's well-known survey [Fog14].

5 Implementation Results

This section compares the speed of our implementation of $\binom{2}{1}$-OT (i.e., $n = 2$) with other similar implementations. We stress that our software is a constant-time one: timing attacks are avoided using the same high-level strategy as [BDL+11].

To show that our speeds for curve operations are competitive, we modify the software to support the function of Diffie-Hellman key exchange and compare the results with existing Curve25519 implementations (our implementation performs scalar multiplications on the twisted Edwards curve, so it is not the same as Curve25519). The experiments are carried out on two machines on the eBACS site for publicly verifiable benchmarks [BL15]: h6sandy (Sandy Bridge) and h9ivy (Ivy Bridge). Since our protocol can serve as the base OTs for an OT extension protocol, we also compare our speed with a base OT implementation presented in [ALSZ13], which is included in the Scapi multi-party computation library; the experiments are made on an Intel Core i7-3537U processor (Ivy Bridge) where each party runs on one core. Note that all experiments are performed with Turbo Boost disabled.

Table 3. DH speeds of our work and existing Curve25519 implementations.

		h6sandy	h9ivy
[MF15]	Average cycles to compute a public key	61828	57612
[BDL+11]	Average cycles to compute a shared secret	194036	182708
this work	Average cycles to generate a public key	61458	60853
	Average cycles to compute a shared secret	182169	180343

Table 4. Timings for per OT in kilocycles. Multiplying the number of kilocycles by 0.5 one can obtain the running time (in μs) on our test architecture.

	m	4	8	16	32	64	128	256	512	1024
this work	Running time of $\mathcal{S}$	548	381	321	279	265	257	246	237	228
	Running time of $\mathcal{R}$	472	366	279	229	205	200	193	184	177
[ALSZ13]	Running time of $\mathcal{S}$	17976	10235	6132	4358	3348	2877	2650	2528	2473
	Running time of $\mathcal{R}$	16968	9261	5188	3415	3382	2909	2656	2541	2462

Comparing with Curve25519 Implementations. Table 3 compares our work with existing Curve25519 implementations. "Cycles to generate a public key" indicates the time to generate the public key given a secret key; the Curve25519 implementation is the implementation by Andrew Moon [MF15]. "Cycles to compute a shared secret" indicates the time to generate the shared secret, given a secret key and a public key; the Curve25519 implementation is from [BDL+11]. Note that since our software runs 4 scalar multiplications in parallel, the numbers in the table are the time for generating 4 public keys or 4 shared secrets divided by 4. In other words, our implementation is optimized for *througput* instead of *latency*.

Comparing with Scapi. Table 4 shows the timings of our implementation for the random OT protocol, along with the timings of a base-OT implementation presented in [ALSZ13]. The paper presents several base-OT implementations; the one we compare with is Miracl-based with "long-term security" using random oracle (cf. [ALSZ13, Section 6.1]). The implementation uses the NIST K-283 curve and SHA-1 for hashing, and it is not a constant-time implementation. It turns out that our work is an order of magnitude faster for $m \in \{4, 8, \ldots, 1024\}$.

Memory Consumption. Our code for public-key generation uses a 284-KB table. For shared-secret computation the table size is 12 KB. For OTs, $\mathcal{S}$ uses a 12-KB table, while $\mathcal{R}$ is *allowed* to use a table of size up to 1344 KB which depends on the parameters given. The current code provides 4 copies of the precomputed points, one for each of the 4 scalar multiplcations, so it is possible to reduce the table sizes by a factor of 4 by broadcasting the precomputed points. Another reason that we have large tables is because of the representation for field elements: each limbs takes 8 bytes, so each field element already takes $12 \cdot 8 = 96$ bytes. The window sizes we use are the same as [BDL+11]. See [BDL+11] for issues related to table sizes.

Acknowledgments. We are very grateful to: Daniel J. Bernstein and Tanja Lange for invaluable comments and suggestions regarding elliptic curve cryptography and for editorial feedback on earlier versions of this paper; Yehuda Lindell for useful comments on our proof of security; Peter Schwabe for various helps on implementation, including providing low-level code for field arithmetic; the anonymous LATINCRYPT reviewer and in particular Gregory Neven.

Tung Chou is supported by Netherlands Organisation for Scientific Research (NWO) under grant 639.073.005. Claudio Orlandi is supported by: the Danish National Research Foundation and The National Science Foundation of China (grant 61361136003) for the Sino-Danish Center for the Theory of Interactive Computation; the Center for Research in Foundations of Electronic Markets (CFEM); the European Union Seventh Framework Programme ([FP7/2007-2013]) under grant agreement number ICT-609611 (PRACTICE).

References

[ABN10] Abdalla, M., Bellare, M., Neven, G.: Robust encryption. In: Micciancio, D. (ed.) TCC 2010. LNCS, vol. 5978, pp. 480–497. Springer, Heidelberg (2010)

[ALSZ13] Asharov, G., Lindell, Y., Schneider, T., Zohner, M.: More efficient oblivious transfer and extensions for faster secure computation. In: Proceedings of the 2013 ACM SIGSAC Conference on Computer Communications Security, pp. 535–548. ACM (2013)

[ALSZ15] Asharov, G., Lindell, Y., Schneider, T., Zohner, M.: More efficient oblivious transfer extensions with security for malicious adversaries. Cryptology ePrint Archive, Report 2015/061 (2015). http://eprint.iacr.org/

[BBJ+08] Bernstein, D.J., Birkner, P., Joye, M., Lange, T., Peters, C.: Twisted edwards curves. In: Vaudenay, S. (ed.) AFRICACRYPT 2008. LNCS, vol. 5023, pp. 389–405. Springer, Heidelberg (2008)

[BCP04] Bresson, E., Chevassut, O., Pointcheval, D.: New security results on encrypted key exchange. In: Bao, F., Deng, R., Zhou, J. (eds.) PKC 2004. LNCS, vol. 2947, pp. 145–158. Springer, Heidelberg (2004)

[BDL+11] Bernstein, D.J., Duif, N., Lange, T., Schwabe, P., Yang, B.-Y.: High-speed high-security signatures. In: Preneel, B., Takagi, T. (eds.) CHES 2011. LNCS, vol. 6917, pp. 124–142. Springer, Heidelberg (2011)

[BDPVA09] Bertoni, G., Daemen, J., Peeters, M., Van Assche, G.: Keccak sponge function family main document. Submission to NIST (Round 2), pp. 3–30 (2009)

[Bea96] Beaver, D.: Correlated pseudorandomness and the complexity of private computations. In: Proceedings of the Twenty-Eighth Annual ACM Symposium on the Theory of Computing, 22–24 May 1996, Philadelphia, Pennsylvania, USA, pp. 479–488 (1996)

[Ber06] Bernstein, D.J.: Curve25519: new Diffie-Hellman speed records. In: Yung, M., Dodis, Y., Kiayias, A., Malkin, T. (eds.) PKC 2006. LNCS, vol. 3958, pp. 207–228. Springer, Heidelberg (2006)

[BL14] Bernstein, D.J., Lange, T.: Safecurves: choosing safe curves for elliptic-curve cryptography. http://safecurves.cr.yp.to. Accessed on 1 December 2014

[BL15]   Bernstein, D.J., Lange, T.: eBACS: ecrypt benchmarking of cryptographic
          systems. http://bench.cr.yp.to. Accessed on 16 March 2015
[BLN+15] Burra, S.S., Larraia, E., Nielsen, J.B., Nordholt, P.S., Orlandi, C., Orsini,
          E., Scholl, P., Smart, N.P.: High performance multi-party computation
          for binary circuits based on oblivious transfer. Cryptology ePrint Archive,
          Report 2015/472 (2015). http://eprint.iacr.org/
[BM89]   Bellare, M., Micali, S.: Non-interactive oblivious transfer and applica-
          tions. In: Brassard, G. (ed.) CRYPTO 1989. LNCS, vol. 435, pp. 547–557.
          Springer, Heidelberg (1990)
[Can01]   Canetti, R.: Universally composable security: a new paradigm for crypto-
          graphic protocols. In: 42nd Annual Symposium on Foundations of Com-
          puter Science, FOCS 2001, 14–17 October 2001, Las Vegas, Nevada, USA,
          pp. 136–145 (2001)
[CF01]   Canetti, R., Fischlin, M.: Universally composable commitments. IACR
          Cryptology ePrint Archive, 2001:55 (2001)
[DNO08]   Damgård, I., Nielsen, J.B., Orlandi, C.: Essentially optimal universally
          composable oblivious transfer. In: Lee, P.J., Cheon, J.H. (eds.) ICISC
          2008. LNCS, vol. 5461, pp. 318–335. Springer, Heidelberg (2009)
[EGL85]   Even, S., Goldreich, O., Lempel, A.: A randomized protocol for signing
          contracts. Commun. ACM 28(6), 637–647 (1985)
[FLPQ13] Farshim, P., Libert, B., Paterson, K.G., Quaglia, E.A.: Robust encryption,
          revisited. In: Kurosawa, K., Hanaoka, G. (eds.) PKC 2013. LNCS, vol.
          7778, pp. 352–368. Springer, Heidelberg (2013)
[Fog14]   Agner Fog. Instruction tables (2014). http://www.agner.org/optimize/
          instruction_tables.pdf
[HL10]   Hazay, C., Lindell, Y.: Efficient Secure Two-Party Protocols - Techniques
          and Constructions. Information Security and Cryptography. Springer,
          Berin (2010)
[HWCD08] Hisil, H., Wong, K.K.-H., Carter, G., Dawson, E.: Twisted edwards curves
          revisited. In: Pieprzyk, J. (ed.) ASIACRYPT 2008. LNCS, vol. 5350, pp.
          326–343. Springer, Heidelberg (2008)
[IKNP03] Ishai, Y., Kilian, J., Nissim, K., Petrank, E.: Extending oblivious transfers
          efficiently. In: Boneh, D. (ed.) CRYPTO 2003. LNCS, vol. 2729, pp. 145–
          161. Springer, Heidelberg (2003)
[IR89]   Impagliazzo, R., Rudich, S.: Limits on the provable consequences of one-
          way permutations. In: Proceedings of the 21st Annual ACM Symposium
          on Theory of Computing, May 14–17, 1989, Seattle, Washigton, USA, pp.
          44–61 (1989)
[Kil88]   Kilian, J.: Founding cryptography on oblivious transfer. In: Proceedings
          of the 20th Annual ACM Symposium on Theory of Computing, 2–4 May
          1988, Chicago, Illinois, USA, pp. 20–31 (1988)
[KOS15]   Keller, M., Orsini, E., Scholl, P.: Actively secure ot extension with optimal
          overhead. In: CRYPTO (2015)
[Lar14]   Larraia, E.: Extending oblivious transfer efficiently, or - how to get active
          security with constant cryptographic overhead. IACR Cryptology ePrint
          Archive, 2014:692 (2014)
[MF15]   Moon, A.: "Floodyberry": implementations of a fast elliptic-curve digi-
          tal signature algorithm. https://github.com/floodyberry/ed25519-donna.
          Accessed on 16 March 2015

[Nie07] Nielsen, J.B.: Extending oblivious transfers efficiently - how to get robust-
         ness almost for free. Cryptology ePrint Archive, Report 2007/215 (2007).
         http://eprint.iacr.org/
[NNOB12] Nielsen, J.B., Nordholt, P.S., Orlandi, C., Burra, S.S.: A new approach
         to practical active-secure two-party computation. In: Safavi-Naini, R.,
         Canetti, R. (eds.) CRYPTO 2012. LNCS, vol. 7417, pp. 681–700. Springer,
         Heidelberg (2012)
[NP01]   Naor, M., Pinkas, B.: Efficient oblivious transfer protocols. In: Proceedings
         of the Twelfth Annual Symposium on Discrete Algorithms, 7–9 January
         2001, Washington, DC, USA, pp. 448–457 (2001)
[PVW08] Peikert, C., Vaikuntanathan, V., Waters, B.: A framework for efficient
         and composable oblivious transfer. In: Wagner, D. (ed.) CRYPTO 2008.
         LNCS, vol. 5157, pp. 554–571. Springer, Heidelberg (2008)
[Rab81] Rabin, M.O.: How to exchange secrets with oblivious transfer. Technical
         report TR-81, Aiken Computation Lab, Harvard University (1981)
[Sch15] Schneider, T.: Personal communication (2015)
[Wie83] Wiesner, S.: Conjugate coding. SIGACT News 15(1), 78–88 (1983)

Foundations

Depth Optimized Efficient
Homomorphic Sorting

Gizem S. Çetin[1]([✉]), Yarkın Doröz[1], Berk Sunar[1], and Erkay Savaş[2]

[1] Worcester Polytechnic Institute, Worcester, USA
{gscetin,ydoroz,sunar}@wpi.edu
[2] Sabanci University, Istanbul, Turkey
erkays@sabanciuniv.edu

Abstract. We introduce a sorting scheme which is capable of efficiently sorting encrypted data without the secret key. The technique is obtained by focusing on the multiplicative depth of the sorting circuit alongside the more traditional metrics such as number of comparisons and number of iterations. The reduced depth allows much reduced noise growth and thereby makes it possible to select smaller parameter sizes in somewhat homomorphic encryption instantiations resulting in greater efficiency savings. We first consider a number of well known comparison based sorting algorithms as well as some sorting networks, and analyze their circuit implementations with respect to multiplicative depth. In what follows, we introduce a new ranking based sorting scheme and rigorously analyze the multiplicative depth complexity as $\mathcal{O}(\log(N) + \log(\ell))$, where N is the size of the array to be sorted and ℓ is the bit size of the array elements. Finally, we simulate our sorting scheme using a leveled/batched instantiation of a SWHE library. Our sorting scheme performs favorably over the analyzed classical sorting algorithms.

Keywords: Homomorphic sorting · Circuit depth · Somewhat homomorphic encryption

1 Introduction

An encryption scheme is *fully homomorphic* (FHE scheme) if it permits the efficient evaluation of any boolean circuit or arithmetic function on ciphertexts [27]. Gentry introduced the first FHE scheme [14,15] based on lattices that supports the efficient evaluation for arbitrary depth circuits. This was followed by a rapid progression of new FHE schemes. van Dijk et al. proposed a FHE scheme based on ideals defined over integers [10]. In 2011, Gentry and Halevi [16] presented the first actual FHE implementation along with a wide array of optimizations to tackle the infamous efficiency bottleneck of FHEs. Further optimizations for FHE which also apply to somewhat homomorphic encryption (SWHE) schemes followed including batching and SIMD optimizations, e.g. see [17,18,29]. Several newer SWHE & FHE schemes appeared in the literature in recent years. Brakerski, Gentry and Vaikuntanathan proposed a new FHE

© Springer International Publishing Switzerland 2015
K. Lauter and F. Rodríguez-Henríquez (Eds.): LatinCrypt 2015, LNCS 9230, pp. 61–80, 2015.
DOI: 10.1007/978-3-319-22174-8_4

scheme (BGV) based on the learning with errors (LWE) problem [5]. To cope with noise the authors propose efficient techniques for noise reduction. While not as effective as Gentry's recryption operation, these lightweight techniques limit the noise growth enabling the evaluation of much deeper circuits using only a depth restricted SWHE scheme. The costly recryption primitive is only used to evaluate extremely complicated circuits. In [18] Gentry, Halevi and Smart introduced a LWE-based FHE scheme customized to achieve efficient evaluation of the AES cipher without bootstrapping. Their implementation is highly optimized for efficient AES evaluation using key and modulus switching techniques [5], batching and SIMD optimizations [29]. Their byte-sliced AES implementation takes about 5 min to homomorphically evaluate an AES block encryption. More recently, López-Alt, Tromer and Vaikuntanathan (LTV) proposed SWHE and FHE schemes based on Stehlé and Steinfeld's generalization of the NTRU scheme [30] that supports inputs from multiple public keys [25]. Bos et al. [3] introduced a variant of the LTV FHE scheme along with an implementation. The authors modify the LTV scheme by adopting a tensor product technique introduced earlier by Brakerski [4] such that the security depends only on standard lattice assumptions. The authors advocate use of the Chinese Remainder Theorem on the message space to improve the flexibility of the scheme. Also, modulus switching is no longer needed due to the reduced noise growth. Doröz, Hu and Sunar propose another variant based on the LTV scheme in [11]. The implementation is batched, bit-sliced and features modulus switching techniques. The authors also specialize the modulus to reduce the key size and report an AES implementation with one minute evaluation time per AES block [18]. More recent FHE schemes displayed significant improvements over earlier constructions in both time complexity and in ciphertext size. Nevertheless, both latency and message expansion rates remain roughly two orders of magnitude higher than those of traditional public-key schemes. Bootstrapping [15], relinearization [6], and modulus reduction [5,6] are indispensable tools for FHEs. In [6, Sect. 1.1], the *relinearization* technique was proposed to re-encrypt quadratic polynomials as linear polynomials under a new key, thereby making their security argument independent of lattice assumptions and dependent only on a standard LWE hardness assumption.

Homomorphic encryption schemes have been used to build a variety of higher level security applications. Lagendijk et al. [22] give a summary of homomorphic encryption and MPC techniques to realize key signal processing operations such as evaluating linear operations, inner products, distance calculation, dimension reduction, and thresholding. Using these key operations it becomes possible to achieve more sophisticated privacy-protected heavy DSP services such as face recognition, user clustering, and content recommendation. Cryptographic tools permitting restricted homomorphic evaluation, e.g. Paillier's scheme, and more powerful techniques such as Yao's garbled circuit [32] have been around sufficiently long to be used in a diverse set of applications. Homomorphic encryption schemes are often used in privacy-preserving data mining applications. Vaidya and Clifton [31] propose to use Yao's circuit evaluation [32] for the comparisons in their privacy-preserving k-means clustering algorithm. The secure comparison protocol by Fischlin [13] uses the GM-homomorphic encryption scheme [19] and the method by Sander et al. [28] to convert the XOR homomorphic encryption in GM scheme into AND homomorphic encryption. The privacy-preserving clustering algorithm for vertically partitioned (distributed) spatio-temporal data [33] uses the Fischlin formulation based on XOR homomorphic secret sharing primitive instead of costly encryption operations. The tools for SWHE developed to achieve FHE have only been around for a few years now and have not been sufficiently explored for use in applications. For instance, in [23] Lauter et al. consider the problems of evaluating averages, standard deviations, and logistical regressions which provide basic tools for a number of real-world applications in the medical, financial, and the advertising domains. The same work also presents a proof-of-concept Magma implementation of a SWHE for the basic operations. The SWHE scheme is based on the ring learning with errors (RLWE) problem proposed earlier by Brakerski and Vaikuntanathan. Later in [24], Lauter et al. show that it is possible to implement genomic data computation algorithms where the patients' data are encrypted to preserve their privacy. They encrypt all the genomic data in the database and able to implement and provide performance numbers for Pearson Goodness-of-Fit test, the D' and r^2-measures of linkage disequilibrium, the Estimation Maximization (EM) algorithm for haplotyping, and the Cochran-Armitage Test for Trend. The authors used a leveled SWHE scheme which is a modified version of [26] where they get rid of the costly relinearization operation. In [2] Bos et al. show how to privately perform predictive analysis tasks on encrypted medical data. They present an implementation of a prediction service running in the cloud. The cloud server takes private encrypted health data as input and returns the probability of cardiovascular disease in encrypted form. The authors use the SWHE implementation of [3] to provide timing results. Graepel et al. in [20] demonstrate that it is possible to execute machine learning algorithms in a service while protecting the confidentiality of the training and test data. The authors propose a confidential protocol for machine learning tasks and design confidential machine learning algorithms using leveled homomorphic encryption. More specifically they implement low-degree polynomial versions of Linear Means Classifier and Fisher's Linear Discriminant Classifier on the Wisconsin Breast Cancer Data set. Finally, they provide benchmarks for small scale data set to show that their scheme is practical. Cheon et al. [9] present a method along with implementation results to compute encrypted dynamic programming algorithms such as Hamming distance, edit distance, and the Smith-Waterman algorithm on genomic data encrypted using a somewhat homomorphic encryption algorithm. The authors design circuits to compute the distances between two genomic strings. The work designs circuits meticulously to reduce their depths to permit efficient evaluation using BGV-type leveled SWHE schemes. In this work, we follow a route very similar to that given in [9] for sorting. In [12], Doröz et al. use an NTRU based SWHE scheme to construct a bandwidth efficient private information retrieval (PIR) scheme. Due to the multiplicative evaluation capabilities of the SWHE, the query and response sizes are significantly reduced compared to earlier PIR constructions. The PIR construction is generic and therefore any SWHE which supports a few multiplicative levels (and many additions) could be used to implement a PIR. The authors also give a leveled and batched reference implementation of their PIR construction including performance figures.

The only homomorphic sorting result we are aware of was reported by Chatterjee et al. in [8]. In this work, for the first time, the authors considered the problem of homomorphically sorting an array using the recently proposed hcrypt FHE library [7]. The authors define a number of FHE functions to realize basic homomorphic comparison and swapping operations and then implement the classical Bubble and Insertion sort algorithms using these homomorphic functions. Noting the exponential rise of evaluation time with the array size, the authors introduce a new approach dubbed **Lazy Sort** which removes the Recrypt operation after additions allowing occasional comparison errors in Bubble Sort. While the array is not perfectly sorted the sorting time is significantly reduced. After Bubble sort the nearly sorted array is then sorted again with a homomorphically evaluated Insertion sort - this time with all Recrypt operations in place. The authors report implementation results with arrays of 5–40 elements (32-bits) which show significant reduction in the evaluation time over direct fully homomorphic evaluation. In the best case, the authors report a 1,399 s evaluation time in contrast to 21,565 s in the fully homomorphic case for an array of size 40. Despite the impressive speed gains, the work opts to alleviate the efficiency bottleneck by relaxing noise management, and by combining classical sorting algorithms instead of targeting the circuit depth of the sorting algorithm. Furthermore, it suffers from the fundamental limitations of the hcrypt library:

- Noise management is achieved by recrypting partial results after every major operation. Recrypt is extremely costly and is considered inferior to more modern noise management techniques such as the **modulus reduction** [5] that yield exponential gains in leveled implementations.
- hcrypt does not take advantage of **batching** or SIMD techniques [29] which greatly improve homomorphic evaluation performance.

Our Contribution. In this work,

- we survey a number of classical sorting algorithms, i.e. Bubble, Insertion, Odd-Even Sort, Merge, Batcher's Odd-even Merge Sort, Bitonic sort, and show that some are more suitable than others for leveled SWHE evaluation, similar to the work for distance computation presented in [9]. Specifically, we characterize them with respect to a new metric, i.e. multiplicative circuit depth. We show that the classical sorting algorithms require deep circuit evaluations and therefore are not ideal for homomorphic evaluation.
- we introduce two new depth optimized sorting schemes: Greedy Sort and Direct Sort. Both algorithms permit shallow circuit evaluation of depth only $\mathcal{O}(\log(N) + \log(\ell))$ for sorting N elements, where ℓ represents the size of the array elements in bits. The Greedy algorithm has slightly lower depth however

requires more multiplications than Direct Sort. Both algorithms improve in the circuit depth metric over classical algorithms by at least *1–3 orders of magnitude.*
- we instantiate a somewhat homomorphic encryption scheme (SWHE) based on NTRU, and present an implementation of the proposed sorting algorithm using this SWHE scheme. Our results, confirm our theoretical analysis, i.e. that the performance of the proposed sorting algorithm scales favorably as N increases.

2 Background

We start by giving a brief summary of the multi-key LTV-FHE scheme and provide a brief explanation on the primitive functions that are proposed by López-Alt, Tromer and Vaikuntanathan. Later, we give details of the DHS FHE library, that is used in the implementation, based on a specialized LTV-FHE version.

2.1 The LTV-SWHE Scheme

In 2012 López-Alt, Tromer and Vaikuntanathan proposed a leveled multi-key FHE scheme (LTV) [25]. The scheme based on a variant of NTRU encryption scheme proposed by Stehlé and Steinfeld [30]. The introduced scheme uses a new operation called relinearization and existing techniques such as modulus switching for noise control. We use the same construction as in [11] which is a single key version of LTV with reduced key size technique. The operations are performed in $R_q = \mathbb{Z}_q[x]/\langle x^n + 1\rangle$ where n is the polynomial degree and q is the prime modulus. The scheme also defines an error distribution χ, which is a truncated discrete Gaussian distribution, for sampling random polynomials that are B-bounded. The term B-bounded means that the coefficients of the polynomial are selected in range $[-B, B]$ with χ distribution. The scheme consists of four primitive functions, namely **KeyGen, Encrypt, Decrypt** and **Eval**. A brief detail of the primitives is as follows:

KeyGen. We choose sequence of primes $q_0 > q_1 > \cdots > q_d$ to use a different q_i in each level. A public and secret key pair is computed for each level: $h^{(i)} = 2g^{(i)}(f^{(i)})^{-1}$ and $f^{(i)} = 2u^{(i)} + 1$, where $\{g^{(i)}, u^{(i)}\} \in \chi$. Later we create evaluation keys for each level: $\zeta_\tau^{(i)}(x) = h^{(i)}s_\tau^{(i)} + 2e_\tau^{(i)} + 2^\tau(f^{(i-1)})^2$, where $\{s_\tau^{(i)}, e_\tau^{(i)}\} \in \chi$ and $\tau = [0, \lfloor\log q_i\rfloor]$.

Encrypt. To encrypt a bit b for the ith level we compute: $c^{(i)} = h^{(i)}s + 2e + b$, where $\{s, e\} \in \chi$.

Decrypt. In order to compute the decryption of a value for specific level i we compute: $m = c^{(i)}f^{(i)} \pmod{2}$.

Eval. The gate level logic operations XOR and AND are done by computing the addition and multiplication of the ciphertexts. In case of $c_1^{(i)} = \mathsf{Encrypt}(b_1)$ and

$c_2^{(i)} = \mathsf{Encrypt}(b_2)$; XOR is equal to $\mathsf{Decrypt}(c_1^{(i)} + c_2^{(i)}) = b_1 + b_2$ and, AND is equal to $\mathsf{Decrypt}(c_1^{(i)} \cdot c_2^{(i)}) = b_1 \cdot b_2$. The multiplication creates a significant noise in the ciphertext and to cope with that we apply Relinearization and modulus switch. The Relinearization computes $\tilde{c}^{(i)}(x)$ from $\tilde{c}^{(i-1)}(x)$ extending $\tilde{c}^{(i-1)}(x)$ as a linear combination of 1-bounded polynomials $\tilde{c}^{(i-1)}(x) = \sum_\tau 2^\tau \tilde{c}_\tau^{(i-1)}(x)$. Then, using the evaluation keys it computes $\tilde{c}^{(i)}(x) = \sum_\tau \zeta_\tau^{(i)}(x) \tilde{c}_\tau^{(i-1)}(x)$ as the new ciphertext. The formula is actually the evaluation of homomorphic product of $c^{(i)}(x)$ and $(f^{(i)})^2$. Later, the modulus switch $\tilde{c}^{(i)}(x) = \lfloor \frac{q_i}{q_{i-1}} \tilde{c}^{(i)}(x) \rceil_2$ decreases the noise by $\log(q_i/q_{i-1})$ bits by dividing and multiplying the new ciphertext with the previous and current moduli, respectively. The operation $\lfloor \cdot \rceil_2$ refers to rounding and matching the parity bits.

2.2 The DHS SWHE Library

We use a customized version of the LTV-SWHE scheme that is previously proposed in [11] by Doröz, Hu and Sunar (DHS). The code is written in C++ using NTL package that is compiled with GMP library. The library contains some special customizations that improve the efficiency in running time and memory requirements. The customizations of the DHS implementation are as follows:

– We select a special m^{th} cyclotomic polynomial $\Psi_m(x)$ as our polynomial modulus. The degree of the polynomial n is equal to the euler totient function of m, i.e. $\varphi(m)$. In each level the arithmetic is performed over $R_{q_i} = \mathbb{Z}_{q_i}[x]/\langle \Psi_m(x) \rangle$, where modulus q_i is equal to p^{k-i}. The value p is a prime number that cuts $(\log_p)$-bits of noise and the value k is equal to the depth plus 1.
– Due to the special structure of the moduli p^{k-i}, the evaluation keys in one level can also be promoted to the next level via modular reduction. For any level we can evaluate the evaluation key as $\zeta_\tau^{(i)}(x) = \zeta_\tau^{(0)}(x) \pmod{q_i}$. This technique reduces the memory requirement significantly and makes it possible to evaluate higher depth circuits.
– The specially selected cyclotomic polynomial $\Psi_m(x)$ is used to batch multiple message bits into the same polynomial for parallel evaluations as proposed by Smart and Vercauteren [17,29] (see also [18]). The polynomial $\Psi_m(x)$ is factorized over $\mathbb{F}_2$ into equal degree polynomials $F_i(x)$ which define the message slots in which message bits are embedded using the Chinese Remainder Theorem. We can batch $\ell = n/t$ number of messages, where t is the smallest integer that satisfies $m|(2^t - 1)$.
– The DHS library can perform 5 main operations; KEYGEN, ENCRYPTION, DECRYPTION, MODULUS SWITCH and RELINEARIZATION. The most time consuming operation is RELINEARIZATION, which is generally the bottleneck. Therefore, the most critical operation for circuit evaluation is RELINEARIZATION. The other operations have negligible effect on the run time.

Modified Relinearization. We modify previously implemented method of relinearization where it uses linear combination of 1-bounded polynomials of the ciphertext $\tilde{c}^{(i-1)}(x) = \sum_\tau 2^\tau \tilde{c}_\tau^{(i-1)}(x)$. Previously, the number of evaluation key polynomials and the number of multiplications in relinearization is $\lceil \log(q) \rceil$. For deep circuits with many levels the bitsize $\lceil \log(q) \rceil$ is two/three orders of magnitude which increase the memory requirements and number of multiplications significantly. In order to achieve a speedup, we group the bits of the ciphertext and use the linear combination of word (r-bits) sized polynomials rather than binary polynomials. Setting the word size as $w = 2^r$, we implement the following changes:

- Compute the evaluation keys as: $\zeta_\tau^{(i)}(x) = h^{(i)} s_\tau^{(i)} + 2 e_\tau^{(i)} + w^\tau (f^{(i-1)})^2$, where $\{s_\tau^{(i)}, e_\tau^{(i)}\} \in \chi$ and $\tau = [0, \lfloor \log q_i / r \rfloor]$.
- Divide the ciphertext into linear combinations of word sized polynomials: $\tilde{c}^{(i-1)}(x) = \sum_\tau w^\tau \tilde{c}_\tau^{(i-1)}(x)$.
- Compute the relinearization as: $\tilde{c}^{(i)}(x) = \sum_\tau \zeta_\tau^{(i)}(x) \tilde{c}_\tau^{(i-1)}(x)$

The changes above decreases the memory requirement by r times. With this change relinearization requires r times fewer multiplications. However this does not yield r times speedup. This is due to the increase of the coefficient size of the linear combination polynomials from 1 to r bits. Thus the cost of a multiplication increases.

3 Basic Circuits

As stated earlier, given level i in a homomorphic circuit, we will have $c_1^{(i)} =$ Encrypt(b_1) and $c_2^{(i)} =$ Encrypt(b_2) where b_1 and b_2 are encrypted by the owner of data. We are allowed to use two fundamental operations on encrypted inputs; bit multiplication (AND, "$\cdot$") and bit addition (XOR, "$\oplus$"). Hence, we can evaluate $c^{(i)} = c_1^{(i)} \oplus c_2^{(i)}$ and $\tilde{c}^{(i)} = c_1^{(i)} \cdot c_2^{(i)}$. Finally, the holder of the secret key can compute and retrieve Decrypt($c^{(i)}$) = $b_1 \oplus b_2$. Decrypt($\tilde{c}^{(i)}$) = $b_1 \cdot b_2$. Using these two, we can define the following circuits. From this point on, we will use $c = E(b)$ instead of $c =$ Encrypt(b) for an arbitrary encryption and $c^{(i)}$ represents the ciphertext of the ith level.

Equality Circuit $\mathcal{C}_{\text{EQ}}$: It compares two encrypted ℓ-bit integers $E(X) = X^{(i)} = (x_{\ell-1})^{(i)} \dots (x_1)^{(i)} (x_0)^{(i)}$ and $E(Y) = Y^{(i)} = (y_{\ell-1})^{(i)}, \dots, (y_1)^{(i)}, (y_0)^{(i)}$, and outputs $\tilde{z}^{(j)}$. Here $(x_k)^{(i)}$ and $(y_k)^{(i)}$ represent the k-th bits of $X^{(i)}$ and $Y^{(i)}$, respectively. Also $D(\tilde{z}^{(j)})$ returns 1 if X equals Y and 0 otherwise. We can formalize the comparison circuit as follows; $\tilde{z}^{(j)} = (E(X) = E(Y)) = \prod_{k \in [\ell]} (E(x_k) = E(y_k)) = \prod_{k \in [\ell]} ((x_k)^{(i)} \oplus (y_k)^{(i)} \oplus 1)$. The product chain of ℓ bits may be evaluated using a binary tree of AND gates which creates a circuit with $\lceil \log(\ell) \rceil$ multiplicative depth. Therefore, $d(\mathcal{C}_{\text{EQ}}) \approx \mathcal{O}(\log(\ell))$.

Less Than Circuit $\mathcal{C}_{\text{LT}}$: In a similar manner, this circuit compares two ℓ-bit integers $E(X)$ and $E(Y)$, and outputs $\tilde{z}^{(j)}$ where $D(\tilde{z}^{(j)}) = 1$ if X is smaller than Y and $D(\tilde{z}^{(j)}) = 0$ otherwise. We can formalize the comparison circuit as follows;

$\tilde{z}^{(j)} = (E(X) < E(Y)) = \sum_{k \in [\ell]} [(E(x_k) < E(y_k)) \prod_{k < t < \ell}(E(x_t) = E(y_t))]$
where $(E(x_k) < E(y_k)) = (y_k)^{(i)} \cdot ((x_k)^{(i)} \oplus 1)$ and $(E(x_t) = E(y_t)) = (y_t)^{(i)} \oplus (x_t)^{(i)} \oplus 1$. The expansion of the formula gives a sum of products expression where the product with the maximum number of bits occurs when $i = 0$, in which case the product chain contains $\ell + 1$ bits, where 2 bits are contributed by the $(E(x_0) < E(y_0))$ term and the rest are from the $(E(x_t) = E(y_t))$ terms. For the product of $\ell + 1$ elements, we may use again a binary tree in which case we achieve the minimum depth of $\lceil \log (\ell + 1) \rceil$. Therefore $d(\mathcal{C}_{\mathrm{LT}}) \approx \mathcal{O}(\log(\ell + 1))$.

Compare and Swap Block $\mathcal{C}_{\mathrm{CS}}$: Since our main goal is the construction of a sorting circuit, we will extensively use the comparators followed by a swap operation. The $\mathcal{C}_{\mathrm{CS}}$ block basically compares two ℓ-bit integers X and Y using previously defined $\mathcal{C}_{\mathrm{LT}}$ circuit and swaps them if $X \not< Y$. Overall circuit can be defined as; $\tilde{X}^{(j+1)} = [\tilde{z}^{(j)} \cdot E(X)] \oplus [(\tilde{z}^{(j)} \oplus 1) \cdot E(Y)]$ and $\tilde{Y}^{(j+1)} = [\tilde{z}^{(j)} \oplus 1) \cdot E(X)] \oplus [\tilde{z}^{(j)} \cdot E(Y)]$, where $\tilde{z}^{(j)} = (E(X) < E(Y))$. Therefore, $d(\mathcal{C}_{\mathrm{CS}}) = d(\mathcal{C}_{\mathrm{LT}}) + 1$.

4   Sorting Algorithms

Sorting is one of the most natural and crucial tasks in computing. Numerous sorting algorithms have been proposed in the literature [21]. These algorithms have been heavily investigated and characterized according to their time and space requirements, as well as to the degree of their suitability for parallelization. As far as homomorphic evaluation is concerned we have another requirement. Since most of the FHE and SWHE schemes are designed to evaluate circuits, and do not scale well when the multiplicative depth of the circuit is high, we need to add another metric; namely multiplicative circuit depth, before we can build a homomorphic sorting scheme. For this we need to first convert the serial sorting algorithm, into a circuit by unrolling loops and eliminating conditional assignments by *arithmetization*. In this paper, the term "circuit depth" is used in lieu of multiplicative depth of the circuit and it should not be confused with "comparison depth", i.e. depth of the circuit measured in terms of comparative levels, which is used in the analysis of classical sorting algorithms in the literature.

A sorting network is a circuit which consists of comparators and swapping operations. The difference between classical comparison-based sorting algorithms and sorting networks is that all operations are set in advance, which means that there is no data dependency in the flow of the algorithm steps in sorting networks. Since we are trying to sort encrypted inputs, we are, in a way, blind in each step of the algorithm. As a result, even though data dependent algorithms may be faster and more efficient over raw data, being independent from the input makes sorting networks the only candidates for FHE sorting. While there are some algorithms specifically designed as a sorting network, some classical sorting algorithms can also be represented as a network, as FHE properties require. Hence we will go over some well known algorithms and give the depth complexity of the corresponding sorting networks.

4.1   Bubble Sort

Bubble Sort is one of the simplest sorting techniques that permits a rather straightforward implementation using only primitive comparison and swap operations. Chatterjee et al. [8] design homomorphic conditional swap circuits to facilitate homomorphic evaluation of the Bubble Sort algorithm. Very briefly the sorting algorithm works by making passes over the array. In each pass the elements are pairwise compared and swapped to move the smaller element to the left (in case of a horizontal array). The average and worst case performance for an array of N elements are the same: $\mathcal{O}(N^2)$. During homomorphic evaluation since we have no way of knowing when the array is sorted for a possible early termination, we need to make $N - 1$ passes over the array always achieving the worst case complexity. Since another element in the rightmost portion is sorted the passes decrease by one in number of elements compared and swapped after each pass. Thus, overall we will have $[(N-1) + (N-2) + \ldots + 1] = (N^2 - N)/2$ $\mathcal{C}_{\text{CS}}$ blocks and the depth of the Bubble Sort circuit will be $[(N^2 - N)/2]d(\mathcal{C}_{\text{CS}})$. Considering ℓ-bit wide array elements, we have $d(\mathcal{C}_{\text{BUBS}}) = \mathcal{O}(N^2 \log(\ell))$. We can gain some economy by not waiting to start the next pass until a pass is finished. We can *overlap* the passes which creates a network version of Bubble Sort, known as Odd Even Sort, detailed in the next section.[1]

4.2   Odd-Even Sort

A trellis shaped circuit arrangement of Bubble sort is known as Odd-Even Sort. The circuit admits N inputs and computes the N sorted output values after N passes. In the first pass, considering a zero-indexed array, every even indexed element is compared and swapped with its right neighbor. In the second pass, every odd indexed element is compared and swapped with its right neighbor. Considering these two steps as a round, the identical operations are applied in each round. The total number of comparisons is $N - 1$ in each round, there are N passes which means $N/2$ rounds and so overall, there are $N(N - 1)/2$ comparators. And the depth of the circuit is $Nd(\mathcal{C}_{\text{CS}})$. Therefore $d(\mathcal{C}_{\text{OES}}) = \mathcal{O}(N \log(\ell))$.

4.3   Insertion Sort

Insertion sort is a simple sorting algorithm that iteratively builds a sorted array from an unsorted one. The sorted array initially holds only the first element. Then each element is one by one added to the sorted list by comparing it from right to left with the elements in the sorted list until an element smaller is encountered. The new element is then inserted into the sorted array next to the

---

[1]   Note that in their implementation Chatterjee et al. [8] perform the comparison using a carry propagate adder based subtraction circuit result in a circuit depth $(N^2 - N)(\ell + 1)/2$. While the computational complexity of the scheme is low, the $\mathcal{O}(N^2)$ circuit depth is prohibitive.

first smaller element when scanning right to left. The average case and the worst case complexity of the algorithm is $\mathcal{O}(N^2)$ while the best case is only $\mathcal{O}(N)$. When considered as a circuit for homomorphic evaluation we need to run the algorithm with the worst case complexity, without making early decisions as in Bubble Sort. We build up the sorted array one by one making increasing number of comparison and conditional swaps. We obtain a circuit depth of $[1 + 2 + \ldots + N - 1]d(\mathcal{C}_{\mathrm{CS}}) = (N^2 - N)/(2)d(\mathcal{C}_{\mathrm{CS}})$. Therefore $d(\mathcal{C}_{\mathrm{INS}}) = \mathcal{O}(N^2 \log(\ell))$. This circuit can be used in a more efficient way by overlapping some comparisons, similar to $\mathcal{C}_{\mathrm{BUBS}}$. Consequently we can see that, Insertion Sort and Bubble Sort reveal the same construction, when they are considered as sorting networks.

In [8] Chatterjee et al. rely on the fact that after the *imperfect* application of Bubble Sort the array is *nearly* sorted. Thus Insertion Sort performs nearly in linear time. But even if the array is *nearly* sorted, the algorithm should run as in the worst case, since we do not have any knowledge of the misplaced elements.

4.4   Merge Sort

Merge Sort is an asymptotically faster algorithm and allows early termination in normal execution, which reduces its complexity. The algorithm is recursively applied by splitting arrays into smaller ones. In the innermost recursion, arrays of two elements are sorted, where only one comparison is needed in one sub-array. In the merging step, which combines two individually sorted arrays into a single sorted array, at most three comparisons are applied in each partition. This eventually requires $\mathcal{O}(N \log(N))$ comparisons in the worst case. But in our case, the merging step requires many more comparisons, due to algorithm's input dependent nature and our lack of input knowledge. For instance, in the classic Merge Sort, to merge two sub-arrays each of size two, as in Fig. 1, we follow one of the paths until all the elements are placed in the sorted sub-array of size four. Let our output array be Z in a merge step. Then, if $\mathcal{C}_{\mathrm{LT}}(E(X_0), E(Y_0))$ output is 1; we can conclude that $Z_0 = X_0$, otherwise $Z_0 = Y_0$. But in homomorphic sorting, we cannot follow any specific path as the output of each $\mathcal{C}_{\mathrm{LT}}(E(X_i), E(Y_j))$ block is also encrypted. Hence, we need to consider every single possible outcome of all comparison operations, i.e. every single path, which eventually necessitates comparing every possible pair.

In summary, we need to perform $(N^2 - N)/2$ comparisons to sort an array of N elements. On the other hand, since there is no swapping, i.e. no data dependency, during the execution of a single merge step, we can compute all of the comparisons in parallel at the beginning of each merge step. Consequently, applying all comparison operations before every merge step simply alters the algorithm and we end up with a totally different scheme from the classical Merge Sort algorithm. Inspired from the analysis of $\mathcal{C}_{\mathrm{MS}}$, we introduce two new sorting circuits, with the same number of comparators $\mathcal{O}(N^2)$ and the total comparison depth of $\mathcal{O}(1)$, in Sect. 5.1.

Indeed, we can reduce the number of comparisons in $\mathcal{C}_{\mathrm{MS}}$ using $\mathcal{C}_{\mathrm{CS}}$ blocks. In the first step, the run of $\mathcal{C}_{\mathrm{CS}}(X_0^{(i)}, Y_0^{(i)})$ will yield $X_0^{(j)}$ and $Y_0^{(j)}$ as the smaller

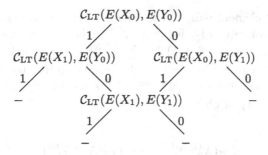

Fig. 1. Merging two individually sorted arrays: $\langle E(X_0), E(X_1)\rangle$ and $\langle E(Y_0), E(Y_1)\rangle$

and larger elements, respectively. Then, we can safely use $Y_0^{(j)}$ in the following step as one of the inputs to $\mathcal{C}_{CS}$ blocks while the second input can be either X_1 or Y_1. But, it is impossible to know which one of them should be used since we do not know the previous comparison result. Therefore, we should apply an additional $\mathcal{C}_{CS}(X_1^{(i)}, Y_1^{(i)})$. This yields $X_1^{(j)}$ as the smaller element. So, as a final step, evaluating $\mathcal{C}_{CS}(X_1^{(j)}, Y_0^{(j)})$ will be sufficient to complete the merging of the two arrays. This algorithm yields to Odd-Even Merge Sort whose details are given in the next section.

4.5 Odd-Even Merge Sort

Odd-Even Merge Sort is a sorting network devised by Batcher [1]. It has a recursive structure similar to Merge Sort. The algorithm considers two already sorted half-lists at each merge step. In a merge step, the merging process is recursively applied to even and odd indexed elements separately while arranging them into two halves. This process continues until there is only one element in each half list in which case a $\mathcal{C}_{CS}$ is applied in order to merge them into an array. Once the even indexed half and the odd indexed half are both internally merged, $\mathcal{C}_{CS}$ are applied to inner adjacent elements only. The merging process is illustrated in Fig. 2.

Assume we have two lists with k elements as input to the merge step. In case $k = 1$, we only need one level of $\mathcal{C}_{CS}$, thus the depth of the circuit $t_1 = 1$. In case $k = 2$, the first step recursively applies the merge step with $k = 1$ twice in parallel. Then, the second step applies inner adjacent comparisons which increment the depth by one, i.e., $t_2 = t_1 + 1 = 2$. For any k, we can conclude that $t_k = t_{k/2} + 1 = \log(k) + 1$. Hence, when we sort N elements, the overall depth can be computed as $\sum_{i=0}^{\log(N)-1} t_k = \sum_{i=0}^{\log(N)-1}(\log(k) + 1)$ where $k = 2^i$ which yields $\sum_{i=0}^{\log(N)-1}(i - 1) = [\log^2(N) + \log(N)]/2$. Therefore, $d(\mathcal{C}_{MS}) = \mathcal{O}(\log^2(N))d(\mathcal{C}_{CS})$. Similarly, let total number of comparison and swaps in a merge step in a single block be c_k. Then $c_1 = 1$ and $c_2 = 2 \cdot c_1 + 1$, since in a single merge block with $k = 2$, there are two merge blocks with $k = 1$ and only one inner adjacent element pair. In the general case, we have $c_k = 2 \cdot c_{k/2} + k - 1$ for arbitrary k. Consequently, for $k = 2^i$, we have $c_i = 2 \cdot c_{i-1} + \log(i) - 1$, which is equal to $c_i = (i+1)2^i - \sum_{j=0}^{i-1} 2^j = i \cdot 2^i + 1$. Since there are $[N/(2k)] = N/2^{i+1}$ parallel blocks in each merge step, the total number of $\mathcal{C}_{\text{CS}}$ will be $\sum_{i=0}^{\log(N)-1} [N 2^{-(i+1)} c_i] = N \sum_{i=0}^{\log(N)-1} [2^{-(i+1)} + i/2]$, which results in an asymptotic complexity of $\mathcal{O}(N \log^2(N))$.

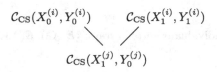

Fig. 2. Odd-Even Merging two individually sorted encrypted arrays: $\langle X_0^{(i)}, X_1^{(i)} \rangle$ and $\langle Y_0^{(i)}, Y_1^{(i)} \rangle$

4.6   Bitonic Sort

Bitonic Sorter is another sorting network created by Batcher [1]. It has similar complexity to Odd-Even Merge Sort, but with slightly different number of comparisons. The algorithm again consists of recursive sort and merge operations. The base case occurs when there are only two elements in the input array in which case only one $\mathcal{C}_{\text{CS}}$ is applied. In order to merge two sorted arrays, first of all their elements are compared and swapped so that all elements of the first subsequence are smaller than the second one. Then the subsequences are individually sorted. The depth is computed as $d(\mathcal{C}_{OEM-SORT}) = (\log^2(N) + \log(N))/2 d(\mathcal{C}_{\text{CS}})$. The depth is $\mathcal{O}(\log^2(N) \log(\ell))$ and a total of $\mathcal{O}(N \log^2(N))$ comparison complexity.

5   Proposed Sorting Algorithms

Given the inadequacies of existing sorting algorithms in permitting shallow circuit evaluation, we develop two new sorting algorithms, Direct Sort and Greedy Sort, optimized for this purpose. Both algorithms take an input vector and compute the sorted vector by evaluating the sorting circuits $\mathcal{C}_{\text{DS}}$ and $\mathcal{C}_{\text{GS}}$. The circuit evaluation makes it easy to apply the SWHE algorithm for homomorphic evaluation. The first circuit $\mathcal{C}_{\text{DS}}$ makes use of the equality check $\mathcal{C}_{\text{EQ}}$ and comparison circuits $\mathcal{C}_{\text{LT}}$ defined in Sect. 3 as building blocks whereas the second $\mathcal{C}_{\text{GS}}$ uses only the comparison circuit $\mathcal{C}_{\text{LT}}$.

Sorting Circuits $\mathcal{C}_{\text{DS}}$, $\mathcal{C}_{\text{GS}}$
Encrypted Input vector: $E(X) = \langle X_0^{(\alpha)}, X_1^{(\alpha)}, \ldots, X_{N-1}^{(\alpha)} \rangle$
Encrypted Output vector: $Y^{(\beta)} = \langle Y_0^{(\beta)}, Y_1^{(\beta)}, \ldots, Y_{N-1}^{(\beta)} \rangle$

The first step, which is mutually used by both of the circuits, constructs a comparison matrix M:

$$M^{(\gamma)} = \begin{pmatrix} m_{0,0}^{(\gamma)} & m_{0,1}^{(\gamma)} & \cdots & m_{0,N-1}^{(\gamma)} \\ m_{1,0}^{(\gamma)} & m_{1,1}^{(\gamma)} & \cdots & m_{1,N-1}^{(\gamma)} \\ \vdots & \vdots & \ddots & \vdots \\ m_{N-1,0}^{(\gamma)} & m_{N-1,1}^{(\gamma)} & \cdots & m_{N-1,N-1}^{(\gamma)} \end{pmatrix}.$$

Each $m_{i,j}^{(\gamma)}$ is computed as follows[2]:

$$m_{ij}^{(\gamma)} = \mathcal{C}_{\mathrm{LT}}(X_i^{(\alpha)}, X_j^{(\alpha)}) = \begin{cases} m_{ij} = 1 \text{ if } X_i < X_j \\ m_{ij} = 0 \text{ else} \end{cases} \tag{1}$$

where $i, j < N$ and $i < j$. The diagonal elements are self comparisons, i.e. $X_i < X_i$, therefore $m_{i,i} = 0$, $\forall i \in [0, N-1]$. The remaining entries in the lower triangular part of M, whose indices satisfy $i > j$, are computed as $m_{ji}^{(\gamma)} = m_{ij}^{(\gamma)} \oplus 1$. Note that the lower triangular part holds the comparison $m_{ji}^{(\gamma)} = (X_i \geq X_j)$. The adopted approach is straightforward as we simply compare every element with every other element in the input vector. But in terms of depth, it has a significant advantage, as performing all comparisons in the beginning reduces the depth by $d(\mathcal{C}_{\mathrm{LT}})$ in each comparison level. In the construction of M we perform $N(N-1)/2$ parallel $\mathcal{C}_{\mathrm{LT}}$ operations. This means the depth of this initial step will be 1 in terms of comparison and $\log(\ell+1)$ in terms of multiplication as stated earlier. By constructing M at the outset we simply avoid further $\mathcal{C}_{\mathrm{LT}}$ computations during the execution of the later steps and the multiplicative depth will thus be minimized.

5.1 Direct Sort

The next step for $\mathcal{C}_{\mathrm{DS}}$ is computing the index vector, σ, which indicates the positions of the vector elements in the sorted output vector, and is computed using the transpose of comparison matrix M^{T} as

$$\sigma^{(\delta)} = \left(\mathrm{HW}\left(M^{(\gamma)^{\mathsf{T}}}[0]\right) \mathrm{HW}\left(M^{(\gamma)^{\mathsf{T}}}[1]\right) \cdots \mathrm{HW}\left(M^{(\gamma)^{\mathsf{T}}}[N-1]\right) \right) \tag{2}$$

where $M^{\mathsf{T}}[i]$ represents the ith row of the transpose matrix and HW computes the Hamming Weight of the given input array. Note that in M, the sum of all elements in a column gives the number of elements, which the element with the index of the column number is greater than. For instance, the sum of all elements in column j is the number of elements, which the element X_j is larger than, as we add 1 to the sum for each such element. Therefore, the sum is also the index

---

[2] Note that when there is no ambiguity we will drop the comma, i.e. write $m_{i,j}^{(\gamma)}$ as $m_{ij}^{(\gamma)}$ in the indices for brevity.

of X_j in the sorted output vector. In other words, if an element is larger than k other elements, then this implies that it is the $k + 1^{st}$ largest element and its index is k in a zero-based output vector. Now, since all data is in an encrypted form, we have no knowledge about the elements of the σ; therefore we cannot use it directly for homomorphic sorting. Here, we simply compare each element of the index vector $\sigma^{(\gamma)}$ (i.e., $\sigma_i^{(\gamma)}$) with each possible index value (which is in the interval $[0, N - 1]$); the equality places the corresponding input element in the current position of the output vector. For this, we make use of C_{EQ} circuit as follows

$$Y_j^{(\beta)} = \sum_{i \in [N]} (\sigma_i^{(\delta)} = j) X_i^{(\alpha)} \quad \text{for } j \in [N]. \tag{3}$$

The overall method for open version C_{DS} is described in Algorithm 1. From the discussions in Sect. 1, we already know that $d(C_{LT}) = \log(\ell + 1)$ and $d(C_{EQ}) = \log(\ell)$. In the computations of the entries of σ we add N bits to form a $\log(N)$-bit sum. In this step full and half adders are used in a Wallace Tree structure, hence the depth of the circuit for the N-bit summation can be given approximately as $d(\sigma) = \mathcal{O}(\log_{3/2}(N))$. Taking into account the parallel C_{LT} and C_{EQ} comparisons and single multiplication in the final summation the total depth becomes $d(C_{LT}) + d(\sigma) + d(C_{EQ}) + 1$. Therefore, we can obtain the following expression for the overall depth of the circuit that implements the proposed algorithm: $d(C_{DS}) = \mathcal{O}(\log(N) + \log(\ell))$.

Algorithm 1. Direct Sorting Algorithm

```
1:  function SORT(X, Y, N)
2:      for i ← 0 to N − 1 do                            ▷ Construct M table
3:          M[i][i] ← 0
4:          for j ← i + 1 to N − 1 do
5:              M[i][j] ← LessThan (X[i], X[j])
6:              M[j][i] ← M[j][i] + 1
7:          end for
8:      end for
9:      M ← Transpose (M)
10:     for i ← i + 1 to N − 1 do                        ▷ Construct σ vector
11:         S[i] ← HammingWeight (M[i], N)
12:     end for
13:     for i ← 0 to N − 1 do                             ▷ Construct Y, output vector
14:         Y[i] ← 0
15:         for j ← 0 to N − 1 do
16:             z ← IsEqual (i, S[j])
17:             Y[i] ← Y[i] + AND (z, X[j])
18:         end for
19:     end for
20: end function
```

5.2   Greedy Sort

In this scheme, we compute every possible permutation of indices for the sorted array. For instance, to determine the smallest element Y_0 in the sorted array we need to check if a candidate element X_i is smaller than all the other elements in X, to be set as the smallest element of the sorted array. We can express the conditions yielding the Y_0 assignment explicitly as in Algorithm 2. Similarly, for Y_1 if an element is smaller than all others except one, then we can conclude that it is the second smallest element. In this case, we compute more possibilities, namely $\binom{N-1}{1}$, in each if-else statement since we have the possibility of an element X_i being larger than any of the other elements. The expression for Y_1, which determines the second smallest element is given in Algorithm 3. Using the comparison matrix $M^{(\gamma)}$, defined in Sect. 5.1, we can convert the if-else statements into logic circuits and compute the sorted elements. The if-else statements give us an exact mutually exclusive partitioning in the output assignments. Therefore, we can use XOR (logical exclusive disjunction $\oplus$) gates to combine each statement. For instance, Y_0 evaluated by the following circuit

$$Y_0^{(\beta)} = \left(m_{0,1}^{(\gamma)} \dots m_{0,N-1}^{(\gamma)}\right) X_0^{(\alpha)} \oplus \left(m_{1,0}^{(\gamma)} \dots m_{1,N-1}^{(\gamma)}\right) X_1^{(\alpha)} \oplus \dots$$
$$\oplus \left(m_{N-1,0}^{(\gamma)} \dots m_{N-1,N-2}^{(\gamma)}\right) X_{N-1}^{(\alpha)}. \tag{4}$$

We can write this equation in a more compact form, if we use a coefficient for each X_i, such as $\theta_{t,i}$, where t stands for the index of Y_t. Using $t = 0$, $\theta_{0,i} = \prod_{\substack{j=0 \\ j \neq i}}^{N-1} m_{ij}$ and the overall equation simply becomes $Y_0 = \theta_{0,0} X_0 \oplus \dots \oplus \theta_{0,N-1} X_{N-1}$.
In condition evaluations we can also convert the OR gates (i.e., logical disjunction $\vee$ in Algorithm 3) to XOR gates. To see why this works, first note that $a \vee b = a \oplus b \oplus (a \cdot b)$ where a and b are bits. If $a \cdot b = 0$ then $a \vee b = a \oplus b$. We can make the following proposition for the conjunction cases of X_i to show that it can either have only one conjunction that outputs 1 or none:

Proposition 1. *In the expression of $\theta_{t,i}$ of the element X_i, for any two distinct conjunctions ρ and ρ' it holds that $\rho\rho' = 0$.*

Proof. In order to evaluate all the combinations we always find $m_{k,l} \in \rho$ and $m_{l,k} \in \rho'$ for some $k, l \in N - 1$. Otherwise $\rho = \rho'$, a contradiction. Since $\rho\rho'$ will contain the conjunction $m_{k,l} m_{l,k}$ we always have $\rho\rho' = 0$ by $m_{k,l} = m_{l,k} \oplus 1$.

Algorithm 2. Finding the minimum element

1: **if** $(X_0 < X_1) \wedge (X_0 < X_2) \wedge \ \dots \ \wedge (X_0 < X_{N-1})$ **then**
2:     $Y_0 = X_0$
3: **else if** $\neg(X_0 < X_1) \wedge (X_1 < X_2) \wedge \ \dots \ \wedge (X_1 < X_{N-1})$ **then**
4:     $Y_0 = X_1$
5: **else if** $\ \dots \ $ **then**

6:     $\vdots$
7: **end if**

Now we can freely convert all occurrences of OR's to $\oplus$ and the circuit for Y_1 becomes $Y_1^{(\beta)} = \theta_{1,0}^{(\gamma)} X_0^{(\alpha)} \oplus \ldots \oplus \theta_{1,N-1}^{(\gamma)} X_{N-1}^{(\alpha)}$ where $\theta_{1,i}^{(\gamma)} = \sum_{\substack{k_1=0 \\ k_1 \neq i}}^{N-1} m_{k_1 i}^{(\gamma)} \prod_{\substack{j=0 \\ j \neq i, k_1}}^{N-1} m_{ij}^{(\gamma)}$. More generally, for other t values, following a similar logical expression, we will have $\binom{N-1}{t}$ possibilities, and $\theta_{t,i}^{(\gamma)}$ will be computed as

$$\theta_{t,i}^{(\gamma)} = \sum_{\substack{k_1=0 \\ k_1 \neq i}}^{N-t} m_{k_1 i}^{(\gamma)} \sum_{\substack{k_2=k_1+1 \\ k_2 \neq i}}^{N-t+1} m_{k_2 i}^{(\gamma)} \cdots \sum_{\substack{k_t=k_{t-1}+1 \\ k_t \neq i}}^{N-1} m_{k_t i}^{(\gamma)} \prod_{\substack{j=0 \\ j \neq i \\ j \neq k_1,\ldots,k_t}}^{N-1} m_{ij}^{(\gamma)} \qquad (5)$$

and the output values of $\mathcal{C}_{\text{GS}}$, Y_t for $t \in [N]$ as $Y_t^{(\beta)} = \sum_{i=0}^{N-1} X_i^{(\alpha)} \theta_{t,i}^{(\gamma)}$. Each output of the circuit $\mathcal{C}_{\text{GS}}$ computes a summation of the input values $X_0, \ldots, X_{N-1}$ where values are weighted with $\theta_{t,i}$. Note that $\theta_{t,i}$ evaluates a logic expression that determines whether X_i ends up in position t, i.e. $Y_t = X_i$, after sorting. The overall depth is $d(\mathcal{C}_{\text{GS}}) = d(\mathcal{C}_{\text{LT}}) + d(\theta_{t,i} X_i)$., where $d(\mathcal{C}_{\text{LT}}) = \log(\ell + 1)$ as given in Sect. 3. During the $\theta_{t,i}$ computations we employ a circuit arranged in a binary tree of depth $d(\theta_{t,i}) = \lceil log(N-1) \rceil$ and $d(\theta_{t,i} X_i) = d(\theta_{t,i}) + 1$. Consequently, the overall circuit depth is found as $d(\mathcal{C}_{\text{GS}}) = \lceil \log(\ell + 1) \rceil + \lceil \log(N-1) \rceil + 1 = \mathcal{O}(\log(N) + \log(\ell))$.

Table 1. The multiplicative depth of different sorting circuits given size N and bit size ℓ

ℓ	8					32				
N	4	8	16	32	64	4	8	16	32	64
$\mathcal{C}_{\text{INS}}\backslash\mathcal{C}_{\text{BUBS}}$	30	140	600	2480	10080	42	196	840	3472	14112
$\mathcal{C}_{\text{OES}}$	20	40	80	160	320	28	56	112	224	448
$\mathcal{C}_{\text{OEMS}}\backslash\mathcal{C}_{\text{BITS}}$	15	30	50	75	105	21	42	70	105	147
$\mathcal{C}_{\text{DS}}$	9	10	11	12	13	11	12	13	14	15
$\mathcal{C}_{\text{GS}}$	7	8	9	10	11	9	10	11	12	13

Algorithm 3. Finding the second minimum element

1: **if** $[(X_0 < X_1) \wedge \ldots \wedge \neg(X_0 < X_{N-1})] \vee \ldots \vee [\neg(X_0 < X_1) \wedge \ldots \wedge (X_0 < X_{N-1})]$ **then**
2:    $Y_1 = X_0$
3: **else if** $[(X_1 < X_0) \wedge \ldots \wedge \neg(X_1 < X_{N-1})] \vee \ldots \vee [\neg(X_1 < X_0) \wedge \ldots \wedge (X_1 < X_{N-1})]$ **then**
4:    $Y_1 = X_1$
5: **else if** $\ldots$ **then**
6:    $\vdots$
7: **end if**

---

6   Implementation Results

We implemented the proposed depth optimized sorting method described in Algorithm 1 using the SWHE scheme of [11] and evaluated C_{DS} for a number of array lengths. Here, we briefly summarize the parameter selection process and present the simulation results (Table 1).

Parameter Selection. According to [11] the NTRU based SWHE Scheme requires Hermite factor $\delta < 1.0066$ to achieve a security level of 80-bit. We set the per level cutting rate $\log p$ depending both on the circuit itself and its total depth, similarly we choose a polynomial degree n according to security threshold and maximum coefficient modulus size. We implemented C_{OES}, C_{OEMS}, C_{DS} and C_{GS} circuits, simulated them for both $\ell = 8$-bit and $\ell = 32$-bit integer inputs and selected array size N.[3] In Table 2, we enumerate the parameters which we used in our experiments for various circuit depths. The largest Hermite factor among our parameter choices is $\delta = 1.0063$, ensuring a security level of 99-bits, which is the lowest security level for all cases.

Table 2. Cutting size $\log p$, maximum coefficient size $\log q_0$, Polynomial degree n, message batching slot size S and Hermite Factor δ for different depths d

Depth d	9	12	15	21	28	42	56
$\log p$	20	20	22	25	25	25	30
$\log q_0$	200	260	352	550	725	1075	1710
n	8190	8190	16384	16384	27000	32768	46656
S	630	630	1024	1024	1800	2048	2592
δ	1.0041	1.0054	1.0037	1.0057	1.0046	1.0056	1.0063

Performance Results. We implemented homomorphic Odd Even Sort, Batcher's Odd Even Merge Sort and both of the proposed algorithms in C++ using DHS-SWHE Library [11]. All simulations were performed on an Intel Xeon @ 2.9 GHz server running Ubuntu Linux 13.10. We compiled our code using Shoup's NTL library version 6.0 and with GMP version 5.1.3. The sorting times for 8 and 32 bit integers are given in Table 3. For instance $N = 64$ elements with $\ell = 32$ bit size, our algorithm Direct Sort, C_{DS}, runs in about 14.15 h whereas the amortized running time, where we use batching with slot size 1024, is about 49.7 s per sort. When the input bit size is $\ell = 8$, the run time is 10.6 h, and with batching slot size of 630, it takes 1 minute. For $N = 4$ the sorting takes as low as 0.20 s per sort. In comparison, the homomorphic Lazy Sort implementation of [8] takes about 976 and 1400 s for array sizes of 10 and 40, respectively. For array sizes $N = 16$ and $N = 64$ our implementation takes 4.28 and 50 s, respectively.

---

[3] Note that N is *not* restricted to a power of two.

Table 3. Amortized execution time of circuits for different array sizes N and input bit sizes ℓ

ℓ	8					32				
N	4	8	16	32	64	4	8	16	32	64
$\mathcal{C}_{OES}$	400 ms	3.45 s	n/a	n/a	n/a	2.4 s	n/a	n/a	n/a	n/a
$\mathcal{C}_{OEMS}$	270 ms	3.30 s	n/a	n/a	n/a	530 ms	5.8 s	31 s	n/a	n/a
$\mathcal{C}_{DS}$	140 ms	690 ms	3.14 s	13.9 s	1 m	200 ms	944 ms	4.28 s	18.6 s	49.7 s
$\mathcal{C}_{GS}$	90 ms	470 ms	2.8 s	13.10 s	52.2 s	500 ms	2.4 s	10.8 s	49.2 s	2.2 m

7 Conclusion

We proposed two depth optimized sorting algorithms for efficient homomorphic evaluation. Circuit depth is intimately related to the parameter sizes in leveled homomorphic encryption implementations and therefore directly affect the overall performance of the homomorphic circuit evaluation. Existing sorting algorithms are not optimized for homomorphic evaluation. To close this gap we presented the depth analysis for several classical sorting algorithms: Bubble sort, Insertion Sort, Odd Even Sort, Odd Even Merge Sort, Merge Sort, and Bitonic Sort. Inspired by the performance of Merge Sort we introduced two new depth-optimized sorting algorithms which achieve a circuit depth of $\mathcal{O}(\log(N)+\log(\ell))$.

To study the real-life performance of our sorting algorithms, we instantiated an NTRU based SWHE scheme in the DHS SWHE library and presented simulation results for selected array lengths. For this we determined the ideal parameter choices, e.g. modulus cutting levels to cope with noise growth and Hermite work factor estimates to ensure reasonable security margins. The implementation performs favorably achieving significant speedup over the proposal in [8] for similar array lengths.

Acknowledgments. Funding for this research was in part provided by the US National Science Foundation CNS Awards #1319130.

References

1. Batcher, K.E.: Sorting networks and their applications. In: Proceedings of the April 30–May 2, 1968, Spring Joint Computer Conference, AFIPS 1968 (Spring), pp. 307–314. ACM, New York (1968). http://doi.acm.org/10.1145/1468075.1468121
2. Bos, J.W., Lauter, K., Naehrig, M.: Private predictive analysis on encrypted medical data. Technical report MSR-TR-2013-81, September 2013. http://research.microsoft.com/apps/pubs/default.aspx?id=200652
3. Bos, J.W., Lauter, K., Loftus, J., Naehrig, M.: Improved security for a ring-based fully homomorphic encryption scheme. In: Stam, M. (ed.) IMACC 2013. LNCS, vol. 8308, pp. 45–64. Springer, Heidelberg (2013). http://dx.doi.org/10.1007/978-3-642-45239-0_4

4. Brakerski, Z.: Fully homomorphic encryption without modulus switching from classical gapSVP. IACR Cryptology ePrint Archive 2012, 78 (2012)
5. Brakerski, Z., Gentry, C., Vaikuntanathan, V.: Fully homomorphic encryption without bootstrapping. Electronic Colloquium on Computational Complexity (ECCC) 18, 111 (2011)
6. Brakerski, Z., Vaikuntanathan, V.: Efficient fully homomorphic encryption from (standard) LWE. In: FOCS, pp. 97–106 (2011)
7. Brenner, M., Perl, H., Smith, M.: libscarab software library. https://hcrypt.com/
8. Chatterjee, A., Kaushal, M., Sengupta, I.: Accelerating sorting of fully homomorphic encrypted data. In: Paul, G., Vaudenay, S. (eds.) INDOCRYPT 2013. LNCS, vol. 8250, pp. 262–273. Springer, Heidelberg (2013). http://dx.doi.org/10.1007/978-3-319-03515-4_17
9. Cheon, J.H., Kim, M., Lauter, K.: Secure dna-sequence analysis on encrypted DNA nucleotides. http://media.eurekalert.org/aaasnewsroom/MCM/FIL_000000001439/EncryptedSW.pdf
10. van Dijk, M., Gentry, C., Halevi, S., Vaikuntanathan, V.: Fully homomorphic encryption over the integers. In: Gilbert, H. (ed.) EUROCRYPT 2010. LNCS, vol. 6110, pp. 24–43. Springer, Heidelberg (2010)
11. Doröz, Y., Hu, Y., Sunar, B.: Homomorphic AES evaluation using NTRU (2014). http://eprint.iacr.org/2014/039.pdf, iACR ePrint Archive
12. Doröz, Y., Sunar, B., Hammouri, G.: Bandwidth efficient PIR from NTRU. In: Böhme, R., Brenner, M., Moore, T., Smith, M. (eds.) FC 2014 Workshops. LNCS, vol. 8438, pp. 195–207. Springer, Heidelberg (2014). http://dx.doi.org/10.1007/978-3-662-44774-1_16
13. Fischlin, M.: A cost-effective pay-per-multiplication comparison method for millionaires (2001)
14. Gentry, C.: A Fully Homomorphic Encryption Scheme. Ph.D. thesis, Stanford University (2009)
15. Gentry, C.: Fully homomorphic encryption using ideal lattices. In: STOC, pp. 169–178 (2009)
16. Gentry, C., Halevi, S.: Implementing Gentry's fully-homomorphic encryption scheme. In: Paterson, K.G. (ed.) EUROCRYPT 2011. LNCS, vol. 6632, pp. 129–148. Springer, Heidelberg (2011)
17. Gentry, C., Halevi, S., Smart, N.P.: Fully homomorphic encryption with polylog overhead. IACR Cryptology ePrint Archive Report 2011/566 (2011). http://eprint.iacr.org/
18. Gentry, C., Halevi, S., Smart, N.P.: Homomorphic evaluation of the AES circuit. IACR Cryptology ePrint Archive 2012 (2012)
19. Goldwasser, S., Micali, S.: Probabilistic encryption & how to play mental poker keeping secret all partial information. In: Proceedings of the Fourteenth Annual ACM Symposium on Theory of Computing, STOC 1982, pp. 365–377. ACM, New York (1982). http://doi.acm.org/10.1145/800070.802212
20. Graepel, T., Lauter, K., Naehrig, M.: ML confidential: machine learning on encrypted data. In: Kwon, T., Lee, M.-K., Kwon, D. (eds.) ICISC 2012. LNCS, vol. 7839, pp. 1–21. Springer, Heidelberg (2013). http://dx.doi.org/10.1007/978-3-642-37682-5_1
21. Knuth, D.E.: The Art of Computer Programming, Fundamental Algorithms, vol. 1, 3rd edn. Addison Wesley Longman Publishing Co., Inc., Redwood City (1998)
22. Lagendijk, R., Erkin, Z., Barni, M.: Encrypted signal processing for privacy protection: conveying the utility of homomorphic encryption and multiparty computation. IEEE Sig. Process. Mag. 30(1), 82–105 (2013)

23. Lauter, K., Naehrig, M., Vaikuntanathan, V.: Can homomorphic encryption be practical. In: Cloud Computing Security Workshop, pp. 113–124 (2011)
24. Lauter, K., Lopez-Alt, A., Naehrig, M.: Private computation on encrypted genomic data. Technical report MSR-TR-2014-93, June 2014. http://research.microsoft.com/apps/pubs/default.aspx?id=219979
25. López-Alt, A., Tromer, E., Vaikuntanathan, V.: On-the-fly multiparty computation on the cloud via multikey fully homomorphic encryption. In: STOC (2012)
26. López-Alt, A., Naehrig, M.: Large integer plaintexts in ring-based fully homomorphic encryption (2014, in preparation)
27. Rivest, R.L., Adleman, L., Dertouzos, M.L.: On data banks and privacy homomorphisms. In: Foundations of Secure Computation, pp. 169–180 (1978)
28. Sander, T., Young, A., Yung, M.: Non-interactive cryptocomputing for nc1. In: 40th Annual Symposium on Foundations of Computer Science, pp. 554–566 (1999)
29. Smart, N.P., Vercauteren, F.: Fully homomorphic SIMD operations. IACR Cryptology ePrint Archive 2011, 133 (2011)
30. Stehlé, D., Steinfeld, R.: Making NTRU as secure as worst-case problems over ideal lattices. In: Paterson, K.G. (ed.) EUROCRYPT 2011. LNCS, vol. 6632, pp. 27–47. Springer, Heidelberg (2011)
31. Vaidya, J., Clifton, C.: Privacy-preserving k-means clustering over vertically partitioned data. In: Proceedings of the Ninth ACM SIGKDD International Conference on Knowledge Discovery and Data Mining, KDD 2003, pp. 206–215. ACM, New York (2003). http://doi.acm.org/10.1145/956750.956776
32. Yao, A.C.: Protocols for secure computations. In: Proceedings of the 23rd Annual Symposium on Foundations of Computer Science, SFCS 1982, pp. 160–164. IEEE Computer Society, Washington, DC (1982). http://dx.doi.org/10.1109/SFCS.1982.88
33. Yildizli, C.B., Pedersen, T., Saygin, Y., Savas, E., Levi, A.: Distributed privacy preserving clustering via homomorphic secret sharing and its application to vertically partitioned spatio-temporal data. Int. J. Data Warehous. Min. 7(1), 46–66 (2011). http://dx.doi.org/10.4018/jdwm.2011010103

The Chain Rule for HILL Pseudoentropy, Revisited

Krzysztof Pietrzak[1] and Maciej Skórski[2(✉)]

[1] IST Austria, Klosterneuburg, Austria
pietrzak@ist.ac.at
[2] University of Warsaw, Warsaw, Poland
maciej.skorski@gmail.com

Abstract. Computational notions of entropy (a.k.a. pseudoentropy) have found many applications, including leakage-resilient cryptography, deterministic encryption or memory delegation. The most important tools to argue about pseudoentropy are chain rules, which quantify by how much (in terms of quantity and quality) the pseudoentropy of a given random variable X decreases when conditioned on some other variable Z (think for example of X as a secret key and Z as information leaked by a side-channel). In this paper we give a very simple and modular proof of the chain rule for HILL pseudoentropy, improving best known parameters. Our version allows for increasing the acceptable length of leakage in applications up to a constant factor compared to the best previous bounds. As a contribution of independent interest, we provide a comprehensive study of all known versions of the chain rule, comparing their worst-case strength and limitations.

1 Introduction

Min-entropy. Various notions of entropy are used to quantify the randomness in a random variable. The most important notion in cryptographic contexts is min-entropy, where a variable X (conditioned on Z) has min-entropy k if one cannot guess X (given Z) with probability better than 2^k.

Definition 1. *The* **min-entropy** *of a variable X is*

$$H_\infty(X) = -\log \max_x \Pr[X = x]$$

More generally, for a joint distribution (X, Z), the **average min-entropy** *of X conditioned on Z is [DRS04]*

$$\widetilde{\mathbf{H}}_\infty(X|Z) = -\log \mathbb{E}_{z \leftarrow Z} \max_x \Pr[X = x | Z = z]$$

$$= -\log \mathbb{E}_{z \leftarrow Z} 2^{-H_\infty(X|Z=z)}.$$

Krzysztof Pietrzak—Research supported by ERC starting grant (259668-PSPC).
Maciej Skórski—Research supported by the Ideas for Poland grant 2/2011 from the Foundation for Polish Science.

K. Lauter and F. Rodríguez-Henríquez (Eds.): LatinCrypt 2015, LNCS 9230, pp. 81–98, 2015.
DOI: 10.1007/978-3-319-22174-8_5

Chain-Rules. Most entropy notions $H(.)$ satisfy a chain rule which roughly capture the fact that when additionally conditioning on a variable Z, the entropy can decrease by at most its length $|Z|$, i.e.,

$$H(X|Y,Z) \geq H(X|Y) - |Z| \tag{1}$$

In particular, average-case min-entropy satisfies such a rule [DRS04]

$$\tilde{\mathbf{H}}_\infty(X|Y,Z) \geq \tilde{\mathbf{H}}_\infty(X|Y) - H_0(Z) \geq \tilde{\mathbf{H}}_\infty(X|Y) - |Z|, \tag{2}$$

where $H_0(Z) \leq |Z|$ denotes the logarithm of the support-size of Z.

Pseudoentropy. Information theoretic entropy notions refer to computationally unbounded parties, e.g., no algorithm can compress a distribution X (given Z) below its Shannon entropy $H(X|Z)$ and no algorithm can guess X (given Z) better than with probability $2^{-\tilde{\mathbf{H}}_\infty(X|Z)}$. Under computational assumptions, in particular in cryptographic settings, one often has to deal with distribution that appear to have high entropy only for computationally bounded parties.

The classical example is a pseudorandom distribution [BM84, Yao82], where $X \in \{0,1\}^n$ is said to be pseudorandom if it cannot be distinguished from the uniform distribution over $\{0,1\}^n$ by polynomial size distinguishers. In this case X appears to have n bits of Shannon and n bits of min-entropy. More generally, $X \in \{0,1\}^n$ has k bits of HILL entropy, if it cannot be distinguished from some distribution Y with k bits of min-entropy. Note that for $k = n$ HILL entropy is simply pseudorandomness, as the only distribution over $\{0,1\}^n$ with n bits of min-entropy is the uniform distribution. HILL entropy was introduced in [HILL99], the more general conditional notion below is from [HLR07].

Definition 2 ([HLR07]). *Let (X,Z) be a joint distribution of random variables. Then X has **conditional HILL entropy** k conditioned on Z, denoted by $\mathbf{H}_{\varepsilon,s}^{\mathrm{HILL}}(X|Z) \geq k$, if there exists a joint distribution (Y,Z) such that $\tilde{\mathbf{H}}_\infty(Y|Z) \geq k$, and $(X,Z) \sim_{\varepsilon,s} (Y,Z)$.[1]*

Computational notions of entropy find important applications in leakage-resilient cryptography [DP08b], deterministic encryption [FOR12], memory delegation [CKLR11], computational complexity [RTTV08a] and foundations of cryptography [HRV10].

Chain Rules for Computational Entropy. When considering chain rules as in as in Eq. (1) for computational notions of entropy, one must not only specify by how much the *quantity* of the entropy decreases, but also its *quality*. For some computational entropy notions like Yao or unpredictability entropy, chain rules are very easy to prove, and have been folklore for a long time (for the short proofs cf.

---

[1] Let us stress that using the same letter Z for the 2nd term in (X,Z) and (Y,Z) means that we require that the marginal distribution Z of (X,Z) and (Y,Z) is the same.

Appendix A in [KPWW14]). For HILL entropy the situation is much more complicated. The first chain rules were found independently by [RTTV08a, DP08a], and several proofs for the chain rule for HILL entropy were given subsequently, often as a corollary of a more general result. The various proofs give different qualitative bounds and are summarised below.

Theorem 1 (Chain Rules for HILL Entropy). *For any joint distribution* $(X, Z) \in \{0,1\}^n \times \{0,1\}^m$ *we have that*

$$\mathbf{H}^{\mathrm{HILL}}_{\varepsilon', s'}(X|Z) \geqslant \mathbf{H}^{\mathrm{HILL}}_{\varepsilon, s}(X) - m - \Delta \qquad (3)$$

where $\varepsilon' = \varepsilon \cdot p(2^\ell, \varepsilon^{-1})$ *and* $s' = s/q(2^\ell, \varepsilon^{-1})$, *for some polynomial functions* $p(.)$ *and* $q(.)$ *as summarised in Table 1 ($\Delta = 0$ except for [DP08b], where $\Delta = 2\log(1/\varepsilon)$).*

Table 1. Qualitative bounds on chain rules for HILL entropy. For simplicity, smaller order terms $\log(1/\epsilon)$, n, m are hidden under the big-O notation.

Reference	Technique	$s' =$	$\epsilon' =$	Meaningful range
(a) [DP08b]	Worst-Case Metric Entropy	$\Omega\left(s \cdot 2^{2m}\epsilon^2\right)$	$O(\sqrt{2^m}\epsilon)$	$s > 2^{-2m}\epsilon^{-2}$, $2^{-m} > \epsilon$
(b) [RTTV08a]	Dense Model Theorem	$\Omega(s \cdot \mathrm{poly}(\epsilon, \min_z$ $(\Pr[Z = z])))$	$O(2^m \epsilon)$	$s > \max_z \frac{1}{\Pr[Z=z]^2} \cdot \epsilon^{-2}$, $2^{-m} > \epsilon$
(c) [FOR12]	Worst-Case Metric Entropy	$\Omega\left(s \cdot 2^{2m}\epsilon^2\right)$	$O(2^m \epsilon)$	$s > 2^{2m}\epsilon^{-2}$, $2^{-m} > \epsilon$
(d) [JP14]	Simulating Auxiliary Inputs	$\Omega\left(s \cdot \frac{\epsilon^2}{2^{3m}} - 2^m\right)$	$O(\epsilon)$	$s > 2^{4m}\epsilon^{-2} + 2^{3m}\epsilon^{-2}$
(e) [VZ13]	Simulating Auxiliary Inputs	$\Omega\left(s \cdot \frac{\epsilon^2}{2^m} - \frac{1}{\epsilon^2} - 2^m\right)$	$O(\epsilon)$	$s > 2^m\epsilon^{-4} + 2^{2m}\epsilon^{-2} + 2^m\epsilon^{-2}$
(f) **This paper** using [GW10]	Relaxed HILL Entropy	$\Omega\left(s \cdot \frac{\epsilon^2}{2^m} - 2^m\right)$	$O(\epsilon)$	$s > 2^{2m}\epsilon^{-2} + 2^m\epsilon^{-2}$
(g) **This paper**	Average Metric Entropy	$\Omega\left(s \cdot \frac{\epsilon^2}{2^m} - 2^m\epsilon^2\right)$	$O(\epsilon)$	$s > 2^m\epsilon^{-2} + 2^{2m}$

As shown in the table, every chain rule losses a factor exponential in m in quality (either in the size s or in the advantage ϵ) and also a factor $\mathrm{poly}(\epsilon)$. The second loss is the reason for poor security bounds in applications, for example in security proofs for leakage resilient stream ciphers (cf. [Pie09] and related papers), but seems unavoidable given the current state of the art. The choice of whether we lose 2^m in size or advantage depends on an application, as we will see later.

All the chain rules in Table 1 can be slightly generalized. Namely, one can opt for a larger s' at the prize of a larger ϵ'. This is possible because the common part of all the corresponding proofs is an approximation argument (typically by the Chernoff Bound). The most general statements can be found in Table 2 below.

Table 1 is recovered from Table 2 by setting the free parameter δ to be of the same order as the other additive term in ϵ' (ϵ or $2^m\epsilon$ in the table), in order to get the smallest possible ϵ', while keeping the number of parameters small. We stress that in later discussions, including applications and our results described in Sect. 1.2, we refer to the general bounds from Table 2.

Table 2. Qualitative bounds on chain rules for HILL entropy, in the most general form with the free parameter δ.

Reference	Technique	$s' =$	$\epsilon' =$
(a) [DP08b]	Worst-Case Metric Entropy	$\Omega\left(s \cdot 2^{2m}\delta^2\right)$	$O(\sqrt{2^m\epsilon} + \delta)$
(b) [RTTV08a]	Dense Model Theorem	$\Omega\left(s \cdot \frac{\delta^2}{\max_z(\Pr[Z=z])^2}\right)$	$O(2^m\epsilon + \delta)$
(c) [FOR12]	Worst-Case Metric Entropy	$\Omega\left(s \cdot \delta^2\right)$	$O(2^m\epsilon + \delta)$
(d) [JP14]	Simulating Auxiliary Inputs	$\Omega\left(s \cdot \frac{\delta^2}{2^{3m}} - 2^m\right)$	$O(\epsilon + \delta)$
(e) [VZ13]	Simulating Auxiliary Inputs	$\Omega\left(s \cdot \frac{\delta^2}{2^m} - \frac{1}{\delta^2} - 2^m\right)$	$O(\epsilon + \delta)$
(f) [GW10]	Relaxed HILL Entropy	$\Omega\left(s \cdot \frac{\delta^2}{2^m} - 2^m\right)$	$O(\epsilon + \delta)$
(g) **This paper**	Average Metric Entropy	$\Omega\left(s \cdot \frac{\delta^2}{2^m} - 2^m\delta^2\right)$	$O(\epsilon + \delta)$

New Chain Rule. We prove the following results

Theorem 2 (Chain Rule for Metric Entropy with Loss in Size). *Let* $X \in \{0,1\}^n$ *and* $Z \in \{0,1\}^m$ *be correlated random variables. Then for any* (ϵ, s) *we have*

$$\mathbf{H}^{\mathrm{Metric,det,[0,1]}}_{\epsilon',s'}(X|Z) \geqslant \mathbf{H}^{\mathrm{Metric,det,[0,1]}}_{\epsilon,s}(X) - m \qquad (4)$$

where $s' = s/2^m - 2^m$ *and* $\epsilon' = \epsilon$.

Corollary 1. *Let* $X \in \{0,1\}^n$ *and* $Z \in \{0,1\}^m$ *be correlated random variables. Then for any* (ϵ, s) *we have*

$$\mathbf{H}^{\mathrm{HILL}}_{\epsilon',s'}(X|Z) \geqslant \mathbf{H}^{\mathrm{HILL}}_{\epsilon,s}(X) - m \qquad (5)$$

where $s' = \Omega\left(\frac{s}{2^m} \cdot \frac{\delta^2}{n+1-k} - 2^m \cdot \frac{\delta^2}{n+1-k}\right)$, $\epsilon' = \epsilon + \delta$, δ *is arbitrary and* $k = \mathbf{H}^{\mathrm{HILL}}_{\epsilon,s}(X)$ *(actually* $k = \mathbf{H}^{\mathrm{Metric,det,[0,1]}}_{(\epsilon,s)}(X|Z)$ *is enough).*

The proofs can be found in Sect. 3. Our new chain rule (g) loses a leakage-dependent factor in s instead in ϵ, and can be viewed as complementary with respect to (c) which loses it only in ϵ. Later we will see that there are settings where both chain rules gives equivalent security (basically when ϵ can be chosen sufficiently small), but for other cases our chain rule might be preferable (when we start with moderate values of ϵ and aim for relatively small ϵ'). We will discuss these applications with practically meaningful numerical examples in Sect. 1.2.

1.1 Proofs Techniques for Chain Rules

Basically, all the previously known chain rules have been obtained by one of the two following ways:

(a) bounding pseuodoentropy for every leakage value separately
(b) using so called relaxed pseudoentropy

The first technique, so called decomposable entropy [FR12], can be viewed as an extension of the dense model theorem which is equivalent when the entropy amount is full (this equivalence holds up to a constant factor as demonstrated in [Sko15a]); this approach yields always an exponential (in m) loss for ϵ. The second way is to use the so called "relaxed" pseudoentropy, which offer an exponential (in m) loss for s, but no loss in ϵ. In this paper we come up with a different approach, namely we first prove a variant of a chain rule for average metric entropy which loses only in s and then use known transformations to convert it back to HILL entropy. As shown in Table 1 our approach yields best possible known loss in s compared to the known chain rules which do not decrease ϵ (Fig. 1).

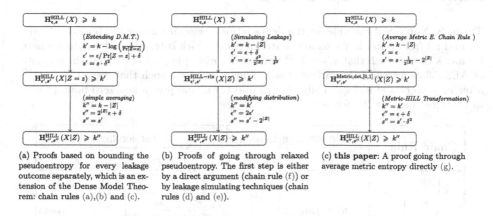

(a) Proofs based on bounding the pseudoentropy for every leakage outcome separately, which is an extension of the Dense Model Theorem: chain rules (a),(b) and (c).

(b) Proofs of going through relaxed pseudoentropy. The first step is either by a direct argument (chain rule (f)) or by leakage simulating techniques (chain rules (d) and (e)).

(c) **this paper:** A proof going through average metric entropy directly (g).

Fig. 1. Chain rules classified by used proof techniques.

1.2 Qualitative Comparison

Table 1 summarizes the known and our new bonds for the HILL entropy chain rule. In the first three bounds (a) to (c) the advantage $\epsilon' = 2^m \epsilon$ degrades exponentially in m (with the best result achieved by (c)), whereas in the bounds (d) to (g) we one have a degradation in the circuit size s', but the distinguishing advantage ϵ' stays the same (up to some small constant hidden on the big-Oh notation, which we'll ignore for the rest of this section).

The degradation in circuit size for all bounds (d) to (g) is of the form $s' = s/\alpha - \beta$, so the circuit size degrades by a factor α and an additive term β.

The best factor $\alpha = 2^m/\epsilon^2$ is achieved by the bounds (e) to (g), and the best additive loss is achieved by (g). Below we give a numerical example showing that for some practical settings of parameters this really matters, and (g) gives meaningful security guarantees whereas (d), (e) and (f) don't. Comparing (g) with (c) is less straight forward, because (c) has a degradation in the advantage whereas (g) does not. To get a meaningful comparison, we consider first settings where we can assume that the running time to advantage ratio s/ϵ is fixed, and then discuss the case when no such a simple tradeoff exists.

Fixed Time-Success Ratio (Application for Weak PRFs). For concreteness, we assume that $X = (x_1, F(K, x_1), \ldots, (x_\ell, F(K, x_\ell))$ consists of input-output pairs of a weak PRF $F(.,.)$ with key K, and we want to know how good the HILL entropy of X is given some m bits of leakage about K. This is the setting in which the chain rule is e.g. used in the security proof of the leakage-resilient stream-cipher from [Pie09]. For example think of $F(.,.)$ as AES256, and assume its security as a weak PRF satisfies $s/\epsilon \approx 2^{256}$, which is the case if bruce force key-search is the best attack.[2] Under this assumption, the degradation in circuit size in [FOR12] and our new bounds (g)) are identical as illustrated with a concrete numerical example in Table 3.

Table 3. Numerical example for the degradation is circuit size for the bound (c), (d), (e) and (f) from Table 1. We assume a distribution which is (ϵ, s) pseudorandom where for any s the ϵ is such that $s/\epsilon = 2^{256}$ (we can for example conjecture that the security of AES256 as a weak PRF satisfies this). Then we chose s such that we get $\epsilon' = 2^{-55}$ after $m = 46$ bits of leakage. Only the (g) and (c) bound give a non-zero bound for s' in this case, i.e. $s' \approx 2^{45}$ for both.

Chain Rule	Before leakage		Leakage	After leakage		
	ϵ	s	m	$s' \approx$		$\epsilon' \approx$
(e) [VZ13]				$s \cdot \frac{\epsilon^2}{2^m} - \frac{1}{\epsilon^2} - 2^m$	< 0	
(d) [JP14]	2^{-55}	2^{201}		$s \cdot \frac{\epsilon^2}{2^{3m}} - 2^m$	< 0	ϵ	2^{-55}
(f) [GW10]			46	$s \cdot \frac{\epsilon^2}{2^m} - 2^m$	< 0	
(g) **this paper**				$s \cdot \frac{\epsilon^2}{2^m} - 2^m \epsilon^2$	2^{45}	
(c) [FOR12]	2^{-101}	2^{155}		$s \cdot 2^{2m} \epsilon^2$	2^{45}	$2^m \epsilon$

More generally, assume that we have a weak PRF that has k bits of security, i.e., it is (s, ϵ) secure for any $s/\epsilon = 2^k$ (see Definition 3). Then, after leaking m bits it is (s', ϵ') secure for any $s'/\epsilon' = 2^t$, where t satisfies the conditions from Table 4 Let us stress that the equivalence of our bounds and the ones

---

[2] We consider the security of AES256 as a weak PRF, and not a standard PRF, because of non-uniform attacks which show that no PRF with a k bit key can have $s/\epsilon \approx 2^k$ security [DTT09], at least unless we additionally require $\epsilon \gg 2^{-k/2}$.

from [FOR12] only holds in the setting where s/ϵ is basically constant. This is a reasonable assumption for secret-key primitives, but certainly not for most other settings like public-key crypto.[3]

Table 4. Consider an (s, ϵ) secure weak PRF where $s/\epsilon = 2^k$ (for any choice of s), then after m bits of leakage on the key, the PRF is $s'/\epsilon' = 2^t$ secure, where depending on the chain rule used, t can take the values as indicated in the table.

Chain Rule	Technique	Security after leakage	Analysis
(e) [VZ13]	Simulating auxiliary inputs	$t = \frac{k}{5} - \frac{m}{5}$	Appendix A.1
(d) [JP14]	Simulating auxiliary inputs	$t = \frac{k}{3} - \frac{4m}{3}$	Appendix A.2
(f) [GW10]	Relaxed HILL Entropy	$t = \frac{k}{3} - \frac{2m}{3}$	Appendix A.3
(c) [FOR12]	Dense Model Theorem	$t = \frac{k}{3} - \frac{m}{3}$	Appendix A.4
(g) **This work**	Average Metric Entropy	$t = \frac{k}{3} - \frac{m}{3}$	Appendix A.5

No Fixed Time-Sucess Ratio (Application for PRGs with Weak Seeds). To be more concrete, consider the problem of generating pseudorandomness from *weak seeds*. Let $\mathsf{PRG} : \{0,1\}^n \rightarrow \{0,1\}^{2n}$ be a length-doubling PRG with known parameters $(\epsilon_{\mathsf{PRG}}, s_{\mathsf{PRG}})$. Suppose that we have a "weak" source X with min-entropy only $n - d$. The output of PRG on X is not guaranteed to be pseudorandom, and in fact it is not secure [DY13]. One way to overcome this problem is so called "expand-extract-reseed" approach [DY13]. Namely, we simply take an extractor $\mathsf{Ext} : \{0,1\}^{2n} \rightarrow \{0,1\}^n$ and output $\mathsf{Ext}(\mathsf{PRG}(X))$. The proof that the output is pseudorandom goes by chain rules. In the second approach the roles of the extractor atincrseed and key X are swapped and then some facts about so called square-friendly applications are used [DY13]. Since the approach based on the square-friendly properties or on chain rules derived by extending the dense model theorem yield a loss in ϵ, for settings where d is relatively big (more specifically when $d > \log \epsilon_{\mathsf{PRG}}^{-1}$) one prefers the use of a chain rule with loss in s. Below in Table 5 we provide corresponding bounds and a numerical example, where only our chain rule guarantees meaningful security.

2   Preliminaries

Security Definitions. Given a cryptographic primitive, we consider the probability ϵ of breaking its security (defined as winning the corresponding security game) by an adversary with running time (circuit size) s. The following definition is the standard way to define the security level.

---

[3] Consider e.g. RSA, here given our current understanding of the hardness of factoring, ϵ goes from basically 0 to 1 as the running time s reaches the time required to run the best factoring algorithms. In any case, it's not reasonable to assume that s/ϵ is almost constant over the entire range of s.

Definition 3 (Security of Cryptographic Primitives, [Lub96]). *We say that a cryptographic primitive has λ bits of security (alternatively: it is 2^λ-secure) if every adversary has time-advantage ratio at least 2^λ.*

We note that for indistinguishability applications, that is when winning the security game is equivalent to distinguishing a given object from the ideal object (like PRFs, PRGs), the advantage is defined as the difference of the winning probability and $\frac{1}{2}$ which corresponds to chances that a random guess succeeds, whereas for unpredictability applications (like one-way functions) the advantage is simply equal the winning probability. In this paper we will consider indistinguishability applications only.

Some Technical Entropy Definitions. We consider several classes of distinguishers. With $\mathcal{D}_s^{\mathrm{rand},\{0,1\}}$ we denote the class of randomized circuits of size at most s with boolean output (this is the standard non-uniform class of distinguishers considered in cryptographic definitions). The class $\mathcal{D}_s^{\mathrm{rand},[0,1]}$ is defined analogously, but with real valued output in $[0,1]$. $\mathcal{D}_s^{\mathrm{det},\{0,1\}}$, $\mathcal{D}_s^{\mathrm{det},[0,1]}$ are defined the corresponding classes for *deterministic* circuits. With $\delta^D(X,Y) = |\mathbb{E}_X[D(X)] - \mathbb{E}_Y[D(Y)]|$ we denote D's advantage in distinguishing X and Y.

Table 5. Security of a PRG fed with weak seeds, by "expand-extract-reseed" technique. We start with a 256-bit PRG output with security parameters $(\epsilon_{\mathsf{PRG}}, s_{\mathsf{PRG}}) = (2^{-40}, 2^{176})$, chosen to exclude best known non-uniform attacks [DTT09] which are of complexity $s > 2^n \epsilon^2$. We aim for $\epsilon' \approx 2^{-39}$.

Technique	Real security for deficiency d		Comments	Numerical example ($n = 256$)				
	ϵ'	s'		ϵ_{PRG}	s_{PRG}	d	ϵ'	s'
square-security [DY13]	$\sqrt{2^d \epsilon_{\mathsf{PRG}}}$	$\Omega(s_{\mathsf{PRG}})$					1	2^{88}
chain rule (c)	$2^d \epsilon_{\mathsf{PRG}} + 2^{-\frac{n-d}{2}} + \delta$	$\Omega(s_{\mathsf{PRG}} \cdot \delta^2)$	δ arbitrary	2^{-40}	2^{176}	50	1	2^{176}
chain rule (f) and (e)	$\epsilon_{\mathsf{PRG}} + 2^{-\frac{n-d}{2}} + \delta$	$\Omega\left(s_{\mathsf{PRG}} \cdot \frac{\delta^2}{2^d}\right)$	δ arbitrary $s_{\mathsf{PRG}} > 2^{2d} \delta^{-2}$				2^{-39}	< 0
chain rule (g)	$\epsilon_{\mathsf{PRG}} + 2^{-\frac{n-d}{2}} + \delta$	$\Omega\left(s_{\mathsf{PRG}} \cdot \frac{\delta^2}{2^d}\right)$	δ arbitrary $s_{\mathsf{PRG}} > 2^{2d}$				2^{-39}	2^{46}

Definition 4 (Metric Pseudoentropy [BSW03,FR12]). *A random variable X has* real deterministic Metric entropy *at least k if*

$$\mathbf{H}_{\epsilon,s}^{\mathrm{Metric},\mathcal{D}_s^{\mathrm{det},[0,1]}}(X) \geqslant k \iff \forall D \in \mathcal{D}_s^{\mathrm{det},[0,1]} \exists Y_D, \ \mathbf{H}_\infty(Y_D) = k \ : \ \delta^D(X,Y_D) \leqslant \epsilon$$

Relaxed Versions of HILL and Metric Entropy. A weaker notion of conditional HILL entropy allows the conditional part to be replaced by some computationally indistinguishable variable.

Definition 5 (Relaxed HILL Pseudoentropy [GW11,Rey11]). *For a joint distribution* (X, Z), *we say that* X *has* relaxed HILL *entropy* k *conditioned on* Z *if*

$$\mathbf{H}_{\epsilon,s}^{\text{HILL-rlx}}(X|Z) \geqslant k$$

$$\iff \exists (Y, Z'), \widetilde{\mathbf{H}}_{\infty}(Y|Z') = k, \forall D \in \mathcal{D}_s^{\text{rand},\{0,1\}}, \;\; : \; \delta^D((X, Z), (Y, Z')) \leqslant \epsilon$$

The above notion of *relaxed* HILL satisfies a chain rule whereas the chain rule for the standard definition of conditional HILL entropy is known to be false. One can analogously define relaxed variants of metric entropy, we won't give these as they will not be required in this paper. The relaxed variant of HILL entropy is also useful because one can convert relaxed entropy into standard HILL entropy, losing in s an additive term exponential in the length of the conditional part.

Lemma 1 (HILL-rlx→HILL, [JP14]). *For any* X *and correlated* Z *of length* m, *we have* $\mathbf{H}_{\epsilon,s'}^{\text{HILL}}(X|Z) \geqslant \mathbf{H}_{\epsilon,s}^{\text{HILL-rlx}}(X|Z)$ *where* $s' = s - 2^m$.

Pseudoentropy Against Different Distinguisher Classes. For randomized distinguishers, it's irrelevant if the output is boolean or real values, as we can replace any $D \in \mathcal{D}_s^{\text{rand},[0,1]}$ with a $D' \in \mathcal{D}^{\text{rand},\{0,1\}}$ s.t. $\mathbb{E}[D'(X)] = \mathbb{E}[D(X)]$ by setting (for any x) $\Pr[D'(x) = 1] = \mathbb{E}[D(x)]$. For HILL entropy (as well as for its relaxed version), it also doesn't matter if we consider randomized or deterministic distinguishers in Definition 2, as we always can "fix" the randomness to an optimal value. This is no longer true for metric entropy,[4] and thus the distinction between metric and metric star entropy is crucial.

3 Main Result

We start with the following recently proven characterization of the distribution maximizing expectations under min-entropy constraints (Sect. 3.1). Based on this auxiliary result, in Sects. 3.2 and 3.3 we prove our chain rules stated in Theorem 2 and Corollary 1.

3.1 An Auxiliary Result on Constrained Optimization

Lemma 2 (Optimizing Expectations Under Entropy Constraints [SGP15,Sko15a]). *Given* D : $\{0, 1\}^{n+m} \times \{0, 1\}^m \to [0, 1]$ *consider the following optimization problem*

$$\max_{Y|Z} \mathbb{E}D(Y, Z)$$
$$\text{s.t. } \widetilde{\mathbf{H}}_{\infty}(Y|Z) \geqslant k \tag{6}$$

The distribution $Y|Z = Y^*|Z$ *satisfying* $\widetilde{\mathbf{H}}_{\infty}(Y^*|Z) = k$ *is optimal for* (6) *if and only if there exist real numbers* $t(z)$ *and a number* $\lambda \geqslant 0$ *such that for every* z

---

[4] It might be hard to find a high min-entropy distribution Y that fools a randomised distinguisher D, but this task can become easy once D's randomness is fixed.

90   K. Pietrzak and M. Skórski

(a) $\sum_x \max(\mathrm{D}(x,z) - t(z), 0) = \lambda$
(b) If $0 < \mathbf{P}_{Y^*|Z=z}(x) < \max_{x'} \mathbf{P}_{Y^*|Z=z}(x')$ then $\mathrm{D}(x,z) = t(z)$.
(c) If $\mathbf{P}_{Y^*|Z=z}(x) = 0$ then $\mathrm{D}(x,z) \leqslant t(z)$
(d) If $\mathbf{P}_{Y^*|Z=z}(x) = \max_{x'} \mathbf{P}_{Y^*|Z=z}(x')$ then $\mathrm{D}(x,z) \geqslant t(z)$

Remark 1. The characterization can be illustrated in an easy and elegant way. First, it says that the area under the graph of $\mathrm{D}(x,z)$ and above the threshold $t(z)$ is the same, no matter what z is (see Fig. 2). Second, for every z the distribution $Y^*|Z = z$ is flat over the set $\{x : \mathrm{D}(x,z) > t(z)\}$ and vanishes for x satisfying $\mathrm{D}(x,z) < t(z)$, see Fig. 3.

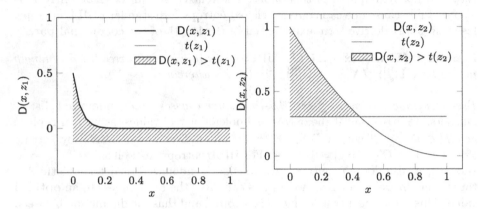

Fig. 2. For every z, the (green) area under $\mathrm{D}(\cdot, z)$ and above $t(z)$ equals λ (Color figure online).

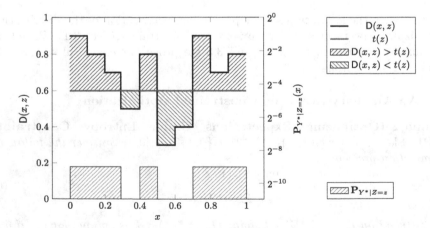

Fig. 3. Relation between distinguisher $\mathrm{D}(x,z)$, threshold $t(z)$ and distribution $Y^*|Z=z$.

Proof (Proof Sketch of Lemma 2). Consider the following linear optimization program

$$
\begin{aligned}
\underset{P_{x,z},a_z}{\text{maximize}} \quad & \sum_{x,z} \mathrm{D}(x,z)P(x,z) \\
\text{subject to} \quad & -P_{x,z} \leqslant 0, \ (x,z) \in \{0,1\}^n \times \{0,1\}^m \\
& \sum_x P_{x,z} - \mathbf{P}_Z(z) = 0, \ z \in \{0,1\}^m \\
& P_{x,z} - a_z \leqslant 0, \ z \in \{0,1\}^m \\
& \sum_z a_z - 2^{-k} \leqslant 0
\end{aligned}
\tag{7}
$$

This problem is equivalent to (6) if we define $\mathbf{P}_{Y,Z}(x,z) = P(x,z)$ and replace the condition $\sum_z \max_x \mathbf{P}_{Y,Z}(x,z) \leqslant 2^{-k}$, which is equivalent to $\tilde{H}_\infty(Y|Z) \geqslant k$, by the existence of numbers $a_z \geqslant \max_x \mathbf{P}_{Y,Z}(x,z)$ such that $\sum_z a_z \leqslant 2^{-k}$. The solutions of (7) can be characterized as follows:

Claim 1. *The numbers $(P_{x,z})_{x,z}, (a_z)_z$ are optimal for (7) if and only if there exist numbers $\lambda^1(x,z) \geqslant 0, \lambda^2(z) \in \mathbb{R}, \lambda^3(x,z) \geqslant 0, \lambda^4 \geqslant 0$ such that*

(a) $\mathrm{D}(x,z) = -\lambda^1(x,z) + \lambda^2(z) + \lambda^3(x,z)$ *and* $0 = -\sum_x \lambda^3(x,z) + \lambda^4$
(b) *We have* $\lambda^1(x,z) = 0$ *if* $P_{x,z} > 0$, $\lambda^3(x,z) = 0$ *if* $P_{x,z} < a_z$, $\lambda^4 = 0$ *if* $\sum_z a_z < 2^{-k}$.

Proof (of Claim). This is a straightforward application of KKT conditions. □

It remains to apply and simplify the last characterization. Let $(P^*_{x,z})_{x,z}, (a^*_z)_z$ be optimal for (7), where $P^*(x,z) = \mathbf{P}_{Y^*,Z}(x,z)$, and $\lambda^1(x,z), \lambda^2(z), \lambda^3(x,z), \lambda^4(x)$ be corresponding multipliers given by the last claim. Define $t(z) = \lambda^2(z)$ and $\lambda = \lambda^4$. Observe that for every z we have $a^*_z \geqslant \max_x \mathbf{P}(x,z) \geqslant 2^{-n}\mathbf{P}_Z(z) > 0$ and thus for every (x,z) we have

$$
\lambda^1(x,z) \cdot \lambda^3(x,z) = 0
\tag{8}
$$

If $P^*(x,z) = 0$ then $P^*(x,z) < a^*(z)$ and $\lambda^3(x,z) = 0$, hence $\mathrm{D}(x,z) \leqslant t(z)$ which proves (c). If $P^*(x,z) = \max_{x'} P^*(x,z)$ then $P^*(x,z) < 0$ and $\lambda^1(x,z) = 0$ which proves (d). Finally observe that (8) implies

$$
\max(\mathrm{D}(x,z) - t(z), 0) = \max(-\lambda^1(x,z) + \lambda^3(x,z), 0) = \lambda^3(x,z)
$$

Hence, the assumption $\sum_x \lambda^3(x,z) = \lambda^4 = \lambda$ proves (a). Suppose now that the characterization given in the Lemma is satisfied. Define $P^*(x,z) = \mathbf{P}_{Y,Z}(x,z)$ and $a_z = \max_z \mathbf{P}_{Y^*,Z}(x,z)$, let $\lambda^3(x,z) = \max(\mathrm{D}(x,z) - t(z), 0)$, $\lambda^1(x,z) = \max(t(z) - \mathrm{D}(x,z), 0)$ and $\lambda^4 = \lambda$. We will show that these numbers satisfy the conditions described in the last claim. By definition we have $-\lambda^1(x,z) +$

$\lambda^2(z) + \lambda^3(x, z) = D(x, z)$, by the assumptions we get $\sum_x \lambda^3(x, z) = \lambda = \lambda^4$. This proves part (a). Now we verify the conditions in (b). Note that $D(x, z) < t(z)$ is possible only if $\mathbf{P}_{Y^*|Z=z}(x) = 0$ and $D(x, z) > t(z)$ is possible only if $\mathbf{P}_{Y^*|Z=z}(x) = \max_{x'} \mathbf{P}_{Y^*|Z=z}(x')$. Therefore, if $\mathbf{P}_{Y,Z}(x, z) > 0$ then we must have $D(x, z) \geqslant t(z)$ which means that $\lambda^1(x, z) = 0$. Similarly if $\mathbf{P}_{Y,Z}(x, z) < \max_z \mathbf{P}_{Y^*,Z}(x, z)$ then $D(x, z) \leqslant t(z)$ and $\lambda^3(x, z) = 0$. Finally, since we assume $\widetilde{H}_\infty(Y^*|Z) = k$ we have $\sum_z a_z = 2^{-k}$ and thus there is no additional restrictions on λ^4. $\qquad\square$

3.2 New Chain Rule for Metric Entropy

We start by sketching the idea of the proof. Assuming contrarily, we have a function D of complexity D' which distinguishes between (X, Z) and all distributions (Y, Z) such that $\widetilde{H}_\infty(Y|Z) \geqslant k - m$. By Lemma 2 we can replace D by a distinguisher D' which is regular conditioned on the second argument, that is $\mathbb{E}\,D(U, z) = \text{const}$ independently on z. This is the key trick in our proof.

Proof (of Theorem 2). Suppose not. There exists real-valued D of size s' such that

$$\mathbb{E}\,D(X, Z) - \mathbb{E}\,D(Y, Z) \geqslant \epsilon, \quad \forall Y : \mathbf{H}_\infty(Y|Z) \geqslant k - m. \qquad (9)$$

The distribution Y^* which minimizes the left-hand side is optimal to the program in (6) (where k is replaced by $k - m$). We start by showing that we can actually assume that D has a very strong property, namely is regular.

Claim (Regular Distinguisher). There exists D' of complexity $\text{size}(D) + 2^m$ which satisfies Eq. 9 in place of D, and is regular, that is $\sum_x D(x, z) = \lambda$ for some λ and every z.

Proof (Proof of Claim). Let $t(z)$ and λ be as in Lemma 2. Define $D'(x, z) = \max(D(x, z) - t(z), 0)$. It is easy to see that Y^* is optimal also when D is replaced by D'. Moreover, we have $\mathbb{E}\,D'(X, Z) \geqslant \mathbb{E}\,D(X, Z) - \lambda$ and $\mathbb{E}\,D'(Y^*, Z) = \mathbb{E}\,D'(Y^*, Z) - \lambda$ and thus $\mathbb{E}\,D'(X, Z) - \mathbb{E}\,D'(Y^*, Z) \geqslant \epsilon$. Therefore,

$$\mathbb{E}\,D'(X, Z) - \mathbb{E}\,D'(Y, Z) \geqslant \epsilon, \quad \forall Y : \mathbf{H}_\infty(Y|Z) \geqslant k - m \qquad (10)$$

note that we have

$$\sum_x D'(x, z) = \lambda, \quad \forall z \qquad (11)$$

which finishes the proof. $\qquad\square$

Having transformed our distinguisher into a more convenient form we define

$$D''(x, z) = \max_z D'(x, z). \qquad (12)$$

Claim. We have $\mathbb{E}\,D''(X) \geqslant \mathbb{E}\,D'(X, Z)$.

Proof. This follows by the definition of D''. $\square$

Claim. For every Y such that $\mathbf{H}_\infty(Y) \geqslant k$ we have $\mathbb{E}\,\mathrm{D}''(Y) \leqslant \mathbb{E}\,\mathrm{D}'(Y^*, Z)$

Proof. We get

$$\mathbb{E}\,\mathrm{D}''(Y) \leqslant 2^{-k} \sum_x \max_z \mathrm{D}'(x, z)$$

$$\leqslant 2^{-k} \sum_{x,z} \mathrm{D}'(x, z)$$

$$= 2^{-k+m} \cdot \lambda = \mathbb{E}\,\mathrm{D}'(Y^*, Z) \qquad (13)$$

where in the last line we have used the fact that D' is regular (see Eq. (11)) and that $HminAv(Y^*|Z) = k - m$ $\square$

Combining the last two claims we get $\mathbb{E}\,\mathrm{D}''(X) - \mathbb{E}\,\mathrm{D}''(Y) \geqslant \epsilon$ for all Y of min-entropy k. It remains to observe that the complexity of D'' equals $s = (s' + 2^m) \cdot 2^m$. $\square$

3.3   The Chain Rule for HILL Entropy

Corollary 1 follows from Theorem 2 by the following result being a tight version of the transformation originally due to [BSW03]

Theorem 3 (Metric$\longrightarrow$HILL Entropy, [Sko15a]). *For any n-bit random variable X and a correlated random variable Z we have*

$$\mathbf{H}^{\mathrm{HILL}}_{(s',\epsilon')}(X|Z) \geqslant \mathbf{H}^{\mathrm{Metric},\mathcal{D}^{\mathrm{det},[0,1]}}_{(s,\epsilon)}(X|Z)$$

where $\delta \in (0,1)$ is an arbitrary parameter, $s' = \Omega\left(s \cdot \delta^2/(\Delta+1)\right)$, $\epsilon' = \epsilon + \delta$ and $\Delta = n - k$ is the entropy deficiency.

A   Time-Success Ratio Analysis

A.1   Chain Rule Given by Vadhan and Zheng

Theorem 4 (Time-success Ratio for Chain Rule (e)). *Suppose that X has n bits of HILL entropy of quality (s, ϵ) for every $s/\epsilon \geqslant 2^k$. Then X conditioned on leakage of m bits has $n - m$ bits of HILL entropy of quality (s', ϵ') for every $s'/\epsilon' \geqslant 2^t$ where*

$$t = \frac{k}{5} - \frac{m}{5} \qquad (14)$$

and this is the best possible bound guaranteed by chain rule (e).

Proof (Proof of Theorem 4). Suppose that we have $s' = s \cdot 2^{-m}\delta^2 - \delta^{-2} - 2^m$ and $\epsilon' = \epsilon + \delta$. We want to find the minimum value of the ratio $\frac{s'}{\epsilon'}$ under the assumption that ϵ, δ, s can be chosen in the possibly most plausible way. Therefore, we want to solve the following min-max problem

$$
\begin{aligned}
&\min_{\epsilon',s'} \max_{s,\epsilon,\delta} \frac{s'}{\epsilon'} \\
&\text{s.t.} \qquad \frac{s}{\epsilon} = 2^k,\ \epsilon + \delta = \epsilon',\ s' = s \cdot 2^{-m}\delta^2 - \delta^{-2} - 2^m
\end{aligned}
\tag{15}
$$

First, we note that

$$
s' = 2^{k-m}(\epsilon' - \delta)\delta^2 - \delta^{-2} - 2^m
$$

Also, since $\delta < \epsilon'$, we need to assume $\epsilon' > 2^{-\frac{k-m}{5}}$ and $\epsilon' > 2^{-\frac{k-2m}{3}}$ to guarantee that $s' > 0$. Now, for $\delta = \Theta(\epsilon')$ we get

$$
\frac{s'}{\epsilon'} = \Omega\left(2^{k-m}\epsilon'^2 - \epsilon'^{-3} - 2^m\epsilon'^{-1}\right) = \Omega\left(2^{\max\left(\frac{3}{5}\cdot(k-m),\frac{k+m}{3}\right)}\right)
\tag{16}
$$

provided that $\epsilon' \gg 2^{-\frac{k-m}{5}}$ and $\epsilon' \gg 2^{-\frac{k-2m}{3}}$.   □

A.2   Chain Rule Given by Jetchev and Pietrzak

Theorem 5 (Time-success Ratio for Chain Rule (d)). *Suppose that X has n bits of HILL entropy of quality (s, ϵ) for every $s/\epsilon \geqslant 2^k$. Then X conditioned on leakage of m bits has $n - m$ bits of HILL entropy of quality (s', ϵ') for every $s'/\epsilon' \geqslant 2^t$ where*

$$
t = \frac{k}{3} - \frac{4m}{3}
\tag{17}
$$

and this is the best possible bound guaranteed by chain rule (d).

Proof (Proof of Theorem 5). Suppose that we have $s' = s \cdot 2^{-3m}\delta^2 - 2^m$ and $\epsilon' = \epsilon + \delta$. We want to find the minimum value of the ratio $\frac{s'}{\epsilon'}$ under the assumption that ϵ, δ, s can be chosen in the possibly most plausible way. Therefore, we want to solve the following min-max problem

$$
\begin{aligned}
&\min_{\epsilon',s'} \max_{s,\epsilon,\delta} \frac{s'}{\epsilon'} \\
&\text{s.t.} \qquad \frac{s}{\epsilon} = 2^k,\ \epsilon + \delta = \epsilon',\ s' = s \cdot 2^{-3m}\delta^2 - 2^m
\end{aligned}
\tag{18}
$$

First, we note that

$$
s' = 2^{k-3m}(\epsilon' - \delta)\delta^2 - 2^m
$$

Also, since $\delta < \epsilon'$, we need to assume $\epsilon' > 2^{-\frac{k-4m}{3}}$ to guarantee that $s' > 0$. Now, setting $\delta = \Theta(\epsilon')$ we have

$$
\frac{s'}{\epsilon'} = \Omega\left(2^{k-m}\epsilon'^2\right) - 2^m\epsilon'^{-1} = \Omega\left(2^{\frac{k-2m}{3}}\right)
\tag{19}
$$

provided that $\epsilon' \gg 2^{-\frac{k-4m}{3}}$.   □

A.3   Chain Rule Given by Gentry and Wichs

Theorem 6 (Time-success Ratio for Chain Rule (f)). *Suppose that X has n bits of HILL entropy of quality (s, ϵ) for every $s/\epsilon \geqslant 2^k$. Then X conditioned on leakage of m bits has $n - m$ bits of HILL entropy of quality (s', ϵ') for every $s'/\epsilon' \geqslant 2^t$ where*

$$t = \frac{k}{3} - \frac{2m}{3} \tag{20}$$

and this is the best possible bound guaranteed by chain rule (f).

Proof (Proof of Theorem 6). Suppose that we have $s' = s \cdot 2^{-m}\delta^2 - 2^m$ and $\epsilon' = \epsilon + \delta$. We want to find the minimum value of the ratio $\frac{s'}{\epsilon'}$ under the assumption that ϵ, δ, s can be chosen in the possibly most plausible way. Therefore, we want to solve the following min-max problem

$$\min_{\epsilon', s'} \max_{s, \epsilon, \delta} \frac{s'}{\epsilon'} \atop \text{s.t.} \quad \frac{s}{\epsilon} = 2^k, \ \epsilon + \delta = \epsilon', \ s' = s \cdot 2^{-m}\delta^2 - 2^m \tag{21}$$

First, we note that

$$s' = 2^{k-m}(\epsilon' - \delta)\delta^2 - 2^m$$

Also, since $\delta < \epsilon'$, we need to assume $\epsilon' > 2^{-\frac{k-2m}{3}}$ to guarantee that $s' > 0$. Now, setting $\delta = \Theta(\epsilon')$ we have

$$\frac{s'}{\epsilon'} = \Omega\left(2^{k-m}\epsilon'^2\right) - 2^m\epsilon'^{-1} = \Omega\left(2^{\frac{k+m}{3}}\right) \tag{22}$$

provided that $\epsilon' \gg 2^{-\frac{k-2m}{3}}$.   □

A.4   Chain Rule Given by Fuller and Reyzin

Theorem 7 (Time-success Ratio for Chain Rule (c)). *Suppose that X has n bits of HILL entropy of quality (s, ϵ) for every $s/\epsilon \geqslant 2^k$. Then X conditioned on leakage of m bits has $n - m$ bits of HILL entropy of quality (s', ϵ') for every $s'/\epsilon' \geqslant 2^t$ where*

$$t = \frac{k}{3} - \frac{m}{3} \tag{23}$$

and this is the best possible bound guaranteed by chain rule (c).

Proof (Proof of Theorem 7). Suppose that we have $s' = s \cdot \delta^2$ and $\epsilon' = 2^m\epsilon + \delta$. We want to find the minimum value of the ratio $\frac{s'}{\epsilon'}$ under the assumption that

ϵ, δ, s can be chosen in the possibly most plausible way. Therefore, we want to solve the following min-max problem

$$\min_{\epsilon', s'} \max_{s, \epsilon, \delta} \frac{s'}{\epsilon'}$$
$$\text{s.t.} \quad \frac{s}{\epsilon} = 2^k, \; 2^m \epsilon + \delta = \epsilon', \; s' = s \cdot \delta^2 \tag{24}$$

First, we note that

$$s' = 2^{k-m}(\epsilon' - \delta)\delta^2$$

Also, since $\delta < \epsilon'$, we need to assume $\epsilon' > 2^{-\frac{k-m}{3}}$ to guarantee that $s' > 1$. Now, setting $\delta = \Theta(\epsilon')$ we have

$$\frac{s'}{\epsilon'} = \Omega\left(2^{k-m}\epsilon'^2\right) = \Omega\left(2^{\frac{k-m}{3}}\right), \tag{25}$$

provided that $\epsilon' > 2^{-\frac{k-m}{3}}$.   □

A.5 Chain Rule in This Paper

Theorem 8 (Time-success Ratio for Chain Rule (g)). *Suppose that X has n bits of HILL entropy of quality (s, ϵ) for every $s/\epsilon \geqslant 2^k$. Then X conditioned on leakage of m bits has $n - m$ bits of HILL entropy of quality (s', ϵ') for every $s'/\epsilon' \geqslant 2^t$ where*

$$t = \frac{k}{3} - \frac{m}{3} \tag{26}$$

and this is the best possible bound guaranteed by chain rule (g).

Proof (Proof of Theorem 8). Suppose that we have $s' = s \cdot 2^{-m}\delta^2 - 2^m\delta^2$ and $\epsilon' = \epsilon + \delta$. We want to find the minimum value of the ratio $\frac{s'}{\epsilon'}$ under the assumption that ϵ, δ, s can be chosen in the possibly most plausible way. Therefore, we want to solve the following min-max problem

$$\min_{\epsilon', s'} \max_{s, \epsilon, \delta} \frac{s'}{\epsilon'}$$
$$\text{s.t.} \quad \frac{s}{\epsilon} = 2^k, \; \epsilon + \delta = \epsilon', \; s' = s \cdot 2^{-m}\delta^2 - 2^m\delta^2 \tag{27}$$

First, we note that

$$s' = 2^{k-m}(\epsilon' - \delta)\delta^2 - 2^m\delta^2$$

Also, since $\delta < \epsilon'$, we need to assume $\epsilon' > 2^{-(k-2m)}$ and $\epsilon' > 2^{-\frac{k-m}{3}}$ to guarantee that $s' > 0$. Setting $\delta = \Theta(\epsilon')$ we obtain

$$\frac{s'}{\epsilon'} = \Omega\left(2^{k-m}\epsilon'^2\right) - 2^m\epsilon' = \Omega\left(2^{k-m}\epsilon'^2\right) \tag{28}$$

provided that $\epsilon' \gg 2^{-(k-2m)}$ and $\epsilon' > 2^{-\frac{k-m}{3}}$. If t is the security level, we must have $t < \min\left(k - 2m, \frac{k-m}{3}\right)$ and $k - m - 2t > t$.   □

References

[BM84]   Blum, M., Micali, S.: How to generate cryptographically strong sequences of pseudorandom bits. SIAM J. Comput. **13**(4), 850–864 (1984)

[BSW03] Barak, B., Shaltiel, R., Wigderson, A.: Computational analogues of entropy. In: Arora, S., Jansen, K., Rolim, J.D.P., Sahai, A. (eds.) RANDOM 2003 and APPROX 2003. LNCS, vol. 2764, pp. 200–215. Springer, Heidelberg (2003)

[CKLR11] Chung, K.-M., Kalai, Y.T., Liu, F.-H., Raz, R.: Memory delegation. Cryptology ePrint Archive, Report 2011/273 (2011). http://eprint.iacr.org/

[DP08a] Dziembowski, S., Pietrzak, K.: Leakage-resilient cryptography. In: FOCS, pp. 293–302 (2008)

[DP08b] Dziembowski, S., Pietrzak, K.: Leakage-resilient cryptography in the standard model. IACR Cryptology ePrint Archive 2008, 240 (2008)

[DRS04] Dodis, Y., Reyzin, L., Smith, A.: Fuzzy extractors: how to generate strong keys from biometrics and other noisy data. In: Cachin, C., Camenisch, J.L. (eds.) EUROCRYPT 2004. LNCS, vol. 3027, pp. 523–540. Springer, Heidelberg (2004)

[DTT09] De, A., Trevisan, L., Tulsiani, M.: Non-uniform attacks against one-way functions and prgs. Electron. Colloquium Comput. Complex. (ECCC) **16**, 113 (2009)

[DY13]   Dodis, Y., Yu, Y.: Overcoming weak expectations. In: Sahai, A. (ed.) TCC 2013. LNCS, vol. 7785, pp. 1–22. Springer, Heidelberg (2013)

[FOR12] Fuller, B., O'Neill, A., Reyzin, L.: A unified approach to deterministic encryption: new constructions and a connection to computational entropy. Cryptology ePrint Archive, Report 2012/005 (2012). http://eprint.iacr.org/

[FR12]   Fuller, B., Reyzin, L.: Computational entropy and information leakage. Cryptology ePrint Archive, Report 2012/466 (2012). http://eprint.iacr.org/

[GW10]   Gentry, C., Wichs, D.: Separating succinct non-interactive arguments from all falsifiable assumptions. Cryptology ePrint Archive, Report 2010/610 (2010). http://eprint.iacr.org/

[GW11]   Gentry, C., Wichs, D.: Separating succinct non-interactive arguments from all falsifiable assumptions. In: STOC 2011, pp. 99–108 (2011)

[HILL99] Hastad, J., Impagliazzo, R., Levin, L.A., Luby, M.: A pseudorandom generator from any one-way function. SIAM J. Comput. **28**(4), 1364–1396 (1999)

[HLR07] Hsiao, C.-Y., Lu, C.-J., Reyzin, L.: Conditional computational entropy, or toward separating pseudoentropy from compressibility. In: Naor, M. (ed.) EUROCRYPT 2007. LNCS, vol. 4515, pp. 169–186. Springer, Heidelberg (2007)

[HRV10] Haitner, I., Reingold, O., Vadhan, S.: Efficiency improvements in constructing pseudorandom generators from one-way functions. In: Proceedings of the 42nd ACM Symposium on Theory of Computing, STOC 2010, pp. 437–446. ACM, New York (2010)

[JP14]   Jetchev, D., Pietrzak, K.: How to fake auxiliary input. In: Lindell, Y. (ed.) TCC 2014. LNCS, vol. 8349, pp. 566–590. Springer, Heidelberg (2014)

[KPWW14] Krenn, S., Pietrzak, K., Wadia, A., Wichs, D.: A counterexample to the chain rule for conditional HILL entropy. IACR Cryptology ePrint Archive 2014, 678 (2014)

[Lub96] Luby, M.: Pseudorandomness and Cryptographic Applications. Princeton Computer Science Notes. Princeton University Press, Princeton (1996)

[Pie09] Pietrzak, K.: A leakage-resilient mode of operation. In: Joux, A. (ed.) EUROCRYPT 2009. LNCS, vol. 5479, pp. 462–482. Springer, Heidelberg (2009)

[Rey11] Reyzin, L.: Some notions of entropy for cryptography (invited talk). In: Fehr, S. (ed.) ICITS 2011. LNCS, vol. 6673, pp. 138–142. Springer, Heidelberg (2011)

[RTTV08a] Reingold, O., Trevisan, L., Tulsiani, M., Vadhan, S.P.: Dense subsets of pseudorandom sets. In: Proceedings of the 2008 49th Annual IEEE Symposium on Foundations of Computer Science, FOCS 2008, pp. 76–85. IEEE Computer Society, Washington, DC (2008)

[SGP15] Skórski, M., Golovnev, A., Pietrzak, K.: Condensed unpredictability. In: Halldórsson, M.M., Iwama, K., Kobayashi, N., Speckmann, B. (eds.) ICALP 2015. LNCS, vol. 9134, pp. 1046–1057. Springer, Heidelberg (2015)

[Sko15a] Skorski, M.: Metric pseudoentropy: characterizations, transformations and applications. In: Lehmann, A., Wolf, S. (eds.) Information Theoretic Security. LNCS, vol. 9063, pp. 105–122. Springer, Heidelberg (2015)

[VZ13] Vadhan, S., Zheng, C.J.: A uniform min-max theorem with applications in cryptography. In: Canetti, R., Garay, J.A. (eds.) CRYPTO 2013, Part I. LNCS, vol. 8042, pp. 93–110. Springer, Heidelberg (2013)

[Yao82] Yao, A.C.-C.: Theory and applications of trapdoor functions (extended abstract). In: FOCS, pp. 80–91 (1982)

Post-Quantum Cryptography

Faster Sieving for Shortest Lattice Vectors Using Spherical Locality-Sensitive Hashing

Thijs Laarhoven[✉] and Benne de Weger

Department of Mathematics and Computer Science,
Eindhoven University of Technology, Eindhoven, The Netherlands
mail@thijs.com, b.m.m.d.weger@tue.nl

Abstract. Recently, it was shown that angular locality-sensitive hashing (LSH) can be used to significantly speed up lattice sieving, leading to a heuristic time complexity for solving the shortest vector problem (SVP) of $2^{0.337n+o(n)}$ (and space complexity $2^{0.208n+o(n)}$. We study the possibility of applying other LSH methods to sieving, and show that with the spherical LSH method of Andoni et al. we can heuristically solve SVP in time $2^{0.298n+o(n)}$ and space $2^{0.208n+o(n)}$. We further show that a practical variant of the resulting SphereSieve is very similar to Wang et al.'s two-level sieve, with the key difference that we impose an order on the outer list of centers.

Keywords: Shortest vector problem (svp) · Sieving algorithms · Nearest neighbor problem · Locality-sensitive hashing (lsh) · Lattice cryptography

1 Introduction

Lattice Cryptography. Lattice-based cryptography has recently received wide attention from the cryptographic community, due to e.g. its presumed resistance against quantum attacks [10], the existence of lattice-based fully homomorphic encryption schemes [18], and efficient cryptographic primitives like NTRU [20] and LWE [40]. An important problem in the study of lattices is the shortest vector problem (SVP): given a lattice, find a shortest non-zero lattice vector. Although SVP is well-known to be NP-hard under randomized reductions [2,30], the computational complexity of finding short(est) vectors is still not well understood, even though it is crucial for applications in lattice-based cryptography [26,38].

Finding Shortest Vectors. Currently the four main methodologies for solving SVP are enumeration [15,23,37], sieving [3], constructing the Voronoi cell of the lattice [31], and a recent method based on discrete Gaussian sampling [1]. Enumeration has a low space complexity, but a time complexity superexponential in the dimension n, which is suboptimal as the other methods all run in single exponential ($2^{\Theta(n)}$) time. Drawbacks of the latter methods are that their space

© Springer International Publishing Switzerland 2015
K. Lauter and F. Rodríguez-Henríquez (Eds.): LatinCrypt 2015, LNCS 9230, pp. 101–118, 2015.
DOI: 10.1007/978-3-319-22174-8_6

complexities are $2^{\Theta(n)}$ as well, and that the hidden constants in the exponents are relatively big. As a result, enumeration (with extreme pruning [17]) is commonly still considered the most practical method for finding shortest vectors in high dimensions [33].

Sieving Algorithms. On the other hand, these other SVP methods are less explored than enumeration, and recent improvements in sieving have considerably narrowed the gap with enumeration. Following the groundbreaking work of Ajtai et al. [3], it was later shown that with sieving one can provably solve SVP in arbitrary lattices in time $2^{2.465n+o(n)}$ [19,35,39]. Heuristic analyses further suggest that with sieving one can solve SVP in time $2^{0.415n+o(n)}$ and space $2^{0.208n+o(n)}$ [7,32,35], or optimizing for time, in time $2^{0.378n+o(n)}$ and space $2^{0.293n+o(n)}$ [7,46,47]. Various papers further studied how to speed up sieving in practice [11,16,22,25,27,28,34,41,42], and currently the highest dimension in which sieving was used to solve SVP is 116 for arbitrary lattices [43], and 128 for ideal lattices [11,22,36].

Locality-sensitive Hashing. Since sieving algorithms store long lists of high-dimensional vectors in memory, and the main procedure of sieving is to go through this list to find vectors nearby a target vector, one might ask whether this can be done faster than with a linear search. This problem is related to the nearest neighbor problem [21], and a well-known method for solving this problem faster is based on locality-sensitive hashing (LSH). Recently, it was shown that the efficient angular LSH technique of Charikar [12] can be used to significantly speed up sieving, both in theory and in practice [24,29], with heuristic time and space complexities of $2^{0.337n+o(n)}$ and $2^{0.208n+o(n)}$ respectively [24]. An open problem of [24] was whether using other LSH techniques would lead to even better results.

Contributions. In this work we answer the latter question in the affirmative. With spherical LSH [5,6] we obtain heuristic time and space complexities for solving SVP of $2^{0.2972n+o(n)}$ and $2^{0.2075n+o(n)}$ respectively, achieving the best asymptotic time complexity for SVP to date. We obtain the space/time trade-off depicted in Fig. 1, and show how the trade-off can be turned into a clean speed-up leading to the blue point in Fig. 1. We further show that a practical variant of our algorithm appears to be very similar to the two-level sieve of Wang et al. [46], with the key difference that the outer list of centers is ordered.

Outline. In Sect. 2 we first provide some background on (spherical) LSH. Section 3 describes how to apply spherical LSH to the NV-sieve [35], and Sect. 4 states the main result. In Sect. 5 we describe a practical variant of our algorithm, and we discuss its relation with Wang et al.'s two-level sieve [46]. In Sect. 6 we discuss practical implications of our results, and remaining open problems for future work.

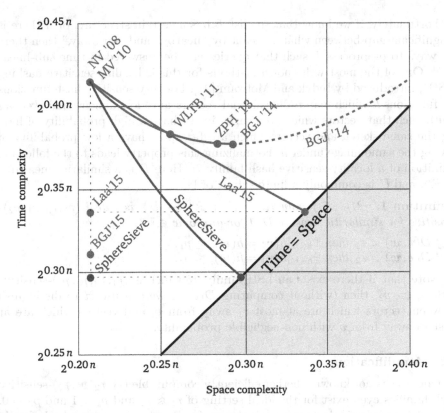

Fig. 1. The space/time trade-offs of various heuristic sieve algorithms from the literature (red), the trade-off for spherical LSH (blue, cf. Proposition 1), and the speedup when using hash tables sequentially rather than in parallel (the point at $(2^{0.208n}, 2^{0.298n})$), cf. Theorem 1). The referenced papers are: NV'08: [35], MV'10: [32], WLTB'11: [46], ZPH'13: [47], BGJ'14: [7], Laa'15: [24], BGJ'15: [8] (Color figure online).

2   Locality-Sensitive Hashing

2.1   Locality-Sensitive Hash Families

The nearest neighbor problem is the following [21]: Given a list of n-dimensional vectors of cardinality N, e.g., $L = \{w_1, w_2, \ldots, w_N\} \subset \mathbb{R}^n$, preprocess L in such a way that given a target vector $v \notin L$, we can efficiently find an element $w \in L$ closest to v. A common variant of this problem is the approximate nearest neighbor problem, where an acceptable solution is a vector "nearby" the target vector (and a solution is unacceptable if it is "far away"). While for low dimensions n there exist ways to answer these queries in time sub-linear or even logarithmic in the list size N, for high dimensions it generally seems hard to do better than with a naive brute-force list search of time $O(N)$. This inability to efficiently store and query lists of high-dimensional data is sometimes referred to as the "curse of dimensionality" [21].

Fortunately, if we know that the list L has a certain structure, or if there is a significant gap between what is meant by "nearby" and "far away," then there are ways to preprocess L such that queries can be answered in time sub-linear in N. One of the most well-known methods for this is locality-sensitive hashing (LSH), introduced by Indyk and Motwani [21]. Locality-sensitive hash functions are functions h which map n-dimensional vectors $\boldsymbol{w}$ to low-dimensional *sketches* $h(\boldsymbol{w})$, such that vectors which are nearby in $\mathbb{R}^n$ have a high probability of having the same sketch and vectors which are far apart have a low probability of having the same image under h. Formalizing this property leads to the following definition of a locality-sensitive hash family $\mathcal{H}$. Here D is a similarity measure[1] on $\mathbb{R}^n$, and U is commonly a finite subset of $\mathbb{N}$.

Definition 1. *[21]   A family $\mathcal{H} = \{h : \mathbb{R}^n \to U\}$ is called (r_1, r_2, p_1, p_2)-sensitive for similarity measure D if for any $\boldsymbol{v}, \boldsymbol{w} \in \mathbb{R}^n$:*

- *if $D(\boldsymbol{v}, \boldsymbol{w}) < r_1$ then $\mathbb{P}_{h \in \mathcal{H}}[h(\boldsymbol{v}) = h(\boldsymbol{w})] \geq p_1$;*
- *if $D(\boldsymbol{v}, \boldsymbol{w}) > r_2$ then $\mathbb{P}_{h \in \mathcal{H}}[h(\boldsymbol{v}) = h(\boldsymbol{w})] \leq p_2$.*

Note that if there exists an LSH family $\mathcal{H}$ which is (r_1, r_2, p_1, p_2)-sensitive with $p_1 \gg p_2$, then (without computing $D(\boldsymbol{v}, \cdot)$) we can use $\mathcal{H}$ to distinguish between vectors which are at most r_1 away from $\boldsymbol{v}$, and vectors which are at least r_2 away from $\boldsymbol{v}$ with non-negligible probability.

2.2   Amplification

In general it is not known whether efficiently computable (r_1, r_2, p_1, p_2)-sensitive hash families even exist for the ideal setting of $r_1 \approx r_2$ and $p_1 \approx 1$ and $p_2 \approx 0$. Instead, one commonly first constructs an (r_1, r_2, p_1, p_2)-sensitive hash family $\mathcal{H}$ with $p_1 \approx p_2$, and then uses several AND- and OR-compositions to turn it into an (r_1, r_2, p_1', p_2')-sensitive hash family $\mathcal{H}'$ with $p_2' < p_2 < p_1 < p_1'$, thereby amplifying the gap between p_1 and p_2.

AND-composition. Given an (r_1, r_2, p_1, p_2)-sensitive hash family $\mathcal{H}$, we can construct an (r_1, r_2, p_1^k, p_2^k)-sensitive hash family $\mathcal{H}'$ by taking a bijective function $\alpha : U^k \to U$ and k functions $h_1, \ldots, h_k \in \mathcal{H}$ and defining $h \in \mathcal{H}'$ as $h(\boldsymbol{v}) = \alpha(h_1(\boldsymbol{v}), \ldots, h_k(\boldsymbol{v}))$. This increases the relative gap between p_1 and p_2 but decreases their absolute values.

OR-composition. Given an (r_1, r_2, p_1, p_2)-sensitive hash family $\mathcal{H}$, we can construct an $(r_1, r_2, 1 - (1 - p_1)^t, 1 - (1 - p_2)^t)$-sensitive hash family $\mathcal{H}'$ by taking $h_1, \ldots, h_t \in \mathcal{H}$, and defining $h \in \mathcal{H}'$ by the relation $h(\boldsymbol{v}) = h(\boldsymbol{w})$ iff $h_i(\boldsymbol{v}) = h_i(\boldsymbol{w})$ for some $i \in \{1, \ldots, t\}$. This compensates the decrease of the absolute values of the probabilities.

Combining a k-wise AND- with a t-wise OR-composition, we can turn an (r_1, r_2, p_1, p_2)-sensitive hash family $\mathcal{H}$ into an (r_1, r_2, p_1^*, p_2^*)-sensitive hash family $\mathcal{H}'$ with $p^* \overset{\text{def}}{=} 1 - (1 - p^k)^t$ for $p = p_1, p_2$. Note that for $p_1 > p_2$ we can always find values k and t such that $p_1^* \approx 1$ and $p_2^* \approx 0$.

---

[1] A similarity measure D may informally be thought of as a "slightly relaxed" metric, which may not satisfy all properties associated to metrics; see e.g. [21] for details.

2.3 Finding Nearest Neighbors

To use these hash families to find nearest neighbors, we can use the following method first described in [21]. First, choose $t \cdot k$ random hash functions $h_{i,j} \in \mathcal{H}$, and use the AND-composition to combine k of them at a time to build t different hash functions $h_1, \ldots, h_t$. Then, given the list L, build t different hash tables $T_1, \ldots, T_t$, where for each hash table T_i we insert $\boldsymbol{w}$ into the bucket labeled $h_i(\boldsymbol{w})$. Finally, given the target vector $\boldsymbol{v}$, compute its t images $h_i(\boldsymbol{v})$, gather all the candidate vectors that collide with $\boldsymbol{v}$ in at least one of these hash tables (an OR-composition), and search this list of candidates for the nearest neighbor.

Clearly, the quality of this algorithm for finding nearest neighbors depends on the quality of the underlying hash family and on the parameters k and t. Larger k and t amplify the gap between the probabilities of finding nearby and faraway vectors as candidates, but this comes at the cost of having to compute many hashes (both during the preprocessing phase and in the querying phase) and having to store many hash tables, each containing all vectors from L. The following lemma shows how to balance k and t such that the overall query time complexity of finding near(est) neighbors is minimized.

Lemma 1. *[21] Suppose there exists a (r_1, r_2, p_1, p_2)-sensitive hash family $\mathcal{H}$. Then, with*

$$\rho = \frac{\log(1/p_1)}{\log(1/p_2)}, \qquad k = \frac{\log(N)}{\log(1/p_2)}, \qquad t = O(N^\rho), \tag{1}$$

with high probability we can find an element $\boldsymbol{w}^* \in L$ with $D(\boldsymbol{v}, \boldsymbol{w}^*) \leq r_2$ or (correctly) conclude that no element $\boldsymbol{w}^* \in L$ with $D(\boldsymbol{v}, \boldsymbol{w}^*) \leq r_1$ exists, with the following costs:

1. *Time for preprocessing the list: $O(kN^{1+\rho})$.*
2. *Space complexity of the preprocessed data: $O(N^{1+\rho})$.*
3. *Time for answering a query $\boldsymbol{v}$: $O(N^\rho)$.*
   (a) *Hash evaluations of the query vector $\boldsymbol{v}$: $O(N^\rho)$.*
   (b) *Candidates to compare to the query vector $\boldsymbol{v}$: $O(N^\rho)$.*

Although Lemma 1 only shows how to choose k and t to minimize the time complexity, we can generally tune k and t to use slightly more time and less space. In a sense this algorithm can be seen as a generalization of the naive brute-force search method, as $k = 0$ and $t = 1$ corresponds to checking the whole list in linear time with linear space. Note that the main costs of the algorithm are determined by the value of ρ, which is therefore often considered the central parameter of interest in LSH literature. The goal is to design $\mathcal{H}$ so that ρ is as small as possible.

2.4 Spherical Locality-Sensitive Hashing

In [24] the family of hash functions that was considered was Charikar's cosine hash family [12] based on angular distances. In the same paper it was suggested that other hash families, such as Andoni and Indyk's celebrated Euclidean LSH family [4], may lead to even better results. The latter method however does not seem to work well in the context of sieving[2], and instead we will focus on yet another LSH family, recently proposed by Andoni et al. [5,6] and coined spherical LSH.

Hash Family. In spherical LSH, we assume[3] that all points in the data set L lie on the surface of a hypersphere $S^{n-1}(R) = \{v \in \mathbb{R}^n : \|x\| = R\}$. In the following description of the hash family we further assume that all vectors lie on $S^{n-1}(1)$, although these definitions can trivially be generalized to the general case $S^{n-1}(R)$.

First, we sample $U = 2^{\Theta(\sqrt{n})}$ vectors $s_1, s_2, \ldots, s_U \in \mathbb{R}^n$ from an n-dimensional Gaussian distribution with average norm $\mathbb{E}\|s_i\| = 1$.[4] This equivalently corresponds to drawing each vector entry from a univariate Gaussian distribution $\mathcal{N}(0, \frac{1}{n})$. To each s_i we associate a hash region H_i:

$$H_i = \{v \in S^{n-1}(1) : \langle v, s_i \rangle \geq n^{-1/4}\} \setminus \bigcup_{j=1}^{i-1} H_j. \qquad (i = 1, \ldots, U) \qquad (2)$$

Since we assume that $v \in S^{n-1}(1)$ and w.h.p. we have $\|s_i\| \approx 1$, the condition $\langle v, s_i \rangle \geq n^{-1/4}$ is equivalent to $\|v - s_i\| \leq \sqrt{2} - \Theta(n^{-1/4})$, i.e., v lies in the *almost-hemisphere* of radius $\sqrt{2} - \Theta(n^{-1/4})$ defined by s_i.

Note that the parts of $S^{n-1}(1)$ that are covered by multiple hash regions are assigned to the *first* region H_i that covers the point. As a result, the size of hash regions generally decreases with i. Also note that the choice of $U = 2^{\Theta(\sqrt{n})}$ guarantees that with high probability, at the end the entire sphere is covered by these hash regions $H_1, H_2, \ldots, H_U$; informally, each hash region covers a $2^{-\Theta(\sqrt{n})}$ fraction of the sphere, so we need $2^{\Theta(\sqrt{n})}$ regions to cover the entire hypersphere. Finally, taking $U = 2^{\Theta(\sqrt{n})}$ also guarantees that computing hashes can trivially be done in $2^{\Theta(\sqrt{n})} = 2^{o(n)}$ time by going through each of the hash regions $H_1, H_2, \ldots, H_U$ and checking whether it contains a given point v.

In our analysis we will use the following result, which is implicitly stated in [5, Lemma 3.3] and [6, Appendix B.1]. Note that in the application of sieving later on, vectors v and w are not assumed to lie on the surface of a sphere, but inside a thin spherical shell with some inner radius γR and outer radius R, with $\gamma = 1 - o(1)$. We can however still apply spherical hashing, due to the observation that $\|\frac{v}{R} - \frac{w}{R}\| - \|\frac{v}{\|v\|} - \frac{w}{\|w\|}\| = O(1 - \gamma) = o(1)$. In other words, by applying the hash method to normalized vectors $\tilde{x} = \frac{x}{\|x\|}$ which all do lie on a

---

[2] Technically speaking, [4] uses the Johnson-Lindenstrauss lemma to project n- to n_0-dimensional vectors with $n_0 = o(n)$, so that single-exponential costs in n_0 $(2^{\Theta(n_0)})$ are sub-exponential in n $(2^{o(n)})$. However, this projection only preserves inter-point distances up to small errors if the length of the list is sufficiently small $(N = 2^{o(n)})$, which is not the case in sieving. Moreover, we estimated the potential improvement using Euclidean LSH to be smaller than the improvement we obtain here.

[3] In Sect. 3 we will justify why this assumption makes sense in sieving.

[4] Note that Andoni et al. sample vectors with average norm $\sqrt{n}$ instead, which means that everything in our description is scaled by a factor $\sqrt{n}$.

Algorithm 1. The Nguyen-Vidick sieve algorithm (sieving step)

1: Compute the maximum norm $R = \max_{v \in L_m} \|v\|$
2: Initialize empty lists L_{m+1} and C_{m+1}
3: **for each** $v \in L_m$ **do**
4:    **if** $\|v\| \leq \gamma R$ **then**
5:       Add v to the list L_{m+1}
6:    **else**
7:       **for each** $w \in C_{m+1}$ **do**
8:          **if** $\|v - w\| \leq \gamma R$ **then**
9:             Add $v - w$ to the list L_{m+1} and continue the outer loop
10:      Add v to the centers C_{m+1}

---

hypersphere, the inter-point distances are preserved up to a negligible additive term $o(1)$, which translates to an $o(1)$ term in the application of LSH.

Lemma 2. *Let $v, w \in \mathbb{R}^n$ with $\|v\|, \|w\| \in [\gamma R, R]$ and $\gamma = 1 - o(1)$, and let θ denote the angle between v and w. Then spherical LSH satisfies:*

$$\mathbb{P}_{h \in \mathcal{H}}[h(v) = h(w)] = \exp\left[-\frac{\sqrt{n}}{2}\tan^2\left(\frac{\theta}{2}\right)(1 + o(1))\right]. \qquad (3)$$

Note that for $\theta_1 = \frac{\pi}{3}$ and $\theta_2 = \frac{\pi}{2}$ this leads to $\rho = \frac{\ln(p_1)}{\ln(p_2)} = \frac{\tan^2(\pi/6)}{\tan^2(\pi/4)}(1 + o(1)) = \frac{1}{3} + o(1)$. This is significantly smaller than the related value of ρ for angular hashing with $\theta_1 = \frac{\pi}{3}$ and $\theta_2 = \frac{\pi}{2}$, which is $\rho' = \log_2(\frac{3}{2}) \approx 0.585$.

3   From the Nguyen-Vidick Sieve to the SphereSieve

We will now describe how spherical LSH can be applied to sieving. More precisely, we will show how spherical LSH can be applied to the heuristic sieve algorithm of Nguyen and Vidick [35]. Applying the same technique to the practically superior GaussSieve [32] seems difficult, and whether this is at all possible is left as an open problem.

3.1   The Nguyen-Vidick Sieve

Initially the Nguyen-Vidick sieve starts with a long list L_0 of long lattice vectors (generated from a discrete Gaussian distribution on the lattice), and it iteratively builds shorter lists of shorter lattice vectors L_{m+1} by applying a sieve to L_m. After $\text{poly}(n)$ applications of the sieve, one hopes to be left with a list L_M containing a shortest non-zero lattice vector. At the heart of the heuristic sieve algorithm of Nguyen and Vidick lies the sieving step, mapping a list L_m to the next list L_{m+1}, and this sieving step is described in Algorithm 1.

The sieving step in Algorithm 1 can be described as follows. We start with an exponentially long list of vectors L_m, and we assume the longest of these vectors has length R; computing R can trivially be done in $\tilde{O}(|L_m|)$ time. Then, for a given parameter $\gamma < 1$ close to 1, we immediately add all vectors of norm less than γR to the next list L_{m+1}; these vectors are not modified in this iteration of the sieve. In the next iteration we want all vectors in L_{m+1} to have norm less than γR, and the remaining vectors in the spherical shell $\mathcal{S} = \{v \in L_m : \gamma R < \|v\| \leq R\}$ do not satisfy this condition, so the main task of the sieving step is to combine lattice vectors in $L_m \cap \mathcal{S}$ to make shorter vectors, which can then be added to L_{m+1}. To do this, we first initialize an empty list of *centers* C_{m+1}, and for each vector $v \in \mathcal{S}$ we do one of the following:

- If v is far away from all center vectors $w \in C_{m+1}$, we add v to C_{m+1};
- If v is close to a center vector $w \in C_{m+1}$, we add $v - w$ to L_{m+1}.

We go through all vectors in L_m one by one, each time deciding whether to add something to L_{m+1} or to C_{m+1}. Note that for each list vector, this decision can be made in $O(|C_{m+1}|) = O(|L_m|)$ time by simply going through all vectors $w \in C_{m+1}$ and checking whether it is close to v. Finally, we obtain a set $C_{m+1} \subset L_m$ which intuitively *covers* $\mathcal{S}$ with balls of radius γR, and we obtain a set L_{m+1} of short vectors. Since the size of C_{m+1} is bounded from above by $2^{\Theta(n)}$, we know that if $|L_m|$ is large enough, many vectors will be included in $|L_{m+1}|$ as well.

To analyze their heuristic sieve algorithm, Nguyen and Vidick used (a slightly stronger version of) the following heuristic assumption.

Heuristic 1. *The angle $\Theta(v, w)$ between two vectors v and w in Line 8 in Algorithm 1 follows the same distribution as the distribution of angles $\Theta(v, w)$ obtained by drawing v and w at random from the unit sphere.*

Using this heuristic assumption, Nguyen and Vidick showed that an initial list of size $|L_0| = (4/3)^{n/2+o(n)} \approx 2^{0.2075n+o(n)}$ suffices to find a shortest vector if $\gamma \approx 1$ [35]. Since the time complexity is dominated by comparing almost every pair of vectors in L_i in each sieving step, this leads to a time complexity quadratic in $|L_i|$. Overall, this means that under the above heuristic assumption, the Nguyen-Vidick sieve solves SVP in time $2^{0.415n+o(n)}$ and space $2^{0.2076n+o(n)}$.

3.2   The SphereSieve

Algorithm 2 describes how we can apply spherical LSH to the sieve step of Nguyen and Vidick's heuristic sieve algorithm, in a similar fashion as how angular LSH was applied to sieving in [24].

To apply spherical LSH to sieving efficiently, there are some subtle issues that we need to consider. For instance, while the angular hashing technique of Charikar considered in [24] is scale invariant, the parameters of spherical LSH slightly change if all vectors in L and the target vector are multiplied by a scalar. This means that for each application of the sieving step, the parameters might change and we must build fresh hash tables. Although this might increase the practical time and space complexities, this does not affect the algorithm's asymptotics.

Algorithm 2. The SphereSieve algorithm (sieving step)

1: Compute the maximum norm $R = \max_{v \in L_m} \|v\|$
2: Initialize an empty list L_{m+1}
3: Initialize t empty hash tables T_i
4: Sample $k \cdot t$ random spherical hash functions $h_{i,j} \in \mathcal{H}$
5: **for each** $v \in L_m$ **do**
6:     **if** $\|v\| \leq \gamma R$ **then**
7:         Add v to the list L_{m+1}
8:     **else**
9:         Obtain the set of candidates $C = \bigcup\limits_{i=1}^{t} T_i[h_i(\pm v)]$
10:        **for each** $w \in C$ **do**
11:            **if** $\|v - w\| \leq \gamma R$ **then**
12:                Add $v - w$ to the list L_{m+1}
13:                Continue the outermost loop
14:        Add v to all t hash tables T_i

More importantly, to justify that we can apply spherical LSH (i.e., to justify the application of Lemma 2), we need to guarantee that $\|v\| \approx \|w\|$ for targets v and (candidate) near neighbors w, i.e., that all these vectors approximately lie on the surface of a sphere. To see why this is true, consider a target vector v and a list vector w. By definition of R, we know that v and w both have norm at most R. Moreover, the case $\|v\| \leq \gamma R$ is handled separately (in polynomial time) in Lines 6–7, and the fact that $w \in C_{m+1}$ implies that $\|w\| > \gamma R$ as well. So when we get to the search in Lines 10–13, we know that the norms of both vectors satisfy $\|v\|, \|w\| \in [\gamma R, R]$. To get the optimal asymptotic time and space complexities, Nguyen and Vidick further let $\gamma \to 1$, which we needed to apply Lemma 2.

4   Theoretical Results

To obtain a first basic estimate of the potential improvements to the time and space complexities using spherical LSH, we first note that in high dimensions "almost everything is orthogonal." In other words, angles close to 90° are much more likely to occur between two random vectors than much smaller angles. So one might guess that for a target vector v and a random list vector w, with high probability their angle is close to 90°. On the other hand, two non-reduced vectors v, w of similar norm for which the if-clause in Line 8 is true (i.e., for which $\|v - w\| \leq \gamma R = R(1 - o(1))$ and $\|v\|, \|w\| \approx R$), always have a common angle of at most 60° + $o(1)$. We therefore expect this angle to be close to 60° with high probability. Under the extreme (and imprecise) assumption that all angles between pairwise reduced vectors are *exactly* 90°, and non-reduced angles are at most 60°, we obtain the following estimate for the optimized time and space complexities using spherical LSH.

Estimate 1. *Assuming that all reduced pairs of vectors are exactly orthogonal, under Heuristic 1 the SphereSieve solves SVP in time and space at most* $(4/3)^{2n/3+o(n)} = 2^{0.2767n+o(n)}$, *using the following parameters:*

$$k = \Theta(\sqrt{n}), \qquad t = (4/3)^{n/6+o(n)} = 2^{0.0692n+o(n)}. \qquad (4)$$

Proof. Assuming all reduced pairs of vectors are orthogonal, we obtain $\rho = \frac{1}{3}$ as described in Sect. 2.4. Since the time complexity is dominated by performing $O(N)$ nearest-neighbor searches on a list of size $O(N)$, with $N = (4/3)^{n/2+o(n)} \approx 2^{0.2075n+o(n)}$, the result follows from Lemma 1.

Of course in practice not all reduced angles are actually 90°, and one should carefully analyze what is the real probability that a vector w whose angle with v is more than 60°, is found as a candidate due to a collision in one of the hash tables. In that sense, Estimate 1 should only be considered a rough estimate, and it gives a lower bound on the best time complexity that we may hope to achieve with this method. Note however that the estimated time complexity is significantly better than the similar estimate obtained for the angular LSH-based HashSieve of Laarhoven [24], for which the estimated time complexity was $2^{0.3289n+o(n)}$. Therefore, one might guess that also the actual asymptotic time complexity, derived after a more precise analysis, is better than that of the HashSieve.

The following proposition shows that this is indeed the case, and it describes exactly what the asymptotic time and space complexities are when the parameters are fully optimized to minimize the asymptotic time complexity. A proof of Proposition 1 and an explanation of the constant 0.2972 can be found in Appendix A.

Proposition 1. *The SphereSieve heuristically solves SVP in time and space* $2^{0.2972n+o(n)}$ *using the following parameters:*

$$k = \Theta(\sqrt{n}), \qquad t = 2^{0.0896n+o(n)}. \qquad (5)$$

By varying k and t, we further obtain the trade-off between the time and space complexities indicated by the solid blue curve in Fig. 1.

Note that the estimated parameters from Estimate 1 are not far off from the main result of Proposition 1. In other words, assuming that reduced vectors are always orthogonal is not entirely realistic, but it provides a reasonable first estimate of the parameters that we have to use.

Finally, note that the space complexity increases by a factor t and thus increases exponentially compared to the Nguyen-Vidick sieve. To get rid of this exponential increase in the memory, instead of storing all hash tables in memory at the same time we may choose to go through the hash tables one by one, as in [24]; we first build one hash table by adding all vectors to their corresponding hash buckets, and then we look for pairs of nearby vectors in each bucket (whose difference has norm less than γR), and add all the found vectors to our new list L_{m+1}. As the number of vectors in each hash bucket is $2^{o(n)}$, comparing all pairs of vectors in a hash bucket can be done in $2^{o(n)}$ time and the cost of processing one hash table is $2^{0.208n+o(n)}$. We then repeat this $t = 2^{0.0896n+o(n)}$ times (each time removing the previous hash table from memory) to finally achieve the following result.

Theorem 1. *The SphereSieve heuristically solves the exact shortest vector problem in time $2^{0.2972n+o(n)}$ and space $2^{0.2075n+o(n)}$.*

5   A Practical SphereSieve Variant and Two-Level Sieving

Let us briefly consider how this algorithm can be made slightly more practical. In particular, note that each spherical hash function requires the use of $U = 2^{\Theta(n)}$ vectors $s_1, \ldots, s_U$, which are (roughly) sampled from the surface of the unit hypersphere. In total, this means that the algorithm uses $t \cdot k \cdot U$ random unit vectors to define hash regions on the sphere, and all these vectors need to be stored in memory. Generating so many random vectors from the surface of the unit hypersphere seems unnecessary, especially considering that we already have a list L_m containing vectors which (almost) lie on the surface of a hypersphere as well.

The above suggests to make the following modification to the algorithm: for building a single hash function $h_{i,j}$, instead of sampling $s_1, \ldots, s_U$ randomly from the surface of the sphere, we randomly sample these vectors from (a scaled version of) L_m. In other words, we use the vectors in L_m to shape the hash regions, rather than sampling and storing new vectors in memory solely for this purpose. According to Heuristic 1 these vectors are also distributed randomly on the surface of the sphere, and so using the same heuristic assumption we can justify that this modification does not drastically alter the behavior of the algorithm. Note that since we need $t \cdot k$ hash functions, we need $t \cdot k$ selections of U vectors from L_m. Fortunately $t \cdot k \cdot U \ll |L_m|$ (cf. Proposition 1), so by independently sampling U random vectors from L_m for each of the $t \cdot k$ hash functions, the hash functions $h_{i,j}$ can practically be considered independent.

Relation with Two-level Sieving. Now, note that for a single hash function, we first use a small set of hash region-defining vectors U (where the radius of each hash region is approximately $(\sqrt{2} - o(1))R$), and then we use the NV-sieve in each of these regions separately to make lists of centers C_{m+1} (where a vector is considered nearby if it is within a radius of approximately $(1-o(1))R$). This very closely resembles the ideas behind Wang et al.'s two-level sieve algorithm [46], where a list $C_1(\cong U)$ of outer centers is built (defining balls of radius $\gamma_1 \cdot R$), and each of the centers w of this outer list contains an inner list $C_2^w(\cong C_{m+1})$ of center vectors (defining a ball of radius $\gamma_2 \cdot R$). In fact, for $t = k = 1$, the SphereSieve is almost identical to the two-level sieve with $\gamma_1 \approx \sqrt{2}$ and $\gamma_2 \approx 1$!

How Order Matters. One difference between the two methods is that the size of U in the SphereSieve is sub-exponential ($2^{\Theta(\sqrt{n})}$), compared to single exponential ($2^{\Theta(n)}$) in the two-level sieve, which means that in our case, one of these hash tables is relatively 'cheap' to build. As a result, the asymptotic exponential overhead in our case only comes from t. However, the key difference that allows us to obtain the improved performance overall seems to be that the analysis of spherical LSH [5,6] (and the closely related analysis of the celebrated Euclidean LSH family [4]) makes crucial use of the fact that the outer list C_1 is *ordered*, and this same order is used each time a vector is assigned to a hash region. Without this observation, Lemma 2 does not hold, and as in [46,47] one would then have to resort to computing intersections of volumes of complicated n-dimensional objects to obtain bounds on the number of points needed to make this method work. One might say that the *order* imposed on C_1 is exactly what makes spherical LSH asymptotically more efficient than the two-level sieve of Wang et al. [46] with $\gamma_1 \approx \sqrt{2}$ and $\gamma_2 \approx 1$.

6   Discussion

Theoretically, Theorem 1 shows that for sufficiently high dimensions n, spherical LSH leads to even bigger speed-ups than angular LSH [24]. With a heuristic time complexity less than $2^{0.2972n+o(n)} < 2^{3n/10+o(n)}$, the SphereSieve is the fastest heuristic algorithm to date for solving SVP in high dimensions. As a result, one might conclude that in high dimensions, to achieve $3k$ bits of security for a lattice-based cryptographic primitive relying on the hardness of exact SVP, one should use a lattice of dimension at least $10k$. As most cryptographic schemes are broken even if a short lattice vector is found (which by using BKZ [44,45] means we can reduce the dimension in which we need to solve SVP), and the time complexity of the SphereSieve is lower than $2^{n/3+o(n)}$, one should probably use lattices of dimension higher than $10k$ to guarantee $3k$ bits of security. So various parameter choices relying on the estimates of e.g. Chen and Nguyen [13] (solving SVP in dimension 200 takes time 2^{111}) would be too optimistic.

Although the leading term $0.2972n$ in the exponent is the best known so far and dominates the complexity in high dimensions, this does not tell the whole story. Especially for the SphereSieve presented in this paper, the $o(n)$-terms in the exponent are not negligible at all for moderate n. Experiments further indicate [16,24,27,28,32,35,41] that the practical time complexity of various sieving algorithms in moderate dimensions n may be higher than quadratic in the list size if we set γ close to 1, while setting $\gamma \ll 1$ makes the use of spherical LSH problematic. Moreover, while the angular LSH method of Charikar [12] considered in [24] is very efficient and hashes can be computed in linear time, with spherical LSH even the cost of computing a single hash value (before amplification) is already sub-exponential (and super-polynomial) in n. So in practice it is not clear whether the SphereSieve will outperform the angular LSH-based sieving algorithm of Laarhoven [24] for any feasible dimension n. Finding an accurate description of the practical costs of finding short(est) vectors in dimension n remains a central problem in lattice cryptography.

An important question for future work remains whether spherical LSH can be made truly efficient. While asymptotic costs are important, lower order terms matter in practice as well, and being able to compute hashes in poly(n)-time would make the SphereSieve significantly faster. As mentioned in Sect. 3, being able to apply spherical LSH to the faster GaussSieve [32] may also lead to a faster sieve. The recent work [9] takes a first step in this direction, showing that the same asymptotics as the SphereSieve can be achieved with efficient hashing.

Appendix

A   Proof of Proposition 1

To prove Proposition 1, we will show how to choose a sequence of parameters $\{(k_n, t_n)\}_{n \in \mathbb{N}}$ such that for large n, the following holds:

1. The probability that a list vector w close[5] to a target vector v collides with v in at least one of the t hash tables is at least constant in n:

$$p_1^* = \mathbb{P}_{\{h_{i,j}\} \subset \mathcal{H}}(v, w \text{ collide } | \theta(v, w) \leq \tfrac{\pi}{3}) \geq 1 - \varepsilon. \quad (\varepsilon \neq \varepsilon(n)) \quad (6)$$

2. The average probability that a list vector w far away (See footnote 5) from a target vector v collides with v is exponentially small:

$$p_2^* = \mathbb{P}_{\{h_{i,j}\} \subset \mathcal{H}}(v, w \text{ collide } | \theta(v, w) > \tfrac{\pi}{3}) \leq N^{-0.5681 + o(1)}. \quad (7)$$

3. The number of hash tables grows as $t = N^{0.4319 + o(1)}$.

This would imply that for each search, the number of candidate vectors is of the order $N \cdot N^{-0.5681} = N^{0.4319}$. Overall we search the list $\tilde{O}(N)$ times, so after substituting $N = (4/3)^{n/2 + o(n)}$ this leads to the following time and space complexities:

-   Time (hashing): $O(N \cdot t) = 2^{0.2972n + o(n)}$.
-   Time (searching): $O(N^2 \cdot p_2^*) = 2^{0.2972n + o(n)}$.
-   Space: $O(N \cdot t) = 2^{0.2972n + o(n)}$.

The next two subsections are dedicated to proving Eqs. (6) and (7).

A.1   Good Vectors Collide with Constant Probability

The following lemma shows how to choose k (in terms of t) to guarantee that (6) holds.

Lemma 3. *Let $\varepsilon > 0$ and let $k = 6n^{-1/2}(\ln t - \ln \ln(1/\varepsilon)) \approx (6 \ln t)/\sqrt{n}$. Then the probability that reducing vectors collide in at least one of the hash tables is at least $1 - \varepsilon$.*

---

[5] Here "close" means that $\|v - w\| \leq \gamma R$, which corresponds to $\theta(v, w) \leq 60° + o(1)$. Similarly "far away" corresponds to a large angle $\theta(v, w) > 60° + o(1)$.

Proof. The probability that a reducing vector $\boldsymbol{w}$ is a candidate vector, given the angle $\Theta = \Theta(\boldsymbol{v}, \boldsymbol{w}) \in (0, \frac{\pi}{3})$, is $p_1^* = \mathbb{E}_{\Theta \in (0, \frac{\pi}{3})}[p^*(\Theta)]$, where we recall that $p^*(\theta) = 1 - (1 - p(\theta)^k)^t$ and $p(\theta) = \mathbb{P}_{h \in \mathcal{H}}[h(\boldsymbol{v}) = h(\boldsymbol{w})]$ is given in Lemma 2. Since $p^*(\Theta)$ is strictly decreasing in Θ, we can obtain a lower bound by substituting $\Theta = \frac{\pi}{3}$ above. Using the bound $1 - x \le e^{-x}$ which holds for all x, and inserting the given expression for k, we obtain $p_1^* \ge p^*\left(\frac{\pi}{3}\right) = 1 - (1 - \exp(\ln \ln(\frac{1}{\varepsilon}) - \ln t))^t = 1 - \left(1 - \frac{\ln(1/\varepsilon)}{t}\right)^t \ge 1 - \varepsilon.$

A.2   Bad Vectors Collide with Low Probability

We first recall a lemma about the density of angles between random vectors. In short, the density at an angle θ is proportional to $(\sin \theta)^n$.

Lemma 4. *[24, Lemma 4] Assuming Heuristic 1 holds, the pdf $f(\theta)$ of the angle between target vectors and list vectors satisfies*

$$f(\theta) = \sqrt{\frac{2n}{\pi}} \, (\sin \theta)^{n-2} [1 + o(1)] = 2^{n \log_2 \sin \theta + o(n)}. \tag{8}$$

The following lemma relates the collision probability p_2^* of (7) to the parameters k and t. Since Lemma 3 relates k to t, this means that only t ultimately remains to be chosen.

Lemma 5. *Suppose $N = 2^{c_n \cdot n}$ with $c_n \ge \gamma_1 = \frac{1}{2} \log_2(\frac{4}{3}) \approx 0.2075$, and suppose $t = 2^{c_t \cdot n}$. Let $k = \frac{6 \ln t}{\sqrt{n}}(1 - o(1))$. Then, for large n, under Heuristic 1 we have*

$$p_2^* = \mathbb{P}_{\{h_{i,j}\} \subset \mathcal{H}}(\boldsymbol{v}, \boldsymbol{w} \text{ collide} \mid \theta(\boldsymbol{v}, \boldsymbol{w}) > \tfrac{\pi}{3}) \le O(N^{-\alpha}), \tag{9}$$

where $\alpha \in (0, 1)$ is defined as

$$\alpha = \frac{-1}{c_n} \left[\max_{\theta \in (\frac{\pi}{3}, \frac{\pi}{2})} \left\{ \log_2 \sin \theta - \left(3 \tan^2\left(\frac{\theta}{2}\right) - 1\right) c_t \right\} \right] + o(1). \tag{10}$$

Proof. First, if we know the angle $\theta \in (\frac{\pi}{3}, \frac{\pi}{2})$ between two bad vectors, then according to Lemma 2 the probability of a collision in at least one of the hash tables is equal to

$$p^*(\theta) = 1 - \left(1 - \exp\left[-\frac{k\sqrt{n}}{2} \tan^2\left(\frac{\theta}{2}\right)(1 + o(1))\right]\right)^t. \tag{11}$$

Letting $f(\theta)$ denote the density of angles θ on $(\frac{\pi}{3}, \frac{\pi}{2})$, we have

$$p_2^* = \mathbb{E}_{\Theta \in (\frac{\pi}{3}, \frac{\pi}{2})}[p^*(\Theta)] = \int_{\pi/3}^{\pi/2} f(\theta) p^*(\theta) d\theta. \tag{12}$$

Substituting $p^*(\theta)$ and the expression of Lemma 4 for $f(\theta)$, noting that $\int_{\pi/3}^{\pi/2} f(\theta)d\theta \approx \int_0^{\pi/2} f(\theta)d\theta = 1$, we get

$$p_2^* = \int_{\pi/3}^{\pi/2} (\sin\theta)^n \left[1 - \left(1 - \exp\left[-3\ln t\tan^2\left(\tfrac{\theta}{2}\right)(1+o(1)) \right] \right)^t \right] d\theta. \quad (13)$$

For convenience, let us write $w(\theta) = [-3\ln t\tan^2\left(\tfrac{\theta}{2}\right)(1+o(1))$. Note that for $\theta \gg \tfrac{\pi}{3}$ we have $w(\theta) \ll -\ln t$ so that $(1 - \exp w(\theta))^t \approx 1 - t\exp w(\theta)$, in which case we can simplify the expression between square brackets. However, the integration range includes $\tfrac{\pi}{3}$ as well, so to be careful we will split the integration interval at $\tfrac{\pi}{3} + \delta$, where $\delta = \Theta(n^{-1/2})$. (Note that any value δ with $\tfrac{1}{n} \ll \delta \ll 1$ suffices.)

$$p_2^* = \underbrace{\int_{\pi/3}^{\pi/3+\delta} f(\theta)p^*(\theta)d\theta}_{I_1} + \underbrace{\int_{\pi/3+\delta}^{\pi/2} f(\theta)p^*(\theta)d\theta}_{I_2}. \quad (14)$$

Bounding I_1. Using $f(\theta) \leq f(\tfrac{\pi}{3}+\delta)$, $p^*(\theta) \leq 1$, and $\sin(\tfrac{\pi}{3}+\delta) = \tfrac{1}{2}\sqrt{3}[1+O(\delta)]$ (which follows from a Taylor expansion of $\sin x$ around $x = \tfrac{\pi}{3}$), we obtain

$$I_1 \leq \text{poly}(n)\sin^n(\tfrac{\pi}{3}+\delta) = \text{poly}(n)(\tfrac{\sqrt{3}}{2})^n (1+O(\delta))^n = 2^{-\gamma_1 n + o(n)}. \quad (15)$$

Bounding I_2. For I_2, our choice of δ is sufficient to make the aforementioned approximation work[6]. Thus, for I_2 we obtain the simplified expression

$$I_2 \leq \text{poly}(n)\int_{\pi/3+\delta}^{\pi/2} (\sin\theta)^n t\exp\left[-3\ln t\tan^2\left(\tfrac{\theta}{2}\right)(1+o(1)) \right] d\theta \quad (16)$$

$$\leq \int_{\pi/3}^{\pi/2} 2^{n\log_2\sin\theta - (3\tan^2(\tfrac{\theta}{2})-1)\log_2 t + o(n)} d\theta. \quad (17)$$

Note that the integrand is exponential in n and that the exponent $E(\theta) = n\log_2\sin\theta + (-3\tan^2\tfrac{\theta}{2}-1)\log_2 t$ is a continuous, differentiable function of θ. So the asymptotic behavior of the entire integral I_2 is the same as the asymptotic behavior of the integrand's maximum value:

$$\log_2 I_2 \leq \max_{\theta \in (\tfrac{\pi}{3},\tfrac{\pi}{2})} \{ n\log_2\sin\theta - (3\tan^2\tfrac{\theta}{2}-1)\log_2 t \} + o(n). \quad (18)$$

Bounding $p_2^* = I_1 + I_2$. Combining (15), (18), and $c_t = \tfrac{1}{n}\log_2 t$, we have

$$\tfrac{\log_2 p_2^*}{n} \leq \max\{ -\gamma_1, \max_{\theta \in (\tfrac{\pi}{3},\tfrac{\pi}{2})} \{\log_2\sin\theta - (3\tan^2\tfrac{\theta}{2}-1)c_t\} \} + o(1). \quad (19)$$

The assumption $c_n \geq \gamma_1$ and the definition of $\alpha \leq 1$ now give $\log_2 p_2^* \leq -\alpha c_n n + o(n)$ which completes the proof.

---

[6] By choosing the order terms in k appropriately, the $o(1)$-term inside $w(\theta)$ may be cancelled out, in which case the δ-term dominates. Note that the $o(1)$-term in $w(\theta)$ can be further controlled by the choice of $\gamma = 1 - o(1)$.

A.3 Balancing the Parameters

Recall that the overall time and space complexities are given by $O(N \cdot t) = 2^{(c_n + c_t)n + o(n)}$ (time for hashing), $O(N^2 \cdot p_2^*) = 2^{(c_n + (1-\alpha)c_n)n + o(n)}$ (time for comparing vectors), and $O(N \cdot t) = 2^{(c_n + c_t)n + o(n)}$ (memory requirement). For the overall time and space complexities $2^{c_{\text{time}}n}$ and $2^{c_{\text{space}}n}$ we find

$$c_{\text{time}} = c_n + \max\{c_t, (1-\alpha)c_n\} + o(1), \quad c_{\text{space}} = c_n + c_t + o(1). \tag{20}$$

Further recall that from Nguyen and Vidick's analysis, we have $N = (4/3)^{n/2+o(n)}$ or $c_n = \gamma_1$. To balance the time complexities of hashing and searching, so that the overall time complexity is minimized, we solve $(1-\alpha)\gamma_1 = c_t$ numerically[7] for c_t to obtain the following corollary. Here θ^* denotes the dominant angle θ maximizing the expression in (10). Note that the final result takes into account the density at $\theta = \theta^*$ as well, and so the result does not simply follow from Lemma 2.

Corollary 1. *Taking $c_t \approx 0.089624$ leads to:*

$$\theta^* \approx 0.42540\pi, \quad \alpha \approx 0.56812, \quad c_{\text{time}} \approx 0.29714, \quad c_{\text{space}} \approx 0.29714. \tag{21}$$

Thus, setting $t \approx 2^{0.08962n}$ and $k = \Theta(\sqrt{n})$, the heuristic time and space complexities of the SphereSieve algorithm are balanced at $2^{0.29714n + o(n)}$.

A.4 Trade-Off Between the Space and Time Complexities

Finally, note that $c_t = 0$ leads to the original Nguyen-Vidick sieve algorithm, while $c_t \approx 0.089624$ minimizes the heuristic time complexity at the cost of more space. One can obtain a continuous trade-off between these two extremes by considering values $c_t \in (0, 0.089624)$. Numerically evaluating the resulting complexities for this range of values of c_t leads to the curve shown in Fig. 1.

References

1. Aggarwal, D., Dadush, D., Regev, O., Stephens-Davidowitz, N.: Solving the shortest vector problem in 2^n time via discrete gaussian sampling. In: STOC (2015)
2. Ajtai, M.: The shortest vector problem in L_2 is NP-hard for randomized reductions (extended abstract). In: STOC, pp. 10–19 (1998)
3. Ajtai, M., Kumar, R., Sivakumar, D.: A sieve algorithm for the shortest lattice vector problem. In: STOC, pp. 601–610 (2001)
4. Andoni, A., Indyk, P.: Near-optimal hashing algorithms for approximate nearest neighbor in high dimensions. In: FOCS, pp. 459–468 (2006)
5. Andoni, A., Indyk, P., Nguyen, H.L., Razenshteyn, I.: Beyond locality-sensitive hashing. In: SODA, pp. 1018–1028 (2014)
6. Andoni, A., Razenshteyn, I.: Optimal data-dependent hashing for approximate near neighbors. In: STOC (2015)

---

[7] Note that α is implicitly a function of c_t as well.

7. Becker, A., Gama, N., Joux, A.: A sieve algorithm based on overlattices. In: ANTS, pp. 49–70 (2014)
8. Becker, A., Gama, N., Joux, A.: Speeding-up lattice sieving without increasing the memory, using sub-quadratic nearest neighbor search. Cryptology ePrint Archive, Report 2015/522 (2015)
9. Becker, A., Laarhoven, T.: Efficient sieving on (ideal) lattices using cross-polytopic LSH. (preprint 2015)
10. Bernstein, D.J., Buchmann, J., Dahmen, E.: Post-Quantum Cryptography. Springer, Heidelberg (2009)
11. Bos, J., Naehrig, M., van de Pol, J.: Sieving for shortest vectors in ideal lattices: a practical perspective. Cryptology ePrint Archive, Report 2014/880 (2014)
12. Charikar, M.S.: Similarity estimation techniques from rounding algorithms. In: STOC, pp. 380–388 (2002)
13. Chen, Y., Nguyen, P.Q.: BKZ 2.0: better lattice security estimates. In: Lee, D.H., Wang, X. (eds.) ASIACRYPT 2011. LNCS, vol. 7073, pp. 1–20. Springer, Heidelberg (2011)
14. Datar, M., Immorlica, N., Indyk, P., Mirrokni, V.S.: Locality-sensitive hashing scheme based on p-stable distributions. In: SOCG, pp. 253–262 (2004)
15. Fincke, U., Pohst, M.: Improved methods for calculating vectors of short length in a lattice. Math. Comput. **44**(170), 463–471 (1985)
16. Fitzpatrick, R., Bischof, C., Buchmann, J., Dagdelen, Ö., Göpfert, F., Mariano, A., Yang, B.-Y.: Tuning gausssieve for speed. In: Aranha, D.F., Menezes, A. (eds.) LATINCRYPT 2014. LNCS, vol. 8895, pp. 288–305. Springer, Heidelberg (2015)
17. Gama, N., Nguyen, P.Q., Regev, O.: Lattice enumeration using extreme pruning. In: Gilbert, H. (ed.) EUROCRYPT 2010. LNCS, vol. 6110, pp. 257–278. Springer, Heidelberg (2010)
18. Gentry, C.: Fully homomorphic encryption using ideal lattices. In: STOC, pp. 169–178 (2009)
19. Hanrot, G., Pujol, X., Stehlé, D.: Algorithms for the shortest and closest lattice vector problems. In: Chee, Y.M., Guo, Z., Ling, S., Shao, F., Tang, Y., Wang, H., Xing, C. (eds.) IWCC 2011. LNCS, vol. 6639, pp. 159–190. Springer, Heidelberg (2011)
20. Hoffstein, J., Pipher, J., Silverman, J.H.: NTRU: a ring-based public key cryptosystem. In: Buhler, J.P. (ed.) ANTS 1998. LNCS, vol. 1423, pp. 267–288. Springer, Heidelberg (1998)
21. Indyk, P., Motwani, R.: Approximate nearest neighbors: towards removing the curse of dimensionality. In: STOC, pp. 604–613 (1998)
22. Ishiguro, T., Kiyomoto, S., Miyake, Y., Takagi, T.: Parallel gauss sieve algorithm: solving the SVP challenge over a 128-dimensional ideal lattice. In: PKC, pp. 411–428 (2014)
23. Kannan, R.: Improved algorithms for integer programming and related lattice problems. In: STOC, pp. 193–206 (1983)
24. Laarhoven, T.: Sieving for shortest vectors in lattices using angular locality-sensitive hashing. In: CRYPTO (2015)
25. Laarhoven, T., Mosca, M., van de Pol, J.: Finding shortest lattice vectors faster using quantum search. Des., Codes Crypt. (2015)
26. Lindner, R., Peikert, C.: Better key sizes (and attacks) for LWE-based encryption. In: Kiayias, A. (ed.) CT-RSA 2011. LNCS, vol. 6558, pp. 319–339. Springer, Heidelberg (2011)
27. Mariano, A., Timnat, S., Bischof, C.: Lock-free gausssieve for linear speedups in parallel high performance SVP calculation. In: SBAC-PAD, pp. 278–285 (2014)

28. Mariano, A., Dagdelen, Ö., Bischof, C.: A comprehensive empirical comparison of parallel listsieve and gausssieve. In: Lopes, L., et al. (eds.) Euro-Par 2014 Workshops. LNCS, pp. 48–59. Springer, Heidelberg (2014)

29. Mariano, A., Laarhoven, T., Bischof, C.: Parallel (probable) lock-free hashsieve: a practical sieving algorithm for the SVP. In: ICPP (2015)

30. Micciancio, D.: The shortest vector in a lattice is hard to approximate to within some constant. In: FOCS, pp. 92–98 (1998)

31. Micciancio, D., Voulgaris, P.: A deterministic single exponential time algorithm for most lattice problems based on voronoi cell computations. In: STOC, pp. 351–358 (2010)

32. Micciancio, D., Voulgaris, P.: Faster exponential time algorithms for the shortest vector problem. In: SODA, pp. 1468–1480 (2010)

33. Micciancio, D., Walter, M.: Fast lattice point enumeration with minimal overhead. In: SODA, pp. 276–294 (2015)

34. Milde, B., Schneider, M.: A parallel implementation of gausssieve for the shortest vector problem in lattices. In: Malyshkin, V. (ed.) PaCT 2011. LNCS, vol. 6873, pp. 452–458. Springer, Heidelberg (2011)

35. Nguyen, P.Q., Vidick, T.: Sieve algorithms for the shortest vector problem are practical. J. Math. Cryptol. **2**(2), 181–207 (2008)

36. Plantard, T., Schneider, M.: Ideal lattice challenge (2014). http://latticechallenge.org/ideallattice-challenge/

37. Pohst, M.E.: On the computation of lattice vectors of minimal length, successive minima and reduced bases with applications. ACM SIGSAM Bull. **15**(1), 37–44 (1981)

38. van de Pol, J., Smart, N.P.: Estimating key sizes for high dimensional lattice-based systems. In: Stam, M. (ed.) IMACC 2013. LNCS, vol. 8308, pp. 290–303. Springer, Heidelberg (2013)

39. Pujol, X., Stehlé, D.: Solving the shortest lattice vector problem in time $2^{2.465n}$. Cryptology ePrint Archive, Report 2009/605 (2009)

40. Regev, O.: On lattices, learning with errors, random linear codes, and cryptography. In: STOC, pp. 84–93 (2005)

41. Schneider, M.: Analysis of gauss-sieve for solving the shortest vector problem in lattices. In: Katoh, N., Kumar, A. (eds.) WALCOM 2011. LNCS, vol. 6552, pp. 89–97. Springer, Heidelberg (2011)

42. Schneider, M.: Sieving for shortest vectors in ideal lattices. In: Youssef, A., Nitaj, A., Hassanien, A.E. (eds.) AFRICACRYPT 2013. LNCS, vol. 7918, pp. 375–391. Springer, Heidelberg (2013)

43. Schneider, M., Gama, N., Baumann, P., Nobach, L.: SVP challenge (2015). http://latticechallenge.org/svp-challenge

44. Schnorr, C.-P.: A hierarchy of polynomial time lattice basis reduction algorithms. Theor. Comput. Sci. **53**(2), 201–224 (1987)

45. Schnorr, C.-P., Euchner, M.: Lattice basis reduction: improved practical algorithms and solving subset sum problems. Math. Program. **66**(2), 181–199 (1994)

46. Wang, X., Liu, M., Tian, C., Bi, J.: Improved nguyen-vidick heuristic sieve algorithm for shortest vector problem. In: ASIACCS, pp. 1–9 (2011)

47. Zhang, F., Pan, Y., Hu, G.: A three-level sieve algorithm for the shortest vector problem. In: Lange, T., Lauter, K., Lisoněk, P. (eds.) SAC 2013. LNCS, vol. 8282, pp. 29–47. Springer, Heidelberg (2014)

FHEW with Efficient Multibit Bootstrapping

Jean-François Biasse$^{(\boxtimes)}$ and Luis Ruiz

Department of Combinatorics and Optimization, University of Waterloo,
200 University Avenue West, Waterloo, ON N2L 3G1, Canada
biasse@lix.polytechnique.fr

Abstract. In this paper, we describe a generalization of the fully homomorphic encryption scheme FHEW described by Ducas and Micciancio [8]. It is characterized by an efficient bootstrapping procedure performed after each gate, as opposed to the HElib of Halevi and Shoup that handles batches of encryptions periodically. While the Ducas-Micciancio scheme was limited to NAND gates, we propose a generalization that can handle arbitrary gates for only one call to the bootstrapping procedure. We also show how bootstrapping can be parallelized and address its performances in a multicore environment.

Keywords: Fully homomorphic encryption · LWE · Bootstrapping · Parallelization

1 Introduction

The first fully homomorphic encryption scheme was proposed by Gentry in 2009 [11]. A fully homomorphic encryption scheme allows anyone to perform computations on an encrypted message, even if they do not know the decryption key. This discovery has a tremendous impact in situations where encrypted data is stored on an untrusted server. This is also seen as a possibility to securely delegate computational power, but Gentry's original solution suffers from a drastic lack of efficiency. Indeed, there is a prohibitive overhead in the computational cost when working on encrypted data. The priority of subsequent work [1,2,5,12–16,18] has been to increase the performances of fully homomorphic encryption schemes to get to the point where industrial applications could be considered.

The security of the majority of the serious proposals for fully homomorphic encryption schemes proposed to this date is based on lattice assumptions. The two main problems on which these schemes rely are the Learning With Errors (LWE) problem, and its variant the Ring Learning With Error (RLWE) problem. The LWE assumption consists of stating that pairs of the form $(a, a \cdot s + e)$ where $s \in \mathbb{Z}_q^n$ is a secret, $a \in \mathbb{Z}_q^n$ is sampled given a certain distribution and $e \in \mathbb{R}/\mathbb{Z}$ is sampled according to another distribution are indistinguishable from randomness. The RLWE assumption is similar to LWE, except that vectors are replaced by ring elements. The LWE problem reduces to finding short vectors

© Springer International Publishing Switzerland 2015
K. Lauter and F. Rodríguez-Henríquez (Eds.): LatinCrypt 2015, LNCS 9230, pp. 119–135, 2015.
DOI: 10.1007/978-3-319-22174-8_7

in lattices [6,21], which is widely admitted to be computationally hard. The error e grows when we perform operations on the ciphertext, and ultimately reaches a threshold above which decryption is no longer possible. The key idea of Gentry [11] was to introduce a bootstrapping procedure that allows us to evaluate the decryption circuit on the encrypted data, thus producing a new and "fresh" ciphertext (provided that the depth of the decryption circuit is reasonable). This requires an extra "circular" security assumption stating that giving away an encryption of the secret key does not compromise the safety of the scheme.

The bootstrapping procedure is the bottleneck of all proposals for fully homomorphic encryption schemes. This is the part of the algorithm that currently receives the most attention from the scientific community. It is both problematic because of the requirements in terms of computational power and in terms of storage (because of the size of the keys). The strategy used in the HElib of Halevi and Shoup [18,19] consists of running an expensive bootstrapping procedure on a batch of encryptions in order to amortize the cost of bootstrapping per homomorphic operation. Ducas and Micciancio [8] opted for the opposite strategy consisting of running a lightweight bootstrapping procedure after each homomorphic operation (limited to NAND gates in their work). Both techniques require a comparable amortized bootstrapping effort, but the Micciancio-Ducas FHEW scheme has the definite advantage of releasing the user from the task of optimizing the homomorphic circuit to offer the best amortized cost. Indeed, they offer bootstrapping at the lowest level of granularity.

Contribution. We improve on the FHEW scheme of Ducas-Micciancio [8] in two different ways. First, following suggestions in the conclusion of [8], we extend the bootstrapping procedure to allow it to refresh encryptions of messages modulo a prime $t > 2$. We also provide arbitrary homomorphic membership tests on these refreshed encryptions. This allows the design of more general gates involving several inputs and outputs (but at the cost of a single bootstrapping procedure). We give the example of the full adder gate, but our method is general and works regardless of the specific gate we use.

Second, we parallelize the bootstrapping procedure and assess its performances in a multicore environment. Fully homomorphic encryption schemes are usually parallelized at the gate level, but this method has the disadvantage of relying on an efficient compiler capable of identifying batches of independent operations. Following the philosophy of FHEW [8], we present a level of parallelization that does not rely on any kind of optimization of the series of homomorphic operations to be performed. In addition, our method is completely orthogonal to the parallelization at the gate level that can be provided by smart compiling method.

2 Mathematical Background

The two main mathematical ingredients to describe fully homomorphic encryption schemes are cyclotomic fields and statistical distributions. As this paper generalizes the methods of [8], we try to stick to the same notations as in [8], and we reuse statements as much as possible. In particular, bold-face lower case letters $\mathbf{a}, \mathbf{b}$ denote column vectors while bold-face upper case letters $\mathbf{A}, \mathbf{B}$ denote matrices. Also, $\mathbf{a} \cdot \mathbf{b}$ denotes the scalar product between $\mathbf{a}$ and $\mathbf{b}$ while $\|\mathbf{a}\|$ is the Euclidean norm of $\mathbf{a}$.

2.1 Distributions

To ensure semantic security, randomized variables are used in encryption. The parameters of the distribution from which these variables are chosen must be adapted to our context. As we see in Sect. 3, LWE encryptions are of the form $(a, a \cdot s + e)$ for some "random" e. In general, e occurs as the result of randomized rounding by a function $\chi : \mathbb{R} \to \mathbb{Z}$ such that $\chi(x + n) = \chi(x) + n$ for all $n \in \mathbb{Z}$. When χ is restricted to $\mathbb{Z}$, we have in fact $\chi(x) = x + \chi(0)$, and all we care about is adding the noise $e = \chi(0)$ to the input. Our random variables follow the subgaussian distribution defined below.

Definition 1 (Subgaussian Variable). *A random variable X over $\mathbb{R}$ is subgaussian with parameter $\alpha > 0$ if the scaled moment-generating function satisfies $E[e^{2\pi t X}] \le e^{\pi \alpha^2 t^2}$.*

As pointed out in [8], if X is subgaussian with parameter $\alpha > 0$, then it satisfies $P(|X| \ge t) \le 2e^{-\pi t^2/\alpha^2}$. Also, any variable X satisfying $|X| \le B$ for some constant B is subgaussian with parameter $B\sqrt{2\pi}$. We refer to [8] for detailed proofs of the statements we use.

2.2 Cyclotomic Fields

A cyclotomic field is an extension of $\mathbb{Q}$ of the form $K = \mathbb{Q}(\zeta_N)$ where $\zeta_N = e^{2i\pi/N}$ is a primitive N-th root of unity. In [8], Ducas and Micciancio restricted themselves to the case where N is a power of 2. To achieve better functionalities in our bootstrapping function, we need to use power-of-p cyclotomics where $p > 2$. The properties of the power-of-two cyclotomics that were used in [8] can be generalized to our case. The ring of integers $\mathcal{R}$ of K is $\mathbb{Z}[X]/(\Phi_N(X))$ where Φ_N is the N-th cyclotomic polynomial. When N is a power of two, $\Phi_N(X) = X^{N/2}+1$, but when $N = p^k$ is a power of $p > 2$, we have $\Phi_N(X) = X^{p^{k-1}(p-1)} + X^{p^{k-1}(p-2)} + \cdots + 1$ (which of course generalizes the case $p = 2$). Elements $a \in \mathcal{R}$ are residues of polynomials in $\mathbb{Z}[X]$ modulo $\Phi_N(X)$, and can be identified with their coefficient vectors $\overrightarrow{a} \in \mathbb{Z}^{\phi(N)}$ where $\phi(N) = p^{k-1}(p-1)$ is the Euler totient of N (and the degree of $\Phi_N(X)$). An element $a \in \mathcal{R}$ can also be uniquely identified by the linear transformation $x \mapsto a \cdot x$ (here the dot denotes the multiplication in $\mathcal{R}$. This linear transformation can in turn be identified uniquely by a matrix denoted by $\overrightarrow{a} \in \mathbb{Z}^{\phi(N) \times \phi(N)}$. In the case of power-of-two cyclotomics, it is an anticirculant matrix while in the case where $N = p^k$, the coefficients of the column of index i are a function of the coefficients of the column of index $i - 1$ (given by a shift and the relation $X^{p^{k-1}(p-1)} \equiv -\sum_j X^{p^{k-1}(p-j)} \bmod \Phi_N(X)$). The first column of $\vec{\vec{a}}$ is $\vec{a}$. The matrices of the form $\vec{\vec{a}}$ are a subring of $\mathbb{Z}^{\phi(N) \times \phi(N)}$, and arithmetic operations on elements in $\mathcal{R}$ correspond to arithmetic operations on the corresponding matrices. Likewise, the trace, norm and spectral norm of an element of $\mathcal{R}$ are that of the corresponding matrix in $\mathbb{Z}^{\phi(N) \times \phi(N)}$.

3   High Level Description of the Algorithm

Our scheme is derived from the Ducas-Micciancio FHEW scheme [8]. The encryption and decryption procedures are based on the standard LWE assumption. The bootstrapping procedure is performed via an accumulator derived from the GSW scheme [17]. Besides reducing the noise via a modulus switch, the bootstrapping procedure can also readjust the message modulus, which is an important component in the FHEW scheme.

3.1   LWE Symmetric Encryption

We keep the encryption-decryption procedure from [8] unchanged. It is a classic symmetric LWE encryption procedure already used in [3,4,21]. It depends on a dimension n, a message modulus $t \geq 2$, a ciphertext modulus $q \in n^{O(1)}$ and a randomized rounding function $\chi : \mathbb{R} \to \mathbb{Z}$. Let $\mathbf{s} \in \mathbb{Z}_q$ be a secret vector. The set of all LWE encryptions of the message $m \in \mathbb{Z}_t$ is given by

$$\mathrm{LWE}_{\mathbf{s}}^{t/q}(m) := \left\{ \left(\mathbf{a}, \chi \left(\mathbf{a} \cdot \mathbf{s} + \frac{mq}{t} \right) \bmod q \right) \right\} \subseteq \mathbb{Z}_q^{n+1}.$$

Correct decryption is ensured when the error distribution satisfies $|\chi(x) - x| < q/2t$. Indeed, in that case let $(\mathbf{a}, b) \in \mathrm{LWE}_{\mathbf{s}}^{t/q}(m)$, we have $|b - \mathbf{a} \cdot \mathbf{s}| = \chi(\mathbf{a} \cdot \mathbf{s} + mq/t) - \mathbf{a} \cdot \mathbf{s} = \chi(mq/t)$. Moreover, $|\chi(mq/t) - mq/t| \leq q/2t$, therefore $\left| \frac{t}{q} \chi(mq/t) - m \right| \leq 1/2$ and

$$m = \lfloor t(b - \mathbf{a} \cdot \mathbf{s})/q \rceil \bmod t.$$

Modulus switching. Changing the ciphertext modulus of an LWE encryption is a classic strategy to reduce the noise. It consists of applying on each coordinate of the ciphertext the randomized rounding function $[\cdot]_{Q:q} : \mathbb{Z}_Q \to \mathbb{Z}_q$ defined as $[x]_{Q:q} = \lfloor qx/Q \rfloor + B$ where $B \in \{0, 1\}$ is a Bernoulli variable with $P(B = 1) = (qx/Q) - \lfloor qx/Q \rfloor \in [0, 1)$. More specifically, the modulus switch procedure is define as

$$\mathrm{ModSwitch}((\mathbf{a}, b), Q : q) := [(\mathbf{a}, b)]_{Q:q} := \left(([a_1]_{Q:q}, \cdots, [a_n]_{Q:q}), [b]_{Q:q} \right).$$

Lemma 1 (Lemma 5 of [8]). For any $\mathbf{s} \in \mathbb{Z}_q^n$, $m \in \mathbb{Z}_t$ and ciphertext $c \in \mathrm{LWE}_{\mathbf{s}}^{t/Q}(m)$ with subgaussian error parameter σ, $c' := \mathrm{ModSwitch}(c, Q : q)$ is of the form $c' \in \mathrm{LWE}_{\mathbf{s}}^{t/q}(m)$ with subgaussian error parameter $\sqrt{(q\sigma/Q))^2 + 2\pi (\|s\|^2 + 1)}$.

Key switching. The bootstrapping procedure involves encryptions with a different key. To get back to an encryption of the message with the original key, one has to perform another classic operation called "Key switching". The technical challenge with key switching is that the new key has to come in an encrypted way (to ensure circular security). Assume the input is an LWE encryption under $\mathbf{z} \in \mathbb{Z}_q^n$. The ciphertext has the form $(\mathbf{a}, \mathbf{a} \cdot \mathbf{z} + mq/t + e)$, and we would like to replace it by one of the form $(\mathbf{a}', \mathbf{a}' \cdot \mathbf{s} + mq/t + e')$ where $\mathbf{s} \in \mathbb{Z}_q^{n'}$ for some integer n'. To do that, we construct $\mathbf{a} \cdot \mathbf{z} + e'$ from the

$$\mathbf{k}_{i,j,v} = \left(\mathbf{a}_{i,j,v}, \mathbf{a}_{i,j,v} \cdot \mathbf{s} + vz_i B^j + e_{i,j,v}\right) \in \mathrm{LWE}_{\mathbf{s}}^{q/q}(vz_i B^j),$$

where B is a base and $v \in \{0, \cdots, B-1\}$ are digits of a base-B number (in our case the digits of the coefficients of $\mathbf{a}$). More specifically, if the $(a_{i,j})_{j \leq \log_B(a_i)}$ are the digits of the coefficients of $\mathbf{a}$, and $\mathfrak{k} = \{\mathbf{k}_{i,j,v}\}$, then the key switching procedure is given by

$$\mathrm{KeySwitch}\,((\mathbf{a}, b), \mathfrak{k}) = (\mathbf{0}, b) - \sum_{i,j} \mathbf{k}_{i,j,a_{i,j}}.$$

Lemma 2 (Lemma 6 of [8]**).** Given a ciphertext $\mathbf{c} \in \mathrm{LWE}_{\mathbf{z}}^{t/q}(m)$ with subgaussian error of parameter α and the key $\left\{\mathbf{k}_{i,j,v} \in \mathrm{LWE}_{\mathbf{s}}^{q/q}(vz_i B)\right\}$ with subgaussian error of parameter σ, and $d = \log_B(q)$, then $\mathrm{KeySwitch}\,(\mathbf{c}, \{\mathbf{k}_{i,j,v}\})$ is in $\mathrm{LWE}_{\mathbf{s}}^{t/q}(m)$ with subgaussian error of parameter $\sqrt{\alpha^2 + nd\sigma^2}$.

3.2   The Bootstrapping Procedure

A tuple of encrypted messages (typically bits, but our scheme allows flexibility on that score) goes through a gate. During the process, the noise increases, and the message modulus may change (in general to contain the growth in noise). The bootstrapping procedure revert back these changes to the original message modulus and noise level. The bootstrapping procedure is divided into three steps. The first one is called the accumulator and is the most time consuming. It produces a new ciphertext with a prescribed message modulus, but under a different encryption key and a larger encryption modulus. The rest of the bootstrapping procedure consists of switching modulus and switching key to restore the original parameters.

Our main contribution is an improvement on the accumulator (denoted by ACC in Fig. 1). We make it more general (it can handle arbitrary numbers) and more efficient in a multicore environment. Since our accumulator handles general gates, we do not want to restrict ourselves to a particular one. However, we would like to illustrate why the message modulus can change after a homomorphic gate. For the sake of completeness, the homomorphic NAND gate was presented in [8]. The change of message modulus comes from the multiplication of two bits occurring in the NAND gate. We may change the modulus after any gate that involves multiplication between bits. Let us see how this occurs in the simplest one: the AND gate (which unlike the NAND gate, is not universal).

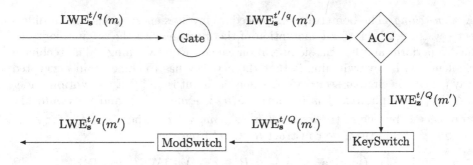

Fig. 1. High level description of the scheme

The multiplication between two bits $m_0, m_1 \in \{0, 1\}$ is linearized by noticing that $(m_0 - m_1)^2 = m_0 + m_1 - 2m_0 m_1$. Let us define the AND gate by

$$\text{AND} : ((\mathbf{a}_0, b_0), (\mathbf{a}_1, b_1)) \longmapsto (-\mathbf{a}_0 - \mathbf{a}_1, q/8 - b_0 - b_1),$$

where $(\mathbf{a}_0, b_0) \in \text{LWE}_{\mathbf{s}}^{4/q}(m_0, q/16)$ and $(\mathbf{a}_1, b_1) \in \text{LWE}_{\mathbf{s}}^{4/q}(m_1, q/16)$ (the second parameter denotes a bound on the error distribution). The resulting ciphertext $(\mathbf{a}_2, b_2)$ is in $\text{LWE}_{\mathbf{s}}^{2/q}(m_0 \cdot m_1, q/4)$. Indeed, we have

$$b_2 - \mathbf{a}_2 \cdot \mathbf{s} - \frac{q}{2} m_0 \cdot m_1 = q/8 - b_0 - b_1 + \mathbf{a}_0 \cdot \mathbf{s} + \mathbf{a}_1 \cdot \mathbf{s}$$

$$- \frac{q}{4}(m_0 + m_1 - (m_0 - m_1)^2)$$

$$= -e_0 - e_1 - \frac{q}{4}\left(\frac{1}{2} - (m_0 - m_1)^2\right) = \pm\frac{q}{8} - (e_0 + e_1).$$

Because of the original bound on the error distribution, we get $|e_2| \leq q/4$.

4   The Accumulator

In this section, we describe a generalization of the homomorphic accumulator scheme described in [8]. It is itself inspired from the work of Alperin-Sheriff and Peikert [2] which identifies the additive group $\mathbb{Z}_q$ with some symmetric group.

4.1   High Level Description of the Accumulator

Just like any bootstrapping procedure, the accumulator homomorphically evaluates a decryption circuit. In our case, the encrypted message has the form

$$\mathbf{c} = (\mathbf{a}, \underbrace{\mathbf{a} \cdot \mathbf{s} + mq/t' + e}_{b}) \in \text{LWE}_{\mathbf{s}}^{t'/q}(m).$$

The decryption circuit consists of extracting m from $b - \mathbf{a} \cdot \mathbf{s}$. When working directly on $\mathbf{c}$, this is just the standard LWE decryption scheme we described earlier. During the bootstrapping procedure, we need to perform that operation homomorphically on $E(b)$ where E is an outer encryption scheme that we describe in Sect. 4.2. Our outer encryption scheme is additive, which means that with $E(\mathbf{a} \cdot \mathbf{s})$, we easily get $E(b - \mathbf{a} \cdot \mathbf{s})$. In Sect. 4.2 we show that there is a $*$ operator satisfying $E(b_1) * E(b_2) = E(b_1 + b_2)$. To create $E(\mathbf{a} \cdot \mathbf{s})$ homomorphically, we use the same base-B_r decomposition technique as in the key switching algorithm (we kept the same notation as in [8], the r stands for refresh). For any digit $c \in \{0, \cdots, B_r - 1\}$, the refreshing key $\mathcal{K}$ is defined as the collection of the $K_{c,i,j} = E(cs_i B_r^j \bmod q)$. Then $E(\mathbf{a} \cdot \mathbf{s})$ is obtained from the base-B_r decomposition of the a_i.

Once $E(b - \mathbf{a} \cdot \mathbf{s}) = E(mq/t' + e)$ is obtained, we need to homomorphically decide which $m \in \mathbb{Z}_{t'}$ is encrypted. We show in Sect. 4.3 how to implement a homomorphic function $f_i(x)$ such that $f_i\left(E(mq/t' + e)\right) \in \mathrm{LWE}_\mathbf{z}^{t/Q}(\mathbb{1}_{m \in \{i\}}(m))$. From the collection of the f_i, we know how to reconstruct an LWE encryption of m just by using $f = \sum_i i \cdot f_i$, but this leaves us absolute freedom to create ciphertexts in $\mathrm{LWE}_\mathbf{z}^{t/Q}(f(m))$ for an arbitrary f. A typical example is the extraction of a particular bit. If the encryption of a word representing several bits is refreshed through the accumulator, we might want to extract LWE (refreshed) encryptions of each individual bits.

Algorithm 1. Homomorphic accumulator

Require: $(\mathbf{a}, b) \in \mathrm{LWE}_\mathbf{s}^{t'/q}(m)$, f, $\mathcal{K} = \{K_{c,i,j}\}$.
Ensure: $c \in \mathrm{LWE}_\mathbf{z}^{t/Q}(f(m))$.
 1: $\mathrm{ACC} \leftarrow E(b)$.
 2: **for** $i \leq n$ **do**
 3:     Compute the base-B_r representation of $-a_i = \sum_j B_r^j a_{i,j} \bmod q$.
 4:     For $j < \log_{B_r}(q)$, $\mathrm{ACC} \leftarrow \mathrm{ACC} * K_{a_{i,j}, i, j}$.
 5: **end for**
 6: $g \leftarrow \sum_{i < t} f(i) f_i$.
 7: **return** $g(\mathrm{ACC})$.

4.2   The Outer Encryption Scheme

Our outer encryption scheme $E(\cdot)$ mainly differs from that of [8] by the use of a different ring. Let $\mathcal{R}$ be the ring of integers of $K = \mathbb{Q}(\zeta_N)$ for N a power of t' where t' is a prime number corresponding to the message modulus of the input of the accumulator $(\mathbf{a}, b) \in \mathrm{LWE}_\mathbf{s}^{t'/q}(m)$. We assume that $t' \mid q$ and that $q \mid N$. Elements $b \in \mathbb{Z}_q$ are encoded as powers of a root of unity Y^b where $Y = X^{N/q}$. Our message space $\mathbb{Z}_q \simeq \langle Y \rangle$ is a subgroup of

$$\langle X \rangle = \left\{ 1, X, \cdots, X^{\frac{N}{t'}(t'-1)}, -\sum_{2 \leq i \leq t'} X^{\frac{N}{t'}(t'-i)}, \cdots, -\sum_{2 \leq i \leq t'} X^{\frac{N}{t'}(t'-i)+(t'-1)} \right\}.$$

The above generalizes the case $t' = 2$, where $\langle X \rangle = \{1, X, \cdots, X^{N/2-1}, -1, \cdots, -X^{N/2-1}\}$. Working in a cyclotomic field with a conductor that is the power of a prime greater than two allows us to define a test function that can test arbitrary subsets of $\mathbb{Z}_{t'}$, as we see in Sect. 4.3.

We hide Y^b by using an encryption scheme working with matrices with coefficients in $\mathcal{R}_{\mathcal{Q}} := \mathcal{R}/\mathcal{Q}\mathcal{R}$ where $\mathcal{Q}$ is a power of a prime h which is a primitive root modulo t'^2 (as in [8], this restriction is for the purpose of a rigorous proof of correctness; in practice it is possible to use a power of 2). To be able to revert back to an LWE encryption with message modulus t, we use an invertible element $u \in \mathbb{Z}_{\mathcal{Q}}$ that is close to $\mathcal{Q}/t't$. More precisely, since $\mathcal{Q}$ is a power of $h \neq t'$, either $\lfloor \mathcal{Q}/t't \rfloor$ or $\lceil \mathcal{Q}/t't \rceil$ is invertible (two consecutive numbers cannot be divisible by q), and $|u - \mathcal{Q}/t't| < 1$. The outer encryption scheme is also parametrized by a basis B_g such that $\mathcal{Q} = B_g^{d_g}$. The coefficients of elements in $\mathcal{R}_{\mathcal{Q}}$ are written in base B_g. We are now ready to define $E(x)$. We remind the reader that $\mathbf{a}$ denotes a column vector and $\mathbf{A}$ a matrix (with coefficients in $\mathcal{R}_{\mathcal{Q}}$).

Definition 2 (Outer Encryption Scheme). *Let $b \in \mathbb{Z}_q$ and a key $z \in \mathcal{R}$. The encryption of b is performed by choosing $\mathbf{a} \in \mathcal{R}_{\mathcal{Q}}^{2d_g}$ uniformly at random and $\mathbf{e} \in \mathcal{R}^{2d_g}$ according to a subgaussian distribution of parameter ς, and computing*

$$E_z(b) := [\mathbf{a}, \mathbf{a} \cdot z + \mathbf{e}] + uY^b \cdot \mathbf{G} \in \mathcal{R}_{\mathcal{Q}}^{2d_g \times 2},$$

where $\mathbf{G} = (\mathbf{I}, B_g\mathbf{I}, \cdots, B_g^{d_g-1}\mathbf{I}) \in \mathcal{R}_{\mathcal{Q}}^{2d_g \times 2}$.

Distribution of the error. To create an error vector $\mathbf{e}$ whose coefficients follow a spherical distribution in the complex domain under the Minkowski embedding

$$x \in \mathbb{Q}(\zeta_N) \longmapsto (\sigma_1(x), \cdots, \sigma_{\phi(N)}(x)),$$

(where the $\sigma_i : \mathbb{Q}(\zeta_N) \to \mathbb{C}$ are the complex embedding of our field), we cannot sample elements coefficient-wise in $\mathbb{Z}[X]/(\Phi_N(X))$ according to a spherical distribution as it can be done in the case where the conductor N is a power of 2. Indeed, the morphism sending the coordinates of a polynomial modulo $\Phi_N(X)$ to its image under the Minkowski embedding is not an isometry. The creation of an error with spherical distribution in a cyclotomic field with arbitrary conductor is dealt with in [7]. Let $N = p^k$ be the conductor we are using. The geometry of the complex-embedding domain relates to the geometry of polynomials modulo $X^N - 1$. By [7, Theorem 5], we know that the map sending $v \in \mathbb{Q}[X]/(X^N - 1)$ to the image of the corresponding elements $v \bmod \Phi_N(X)$ of $\mathbb{Q}(\zeta_N)$ under the Minkowski embedding preserves the sphericity of the distribution. Therefore, if the coefficients of v are distributed as ψ_s^N (the spherical distribution of variance s) in the power basis, then it is distributed as $\psi_{s\sqrt{N}}^{\phi(N)}$ under the Minkowski embedding. We therefore sample the error in $\mathbb{Q}[X]/(X^N - 1)$ according to a spherical distribution ψ_ς and then reduce modulo Φ_N, thus getting an error in $\mathbb{Q}(\zeta_N)$. As pointed out in [7], the resulting distribution is subgaussian with parameter $2p\varsigma$ (Instead of ς in [8]). We need to account for this increase of the subgaussian parameter of the error in our analysis.

Homomorphic additions. The outer encryption scheme has to be additively homomorphic to ensure that the bootstrapping procedure gives us a fresh encryption of the message. Additive operations on ciphertexts encrypted with $E_z(\cdot)$ are referred to as accumulator incrementation in [8]. Let $ACC = E_z(b_1)$ and $\mathbf{C} = E_z(b_2)$, then the homomorphic addition is done by first decomposing $u^{-1} ACC = \sum_{i \le d_g} B_g^{i-1} \mathbf{D}_i$ where each $\mathbf{D}_i \in \mathcal{R}^{2d_g \times 2}$ has entries with coefficients in $\left\{ \frac{1-B_g}{2}, \cdots, \frac{B_g-1}{2} \right\}$. We define the $*$ operator by

$$E_z(b_1) * E_z(b_2) := [\mathbf{D}_1, \cdots, \mathbf{D}_{d_g}] \cdot \mathbf{C}.$$

Proposition 1. *Let $b_1, b_2 \in \mathbb{Z}_q$, $E_z(b_1) = [\mathbf{a}, \mathbf{a} \cdot z + \mathbf{e}] + uY^{b_1} \cdot \mathbf{G}$, and $E_z(b_2) = [\mathbf{a}', \mathbf{a}' \cdot z + \mathbf{e}'] + uY^{b_2} \cdot \mathbf{G}$. Then $E_z(b_1) * E_z(b_2)$ has the form $[\mathbf{a}'', \mathbf{a}'' \cdot z + \mathbf{e}''] + uY^{b_1+b_2} \cdot \mathbf{G}$ for $\mathbf{e}'' = Y^{b_2} \cdot \mathbf{e} + [\mathbf{C}_1, \cdots, \mathbf{C}_{d_g}] \cdot \mathbf{e}'$, and is therefore an encryption of $b_1 + b_2$.*

4.3   The Test Functions

Our test functions are applied on $E_z(mq/t' + e)$ to decide (homomorphically) if m belongs to a given subset of $\mathbb{Z}_{t'}$. It is a generalization of the methods of [8] which allow us to compute an encryption of the most significant bit of $mq/t' + e$. In the case where $t' = 2$ and $m \in \{0, 1\}$, this gives an LWE encryption of m. The test function returning the most significant bit is a membership test. Indeed, the most significant bit is 0 if $mq/t' + e \in [0, q/2 - 1]$ and it is 1 if $mq/t' + e \in [q/2, q - 1]$. According to [8], this is likely to generalize to any anti-symmetric subsets of $\mathbb{Z}_q$. It is however harder to see how this can generalize to arbitrary subsets of $\mathbb{Z}_q$ when $\mathcal{R}$ is the ring of integers of cyclotomic fields whose conductor is a power of two.

We turn $ACC = E_z(mq/t' + e)$ into an LWE encryption of $f(m)$ by a linear algebra operation on $\overrightarrow{ACC} \in \mathbb{Z}^{Nd_g \times N}$. More specifically, let $[a, b'] \in \mathcal{R}^{1 \times 2}$ be the second row of $ACC \in \mathcal{R}^{2d_g \times 2}$. The corresponding vectors satisfy $\overrightarrow{b'} = \overrightarrow{a} \cdot \overrightarrow{z} + u\overrightarrow{Y^v} + \mathbf{e}$ where $v = mq/t' + e$. Let $\mathbf{t} \in \mathbb{Z}^N$ and $\mathbf{t}^T$ be the corresponding row vector. Then we have

$$\mathbf{t}^T \left[\overrightarrow{a}, \overrightarrow{b'} \right] = (\mathbf{a}, \mathbf{a} \cdot \overrightarrow{z} + \mathbf{t} \cdot \mathbf{e} + u \cdot \mathbf{t}^T \cdot \overrightarrow{Y^v}), \text{ where } \mathbf{a} = \mathbf{t}^T \cdot \overrightarrow{a}.$$

Depending on our choice of $\mathbf{t}$ (and on the size of u), the right hand side of the equation can be in $LWE_{\mathbf{z}}^{t/Q}(\mathbb{1}_{m \in \{i\}}(m))$. In the following, we give proofs for the simpler case $N = q$.

Proposition 2. *Let $u = \lfloor Q/t't \rfloor$ or $\lceil Q/t't \rceil$ (whichever is invertible modulo Q), $v \in \mathbb{Z}_q$ and $0 \le i \le t' - 2$. We assume $N = q$. We define $\mathbf{t}_i^T$ by*

$$\mathbf{t}_i^T = (\underbrace{0, \cdots, 0}_{iq/t'}, \underbrace{t', \cdots, t'}_{q/t'}, 0, \cdots, 0) - (1, \cdots, 1) \in \mathbb{Z}^{\phi(N)}.$$

Let $S_i = \left\{\frac{iq}{t'}, \cdots, \frac{(i+1)q}{t'} - 1\right\}$, *and* $\mathbf{t}_i^T\left[\overrightarrow{a}, \overrightarrow{b'}\right] = (\mathbf{a}, \mathbf{a} \cdot \overrightarrow{z} + \mathbf{t}_i \cdot \mathbf{e} + u \cdot \mathbf{t}_i^T \cdot \overrightarrow{Y^v}) = (\mathbf{a}, b_0)$, *then*

$$\mathbf{c} = (\mathbf{a}, b_0 + u) = (\mathbf{a}, \mathbf{a} \cdot \overrightarrow{z} + \mathbf{t} \cdot \mathbf{e} + tu \cdot \mathbb{1}_{v \in S_i}(v)) \in \mathrm{LWE}_{\mathbf{z}}^{t/Q}(\mathbb{1}_{v \in S_i}(v)).$$

Proof. Let $\overrightarrow{Y^v} = (v_0, \cdots, v_{k-1})$ where $k = \frac{q}{t'}(t' - 1)$. Then $\sum_{j \in S_i} v_j$ equals 1 if $v \in S_i$, 0 if $v \in S_j$ for $j \le t' - 2$ and -1 if $v \in S_{t'-1} = \{\phi(q), \cdots q - 1\}$. This information does not allow us to distinguish between S_i and $S_{t'-1}$. For that, we notice that $(\sum_j v_j) - 1$ equals 0 if $v \in S_j$ for $i \ne j \le t' - 2$ or $-t'$ if $v \in S_{t'-1} =$. Therefore

$$\mathbf{t}_i^T \cdot \overrightarrow{Y^v} + 1 = t' \cdot \sum_{j \in S_i} v_j - \left(\left(\sum_j v_j\right) - 1\right) = t' \cdot \mathbb{1}_{v \in S_i}(v).$$

The fact that we end up with an LWE encryption of $\mathbb{1}_{v \in S_i}(v)$ with message modulus t due to the fact that $u \approx Q/t't$.

The above allows us to decide homomorphically if $v \in S_i$ for $0 \le i \le t' - 2$. Deciding if $v \in S_{t'-1}$ must be done with a different test vector due to the special form of Y^v for $v \in S_{t'-1}$. We provide the following statement without a proof, since the method we use is similar.

Proposition 3. *Let* $u = \lfloor Q/t't \rfloor$ *or* $\lceil Q/t't \rceil$ *(whichever is invertible modulo Q) and* $v \in \mathbb{Z}_q$. *We assume* $N = q$. *We define* $\mathbf{t}_{t'-1}^T$ *by*

$$\mathbf{t}_{t'-1}^T = (-1, \cdots, -1).$$

Let $S_{t'-1} = \{\phi(q), \cdots q - 1\}$ *and* $\mathbf{t}_{t'-1}^T\left[\overrightarrow{a}, \overrightarrow{b'}\right] = (\mathbf{a}, \mathbf{a} \cdot \overrightarrow{z} + \mathbf{t}_{t'-1} \cdot \mathbf{e} + u \cdot \mathbf{t}_{t'-1}^T \cdot \overrightarrow{Y^v}) = (\mathbf{a}, b_0)$, *then*

$$\mathbf{c} = (\mathbf{a}, b_0 + u) = (\mathbf{a}, \mathbf{a} \cdot \overrightarrow{z} + \mathbf{t} \cdot \mathbf{e} + t'u \cdot \mathbb{1}_{v \in S_{t'-1}}(v)) \in \mathrm{LWE}_{\mathbf{z}}^{t/Q}(\mathbb{1}_{v \in S_{t'-1}}(v)).$$

LWE encryptions of functions of the form $\mathbb{1}_{v \in S_i}$ readily give us the encryption of $\mathbb{1}_{m \in \{i\}}$ when $v = mq/t' + e$ for $|e| < q/2t'$ since in this case $v \in S_i \Leftrightarrow m = i$. The last parameter we need to estimate is the distribution of the error in $\mathrm{LWE}_{\mathbf{z}}^{t/Q}(\mathbb{1}_{m \in \{i\}}(m))$. We follow the approach of [8] which consists of modeling the $\mathbf{D}^{(i)}$ as independent and uniformly distributed. The only difference with [8, Lemma 11] is that our test vector has some entries greater than 1.

Lemma 3. *Let* l *be the number of incrementation of the accumulator were* $l \ge \omega(\sqrt{\log(n)})$ *leading to* $E_z(v)$, *then the ciphertext* $\mathbf{c} \in \mathrm{LWE}_{\mathbf{z}}^{t/Q}(\mathbb{1}_{v \in S_i}(v))$ *computed as described in Propositions 2 and 3 has a subgaussian error distribution of parameter* β *where* $\beta \in O\left(\varsigma p^2 B_g t' \sqrt{qNd_g l}\right)$.

The proof of this statement is essentially similar to that of Lemma 11 of [8]. The main difference is that the size of the coefficients involved in matrix multiplications are slightly worse in our case. The test vector used to produce the encryption of $\mathbb{1}_{v \in S_i}(v)$ has some coefficients equal to $t' - 1$. When evaluating an arbitrary function of the form $f = \sum_i f(i)\mathbb{1}_{v \in S_i}$, one has to take into account the Euclidean norm of the corresponding test vector in the estimation of the error distribution. Moreover, we need to adapt [8, Fact 1] and [8, Fact 12] to account for the growth in the subgaussian parameter (by a factor $O(p)$ occurring when going from $\mathbb{Q}[X]/(X^N - 1)$ to $\mathbb{Z}[X]/(\Phi_N(X))$ by reduction modulo $\Phi_N(X)$). After the accumulator step, the bootstrapping procedure modifies the error distribution when applying the key switching and modulus switching procedures. The following corresponds to [8, Theorem 10], and the main difference is that t' occurs in the error parameter.

Proposition 4. *After the bootstrapping procedure, we obtain* $c \in \text{LWE}_s^{t/q}$ *$(\mathbb{1}_{m \in \{i\}}(m))$ where the error parameter is*

$$\sqrt{\frac{q^2}{Q^2}\left(\varsigma^2 p^4 B_g^2 t'^2 \cdot l \cdot q \cdot N d_g + \sigma^2 N d_{ks}\right) + \|s\|^2 \cdot \omega\left(\sqrt{\log(n)}\right)}.$$

5   Numerical Results

In this section, we illustrate the performances our the modified FHEW scheme. Better timings are obtained by allowing more general gates (and thus diminishing the number of calls to the bootstrapping procedure), and by computing in parallel. As pointed out before, we only seek to evaluate the impact of parallelism at the gate level. Another layer of parallelism can be achieved by using optimized compilation techniques that gather independent gate evaluations. These techniques are essential too, but they are completely orthogonal to the work that is presented in this paper. Our timings were obtained on a node with a 12 core AMD Opteron 1.8 GHz and 256 GB of memory. We modified the C++ code [9] that was used for the experiments of the original paper of Ducas and Micciancio [8]. This has the obvious advantage of minimizing the amount of extra work to implement the generalization of the techniques of [8]. It also allows a better comparison between the two methods by minimizing the changes. Indeed, it is harder to isolate the causes for a difference in terms of performances between the original method and a new implementation from scratch.

5.1   Arithmetic Operations in $\mathbb{Q}(\zeta_{p^k})$

Homomorphic operations with the outer encryption scheme are carried on elements in $\mathbb{Q}(\zeta_{p^k})$. We use the correspondence between $\mathbb{Z}[X]/(\Phi_N(X))$ and $\mathbb{C}[X]/(X^N - 1)$ and the fact that we can identify a polynomial in $\mathbb{C}[X]/(X^N - 1)$ by its values at the ζ_N^i for all $0 \leq i < N$. We in fact only need to keep the values of our polynomials at the primitive roots of unity, but it would have required a significant extra effort in terms of implementation. Instead, we use the standard FFT tools publicly available [10]. We carry all operations between elements in

$\mathbb{Q}(\zeta_{p^k})$ in FFT format, and once we apply the FFT backward, we get a polynomial in $\mathbb{Z}[X]/(X^N - 1)$ that we further reduce modulo $\Phi_N(X)$. To optimize even further this extra reduction step, we compute the reductions of $X^i \mathrm{mod} \Phi_N(X)$ for $i \geq \phi(N)$. These are given by the relation $X^{p^{k-1}(p-1)} = -\sum_j X^{p^{k-1}(p-j)}$. According to our experiments, the extra reduction step induces an overhead of less than 5 % when working with elements of $\mathbb{Q}(\zeta_{5^4})$. Note that the FFT is also slowed down by the fact that we work with $p > 2$. This induces a slow down compared to the techniques of [8] that exploit the optimizations provided by [10] for the specific case $p = 2$ (we provide a comparison in Sect. 5.3). We have not investigated alternative methods to speed-up arithmetic in $\mathbb{Q}(\zeta_{p^k})$ such as using a different polynomial multiplication algorithm (such as Toom-Cook for example). The impact of a different polynomial multiplication algorithm would have to be put into perspective with its effect on the original FHEW scheme working in $\mathbb{Q}(\zeta_{2^k})$. It would be interesting, but it would require a careful assessment.

5.2   Parallel Computations

A profiling of the execution of the code indicates that 99 % of the time is spent in the accumulator. This is not a surprise as bootstrapping is known as the most expensive step in homomorphic operations. Any optimization on the bootstrapping procedure automatically induces a comparable speedup on the overall time. The accumulator is a series of operations that could be executed in parallel. We used the openMP tool to run independent threads sharing memory. Our early experiments indicated that using the built-in functions to parallelize the `for` loops involved was completely counter-productive due to the overhead in thread management. Indeed, the individual cost of each elementary step is too low. Therefore, the code had to be reorganized, and it occurred that the separation of the accumulator into independent rows was the optimal solution. Memory is allocated to each row, and they are treated independently by each thread. The speedup that we can obtain this way is limited by the parameters of the accumulator. We used the parameters of [8, Sect. 6.2]

LWE parameters:	$n = 500, Q = 2^{32}, \sigma = 2^{17}, q = 2^9$.
Outer scheme parameters:	$\phi(N) = 2^{10}, \varsigma = 1.4$.
Gadget matrix $\mathbf{G}$:	$B_g = 2^{11}, d_g = 3, u = \frac{Q}{8} + 1$.
bootstrapping key parameters:	$B_r = 23, d_r = 2$.
Key switching parameters:	$B_{ks} = 24, d_{ks} = 7$.

In this case, we deal with an accumulator that has $2d_g = 6$ rows. Table 1 summarizes the effect on the use of multiple cores for the homomorphic evaluation of one NAND gate with these parameters. We took the average over 100 evaluations, and we obtained a speedup factor of almost 4 when using 6 cores. Because of the dimensions of the accumulator with these parameters, the speedup decreases when we add more cores. Note that it seems that the architecture we used for the timings is slower than what was used in [8] on which bootstrapping took less of a second. We assume that the speedup factor we achieved would have been the same on a faster machine.

Table 1. Influence of parallel computation on the NAND gate

Number of cores	1	2	3	4	5	6
CPU time (sec)	1.5426	0.8848	0.6845	0.6672	0.5853	0.4142
Speedup	1.00	1.74	2.25	2.31	2.64	3.72

5.3   Benchmarks with the Full Adder

The other improvement we made on the accumulator scheme is that it is more general than the one presented in [8]. This allows to reduce the number of gates in a circuit, thus making its homomorphic evaluation more efficient (since it reduces the number of call to the refresh function). As we are trying to be as general as possible, choosing a particular gate to illustrate the impact of our scheme is a little delicate. In particular, it is probably possible to design a gate whose implementation with NAND gates is unusually complicated, in which case the speedup obtained by using a dedicated gate would be quite substantial. We believe that such an illustration would not make the best case for our scheme. The "add-with-carry" gate was mentioned as an example for potential generalizations in [8, Sect. 7]. It is unclear how to implement it with a message modulus $t = 6$ and an outer encryption scheme working with $\mathcal{R} = \mathbb{Z}[X]/(X^{2^k} + 1)$ as suggested in [8], but our scheme supports it with a ring of the form $\mathcal{R} = \mathbb{Z}[X]/(\Phi_{5^k}(X))$. Moreover, this gate is undeniably useful for homomorphic computations.

The full adder takes bits b_1, b_2 and a carry c as input and returns a bit b_3 and carry c' corresponding to the addition of b_1, b_2 and c. To implement it, we choose $t = 5$ and we simply add the LWE encryptions of b_1, b_2 and c, thus getting an encryption of $m := b_1 + b_2 + c \in \{0, 1, 2, 3\}$ (note that we do not use all the available message space). Then we refresh the encryption of m through the accumulator and apply two membership test functions: $\mathbb{1}_{m \in \{2,3\}}(m) = \mathbb{1}_{m \in \{2\}}(m) + \mathbb{1}_{m \in \{3\}}(m)$ which gives us the most significant bit of m (and thus c'), and $\mathbb{1}_{m \in \{1,3\}}(m) = \mathbb{1}_{m \in \{1\}}(m) + \mathbb{1}_{m \in \{3\}}(m)$ which gives us the least significant bit of m (and thus b_3). The procedure describing our full adder gate is given by Algorithm 2. We compare our dedicated full adder gate to a full adder implemented with homomorphic binary gates using the FHEW scheme of Ducas and Micciancio [8]. In [8], only the NAND gate (which is universal) was described. A full adder takes 9 NAND gates, as shown in Fig. 2. Limiting ourselves to NAND gates does not give a fair comparison with the FHEW scheme presented in [8] since it is straightforward to deduce homomorphic AND, OR and NOR gates at a comparable cost. Moreover, the NOT gate comes for free. We compare our full adder gate to an implementation using 7 NOR gates and 5 NOT gates presented in Fig. 3.

Table 2 summarizes the numerical results obtained with the full adder gate. A fair comparison between the two methods is made difficult by the fact that augmenting the message modulus decreases the granularity in terms of security level. The higher the security level, the worse are the performances, and at the same time, when the security level is given in powers of 2, it is hard to

Algorithm 2. Homomorphic full adder gate

Require: $\mathrm{LWE}_s^{t/q}(b_1)$, $\mathrm{LWE}_s^{t/q}(b_2)$, $\mathrm{LWE}_s^{t/q}(c)$ with error $E < q/6t$.
Ensure: $\mathrm{LWE}_s^{t/q}(b_3)$, $\mathrm{LWE}_s^{t/q}(c')$, digit and carry of $b_1 + b_2 + c$.
1: $(\mathbf{a}, b) \leftarrow \mathrm{LWE}_s^{t/q}(b_1) + \mathrm{LWE}_s^{t/q}(b_2) + \mathrm{LWE}_s^{t/q}(c)$.
2: Call Algorithm 1 on $(\mathbf{a}, b) = \mathrm{LWE}_s^{t/q}(m)$
3: $\mathrm{LWE}_z^{t/Q}(c) \leftarrow \mathrm{LWE}_z^{t/Q}(\mathbb{1}_{m \in \{2,3\}}(m))$.
4: $\mathrm{LWE}_z^{t/Q}(b_3) \leftarrow \mathrm{LWE}_z^{t/Q}(\mathbb{1}_{m \in \{1,3\}}(m))$.
5: Switch key and modulus on $\mathrm{LWE}_z^{t/Q}(b_3), \mathrm{LWE}_z^{t/Q}(c)$.
6: **return** $\mathrm{LWE}_s^{t/q}(b_3), \mathrm{LWE}_s^{t/q}(c)$.

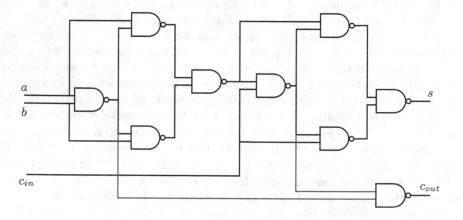

Fig. 2. Full adder with 9 NAND gates

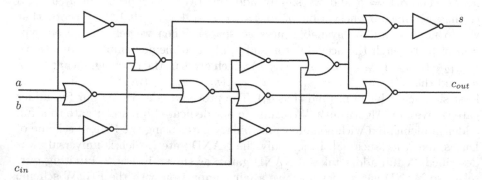

Fig. 3. Full adder with 7 NOR gates and 5 NOT gates

make it coincide with a power of 5. The timings we quote for our full adder gate are obtained with a ciphertext modulus $q = 5^4 = 625$ and a degree for the FFT operations of $N = q^5 = 3125$. This corresponds to a higher security level (for both the LWE scheme and the GSW-based outer scheme) than the NAND gates of [8] which were used with a ciphertext modulus $q = 2^9 = 512$ and a ring degree degree (for the outer encryption scheme) of $\phi(2^{11}) = 2^{10} = 1024$.

To provide the most accurate comparison, we modified the parameters from [8] to work with a ring of degree $\phi(2^{12}) = 2048$, while our schemes runs with a ring of degree $\phi(5^5) = 2500$. This is still a conservative choice since our scheme still offers a higher level of security with these parameters. Note that our definition of N corresponds to $2N$ in [8] where the relation between 2^k and $\phi(2^k)$ is a simple division by 2.

	NOR parameters	Full adder gate parameters
LWE	$n = 500, Q = 2^{32}, \sigma = 2^{17}, q = 2^9.$	$n = 500, Q = 2^{32}, \sigma = 2^{17}, q = 5^4$
Outer scheme	$\phi(N) = 2048, \varsigma = 1.27.$	$\phi(N) = 2500, \varsigma = 1.4.$
Gadget matrix	$B_g = 2^{11}, d_g = 3, u = \frac{Q}{8} + 1.$	$B_g = 2^{11}, d_g = 3, u = \left\lceil \frac{Q}{25} \right\rceil.$
bootstrapping key	$B_r = 23, d_r = 2.$	$B_r = 23, d_r = 2.$
Key switching	$B_{ks} = 24, d_{ks} = 7.$	$B_{ks} = 24, d_{ks} = 7.$

We follow the same approach as in [8, Sect. 6.1] for the security analysis for this set of parameters. We use the methodology following from the analysis of [20]. An attack creating an ϵ-distinguisher for LWE with dimension n, message modulus q, and error with standard deviation σ must reach a root Hermite factor of

$$\delta = \delta\text{-}\mathrm{LWE}(n, q, \sigma, \epsilon) = 2^{(\log_2^2(\rho))/(4n\log_2(q))} \text{ where } \rho = (q/\sigma)\sqrt{2\log(1/\epsilon)}.$$

This quantifies the security of the LWE ciphertexts coming out of the accumulator. After the key switching procedure, we have a binary secret. As pointed out in [8], we have to account for the possibility of an attack consisting of changing the message modulus to decrease the error. Taking the maximum over all the possible δ-LWE, the attack should reach a root Hermite factor of

$$\delta\text{-}\mathrm{binLWE}(n, q, \sigma, \epsilon) = \max_{q' \leq q} \left\{ \delta\text{-}\mathrm{LWE}\left(n, q', \sigma' = \sqrt{(q'/q)^2\sigma^2 + n/24}, \epsilon\right) \right\}.$$

	NOR parameters	Full adder gate parameters
$\delta\text{-}\mathrm{LWE}(\phi(N), Q, \varsigma, 2^{-64})$	1.0032	1.0026
$\delta\text{-}\mathrm{binLWE}(n, Q, \sigma, 2^{-64})$	1.0064	1.0064

The security of the LWE ciphertexts with binary secret key is the same with both sets of parameters and equal to the choices made in [8]. As expected, we have increased the level of security of the LWE ciphertext coming out of the accumulator with respect to what it was in [8] (for the sake of a more accurate comparison between the two schemes). Also, the security of our accumulator is higher than that of the homomorphic NOR gates, thus confirming that our choice of parameters for the comparison was conservative.

Table 2 presents the comparative timings between our full adder gate and an implementation with NOR and NOT gates using the original FHEW scheme (and its parallel variants). Our full adder gate is approximately 1.8 times slower than a single NOR (or NAND) gate. Therefore, on a single core, the homomorphic evaluation of the full adder procedure is almost 4 times faster than with NOR gates. Combined with the effect of parallelization, our dedicated full adder gate is 14 times faster on 6 cores than the same circuit evaluated with NOR gates on a single core as in [8].

Table 2. Benchmark on the homomorphic full adder

Number of cores	1	2	3	4	5	6
Full adder with 7 NOR gates	25.003	13.528	10.309	10.247	10.061	6.629
Full adder gate	6.429	3.633	2.728	2.699	2.430	1.754
Speedup	3.89	3.72	3.78	3.79	4.14	3.79

6   Conclusion

We described a generalization of the FHEW efficient bootstrapping procedure described by Ducas and Micciancio [8]. Our method allows the treatment of arbitrary gates, and we provided an efficient implementation that takes advantage of a multicore environment at the gate level (circuits can be further parallelized by efficient compiling methods that we did not describe). We provided a benchmark on the full adder gate, whose practical interest is undeniable. The combined effect of using a dedicated gate and exploiting a multicore environment resulting in a speedup of a factor 14 (our experiments could not compare timings on the exact same security level, therefore we made the conservative choice of quoting our full adder gate with a higher security level). This is a significant step towards efficient fully homomorphic encryption with practical applications. Further work needs to be done to assess which (useful) gates benefit from our methods. In particular, since we provide a very general bootstrapping technique, it would be interesting to see how our scheme performs on a homomorphic implementation of the AES.

Acknowledgments. We thank Leo Ducas for taking the time to explain us the technical details of [8] and of the corresponding open source implementation [9]. We also thank the reviewers for their many helpful comments, especially about the error distribution in arbitrary cyclotomic fields.

References

1. Alperin-Sheriff, J., Peikert, C.: Practical bootstrapping in quasilinear time. In: Canetti, R., Garay, J.A. (eds.) CRYPTO 2013, Part I. LNCS, vol. 8042, pp. 1–20. Springer, Heidelberg (2013)
2. Alperin-Sheriff, J., Peikert, C.: Faster bootstrapping with polynomial error. In: Garay, J.A., Gennaro, R. (eds.) CRYPTO 2014, Part I. LNCS, vol. 8616, pp. 297–314. Springer, Heidelberg (2014)
3. Applebaum, B., Cash, D., Peikert, C., Sahai, A.: Fast cryptographic primitives and circular-secure encryption based on hard learning problems. In: Halevi, S. (ed.) CRYPTO 2009. LNCS, vol. 5677, pp. 595–618. Springer, Heidelberg (2009)
4. Blum, A., Furst, M.L., Kearns, M., Lipton, R.J.: Cryptographic primitives based on hard learning problems. In: Stinson, D.R. (ed.) CRYPTO 1993. LNCS, vol. 773, pp. 278–291. Springer, Heidelberg (1994)

5. Brakerski, Z., Gentry, C., Halevi, S.: Packed ciphertexts in LWE-based homomorphic encryption. In: Kurosawa, K., Hanaoka, G. (eds.) PKC 2013. LNCS, vol. 7778, pp. 1–13. Springer, Heidelberg (2013)

6. Brakerski, Z., Langlois, A., Peikert, C., Regev, O., Stehlé, D.: Classical hardness of learning with errors. In: Boneh, D., Roughgarden, T., Feigenbaum, J. (eds.) Symposium on Theory of Computing Conference, STOC 2013, Palo Alto, CA, USA, June 1–4, pp. 575–584. ACM (2013)

7. Ducas, L., Durmus, A.: Ring-LWE in polynomial rings. In: Fischlin, M., Buchmann, J., Manulis, M. (eds.) PKC 2012. LNCS, vol. 7293, pp. 34–51. Springer, Heidelberg (2012)

8. Ducas, L., Micciancio, D.: FHEW: Bootstrapping homomorphic encryption in less than a second. Cryptology ePrint Archive, Report 2014/816 (2014). http://eprint.iacr.org/

9. Ducas, L., Micciancio, D.: Implementation of FHEW (2014). https://github.com/lducas/FHEW

10. Frigo, M., Johnson, S.: The design and implementation of FFTW3. In: Proceedings of the IEEE, 93(2):216–231. Special issue on "Program Generation, Optimization, and Platform Adaptation" (2005)

11. Gentry, C.: Fully homomorphic encryption using ideal lattices. In: Proceedings of the Forty-first Annual ACM Symposium on Theory of Computing, STOC 2009, pp. 169–178. ACM, New York (2009)

12. Gentry, C., Halevi, S.: Implementing gentry's fully-homomorphic encryption scheme. In: Paterson, K.G. (ed.) EUROCRYPT 2011. LNCS, vol. 6632, pp. 129–148. Springer, Heidelberg (2011)

13. Gentry, C., Halevi, S., Peikert, C., Smart, N.P.: Ring switching in BGV-style homomorphic encryption. In: Visconti, I., De Prisco, R. (eds.) SCN 2012. LNCS, vol. 7485, pp. 19–37. Springer, Heidelberg (2012)

14. Gentry, C., Halevi, S., Smart, N.P.: Better bootstrapping in fully homomorphic encryption. In: Fischlin, M., Buchmann, J., Manulis, M. (eds.) PKC 2012. LNCS, vol. 7293, pp. 1–16. Springer, Heidelberg (2012)

15. Gentry, C., Halevi, S., Smart, N.P.: Fully homomorphic encryption with polylog overhead. In: Pointcheval, D., Johansson, T. (eds.) EUROCRYPT 2012. LNCS, vol. 7237, pp. 465–482. Springer, Heidelberg (2012)

16. Gentry, C., Halevi, S., Smart, N.P.: Homomorphic evaluation of the AES circuit. In: Safavi-Naini, R., Canetti, R. (eds.) CRYPTO 2012. LNCS, vol. 7417, pp. 850–867. Springer, Heidelberg (2012)

17. Gentry, C., Sahai, A., Waters, B.: Homomorphic encryption from learning with errors: conceptually-simpler, asymptotically-faster, attribute-based. In: Canetti, R., Garay, J.A. (eds.) CRYPTO 2013, Part I. LNCS, vol. 8042, pp. 75–92. Springer, Heidelberg (2013)

18. Halevi, S., Shoup, V.: Algorithms in HElib. In: Garay, J.A., Gennaro, R. (eds.) CRYPTO 2014, Part I. LNCS, vol. 8616, pp. 554–571. Springer, Heidelberg (2014)

19. Halevi, S., Shoup, V.: Algorithms in HElib. IACR Cryptology ePrint Archive 2014:106 (2014)

20. Lindner, R., Peikert, C.: Better key sizes (and Attacks) for LWE-based encryption. In: Kiayias, A. (ed.) CT-RSA 2011. LNCS, vol. 6558, pp. 319–339. Springer, Heidelberg (2011)

21. Regev, O.: On lattices, learning with errors, random linear codes, and cryptography. In: Gabow, H., Fagin, R. (eds.) Proceedings of the 37th Annual ACM Symposium on Theory of Computing, Baltimore, MD, USA, May 22–24, pp. 84–93. ACM (2005)

Symmetric Key Cryptanalysis

Improved Top-Down Techniques in Differential Cryptanalysis

Itai Dinur[1]([✉]), Orr Dunkelman[2,3], Masha Gutman[3], and Adi Shamir[3]

[1] Département d'Informatique, École Normale Supérieure, Paris, France
itai.dinur@ens.fr
[2] Computer Science Department, University of Haifa, Haifa, Israel
[3] Computer Science Department, The Weizmann Institute, Rehovot, Israel

Abstract. The fundamental problem of differential cryptanalysis is to find the highest entries in the Difference Distribution Table (DDT) of a given mapping F over n-bit values, and in particular to find the highest diagonal entries which correspond to the best iterative characteristics of F. The standard bottom-up approach to this problem is to consider all the internal components of the mapping along some differential characteristic, and to multiply their transition probabilities. However, this can provide seriously distorted estimates since the various events can be dependent, and there can be a huge number of low probability characteristics contributing to the same high probability entry. In this paper we use a top-down approach which considers the given mapping as a black box, and uses only its input/output relations in order to obtain direct experimental estimates for its DDT entries which are likely to be much more accurate. In particular, we describe three new techniques which reduce the time complexity of three crucial aspects of this problem: Finding the exact values of all the diagonal entries in the DDT for small values of n, approximating all the diagonal entries which correspond to low Hamming weight differences for large values of n, and finding an accurate approximation for any DDT entry whose large value is obtained from many small contributions. To demonstrate the potential contribution of our new techniques, we apply them to the SIMON family of block ciphers, show experimentally that most of the previously published bottom-up estimates of the probabilities of various differentials are off by a significant factor, and describe new differential properties which can cover more rounds with roughly the same probability for several of its members.

Keywords: Differential cryptanalysis · Difference distribution tables · Iterative characteristics · SIMON

1 Introduction

Differential cryptanalysis, which was first proposed in [6], is one of the best known and most widely used tools for breaking the security of many types

O. Dunkelman—The second author was supported in part by the Israel Science Foundation through grants No. 827/12 and No. 1910/12.

© Springer International Publishing Switzerland 2015
K. Lauter and F. Rodríguez-Henríquez (Eds.): LatinCrypt 2015, LNCS 9230, pp. 139–156, 2015.
DOI: 10.1007/978-3-319-22174-8_8

of cryptographic schemes (including block ciphers, stream ciphers, keyed and unkeyed hash functions, etc.). Its main component is a *Difference Distribution Table* (abbreviated as DDT) which describes how many times each input difference is mapped to each output difference by a given mapping F over n-bit values. The DDT table has exponential size (with 2^n rows and 2^n columns), but we are usually interested only in its large entries: When we try to attack an existing scheme we try to find the largest DDT entry, and when we develop a new cryptographic scheme we try to demonstrate that all the DDT entries are smaller than some bound.

For large value of n such as 128, it is impractical to find the exact value of even a single entry in the table, but in most cases we are only interested in finding a sufficiently good approximation of its large values. There are many proposed algorithms for computing such approximations, but almost all of them are bottom-up techniques which start by analyzing the differential properties of small components such as a single S-box, and then combine them into large components such as a reduced-round version of the full scheme. To find the best differential attack, they use the detailed description of the scheme in order to identify a consistent collection of high probability differential properties of all the small components, and then multiply all these probabilities. In order to claim that there are no high probability differentials, they lower bound the number of multiplied probabilities, e.g., by showing that any differential characteristic has a large number of active S-boxes.

A second problem is that in most cases, this bottom-up approach concentrates on a single differential characteristic and describes one particular way in which the given input difference can give rise to the given output difference by specifying all the intermediate differences. Moreover, this approach is also more susceptible to deviations from the Markov cipher model [16], where dependence between different rounds can lead to an estimation of probability which is far from the correct value.

In this paper we follow a different top-down approach, in which we consider the given mapping as a black box and ignore its internal structure. In particular, we do not multiply or add a large number of probabilities associated with its smallest components, and thus we do not suffer from the three methodological problems listed above. Our goal is to use the smallest possible number of evaluations of the given mapping in order to compute either the precise value or a sufficiently good approximation of the most interesting entries in its DDT.

A straightforward black box algorithm can calculate the exact value of any particular entry in the DDT table in 2^n time by evaluating the mapping for all the pairs of inputs with the desired input difference, and counting how many times we got the desired output difference. When we want to compute a set of k entries in the DDT, we can always repeat the computation for each entry separately and thus get a $k2^n$ upper bound on the time complexity. However, for some large sets of entries we can do much better. In particular, we can compute all the $k = 2^n$ entries in a single row (which corresponds to a fixed input difference and arbitrary output differences) with the same 2^n time complexity by using the same algorithm. This also implies that the whole DDT can be computed in 2^{2n} time, whereas a naive algorithm which computes each one of the 2^{2n} entries separately would require 2^{3n} time. If the mapping is a permutation and we are also given its inverse as a black box, we can similarly compute each column in the DDT (which corresponds to a fixed output difference and arbitrary input differences) in 2^n time by applying the inverse black box to all the pairs with the desired output difference.

Which other sets of entries in the DDT can be simultaneously computed faster than via the naive algorithm? The first result we show in this paper is a new technique called the *diagonal algorithm*, which can calculate the exact values of all the 2^n diagonal entries in the DDT (whose input and output differences are equal) with a total time complexity of about 2^n. These entries in the DDT are particularly interesting in differential cryptanalysis, since they describe the probabilities of all the possible iterative characteristics which can be concatenated to themselves an arbitrarily large number of times in a consistent way. For many well known cryptosystems (such as DES), the best known differential attack on the scheme is based on such iterative characteristics. We then extend the diagonal algorithm to generalized diagonals which are defined as sets of 2^n DDT entries in which the input difference and output difference are linearly related (with respect to the same linear operation used to calculate differences in the DDT) rather than equal. This can be particularly useful in schemes such as Feistel structures, in which we are often interested in output differences which are equal to the input differences but with swapped halves.

In many applications of differential cryptanalysis, we can argue that only rows in the DDT which correspond to input differences with low Hamming weight can contain large values (and thus lead to efficient attacks). Our next result is a new top-down algorithm which we call the *Hamming Ball algorithm*, which can efficiently identify all the large diagonal entries in the DDT whose input and output differences have a low Hamming weight, and approximate their values.

Our third result is a new *bins-in-the-middle(BITM) algorithm* for computing in a more efficient way an improved approximation for any particular DDT entry whose high value may be accumulated from a large number of differential characteristics which have much smaller probabilities. In this algorithm we assume that the given mapping is only quasi black box in the sense that it is the concatenation of two black boxes which can be computed separately. A typical example of such a situation is a cryptographic scheme which consists of many rounds, where we can choose in our analysis how many rounds we want to evaluate in the first black box, and then define the remaining rounds as the second black box.

In our complexity analysis, we assume that most of the DDT entries are distributed as if the mapping is randomly chosen, but a small number of entries have unusually large values which we would like to locate and to estimate by evaluating the mapping on the smallest possible number of inputs. This is analogous to classical models of random graphs in which we try to identify some planted structure such as a large clique which was artificially added to the random graph.

The algorithms we develop analyze keyless functions, but can be easily extended to keyed functions (such as block ciphers) assuming the existence of high probability DDT entries which are common to a large fraction of the keys. Such entries are typically the result of high probability differential characteristics, and can be estimated by independently running our algorithms for a few keys selected at random. Then, we look for DDT entries of high probability which are common to several keys.

To demonstrate the power of our new techniques, we used the relatively new but extensively studied proposal of the SIMON family of lightweight block ciphers, which was developed by a team of experienced cryptographers from the NSA. Several previous papers [1,2,9,24] tried to find the best possible differential properties of reduced-round variants of SIMON with the bottom-up approach by analyzing its individual components. By using our new top-down techniques, we can provide strong experimental evidence that the previous probability estimates were inaccurate, and in fact we found new differential properties which are either longer by two rounds or have better probabilities for the same number of rounds compared to all the previously published results.

The paper is organized as follows. After introducing our notation in Sect. 2, we survey in Sect. 3 the main bottom-up techniques for estimating differential probabilities which were proposed in the literature. Our three new top-down techniques are described in Sects. 4, 5 and 6. We describe the application of our new techniques to the SIMON family of block ciphers in Sect. 7. In the full version of this paper [13] we show how to use these top-down techniques in order to analyze the differential properties of the Even-Mansour scheme (whose random permutation is only given in the form of a black box), and to find the first generic attack on its 4-round 1-key version in the related key setting.

2   Notations

In this section, we describe the notations used in the rest of this paper.

Given a function $F : \mathrm{GF}(2)^n \to \mathrm{GF}(2)^n$, the *difference distribution table* (DDT) is a $2^n \times 2^n$ table, where $DDT[\Delta_I][\Delta_O]$ counts the number of input pairs to F with an n-bit difference of Δ_I whose n-bit output difference is Δ_O. More formally we define $DDT[\Delta_I, \Delta_O] \triangleq |\{x \in \mathrm{GF}(2)^n : F(x) \oplus F(x \oplus \Delta_I) = \Delta_O\}|$.

We define the *diagonal* $(DIAG)$ of the DDT as a vector of length 2^n which contains only the $[\Delta_I, \Delta_O]$ entries for which $\Delta_O = \Delta_I$, namely $DIAG[\Delta] \triangleq DDT[\Delta, \Delta]$. Given an auxiliary function $L : \mathrm{GF}(2)^n \to \mathrm{GF}(2)^n$, we define the *generalized diagonal* $(GDIAG)$ of the DDT as a table of size 2^n, which contains only the $[\Delta_I, \Delta_O]$ entries for which $\Delta_O = L(\Delta_I)$, namely $GDIAG_L[\Delta] \triangleq DDT[\Delta, L(\Delta)]$. Thus, the diagonal is a particular case of the generalized diagonal for which the auxiliary function L is the identity. In this paper, we are mostly interested in generalized diagonals for linear functions L (over $\mathrm{GF}(2)^n$), which can be computed efficiently using our algorithms.

Given an n-bit word x, we denote by $ham(x)$ its Hamming weight. Given two n-bit words x, y, we denote by $dist(x, y)$ their Hamming distance, i.e. $ham(x \oplus y)$.

For an integer $0 \leq r \leq n$, we denote by $B_r(y)$ the Hamming ball of radius r centered at c, namely $B_r(c) \triangleq \{x | dist(x, c) \leq r\}$. The number of points in $B_r(c)$ is denoted as $M_r^n \triangleq |B_r(c)| = \sum_{i=0}^{r} \binom{n}{i}$.

We denote the n-bit word with bits $i_1, ..., i_k$ set to 1 and the rest set to 0 by $e_{i_1, ..., i_k}$.

3   Previous Work

3.1   Bottom-Up Differential Characteristic Search

Since the early works on differential cryptanalysis (including the original work of [6]), there was a need to find good differential characteristics. This need was usually answered in the bottom-up approach: In [18] Matsui described the first general purpose differential characteristic search algorithm, which uses "bound-and-branch" approach. Matsui's algorithm is assured to find the best characteristic, but its running time may be unbounded. Later works in the field was sometimes applied to specific ciphers (e.g., analyzing FEAL in [3]), or extending Matsui's approach using basic properties of the block cipher (notably, the byte-oriented ciphers studied in [7, 8, 14, 21] or the ARX constructions studied in [10, 12, 17, 19]).

Offering an upper bound on the probability of differential characteristics dates back to the early works of [22], which suggested bounds for Feistel constructions, based on bounds on the probability of differential characteristics through the round function. This method is the basis of the approach of counting the number of active S-boxes (introduced in [11]), which is widely used today. Another approach introduced in [20] is the transformation of the problem into a linear-programming problem, and solving it for constraints. This technique was later extended in [23, 24].

Finally, we note that [10] also explored the concept of computing a subset of the DDT entries in the context of ARX constructions. The value of each DDT entry is computed analytically by using properties of the ARX operations rather than running actual experiments (as in our approach).

3.2   Top-Down Algorithms

The first top-down algorithm which we are aware of is due to [5] — the "Shrinking" algorithm that searches for impossible differentials. The main idea behind the shrinking algorithm is to take a scaled-down version of the cipher (e.g., with reduced word sizes and S-boxes). Such a scaled-down version allows evaluating the full difference distribution table, which in turn can be used to automatically identify impossible differentials. However, we note that many cryptosystems cannot be scaled down in an obvious way while maintaining properties of their DDT, and therefore the applicability of this algorithm is limited.

4   The Diagonal Algorithm and Its Extensions

4.1   The Diagonal Algorithm

We begin by describing our basic algorithm for calculating the exact values of all the diagonal entries in the difference distribution table with about the same time complexity as computing a single entry. The algorithm is given black box access to a function $F : \mathbb{GF}(2)^n \to \mathbb{GF}(2)^n$, and outputs the diagonal of the difference distribution table $DIAG[\Delta] \triangleq DDT[\Delta, \Delta]$. The algorithm is based on the simple property that the equality $x \oplus y = F(x) \oplus F(y)$ along the diagonal is equivalent to the equality $x \oplus F(x) = y \oplus F(y)$. Therefore, we can efficiently identify all the (x, y) pairs with equal input and output differences $\Delta = x \oplus y$ (which contribute to the $DIAG$ table) by searching for all the collisions between values of $x \oplus F(x)$ and $y \oplus F(y)$.

1. Initialize all the entries of the table $DIAG$ to zero, and set $DIAG[0]$ to 2^n.
2. For each n-bit value x:
   (a) Compute $x \oplus F(x)$, and store the pair $(x \oplus F(x), x)$ in a hash table H, i.e., add x to the set of values stored at $H[x \oplus F(x)]$.
3. For each n-bit value b:
   (b) For each pair (x, y) of distinct values such that $x, y \in H[b]$, increment $DIAG[x \oplus y]$ by 1.

The time complexity of Steps 1 and 2 is 2^n each, and the time complexity of Step 3 is proportional to D, which denotes the total number of pairs (x, y) such that $x \oplus y = F(x) \oplus F(y)$ (which is the same as the sum of all the entries in $DIAG$). Note that for a random function F, the expected value of D is about $2^{2n-n} = 2^n$ as we have about 2^{2n} (x, y) pairs, and the probability that a pair satisfies the n-bit equality is 2^{-n}. Consequently, the expected time complexity of the algorithm for a random function is about 2^n, and the total memory complexity is also 2^n, which is the size of the hash table H and the output table $DIAG$.

We note that there are several previous algorithms whose general structure resembles the diagonal algorithm. One such algorithm is impossible differential cryptanalysis of Feistel structures [15] and its various extensions, which use a data structure similar to H to iterate over pairs with related input and output differences. However, in these algorithms H is used in order to filter pairs required to attack specific cryptosystems, and not to explicitly calculate the DDT (as we do in Step 3.(a)).

4.2   The Generalized Diagonal Algorithm

We now extend the diagonal algorithm to compute a generalized diagonal $GDIAG_L$ for any given linear function L over $\mathbb{GF}(2)^n$. In this case, we are interested in (x, y) pairs such that $L(F(x) \oplus F(y)) = x \oplus y$, which is equivalent to the equality $x \oplus L(F(x)) = y \oplus L(F(y))$, since L is linear. Therefore, the generalized diagonal algorithm is very similar to the diagonal algorithm above, and only differs in Step 2.(a), where we store the pair $(x \oplus L(F(x)), x)$ in the hash table H (instead of storing the pair $(x \oplus F(x), x)$). The complexity analysis of the generalized diagonal algorithm is essentially identical to the basic diagonal algorithm.

5 The Hamming Ball Algorithm

The (generalized) diagonal algorithm computes the exact value of the (generalized) diagonal of the DDT of the function F in about 2^n time, which is practical for $n = 32$ but marginal for $n = 64$. In fact, it is easy to show that information theoretically, the only way to compute the precise value of a single DDT entry is to test all the 2^n relevant pairs of inputs or outputs. However, if we assume that we only want to find large entries on the diagonal and to approximate their values, we can do much better.

Assume that there exists some entry $DDT[\Delta, L(\Delta)]$ with a value of $p \cdot 2^n$ (where $0 < p \leq 1$ is the probability of an input pair with difference Δ to have an output difference of $L(\Delta)$) for a fixed linear function L. A trivial adaptation to the (generalized) diagonal algorithm evaluates and stores the pairs $(x \oplus L(F(x)), x)$ for only $0 < C \leq 2^n$ random values of x. Clearly, we do not expect to generate a non-zero value in entry $DDT[\Delta, L(\Delta)]$ before evaluating at least p^{-1} $(x, x \oplus \Delta)$ pairs. This gives a lower bound on C and on the complexity of the algorithm, since after the evaluation of C arbitrary values x, we expect to have about $C^2 \cdot 2^{-n}$ pairs with randomly scattered input differences, and thus we require $C^2 \cdot 2^{-n} \geq p^{-1}$ or $C \geq 2^{n/2} \cdot p^{-1/2}$. Therefore, the time and memory complexity of our adaptation are still somewhat large for big domains, and in particular it is barely practical for $n = 128$ even when p is close to 1 (as $C \geq 2^{n/2} = 2^{64}$).

We now describe a more efficient adaptation that requires the stronger assumption that the high probability entries $DDT[\Delta, L(\Delta)]$ occur at Δ's which have (relatively) low Hamming weight. The motivation behind this assumption is that we are interested in applying our algorithms to concrete cryptosystems in which a high probability entry $DDT[\Delta_I, \Delta_O]$ typically indicates the existence of a high probability differential characteristic with the corresponding input-output differences. Such high probability characteristics in SP networks are likely to have a small number of active Sboxes, and thus Δ_I and Δ_O are likely to have low Hamming weights.

In order to consider only $DDT[\Delta, L(\Delta)]$ entries where Δ is of small Hamming weight, we pick an arbitrary center c and a small radius r, and evaluate F only for inputs inside the Hamming ball $B_r(c)$. All the pairs of points inside the Hamming ball have a small Hamming distance, and thus for a carefully chosen value of r, we will obtain a quadratic number of relevant pairs from a linear number of values which have small Hamming distances d.

It is easy to see that the raw estimates we get with this approach for the entries in the DDT are biased, since the Hamming ball has more pairs which differ only in their least significant bit than pairs which differ in their d least significant bits for $d > 1$.[1] Given a difference Δ such that $ham(\Delta) = d$, an important measure which is used by our Hamming ball algorithm is the number of pairs with difference Δ in $B_r(c)$. This measure, which we denote[2] by $P_{r,d}^n$ (it does not depend on the actual values of c or Δ), is used in order to create from the experimental data unbiased estimates for the values of the entries $DDT[\Delta_I, \Delta_O]$, as described below.

1. Initialize the entries of the table $GDIAG_L$ to zero.
2. For each n-bit value $x \in B_r(c)$:
   (a) Compute $x \oplus L(F(x))$, and store the pair $(x \oplus L(F(x)), x)$ in a hash table H, i.e., add x to the set of values stored at $H[x \oplus L(F(x))]$.
3. For each n-bit value b such that $H[b]$ contains at least 2 values:
   (a) For each pair (x, y) such that $x, y \in H[b]$, increment $GDIAG_L[x \oplus y]$ by 1.
4. For each n-bit value Δ such that $GDIAG_L[\Delta] > 0$:
   (a) Denote $ham(\Delta) = d$ and normalize the entry $GDIAG_L[\Delta]$ by setting $GDIAG_L[\Delta] \leftarrow GDIAG_L[\Delta] \cdot (2^n / P_{r,d}^n)$.

The time and memory complexities of Step 2 are M_r^n. The time and memory complexities of steps 3 and 4 are determined by the number of collisions in the hash table H, which depends on F. For a random function, we expect to have $(M_r^n)^2 \cdot 2^{-n} \leq M_r^n$ such collisions, and therefore we generally do not expect steps 3 and 4 to dominate the time or memory complexities of the attack (especially for large domains where we select a small r implying that $M_r^n \ll 2^n$ and thus $(M_r^n)^2 \cdot 2^{-n} \ll M_r^n$).

In order to detect an entry $DDT[\Delta, L(\Delta)]$ with $ham(\Delta) = d$ whose probability is p, the most efficient method (assuming that we have sufficient memory) is to select a r such that $B_r(c)$ contains about p^{-1} pairs of points with input different Δ, or $P_{r,d}^n \geq p^{-1}$.

The efficiency of our algorithm for low Hamming weights is derived from the fact that Hamming balls are relatively closed under XOR's - pairs of points which are close to the origin are also close to each other. Similar efficiencies can be obtained for other sets with similar closure properties, such as arbitrary linear subspaces and sets of points which have short Hamming distance to linear subspaces.

5.1   Analyzing Keyed Functions

The algorithms described so far analyze a keyless function F. In order to obtain meaningful results for a keyed function F_K, we assume the existence of high probability entries $DDT[\Delta_I, \Delta_O]$, which are common to a large fraction of the keys. Such common high probability entries are typically the result of high probability

---

[1] This claim can be easily supported by the fact that as more bits are changed, the probability that the new computed value is outside the ball increases.
[2] The computation of $P_{r,d}^n$ is discussed in Appendix A.

differential characteristics (with the corresponding input-output differences) in iterated block ciphers where the round keys are XORed into the state.[3]

Based on this assumption, we can select a few keys K_i at random, and independently run our algorithms on F_{K_i} for each K_i. Then, we look for high probability entries $DDT[\Delta_I, \Delta_O]$ which are common to several keys. An additional possibility is to first run our algorithms on F_{K_1}, and then to test the obtained high probability entries $DDT[\Delta_I, \Delta_O]$ on F_{K_i} for $i > 1$, by encrypting sufficiently many pairs with input difference Δ_I for each key.

6   Improved Approximation of a Single Large DDT Entry

We now turn our attention to a related problem. Assume that we found a pair of input/output differences (Δ_I, Δ_O) which are somehow related. For example, this can occur when an iterative characteristic is repeated several times. Given (Δ_I, Δ_O), we wish to estimate the probability of the transition $\Delta_I \xrightarrow{r} \Delta_O$ (where r is the number of rounds in the differential). The standard method to estimate this probability is to take many pairs with input difference Δ_I and check how many of them have output difference Δ_O (again, trying multiple keys). If the probability of the differential is p, a good estimation requires $O(p^{-1})$ queries to the encryption algorithm.

Now, assume that the cipher (or the rounds) for which we analyze this transition, can be divided into two (roughly equal) parts, where M denotes the state in the middle. In such a case, we can discuss the transition from Δ_I to some Δ_M after about $r/2$ rounds, and from Δ_M to Δ_O in the remaining rounds. In other words, we look at Δ_M after r' rounds, and use the fact that:

$$\Pr[\Delta_I \xrightarrow{r} \Delta_O] = \sum_{\Delta_M} \Pr[\Delta_I \xrightarrow{r'} \Delta_M \xrightarrow{r-r'} \Delta_O] \tag{1}$$

which by the stochastic equivalence assumption (see [16]) we can re-write as

$$\Pr[\Delta_I \xrightarrow{r} \Delta_O] = \sum_{\Delta_M} \Pr[\Delta_I \xrightarrow{r'} \Delta_M] \cdot \Pr[\Delta_M \xrightarrow{r-r'} \Delta_O] \tag{2}$$

To correctly evaluate the probability suggested by Eq. (2), one needs to go over all possible Δ_M values (which is usually infeasible for common block sizes), and for each one of them evaluate the probability of two shorter differentials, $\Delta_I \xrightarrow{r'} \Delta_M$ and $\Delta_M \xrightarrow{r-r'} \Delta_O$ (which in itself may be a hard task).

Luckily, it was already observed in [6] that (in most cases) a high probability differential characteristic has several "close" high probability neighbors. This is explained by taking slightly different transitions through the active S-boxes with probability which is only slightly lower than the highest possible probability (used in the high probability characteristic). Similar behavior sometimes happen

---

[3] In such cases, the probability of the characteristic can be estimated independently of the round keys, assuming the input values are selected at random.

for differentials (especially for differentials which are based on a few "strong" characteristics, each having a few high probability "neighbors").

Hence, to give a lower bound on the value suggested by Eq. (2), we can use the following computation:

$$\Pr[\Delta_I \xrightarrow{r} \Delta_O] \geq \sum_{\Delta_M \in S} \Pr[\Delta_I \xrightarrow{r'} \Delta_M] \cdot \Pr[\Delta_M \xrightarrow{r-r'} \Delta_O] \qquad (3)$$

where the set S contains all the Δ_M values for which the differentials $\Delta_I \xrightarrow{r'} \Delta_M$ and $\Delta_M \xrightarrow{r-r'} \Delta_O$ have a sufficiently high probability.[4]

Obviously, this approximation relies on the fact that the two parts of the cipher are independent of each other. When taking into consideration a Markov-cipher assumption or the Stochastic Equivalence assumption (see [16] for more details), then the independence assumption immediately holds. However, in real life, one needs to verify it.

One advantage of the Bins-in-the-Middle algorithm which is presented next over the standard analytical approach is the fact that we "reduce" the independence assumption only to the transition between the two parts of the cipher. This is to be compared with an analytical approach that computes the probability of each round independently, and then simply multiplies the probabilities of each round (i.e., approaches that assume that each round is independent of others). In the Bins-in-the-Middle algorithm, the probabilities which are multiplied are the sampled probabilities of differentials, i.e., probabilities that were experimentally verified.[5]

6.1   The Bins-in-the-Middle (BITM) Algorithm

We now present an algorithm that finds all the "good" Δ_M values in the set S and experimentally estimates the probability of the two differentials $\Delta_I \xrightarrow{r'} \Delta_M$ and $\Delta_M \xrightarrow{r-r'} \Delta_O$. The algorithm requires that the last $r - r'$ rounds are invertible (and thus, can be used only on permutations).

The algorithm's basic idea is to actually produce a list of plausible Δ_M by sampling random pairs with input difference Δ_I (for the first r' rounds) and a corresponding list by sampling random pairs with output difference Δ_O (for the last $r - r'$) rounds. We shall denote the two lists, L_1 and L_2, respectively. The first list, L_1, contains pairs of the form (Δ_{M_i}, p_i) (i.e., the difference Δ_{M_i} appears with probability p_i given an input difference Δ_I). Similarly, the second list, L_2, contains pairs of the form (Δ_{M_j}, q_j).

Given these two lists, we can define the set S as all the differences which appear both in L_1 and L_2 with sufficiently high probability (which we denote

---

[4] When using $BITM$ to calculate the probability of a differential, one can choose the meeting round in a variety of ways. Usually setting $r' \approx r/2$ gives the optimal results.

[5] Of course, we still need to assume independence between the two parts of the cipher.

by p_b). Then, by using Eq. (3), and the estimations for the p_i's and q_j's, we can compute an estimation for the probability of the differential $\Delta_I \xrightarrow{r} \Delta_O$:

1. Pick N plaintext pairs[6] of the form $(x, x \oplus \Delta_I)$, and obtain their partial encryption after r' rounds, (z, z').
2. Collect the differences $z \oplus z'$, and produce L_1.
3. Pick N ciphertext pairs of the form $(y, y \oplus \Delta_O)$, and obtain their partial decryption after $r - r'$ rounds, (w, w').
4. Collect the differences $w \oplus w'$, and produce L_2.
5. For all the differences that appear with probability above some bound p_b in both L_1 and L_2, compute the sum of all products $(p_i \cdot q_j)$.

First, it is easy to see that both L_1 and L_2 contain two types of differences: High-probability differences (e.g., differences that appear with probability higher than p_b) as well as low-probability differences that got sampled by chance. For an n-bit block cipher, after sampling N pairs, we expect low probability differences Δ_M to be encountered only once[7] (both in L_1 and in L_2) as long as $N < 2^{n/2}$. Moreover, as we later discuss, estimating the probabilities p_i's and q_j's can be done over many keys, offering a better estimation.

Now, given p_b, we wish to assure that we sample the high probability differences. This can be done, by looking for differences that appear at least twice during Steps 1–2 (for L_1) or Steps 3–4 (for L_2). Given that the number of "appearances" of an output difference follows the Poisson distribution, we need to take $N = \alpha/p_b$ pairs, where α determines the quality of our sampling. For example, if we pick $\alpha = 4$, i.e., we expect 4 pairs that follow the differential $\Delta_I \xrightarrow{r'} \Delta_M$, then with probability of 90%, Δ_M would appear at least twice in Steps 1–2. Increasing the value of α (and/or sampling using more keys) improves the quality of the values in L_1 and L_2. For example, for $\alpha = 10$, the probability that a good Δ_M will not appear at least twice is less than 0.5%.

It is important to note that differences of low probability do not affect the overall estimation. This follows from the fact that we count only differences that appear in *both* lists L_1 and L_2. Hence, even though there are some low probability differences in each list, it is extremely unlikely that the same low probability difference will appear in both lists simultaneously. Even in the extreme case that there are N low probability Δ_M values in each list, the expected number of low probability Δ_M appearing in both lists is $N^2/2^n$, which is less than 1.

We recall that similarly to all approaches that estimate the probability of differentials, we need to rely on some randomness assumptions. A round-by-round

---

[6] The value of N is discussed later.

[7] We note that one can take more pairs, but as we later show, $N = O(1/p_b)$, i.e., as long as p_b is above $2^{-n/2}$ the algorithm is expected to work. Moreover, if both $\Delta_I \xrightarrow{r'} \Delta_M$ and $\Delta_M \xrightarrow{r-r'} \Delta_O$ have probability lower than p_b, the overall contribution of the characteristic $\Delta_I \xrightarrow{r'} \Delta_M \xrightarrow{r-r'} \Delta_O$ to the probability we estimate is at most p_b^2. Picking $p_b < 2^{-n/2}$ suggests that the contribution is less than 2^{-n}. Such a low probability is usually of little interest in cryptanalysis, and requires a very careful analysis.

approach relies on the cipher being Markovian, whereas an experimental verification of the full differential does not require any assumption. The independence assumption needed by the $BITM$ algorithm lies between these two extremes. We need to assume that the transition between the two parts of the cipher does not affect the probability estimations. In other words, even though the actual pairs in L_1 and L_2 are different, we can use a (reduced) Markov-cipher assumption to obtain an estimate for the total probability of the differential $\Delta_I \xrightarrow{r} \Delta_O$.

As mentioned earlier, as α increases (or if the probability of the difference we check is higher than p_b) the quality of the estimation of the probabilities in L_1 and L_2 improves. This is explained by the fact that we estimate the probability of an event which follows a Poisson distribution. If $X \sim Poi(\lambda)$, then $E[X] = Var[X] = \lambda$, so the larger λ is, the closer X is to its mean.

Moreover, we note that the use of multiple keys can significantly improve the quality of the estimation. If we repeat the experiment with t different keys, the expected number of times Δ_M appeared in all t experiments is increased by a factor t. As the sum of Poisson random variables is itself a Poisson random variable, we obtain a significantly better estimate for the actual probability of the difference.[8]

Hence, after sampling sufficiently many keys, one can obtain a better estimation of the actual probabilities of the various differences in L_1 and L_2, and discard the low probability differences. These probabilities can then be combined to offer a higher quality estimate of the probability of the differential $\Delta_I \xrightarrow{r} \Delta_O$.

We note that while the $BITM$ algorithm is superficially similar to the meet in the middle (MITM) algorithm, it is quite different. In the MITM algorithm, we typically try to find some common value between the two parts of a cipher, and use this value to find the key (depending on the cryptanalytic task at hand, we may search for all the common values). In the BITM algorithm our goal is not to find these values, but to estimate the probability that they exist, in order to choose the best differential attack on the scheme. Finally, some improvements for the $BITM$ algorithm are suggested in Appendix B.

6.2 The Advantages of the $BITM$ Algorithm

The main advantage of the $BITM$ algorithm over a pure top-down algorithm which evaluates the full mapping is its greatly improved efficiency. Indeed, in order to estimate the differential $\Delta_I \xrightarrow{r} \Delta_O$ requires $O(p^{-1})$ pairs. However, if we pick $r' \approx r/2$, and under the assumption that both parts are roughly of the same strength, we obtain $p_b = O(\sqrt{p})$. This is extremely important for the

---

[8] For $\alpha = 4$ and $t = 32$ (expecting four pairs in 32 experiments), the total number of times Δ_M appears in all experiments follows a Poisson distribution with a mean of 128. Hence, with probability 95 %, counting over all experiments will suggest Δ_M somewhere between 105 and 151 times (in all 32 experiments). In other words, taking the number of times Δ_M appears (divided by $32N$) as an estimate for the actual probability will be accurate within 18 % of the correct probability with probability 95 %.

cases where a time complexity of p_b^{-1} is still feasible but p_b^{-2} is not (e.g., when $p_b \approx 2^{-40}$).

Another advantage of the $BITM$ algorithm over bottom-up algorithms is that we take into account all the high probability differential characteristics simultaneously. Hence, the estimation for the differential probability is closer to the actual probability than an estimation which is based on the multiplication of many probabilities along a single differential characteristic.

Finally, this method offers some experimental verification of the stochastic equivalence assumption. Indeed, for most ciphers (and most of the keys), the stochastic equivalence assumption tends to hold (or seem to "work" most of the time). However, when we discuss a single long characteristic, we may encounter some inconsistencies between different parts of the characteristic. In these situations, the real probability and computed probability will differ. Once we take into consideration multiple differential characteristics, the estimation becomes more resilient (though not 100 % foolproof), as a "failure" of one of the longer characteristics does not invalidate the full differential. In addition, by running the algorithm with several different r' can also help in validating the probability of the transition between the top half and the bottom half.

7 Applying Our New Algorithms to the SIMON Family of Block Ciphers

The SIMON family of lightweight block ciphers, presented in [4], is implemented using a balanced Feistel structure. The SIMON round function is very simple and consists of only three operations: AND, XOR and constant rotations. All the ciphers in the SIMON family use the same round function and differ only by the key size, the block size (which ranges from 32 to 128 bits) and the number of Feistel rounds which is dependant on the former two. As in any Feistel structure, the plaintext is divided into two blocks of size n: $P = (L_0, R_0)$ and then every round $1 \leq i \leq r$:

$$L_i = R_{i-1} \oplus F(L_{i-1}) \oplus K_{i-1}; \quad R_i = L_{i-1}$$

where the ciphertext is $C = (L_r, R_r)$ and F is the SIMON round function:

$$F(x) = ((x \lll 1) \wedge (x \lll 8)) \oplus (x \lll 2)$$

In this section we present the best differentials for SIMON64 SIMON96 and SIMON128 we found using our various diagonal estimation algorithms. We also describe more accurate $BITM$-based estimates for the differential probabilities of previously presented SIMON characteristics, which are substantially different from the original estimates.

7.1 Applying $BITM$ to Previously Known SIMON Differentials

The discussion of previous known SIMON differentials is given in the full version of this paper [13]. Table 1 compares all the previously published bottom-up

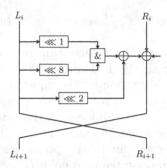

Fig. 1. The SIMON round function

estimates of the probabilities of various differential transitions with our experimentally obtained top-down results (where the numbers are the log to the base 2 of the probabilities). All the results (including those presented in the next subsection) were obtained by the same method as described for the SIMON128 differential. Namely, we performed experiments with dozens of arbitrary keys, where in each experiment as many as 2^{35} balls were thrown from each side with a narrow result range, and the average value was taken as the final probability (Fig. 1).

Table 1. The original and our improved estimates of the probabilities of the best previously published differentials

Cipher	Rounds	Presented prob	$BITM$ prob	Source
SIMON64	21	-60.53	-56.05	[9]
SIMON64	21	-61.01	-56.05	[1]
SIMON64	21	-60.21	-59	[23]
SIMON96	30	-92.20	-88.5	[1]
SIMON128	41	-124.6	-118.6	[1]

This table shows that the previous estimates were too pessimistic (sometimes by a significant factor of $2^6 = 64$) since they considered only a limited number of differential characteristics. Since some of the differential probabilities that appear in the mentioned papers are significantly lower than the probabilities estimated by our $BITM$ algorithm, we can extend the differentials to a larger number of rounds, while maintaining the probability above 2^{-n}.[9] However, even without extending the characteristics, the results of Table 1 automatically translate into

---

[9] All these differential characteristics could be theoretically extended to cover more rounds, but in order to break an n-bit block cipher, the probabilities generally need to be higher than 2^{-n} (otherwise we do not expect to find more than a single accidental pair, even when we try the full code book).

better key recovery attacks on the SIMON members, as the previous attacks only depended on differentials for SIMON (and not on the internal characteristics).

Verifying the Accuracy of BITM on SIMON32. In order to verify that BITM estimates well the actual probability of differentials, we performed experiments on SIMON32, for which we can calculate the actual differential probability for a given key in 2^{32} time. For this purpose, we picked a 9-round differential[10] that has probability of about 2^{-25} and compared the results of BITM with the actual probability in more than 300 experiments, each with a different key picked at random. In each BITM experiment, 2^{26} balls were thrown from each side. The actual probability had a big dependency on the key and ranged in the interval $[2^{-27.2}, 2^{-18.7}]$, a fact that we attribute to the small block size, as such fluctuations do not seem to occur in larger SIMON variants (although for them we cannot calculate actual probabilities precisely). Despite the large fluctuations in the actual probability, BITM estimated it within a factor of 2 for more than 60 % of the keys, and we expect even better estimates for larger SIMON variants for which the actual probabilities are less dependant on the key.

7.2   Improved SIMON Differentials Found Using $GDIAG_L$

Due to space limitations, we now give the improved differentials that we found using the $GDIAG_L$ algorithm, whose probabilities were estimated using $BITM$. The result is an improvement by two rounds of the best previously known differential from [1] while maintaining roughly the same probability. The full details are given in the full version of this paper [13] (Table 2).

Table 2. Summary of the $GDIAG_L$ results for SIMON

Cipher	Differential family	Rounds	Prob. ($\log_2$)
SIMON64	$(e_{i,i+4}, e_{i+6}) \rightarrow (e_{i,i+4}, e_{i+2})$	22	-59.9
SIMON64	$(e_{i+2}, e_{i,i+4}) \rightarrow (e_{i+6}, e_{i,i+4})$	22	-59.3
SIMON64	$(e_{i+6}, e_{i,i+4,i+8}) \rightarrow (e_{i,i+4}, e_{i+2})$	23	-61.9
SIMON64	$(e_{i+2}, e_{i,i+4}) \rightarrow (e_{i,i+4,i+8}, e_{i+6})$	23	-61.3
SIMON128	$(e_{i+6}, e_{i,i+4,i+8}) \rightarrow (e_{i,i+8,i+12}, e_{i+2,i+10})$	43	-125.6
SIMON128	$(e_{i+2,i+10}, e_{i,i+8,i+12}) \rightarrow (e_{i,i+4,i+8}, e_{i+6})$	43	-124.4

8   Conclusions

In this paper we described and motivated the top-down approach to differential cryptanalysis, which tries to compute or approximate certain DDT values

---

[10] The results for other differentials do not seem to differ significantly.

without looking at the internal structure of the given mapping. We introduced three novel techniques which can compute three types of interesting entries (on the diagonal, in low Hamming weight entries on the diagonal, and arbitrarily located entries with large values) with improved efficiency. We then applied the new BITM technique to SIMON in order to obtain more accurate estimates of the probabilities of all the previously published differentials and combined it with the generalized diagonal algorithm to find better differentials for a larger number of rounds. This improves the best known cryptanalytic results for this scheme and demonstrates the power and versatility of our new top-down techniques. Finally, in the full version of this paper [13] we describe how to use our new algorithms to efficiently locate the highest diagonal entry in any given incarnation of the four round version of Even-Mansour scheme, in order to break the scheme with a related key attack which is faster than exhaustive search.

A   Calculating

$P_{r,d}^n$ The Hamming ball algorithm of Sect. 5 relies on the value of $P_{r,d}^n$. We compute this value by distinguishing between two cases: when $d > 2r$, then $P_{r,d}^n = 0$, as the largest Hamming distance between points in $B_r(c)$ is $2r$. Otherwise, $d \leq 2r$, and we consider the conditions on a point x such that both $x \in B_r(c)$ and $x \oplus \Delta \in B_r(c)$. We partition the coordinates of $x \oplus c$ which are set to 1 into two groups: the $d_1 \leq min(r,d)$ coordinates which are common to $x \oplus c$ and $\Delta \oplus c$, and the remaining $d_2 \leq min(r, n-d)$ coordinates. Thus, we have $dist(x,c) = d_1 + d_2$ and $dist(x \oplus \Delta, c) = d + d_2 - d_1$, implying that $d_1 + d_2 \leq r$ and $d + d_2 - d_1 \leq r$. In particular, the last equality implies that $d_1 \geq max(d-r, 0)$, and so $max(d-r,0) \leq d_1 \leq min(r,d)$, while $0 \leq d_2 \leq min(r-d_1, r+d_1-d, n-d)$. Therefore, we obtain

$$P_{r,d}^n = \sum_{d_1=m_1}^{m_2} \sum_{d_2=0}^{m_3} \binom{d}{d_1} \binom{n-d}{d_2}$$

where $m_1 = max(d-r, 0)$, $m_2 = min(r,d)$ and $m_3 = min(r-d_1, r+d_1-d, n-d)$.

B   Improving the BITM Algorithm

We first note that there is no need to actually store L_2. One can generate L_1, and for each $w \oplus w'$ value of Steps 3–4, to increment the counter if $w \oplus w'$ happens to be in L_1.

We now turn our attention to the generation of L_1. It is easy to see that L_1 can take at most $O(N)$ memory cells. As N increases this may be a practical bottleneck. Hence, once the used memory reaches the machine's limit (or the process' limit), we suggest to "extract" all the high probability differences encountered so far into a shorter list L_1'. Then, we sample more random pairs, but this time, we only deal with those pairs whose "output" difference is in the short list L'_1. The main advantage is now that we use almost no memory (as L'_1 tends to be small), we can actually increase the number of queries, thus obtaining a more accurate estimate.

The final improvement in this front is to perform the previous idea in steps. We first sample many pairs, and store the differences $z \oplus z'$ in a hash table (with less than N bins). After finding the bins which were suggested more than others, we can dive into them by re-sampling more pairs.

References

1. Abed, F., List, E., Lucks, S., Wenzel, J.: Differential cryptanalysis of round-reduced simon and speck. In: Cid, C., Rechberger, C. (eds.) FSE 2014. LNCS, vol. 8540, pp. 525–545. Springer, Heidelberg (2015)
2. Alkhzaimi, H.A., Lauridsen, M.M.: Cryptanalysis of the SIMON Family of Block Ciphers. Cryptology ePrint Archive, Report 2013/543 (2013)
3. Aoki, K., Kobayashi, K., Moriai, S.: Best differential characteristic search of FEAL. In: Biham, E. (ed.) FSE 1997. LNCS, vol. 1267, pp. 41–53. Springer, Heidelberg (1997)
4. Beaulieu, R., Shors, D., Smith, J., Treatman-Clark, S., Weeks, B., Wingers, L.: The SIMON and SPECK Families of Lightweight Block Ciphers. Cryptology ePrint Archive, Report 2013/404 (2013)
5. Biham, E., Biryukov, A., Shamir, A.: Cryptanalysis of skipjack reduced to 31 rounds using impossible differentials. In: Stern, J. (ed.) EUROCRYPT 1999. LNCS, vol. 1592, pp. 12–23. Springer, Heidelberg (1999)
6. Biham, E., Shamir, A.: Differential cryptanalysis of DES-like cryptosystems. J. Cryptol. **4**(1), 3–72 (1991)
7. Biryukov, A., Nikolić, I.: Automatic search for related-key differential characteristics in byte-oriented block ciphers: application to aes, camellia, khazad and others. In: Gilbert, H. (ed.) EUROCRYPT 2010. LNCS, vol. 6110, pp. 322–344. Springer, Heidelberg (2010)
8. Biryukov, A., Nikolić, I.: Search for related-key differential characteristics in DES-like ciphers. In: Joux, A. (ed.) FSE 2011. LNCS, vol. 6733, pp. 18–34. Springer, Heidelberg (2011)
9. Biryukov, A., Roy, A., Velichkov, V.: Differential analysis of block ciphers SIMON and SPECK. In: Cid, C., Rechberger, C. (eds.) FSE 2014. LNCS, vol. 8540, pp. 546–570. Springer, Heidelberg (2015)
10. Biryukov, A., Velichkov, V.: Automatic search for differential trails in ARX ciphers. In: Benaloh, J. (ed.) CT-RSA 2014. LNCS, vol. 8366, pp. 227–250. Springer, Heidelberg (2014)
11. Daemen, J., Govaerts, R., Vandewalle, J.: A new approach to block cipher design. In: Anderson, R. (ed.) FSE 1993. LNCS, vol. 809. Springer, Heidelberg (1994)
12. De Cannière, C., Rechberger, C.: Finding SHA-1 characteristics: general results and applications. In: Lai, X., Chen, K. (eds.) ASIACRYPT 2006. LNCS, vol. 4284, pp. 1–20. Springer, Heidelberg (2006)
13. Dinur, I., Dunkelman, O., Gutman, M., Shamir, A.: Improved top-down techniques in differential cryptanalysis. IACR Cryptol. ePrint Arch. **2015**, 268 (2015)
14. Fouque, P.-A., Jean, J., Peyrin, T.: Structural evaluation of AES and chosen-key distinguisher of 9-round AES-128. In: Canetti, R., Garay, J.A. (eds.) CRYPTO 2013, Part I. LNCS, vol. 8042, pp. 183–203. Springer, Heidelberg (2013)

15. Knudsen, L.: DEAL - A 128-bit Block Cipher. NIST AES Proposal (1998)
16. Lai, X., Massey, J.L.: Markov ciphers and differential cryptanalysis. In: Davies, D.W. (ed.) EUROCRYPT 1991. LNCS, vol. 547, pp. 17–38. Springer, Heidelberg (1991)
17. Leurent, G.: Analysis of differential attacks in ARX constructions. In: Wang, X., Sako, K. (eds.) ASIACRYPT 2012. LNCS, vol. 7658, pp. 226–243. Springer, Heidelberg (2012)
18. Matsui, M.: On correlation between the order of s-boxes and the strength of DES. In: De Santis, A. (ed.) EUROCRYPT 1994. LNCS, vol. 950, pp. 366–375. Springer, Heidelberg (1995)
19. Mendel, F., Nad, T., Schläffer, M.: Finding SHA-2 characteristics: searching through a minefield of contradictions. In: Lee, D.H., Wang, X. (eds.) ASIACRYPT 2011. LNCS, vol. 7073, pp. 288–307. Springer, Heidelberg (2011)
20. Mouha, N., Wang, Q., Gu, D., Preneel, B.: Differential and linear cryptanalysis using mixed-integer linear programming. In: Wu, C.-K., Yung, M., Lin, D. (eds.) Inscrypt 2011. LNCS, vol. 7537, pp. 57–76. Springer, Heidelberg (2012)
21. Nikolić, I.: Tweaking AES. In: Biryukov, A., Gong, G., Stinson, D.R. (eds.) SAC 2010. LNCS, vol. 6544, pp. 198–210. Springer, Heidelberg (2011)
22. Nyberg, K., Knudsen, L.R.: Provable security against differential cryptanalysis. In: Brickell, E.F. (ed.) CRYPTO 1992. LNCS, vol. 740, pp. 566–574. Springer, Heidelberg (1993)
23. Sun, S., Hu, L., Wang, M., Wang, P., Qiao, K., Ma, X., Shi, D., Song, L.: Automatic Enumeration of (Related-key) Differential and Linear Characteristics with Predefined Properties and Its Applications (2014)
24. Sun, S., Hu, L., Wang, P., Qiao, K., Ma, X., Song, L.: Automatic Security Evaluation and (Related-key) Differential Characteristic Search: Application to SIMON, PRESENT, LBlock, DES(L) and Other Bit-oriented Block Ciphers. Cryptology ePrint Archive, Report 2013/676 (2013). Accepted to ASIACRYPT 2014

Algebraic Analysis of the Simon Block Cipher Family

Håvard Raddum[✉]

Simula Research Laboratory, Bergen, Norway
haavardr@simula.no

Abstract. This paper focuses on algebraic attacks on the Simon family of block ciphers. We construct equation systems using multiple plaintext/ciphertext pairs, and show that many variables in the cipher states coming from different plaintexts are linearly related. A simple solving algorithm exploiting these relations is developed and extensively tested on the different Simon variants, giving efficient algebraic attacks on up to 16 rounds of the largest Simon variants.

Keywords: Block cipher · Algebraic attack · Equation system · Simon

1 Introduction

The Simon and Speck families of block cipher were published by the National Security Agency in June 2013 [1]. Both families consist of lightweight block ciphers, where the Simon ciphers are optimized for hardware and the Speck ciphers are optimized for software. The ciphers use very little area and have high throughput.

In the publication from the NSA all ciphers were specified, but there was no security analysis. In the relatively short time since the Simon and Speck ciphers became known the cryptographic community has spent a lot of effort cryptanalyzing them. Some work focused on differential cryptanalysis has been published [2,3], as well as a paper on linear cryptanalysis of Simon [4]. Several other papers on cryptanalysis of Simon have also been posted to the IACR eprint archive [5–11].

The work in [10] is focused on cube attacks, but other than that we have not seen any published work on Simon's resistance to algebraic attacks. This paper investigates to what extent algebraic attacks may be applied to the Simon family of ciphers and helps to fill this gap in the cryptanalysis of Simon. The main contribution of this work is the analysis that finds linear relations between variables coming from different plaintext/ciphertext pairs. Using the linear relations we may reduce the number of variables a lot when constructing an equation system based on several chosen plaintexts. We use a simple and straight-forward solving algorithm to test how many rounds of different Simon variants that can be broken when the linear relations are exploited. For instance, for Simon with a 128-bit block we can solve a system representing 16 rounds of the cipher using

© Springer International Publishing Switzerland 2015
K. Lauter and F. Rodríguez-Henríquez (Eds.): LatinCrypt 2015, LNCS 9230, pp. 157–169, 2015.
DOI: 10.1007/978-3-319-22174-8_9

20 chosen plaintexts in three and a half minutes on a MacBook Air, even though the equation system initially contains 15488 variables.

The paper is organized as follows. In Sect. 2, we give a brief description of the Simon ciphers. In Sect. 3 we construct equation systems representing Simon, and in particular look at reusing variables across different plaintext/ciphertext pairs. In Sect. 4 we introduce a simple way to eliminate redundant variables, which turns into a solving algorithm when all variables get eliminated. Extensive experiments on solving equation systems representing five different Simon variants are also reported here. Section 5 concludes the paper.

2   Description of the Simon Family of Ciphers

The Simon family of block ciphers consists of 10 members. There are two parameters that can be varied: the block size and the key size. For each of the allowed choices of block and key sizes the number of rounds is given. The variants of Simon ciphers that have been defined are listed in Table 1.

Table 1. Parameters for the Simon variants.

Block size	Key size	Rounds
32	64	32
48	72	36
48	96	36
64	96	42
64	128	44
96	96	52
96	128	54
128	128	68
128	192	69
128	256	72

Each cipher is a traditional Feistel cipher with a simple round function, depicted in Fig. 1. Two copies of the bitstring input to the round function are cyclically shifted by 1 and 8 positions to the left, respectively. These two words are joined together with bit-wise AND. Then the input word rotated by 2 positions to the left and the round key are XORed onto the output from the AND-operation and the result forms the output of the round function.

Following the specification in [1], we let the left half of the cipher block be the input to the round function, and we include the swap also after the last round. The left and right halves of the plaintext are denoted by P_1 and P_0, respectively. The left and right halves of the ciphertext are similarly denoted C_1 and C_0. We start the numbering of the rounds at 0, so the first and last rounds in an r-round

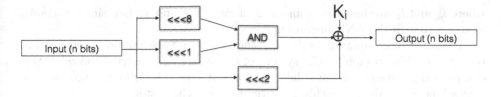

Fig. 1. Round function of the Simon ciphers

variant of Simon is round 0 and round $r - 1$, respectively. The size of one half of the cipher block is denoted by n, and the key size is denoted by k. A member of the Simon family having block size $2n$ and key size k will be referred to as Simon$2n/k$.

The key schedule for Simon comes in three different variants, depending on which member of the Simon family is used. We omit the details here (they can be found in [1]), but note that all three variants are linear. The bits of each round key can be expressed as a linear combination of the user-selected key bits.

3   Constructing Simon Equation Systems Using Multiple Plaintext/Ciphertext Pairs

We proceed to create equation systems representing instances of a Simon cipher. First we create an equation system using a single plaintext and its corresponding ciphertext. In all systems we work with, all variables will represent the values of single bits, and all additions and multiplications will be over the binary field.

3.1   Equations Representing the Encryption of One Plaintext

The only non-linear part of Simon is the AND-operation in the round function. This operation works on individual bits in parallel, taking 2 input bits and producing 1 output bit. For two input variables x and y producing the output z the equation is simply $x \cdot y = z$.

To keep equations simple and quadratic, we introduce variables representing the cipher state at the output of the AND-operations. In addition, the user-selected key bits are also variables. As noted above, the bits of each round key can be expressed as linear combinations of the user-selected key bits. All other operations in Simon are linear, so every input and output bit to each AND-operation can be described as linear combinations of these variables.

Note that there is no key material introduced before the AND-operation in the round function, so in the first and last rounds we do not need to introduce any variables. The outputs of the AND-operations in these rounds are simply constants depending on the left half of the (known) plaintext and the right half of the ciphertext.

For an r-round version of Simon using block size $2n$ and key size of k bits we will get $n(r - 2)$ basic equations in $n(r - 2) + k$ variables. Each equation will have the form

$$l_a \cdot l_b = x_c, \tag{1}$$

where l_a and l_b are linear combinations of variables and x_c is a single variable in the output of the AND-operation.

The set of basic equations will represent an r-round encryption of some plaintext using a Simon cipher. The system of equations we construct following the description given so far is underdefined by k variables. The following proposition shows how to eliminate another $2n$ variables from the system.

Proposition 1. *Let $\mathcal{F}$ be an r-round Feistel cipher, where $(C_1, C_0) = \mathcal{F}(P_1, P_0)$. Denote the output of the round function in round j by u_j, for $0 \leq j \leq r - 1$. Then both XOR sums*

$$\sum_{j=0}^{\lfloor \frac{r-1}{2} \rfloor} u_{2j} \ and \ \sum_{j=0}^{\lfloor \frac{r-1}{2} \rfloor} u_{2j+1}$$

can be expressed as the XOR of one half of P and one half of C.

Proof: Denote the input to round j by i_j for $0 \leq j \leq r - 1$, and assume $\mathcal{F}$ has a swap after the last round. Then $i_0 = P_1$ and $i_{r-1} = C_0$. Define $i_{-1} = P_0$ and $i_r = C_1$. In the Feistel cipher $\mathcal{F}$, i_j is given as $i_j = u_{j-1} + i_{j-2}$ for $1 \leq j \leq r$, or equivalently

$$u_j = i_{j-1} + i_{j+1}, \ 0 \leq j \leq r - 1.$$

The sum of u's with even indices then becomes

$$\sum_{j=0}^{\lfloor \frac{r-1}{2} \rfloor} u_{2j} = (i_{-1} + i_1) + (i_1 + i_3) + (i_3 + i_5) + \dots,$$

where the last term will be i_{r-1} when r is even and i_r when r is odd. All terms in the sum except for the first and last will cancel, so the sum just becomes the sum of P_0 and C_0 or C_1.

The sum $\sum u_{2j+1}$ similarly becomes a telescope sum where all terms except for the first and last will cancel. The first term will be P_1 and the last term will be C_0 when r is odd and C_1 when r is even. $\square$

All outputs from the round function in Simon can be expressed as linear combinations of the variables we have introduced, and using Proposition 1 we can derive $2n$ independent linear equations (assuming known plaintext) that can be used to eliminate $2n$ variables. After doing this the number of variables in the basic system becomes $n(r-4)+k$, and the system will be underdefined by $k-2n$ variables.

When the block and key sizes are equal one known plaintext is expected to be enough to uniquely determine the key in Simon, otherwise we will need at least two known plaintexts to have enough information to uniquely determine the user-selected key.

3.2 Equation System Representing Encryptions of Two Plaintexts

We call the set of equations and variables representing the encryption of plaintext P into ciphertext C for a (P, C)-*instance*.

Above we constructed the equations representing the encryption of the single plaintext P into C, introducing variables for the cipher state. If we repeat the construction for another plaintext/ciphertext pair P' and C' we must in general make new variables for the cipher state in the (P', C')-instance. However, we may construct P and P' in some particular way to try to make the cipher states in the first few rounds similar, or related. This will allow us to not introduce new variables for some of the cipher state bits in the (P', C')-instance, but rather re-use the variables from the (P, C)-instance.

The benefit will be a more overdefined total system, with fewer variables, which should be easier to solve. Two questions are: How many rounds of Simon must be executed before the cipher states of the two instances become unrelated, and how many new variables can be avoided in the (P', C')-instance?

Constructing Plaintexts. The plaintexts P and P' may be constructed as follows. Let P_1 be a string of n 0-bits. Let the rightmost bit of P_1' be 1, and the rest of the bits in P_1' be 0. The output of the AND-operation for P_1 and P_1' will be the all-zero word in both cases. After adding the input word shifted by 2 positions to the left, the difference in the output of the round function will be 4, hexadecimally. Choosing P_0 to be anything, and $P_0' = P_0 + 4$, the differences will cancel in the addition at the end of the first round and the inputs to the second round will be the same for both texts. Hence the variables to be introduced in the output of the AND-operation in the second round of the (P', C')-instance are exactly the same as in the (P, C)-instance. The situation is depicted in Fig. 2.

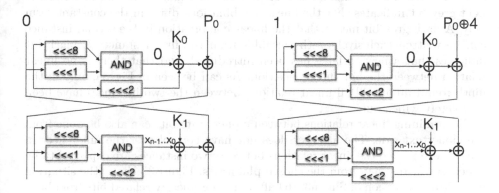

Fig. 2. Chosen plaintexts allowing reuse of all variables in second round

Tracing Relations Further. The inputs to the third round will differ in the rightmost bit due to the XOR of P_1 and P_1' onto the outputs from the second round. This difference means that two of the bits in the output of the AND-operation may be unequal in the two instances. The other $n - 2$ bits will still be the same, so $n - 2$ of the third-round variables in the (P, C)-instance carry over and can be used in the (P', C')-instance directly. However, we do not need to introduce new variables for the two bits that differ, either.

Let l_a be the linear combination representing the rightmost bit in the input of the third round of the (P, C)-instance. Then the same bit will be represented as $l_a + 1$ in the (P', C')-instance. One equation using l_a in the first instance will be $l_a \cdot l_b = x_c$ for some l_b and x_c. In the second instance the same equation will be $(l_a + 1) \cdot l_b = x_d$, where x_d is a (temporary) new variable in the second instance. Adding these two equations will cause the quadratic terms to cancel, and we get the relation $x_d = x_c + l_b$, which can be used to express x_d as a linear combination of variables only from the (P, C)-instance.

In general, when the two sets of input bits (represented as linear combinations) to the AND-operations only differ in the constant terms, the output bits are related through a linear equation. We can then reuse variables across the instances. Only when inputs in two instances differ in more than the constant term will we have to make new variables in the second instance.

Complete Mapping of Linear Relations. We have made a computer program for tracing the relations between two (P, C)- and (P', C')-instances in Simon$2n/k$ for all different choices for n. Figure 3a shows how the individual bits in the input to the round functions in Simon32/64 will be related between two plaintexts prepared as explained above. The bits in the input to the round function can be either white, light grey, dark grey or black. A white bit indicates that the linear combinations for this bit are equal in the two instances. A light grey bit indicates that the linear combinations differ in the constant term only. A dark grey bit means that the linear combination in the second instance can be written exclusively with variables from the first instance. A black bit indicates that a new variable has been introduced, and that there is no linear relation between the bits in this position. As can be seen in Fig. 3a, only in the ninth round input will all linear relations between the two instances have been completely wiped out.

Determining linear relations between cipher state bits can also be done from the ciphertext side. The attacker does not have control over the differences in the ciphertexts, so the linear relations between two instances are expected to be wiped out faster than from the chosen plaintexts. Figure 3b shows the situation in a 10-round version of Simon32/64 after tracing linearly related bits from both the plaintext and the ciphertext sides. In the figure, the colour of a bit has the lightest colour found when tracing from both sides.

In the example of Fig. 3b there are only 21 bits of internal cipher state that must be represented by new variables in the (P', C')-instance. The number of new variables we have to introduce from a new (P', C')-instance will vary slightly according to how "lucky" we are with the ciphertexts. However, if we had created the second instance directly from the description of Sect. 3.1 we would have had to introduce 96 new variables in a 10-round version of Simon32/64.

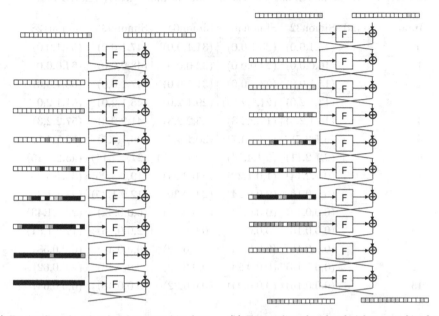

(a) Tracing linearly related cipher state bits from chosen plaintexts only.

(b) Tracing linearly related cipher state bits from both plaintext and ciphertext sides.

Fig. 3. Tracing linearly related cipher state bits in Simon32/64

We have traced and mapped the linearly related cipher state bits for all variants of Simon using plaintexts prepared as described above. The rotations in Simon shift the words with a constant number of positions, independent og the block size, so we can expect the number of rounds needed to wipe out all linear relations between two instances to increase in the larger variants. Figures for the larger Simon variants will not fit in the paper in a readable way, but in Table 2 we report on the number of white, light grey, dark grey and black bits in the input to each round for the different Simon variants.

3.3 Equation System Using Multiple (P, C)-Instances

We can make an equation system using as many (P, C)-instances as we like. For each new instance added, we can trace the linear relations the new instance has to all previous instances. We expect to add fewer and fewer variables (up to some limit) for each new instance added. The number of equations we get from each new instance is initially $n(r-2)$, but we can expect many of the new equations to be expressible through equations from other instances and therefore redundant.

Table 2. Numbers and types of linearly related bits in Simon using two chosen plaintexts. In each 4-tuple the first entry is the number of equal linear combinations, the second entry is the number of linear combinations only differing in the constant term, the third entry is the number of linear combinations differing in more than only the constant term, and the fourth entry is the number of non-linearly related bits

Input to round	Simon32	Simon48	Simon64	Simon96	Simon128
0	(15,1,0,0)	(23,1,0,0)	(31,1,0,0)	(47,1,0,0)	(63,1,0,0)
1	(16,0,0,0)	(24,0,0,0)	(32,0,0,0)	(48,0,0,0)	(64,0,0,0)
2	(15,1,0,0)	(23,1,0,0)	(31,1,0,0)	(47,1,0,0)	(63,1,0,0)
3	(13,1,2,0)	(21,1,2,0))	(29,1,2,0)	(45,1,2,0)	(61,1,2,0)
4	(10,1,2,3)	(17,2,2,3)	(25,2,2,3)	(41,2,2,3)	(57,2,2,3)
5	(6,1,2,7)	(12,1,4,7)	(20,1,4,7)	(36,1,4,7)	(52,1,4,7)
6	(2,1,2,11)	(5,1,4,14)	(12,1,4,15)	(27,2,4,15)	(43,2,4,15)
7	(0,0,2,14)	(1,0,1,22)	(6,0,2,24)	(19,1,3,25)	(35,1,3,25)
8	(0,0,0,16)	(0,0,0,24)	(2,0,0,30)	(12,0,1,35)	(27,1,1,35
9	(0,0,0,16)	(0,0,0,24)	(0,0,0,32)	(6,0,0,42)	(20,0,1,43)
10	(0,0,0,16)	(0,0,0,24)	(0,0,0,32)	(2,0,0,46)	(12,0,1,51)
11	(0,0,0,16)	(0,0,0,24)	(0,0,0,32)	(0,0,0,48)	(6,0,0,58)
12	(0,0,0,16)	(0,0,0,24)	(0,0,0,32)	(0,0,0,48)	(2,0,0,62)
13	(0,0,0,16)	(0,0,0,24)	(0,0,0,32)	(0,0,0,48)	(0,0,0,64)

However, the variables for the user-selected key remains the same, so the degree of overdefinedness in the total system will increase with the number of (P,C)-instances used. This is also confirmed in the experiments reported in Sect. 4.

When exploiting the fact that many of the cipher state bits from different (P,C)-instances are linearly related, the equations in the total system will be a lot more connected than if only the variables for the key were common. In the next section we develop a very simple solving algorithm that exploits the linear relations that have been shown to exist.

4   Simple Solving Algorithm and Experiments

By creating variables as described in Sect. 3.1 we can easily construct the basic equations in one (P,C)-instance. It is also straight-forward to construct a system from multiple (P,C)-instances by introducing new variables in each instance. Identifying and using all the linear relations we know exist in the first rounds of the Simon ciphers are a bit more complicated.

4.1   Identifying Linear Relations

Assume we have constructed all basic Eq. (1) as described in Sect. 3.1, including the use of Proposition 1, and say we used t chosen (P, C)-instances. Then we will get a system of $tn(r-2)$ equations in $tn(r-4) + k$ variables.

The form of each equation is $l_a \cdot l_b + x_c = 0$ for some linear combinations l_a and l_b. If we have two such equations

$$l_a \cdot l_b + x_c = 0$$
$$l_d \cdot l_e + x_f = 0$$

where either

$$\deg(l_a + l_d) = \deg(l_b + l_e) = 0 \text{ or } \deg(l_a + l_e) = \deg(l_b + l_d) = 0 \qquad (2)$$

we know that adding the two equations together will yield a linear equation. This linear equation can be used to eliminate one variable. We could use diagrams like those in Fig. 3 to find exactly which pairs of equations that will yield linear relations when added together. A simpler but more time consuming method is just to try adding all pairs of equations and see which additions that become linear.

To get a fast check of whether two equations will give a linear relation we do not actually multiply out the product $l_a \cdot l_b$ in (1), but rather just check if condition (2) holds.

4.2   Solving Algorithm

After finding all linear relations in a set of equations representing Simon, we can use them to eliminate variables. The whole process can then be repeated, to see if some other pairs of equations will yield new linear equations after the first substitution of variables. The pseudocode for the whole algorithm is given in Algorithm 1, and is nothing more than a simplified version of ElimLin [12].

Algorithm 1. Solving system of equations on the form $l_a \cdot l_b + l_c = 0$.

---
repeat
    for every pair of equations f, g **do**
        if $f + g$ is linear **then**
            Eliminiate one variable from system
        end if
    end for
until No more linear equations found

The complexity of Algorithm 1 on a system with m equations in n variables can be estimated as follows. The inner for-loop iterates $\binom{m}{2} = \mathcal{O}(m^2)$ times.

Table 3. Number of times an equation system representing Simon32/64 was solved out of 256 tries

r \ t	4	5	6	8	10	12	16	20
7	0	0	0	0	37	42	44	198
8	0	68	109	164	250	251	252	254
9	26	213	240	253	256	255	256	256
10	0	2	152	255	255	255	255	256
11	0	0	0	0	0	3	35	50

Table 4. Number of times an equation system representing Simon48/72 was solved out of 256 tries

r \ t	4	5	6	8	10	12	16	20
8	6	159	185	231	255	256	256	256
9	74	229	247	256	256	256	256	256
10	0	44	231	256	256	256	256	256
11	0	0	0	0	1	27	220	250

At least one variable has to be eliminated from the system each time the for-loop is run, otherwise the algorithm stops. As there are n variables in the system the for-loop can be run at most n times, so the worst-case complexity for the whole algorithm is $\mathcal{O}(m^2 n)$.

At the beginning of the research for this paper, Algorithm 1 was only considered to be a preprocessing step for the equation system before applying a more advanced solver. We were a bit surprised to find that the algorithm actually solved systems, by eliminating *all* variables. If we store all the linear equations used to eliminate variables we get a uniquely defined linear system of equations, which is easily solved to give the variables for the user-selected key. We have verified that the returned solution from the solver actually gives the correct key.

4.3 Experiments

We have investigated how often our algorithm is able to solve the various Simon systems, using different number of rounds r and different numbers t of (P, C)-instances. The results are reported in the tables below.

For each choice of r and t parameters we have generated 256 systems and tried to solve them using Algorithm 1. The number of times we succeeded is indicated in each cell. For each system, the P_1-half of the plaintexts were chosen to be the bit string representing i, for $0 \leq i \leq t - 1$. When t is less than 128 we know that the output of the AND-operation in the first round will be 0 for all P_1's. An initial value v was chosen at random for the other half of the plaintext, and the P_0-values were given as $v \oplus (i << 2)$ for each of the t plaintexts. The key was chosen at random for every system.

Attempts to solve systems representing more rounds than reported in the tables were unsuccessful in the experiments.

When Algorithm 1 fails, we have tried to apply the CRHS method [13] for solving the remaining system Table 3. This has only been successful in the smaller cases of Simon32/64 and Simon48/72 in seven to nine round variants, using relatively few (i.e. three or four) plaintext/ciphertext pairs Table 4. When Algorithm 1 stops without returning a solution we are left with a non-linear equation systems containing several thousand variables in the larger cases Table 5. These systems were too big to handle for the CRHS method on an ordinary computer.

Table 5. Number of times an equation systemrepresenting Simon64/96 was solved out of 256 tries

t / r	4	5	6	8	10	12	16	20
8	0	7	20	54	211	226	242	256
9	1	115	177	229	255	256	256	256
10	5	135	242	256	256	256	256	256
11	0	0	38	253	256	256	256	256
12	0	0	0	0	5	137	255	256

Table 6. Number of times an equation system representing Simon96/96 was solved out of 256 tries

t / r	4	5	6	8	10	12	16	20
9	8	165	222	248	256	256	256	256
10	2	89	234	253	256	256	256	256
11	0	5	66	231	256	256	256	256
12	0	1	23	143	233	248	256	255
13	0	0	0	82	189	233	251	255
14	0	0	0	0	2	62	225	250

Table 7. Number of times an equation system representing Simon128/128 was solved out of 256 tries

t / r	4	5	6	8	10	12	16	20
10	1	63	226	247	256	256	256	256
11	0	7	71	236	255	256	256	256
12	0	0	15	153	227	250	256	256
13	0	0	12	77	166	220	250	254
14	0	0	1	35	104	164	220	243
15	0	0	0	15	51	106	172	214
16	0	0	0	0	0	13	106	182

It is still possible that deeper insight into the particular equations for Simon systems may give a better solving algorithm for these special systems Table 6. This could be a topic for future work.

5 Conclusions

Simon's resistance against differential and linear cryptanalysis has been thoroughly investigated in [2–4]. In this paper we have looked at Simon's resistance against algebraic attacks, contributing to the cryptanalysis of the Simon family of ciphers.

We have traced how two (P, C)-instances of Simon, differing in a single bit in the plaintexts, remain related through the first rounds of the cipher. Two instances of Simon32/64 are still linearly connected in two bits in the input of the eighth round, and two instances of Simon128/128 are linearly related up to and including the thirteenth round Table 7. Using these relation mappings one can reduce the number of variables a lot when creating equation systems representing Simon.

We developed a simple solving algorithm that exploits the many linear relations found in Simon. Using several chosen plaintexts, we tested how many rounds of Simon that could be broken with this method. The experiments verified the findings of Sect. 3, in that we always could solve a few more rounds of Simon after the linear relations from the plaintext side had completely disappeared. It is unusual that one can solve an equation system representing 16 rounds of a real cipher with 128-bit block and key sizes. Even though the largest systems tested initially contained 15488 variables, and 4839 variables after the first iteration of finding linearly related cipher state bits, they could be solved in three and a half minutes on a MacBook Air 2013.

To compensate for its very lightweight structure, the Simon ciphers have rather many rounds. The attack reported here does not threaten the security of Simon, but it is nevertheless interesting to see that large non-linear equation systems representing some cipher can still be efficiently solved when the cipher has very simple operations.

References

1. Beaulieu, R., Shors, D., Smith, J., Treatman-Clark, S., Weeks, B., Wingers, L.: The SIMON and SPECK Families of Lightweight Block Ciphers, Cryptology ePrint Archive, Report 2013/404 (2013). http://eprint.iacr.org/2013/404
2. Biryukov, A., Roy, A., Velichkov, V.: Differential analysis of block ciphers SIMON and SPECK. In: Cid, C., Rechberger, C. (eds.) FSE 2014. LNCS, vol. 8540, pp. 546–570. Springer, Heidelberg (2015)
3. Abed, F., List, E., Lucks, S., Wenzel, J.: Differential cryptanalysis of round-reduced Simon and Speck. In: Cid, C., Rechberger, C. (eds.) FSE 2014. LNCS, vol. 8540, pp. 525–545. Springer, Heidelberg (2015)
4. Wang, Q., Liu, Z., Varici, Y., Sasaki, V., Rijmen, Y.: Cryptanalysis of reduced-round SIMON32 and SIMON48. In: Meier, W., Mukhopadhyay, D. (eds.) Progress in Cryptology – INDOCRYPT 2014. LNCS, pp. 143–160. Springer, Switzerland (2014)
5. Alkhzaimi, H.A., Lauridsen, M.M.: Cryptanalysis of the SIMON Family of Block Ciphers, Cryptology ePrint Archive, Report 2013/543 (2013). http://eprint.iacr.org/2013/543
6. Alizadeh, J., Bagheri, N., Gauravaram, P., Kumar, A., Sanadhya, S.K.: Linear Cryptanalysis of Round Reduced SIMON, Cryptology ePrint Archive, Report 2013/663 (2013). http://eprint.iacr.org/2013/663
7. Wang, N., Wang, X., Jia, K., Zhao, J.: Improved Differential Attacks on Reduced SIMON Versions, Cryptology ePrint Archive, Report 2014/448 (2014). http://eprint.iacr.org/2014/448
8. Alizadeh, J., Alkhazaimi, H.A., Aref, M.R., Bagheri, N., Gauravaram, P., Lauridsen, M.M.: Improved Linear Cryptanalysis of Round Reduced SIMON, Cryptology ePrint Archive, Report 2014/681 (2014). http://eprint.iacr.org/2014/681
9. Shi, D., Hu, L., Sun, S., Song, L., Qiao, K., Ma, X.: Improved Linear (hull) Cryptanalysis of Round-reduced Versions of SIMON, Cryptology ePrint Archive, Report 2014/973 (2014). http://eprint.iacr.org/2014/973
10. Ahmadian, Z., Rasoolzadeh, S., Salmasizadeh, M., Aref, M.R.: Automated Dynamic Cube Attack on Block Ciphers: Cryptanalysis of SIMON and KATAN, Cryptology ePrint Archive, Report 2015/040 (2015). http://eprint.iacr.org/2015/040

11. Kölbl, S., Leander, G., Tiessen, T.: Observations on the SIMON block cipher family, Cryptology ePrint Archive, Report 2015/145 (2015). http://eprint.iacr.org/2015/145
12. Courtois, N.T., Bard, G.V.: Algebraic cryptanalysis of the data encryption standard. In: Galbraith, S.D. (ed.) Cryptography and Coding 2007. LNCS, vol. 4887, pp. 152–169. Springer, Heidelberg (2007)
13. Schilling, T.E., Raddum, H.: Solving compressed right hand side equation systems with linear absorption. In: Helleseth, T., Jedwab, J. (eds.) SETA 2012. LNCS, vol. 7280, pp. 291–302. Springer, Heidelberg (2012)

Cryptanalysis of the Full 8.5-Round REESSE3+ Block Cipher

Jorge Nakahara Jr.[(✉)]

Jose Benedito de Moraes Leme, 13, Jaguare', Sao Paulo 05336-060, Brazil
jorge_nakahara@yahoo.com.br

Abstract. This paper describes the first independent cryptanalysis of the full 8.5-round REESSE3+ block cipher, a large-block variant of the IDEA cipher. We show that large classes of weak keys exist in REESSE3+, just like in IDEA, under differential and linear attacks. Moreover, doubling the number of rounds is not enough to avoid weak keys. The existence of weak keys jeopardizes the use of REESSE3+ as a building block in the construction of other cryptographic primitives such as hash functions in modes such as Davies-Meyer's. We also describe square and impossible differential attacks on reduced-round versions.

Keywords: Cryptanalysis · IDEA · Weak keys · Block cipher design

1 Introduction

The REESSE3+ cipher, designed by Su and Lu, was described in [28] and is an *ad hoc* 128-bit block variant of the IDEA cipher. Another new feature of REESSE3+ is the fact that its MA-box is a fixed, unkeyed permutation (ressembling an iterated Even-Mansour cipher).

The IDEA block cipher [20,21] dates back to 1991 and has been subject to extensive cryptanalysis ever since: an attack by Meier in [23] on up to 2.5 rounds; a differential-linear attack by Borst *et al.* [7] on 3 rounds; a truncated-differential attack by Borst *et al.* [7] on 3.5 rounds; a square attack by Demirci [13] on up to 4 rounds; an impossible differential attack by Biham *et al.* [2] on up to 4.5-rounds; a meet-in-the-middle attack by Demirci *et al.* [14] on 5 rounds, and so on. There are attacks on the full 8.5-round IDEA, such as linear cryptanalysis [10,29], differential-linear [17], and boomerang attacks [5], but they all operate under weak-key assumptions, namely, the (unknown) user key should cause multiplicative subkeys equal to 0 or 1. There are other attacks such as the biclique technique [18] that do not require weak keys.

IDEA became well-known probably because it was embedded in the widely distributed Pretty Good Privacy (PGP) software package from Phil Zimmerman [16], for general-purpose file encryption.

We describe the first independent analysis of the full 8.5-round REESSE3+: differential and linear attacks (under weak-key assumptions) as well as square and impossible-differential analyses of reduced-round versions (without weak-key

© Springer International Publishing Switzerland 2015
K. Lauter and F. Rodríguez-Henríquez (Eds.): LatinCrypt 2015, LNCS 9230, pp. 170–186, 2015.
DOI: 10.1007/978-3-319-22174-8_10

assumptions). We also briefly discuss REESSE3+ as an iterated Even-Mansour construction.

This paper is organized as follows: Sect. 2 describes the REESSE3+ block cipher; Sect. 3 describes the key schedule of REESSE3+; Sect. 4 compares the relative costs of implementing REESSE3+ and IDEA; Sect. 5 describes weak keys of REESSE3+ and its use in differential and linear attacks; Sect. 6 describes square attacks; Sect. 7 describes impossible-differential attacks; Sect. 9 concludes the paper.

2   The REESSE3+ Block Cipher

REESSE3+ operates on 128-bit text blocks, under a 256-bit key and iterates eight rounds plus an output transformation [28]. Figure 1 depicts the computational graph for encryption.

Both IDEA and REESSE3+ employ only three group operations on 16-bit words: exclusive-or (denoted $\oplus$), addition in $\mathbb{Z}_{2^{16}}$ (denoted $\boxplus$), and multiplication in $GF(2^{16}+1)$ (denoted $\odot$), with the exception that $0 \equiv 2^{16}$. The condition $0 \equiv 2^{16}$ has at least two important consequences: (i) every element in $GF(2^{16}+1)$ has a multiplicative inverse, even 0 has a multiplicative inverse, which is 0 itself; (ii) all elements in $GF(2^{16}+1)$ fit in 16 bits.

A common feature of both ciphers is that the same group operation is never used twice in succession across the entire cipher. Also, both ciphers contain a so called MA-box (Multiplication-Addition box), a bijective mapping composed of addition and multiplication operations alternatedly and in a zig-zag order. But, IDEA uses a 32-bit keyed MA-box, while REESSE3+'s 64-bit MA-box is not keyed. Nonetheless, the authors in [28] did not list explicitly all the design criteria for the round function nor for the MA-box of REESSE3+.

One full round of REESSE3+ can be split into a Key-Mixing (KM) and a Multiplication-Addition (MA) half-rounds, in this order. In the i-th KM half-round, the 8-word input $(X_1, X_2, X_3, X_4, X_5, X_6, X_7, X_8)$ is combined via keyed-addition or keyed-multiplication with round subkeys resulting in the following output:

$$(Y_1, Y_2, \ldots, Y_8) = (X_1 \odot Z_1^{(i)}, X_2 \boxplus Z_2^{(i)}, X_3 \boxplus Z_3^{(i)},$$
$$X_4 \odot Z_4^{(i)}, X_5 \boxplus Z_5^{(i)}, X_6 \odot Z_6^{(i)}, X_7 \odot Z_7^{(i)}, X_8 \boxplus Z_8^{(i)}),$$

where $Z_j^{(i)}$ denotes the j-th subkey of the i-the round, $1 \leq j \leq 8$, $1 \leq i \leq 9$. Subkey which are combined via $\boxplus$ are called additive subkeys. Subkeys which are combined via $\odot$ are called multiplicative subkeys.

The MA half-round contains a Multiplication-Addition box (MA-box) whose input is the 4-tuple

$$(T_1, T_2, T_3, T_4) = (Y_1 \oplus Y_3, Y_2 \oplus Y_4, Y_5 \oplus Y_7, Y_6 \oplus Y_8).$$

The MA-box of REESSE3+ is a key-independent, fixed permutation mapping, MA: $\mathbb{Z}_{2^{16}}^4 \rightarrow \mathbb{Z}_{2^{16}}^4$. It can be denoted $MA(T_1, T_2, T_3, T_4) = (W_1, W_2, W_3, W_4)$. See Fig. 1.

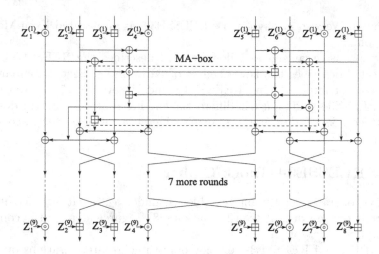

Fig. 1. Computational graph of REESSE3+ for encryption.

The output of the MA-box is (W_1, W_2, W_3, W_4), where $\odot$ has higher priority than $\boxplus$:

$$W_1 = (T_1 \odot T_2) \boxplus ((T_1 \odot T_2) \boxplus T_3) \odot T_4;$$

$$W_2 = T_1 \boxplus ((T_1 \odot T_2) \boxplus T_3) \odot T_4;$$

$$W_3 = T_4 \odot ((T_1 \odot T_2) \boxplus ((T_1 \odot T_2) \boxplus T_3) \odot T_4)$$

and

$$W_4 = ((T_1 \odot T_2) \boxplus T_3) \odot T_4.$$

Note that W_1 already depends on all four input words (T_1, T_2, T_3, T_4), which depend on all Y_i's, $1 \leq i \leq 8$. Consequently, W_2, W_3 and W_4 also depend on all eight eight Y_i, although under different algebraic relations. Nonetheless, there is complete text diffusion across a single MA-box, and this dependence will spread further to the entire round.

Next, the W_i values are combined with the original input to the MA half-round:

$$U = (U_1, U_2, U_3, U_4, U_5, U_6, U_7, U_8) = (Y_1 \oplus W_1, Y_2 \oplus W_2, Y_3 \oplus W_1,$$
$$Y_4 \oplus W_2, Y_5 \oplus W_3, Y_6 \oplus W_4, Y_7 \oplus W_3, Y_8 \oplus W_4).$$

Finally, there is a mapping σ that swaps words at the end of the round: $\sigma(U) = (U_1, U_3, U_2, U_5, U_4, U_7, U_6, U_8)$. Thus, σ is an involution, that is, $\sigma \circ \sigma(U) = U$, $\forall U$.

The i-th round output is:

$$\sigma(U) = (Y_1 \oplus W_1, Y_3 \oplus W_1, Y_2 \oplus W_2, Y_5 \oplus W_3, Y_4 \oplus W_2, Y_7 \oplus W_3, Y_6 \oplus W_4, Y_8 \oplus W_4).$$

This procedure for a full round is repeated eight times. The output transformation (OT) consists of σ plus a KM half-round. Thus, the OT counts as half a round. Moreover, since σ is an involution, the swap in the last round is effectively removed by the one in the OT, and the decryption framework of REESSE3+ becomes very similar to the one for encryption except for the order and value of the subkeys in the KM half-rounds: subkeys added in encryption are subtracted for decryption, and subkeys multiplied during encryption are divided during decryption.

3   The Key Schedule of REESSE3+

There are eight subkeys per round in REESSE3+, and all of them are in the KM half-rounds. In total, 72 round subkeys are required for either encryption or decryption.

Let the 256-bit user key be denoted $K = (k_0, k_1, k_2, \ldots, k_{254}, k_{255})$.

The key schedule algorithm of REESSE3+ is as follows (starting with $i = 1$):

(1) the 128 most significant bits K, $(k_0, k_1, k_2, \ldots, k_{126}, k_{127})$, is partitioned into eight 16-bit contiguous pieces, and each piece is assigned to the eight round subkeys of the i-th KM half-round: $Z_1^{(i)}$, $Z_2^{(i)}$, $Z_3^{(i)}$, $Z_4^{(i)}$, $Z_5^{(i)}$, $Z_6^{(i)}$, $Z_7^{(i)}$ and $Z_8^{(i)}$.
(2) K is left rotated by 25 bits and $i = i + 1$.
(3) steps (1) and (2) are repeated until all 72 round subkeys are generated.

Notice that key diffusion is very slow: only 25 additional bits, out of 256, are added per round, due to the fixed rotation by 25 bits. In total, $\lceil \frac{256}{25} \rceil = 6$ rounds will be required until all 256 key bits are used at least once during the encryption process. This key schedule can generate subkeys beyond 8.5 rounds, i.e. create additional round subkeys for an arbitrary number of rounds. This fact will be exploited in the following sections.

4   Relative Performance

From Sect. 2, REESSE3+ contains seven $\odot$, seven $\boxplus$ and twelve $\oplus$ per round, while IDEA uses four $\odot$, four $\boxplus$ and six $\oplus$ per round. But, REESSE3+ operates on 128-bit blocks, while IDEA operates on 64-bit blocks.

Note that both ciphers: (i) iterate the same number of rounds, 8.5; (ii) use the same three group operations; (iii) have very similar key schedule algorithms. Thus, it is expected that an implementation (in software or hardware) of REESSE3+ be faster than IDEA (per 128-bit block encrypted), as long as it in a chained mode (not ECB nor CTR, otherwise, two IDEA instances could be run in parallel). This fact motivated the cryptanalysis of REESSE3+: since there are less operations per round in REESSE3+ compared to (two instances of) IDEA, how the REESSE3+ design resists the cryptanalysis techniques that were applied to IDEA, such as differential and linear attacks?

5   Weak Keys of REESSE3+

From the key schedule described in Sect. 3, it is clear that each subkey contains an exact copy of (some) bits of the user key K. These bits may or may not be contiguous, but recovery of any subkey immediately reveals the same number of bits of the user key K. REESSE3+'s key schedule is actually very similar to the one used in IDEA [20], and thus, the same weaknesses [10] are inherited by REESSE3+.

An important issue concerning key schedule algorithms is the existence (and the number) of weak keys. A weak key for REESSE3+ is a 256-bit key K which, under its key schedule algorithm (Sect. 3), leads to multiplicative subkeys with value 0 or 1 [10]. The existence of weak keys is straightforward to demonstrate. Consider the user key consisting of 256 zero bits. From the key schedule in Sect. 3, all round subkeys will be zero. Consequently, all multiplicative subkeys will be zero as well, satisfying any condition dictated by an attack trail.

Now, let's discuss how weak keys help differential and linear attacks separately.

5.1   Differential Cryptanalysis Under Weak Keys

For differential cryptanalysis (DC) [4], we assume the difference operator is bitwise exclusive-or, $\oplus$. The DC analysis to REESSE3+, under weak keys, is similar to the analysis applied to IDEA [10].

Let $\Delta X = X \oplus X*$ denote a wordwise difference. Under multiplication with a subkey Z, the input difference 8000_x causes the output difference: $(X \odot Z) \oplus ((X \oplus \Delta) \odot Z) = (X \odot Z) \oplus ((X \oplus 8000_x) \odot Z)$.

Note that $-X = X \oplus 8000_x$, and $0 \equiv 2^{16} \equiv -1 \mod (2^{16} + 1)$.

Under a weak subkey ($Z \in \{0, 1\}$), this output difference becomes either:

- $(X \odot 1) \oplus ((X \oplus 8000_x) \odot 1) = X \oplus X \oplus 8000_x = 8000_x$,
- $(X \odot 0) \oplus ((X \oplus 8000_x) \odot 0) = (X \odot (-1)) \oplus (-X \odot (-1)) = (-X) \oplus X = 8000_x$.

So, in both cases, the difference 8000_x is a fix-point difference for the multiplication operation under a weak subkey.

Selecting a differential trail or characteristic for REESSE3+ involves balancing several criteria:

(i)   covering as many rounds as possible
(ii) finding a *narrow trail*; the final objective is that it holds with high probability
(iii) minimize the number of weak-subkey assumptions

Balancing these criteria typically requires some kind of trade-off. Our preliminary analysis is based on narrow trails and started with low Hamming Weight differences, e.g. one single nonzero word difference, following (ii). Such difference would initially be placed on top of an additive subkeys entry, to satisfy (iii), such as $(0, \Delta, 0, 0, 0, 0, 0, 0)$. Soon, it became clear that such differences would have to cross the MA-box, leading to very low overall probability and unpredictable differences at the round output (after a single round), because of the unkeyed multiplications in the MA-box. Moreover, imposing weak-key conditions would not help control the output difference since there are no subkeys in the MA-box (unlike in IDEA). It also became clear that the round output difference would hardly be the same as the input difference (and neither of low Hamming Weight).

Further, still using narrow trails but with nonzero differences in two words, we exploited differences that avoid the MA-box. Such differences cause the input difference to the MA-box to be $(0, 0, 0, 0)$. This means that all pairs of words that form the input to the MA-box have either zero difference or the same wordwise difference. For instance, the round input difference $(\Delta, 0, \Delta, 0, 0, 0, 0, 0)$ would remain the same after the KM half-round assuming $Z_1^{(i)}$ is a weak subkey, and where $\Delta = 8000_x$. Further, the input difference to the MA-box is $(0, 0, 0, 0)$, that is, we bypass the MA-box altogether. The round output difference is $(\Delta, \Delta, 0, 0, 0, 0, 0, 0)$, which does not allow to bypass the MA-box in the next round. In summary, it is not an iterative characteristic.

We also exploited the fact that the MA half-round is mirror symmetric, that is, the left half of the block is a mirror image of the right half. This symmetry motivates the use of mirror symmetric differences such as $(\Delta, 0, \Delta, 0, 0, \Delta, 0, \Delta)$, but the round output difference becomes $(\Delta, \Delta, 0, 0, 0, 0, \Delta, \Delta)$. The arrow (topped by '1r') indicates that the difference on the left-hand side propagates (or 'causes') the difference on the right-hand side after one full round.

Similarly,

$$(0, \Delta, 0, \Delta, \Delta, 0, \Delta, 0) \xrightarrow{1r} (0, 0, \Delta, \Delta, \Delta, \Delta, 0, 0).$$

Neither of them is iterative, and the output differences would not avoid the MA-box in the following round. Building characteristics in this fashion, round-by-round, would not allow one to bound the total probability of the characteristic (for increasing number of rounds), and would hardly lead to iterative characteristics. We concluded that the best overall strategy is to focus on iterative characteristics from the beginning.

In our analysis, we arrived at the following one-round iterative characteristic, as the best trade-off:

$$(\Delta, \Delta, \Delta, \Delta, \Delta, \Delta, \Delta, \Delta) \xrightarrow{1r} (\Delta, \Delta, \Delta, \Delta, \Delta, \Delta, \Delta, \Delta), \qquad (1)$$

where '$\Delta = 8000_x$' means the difference between 16-bit words, i.e. if X and X^* are two words in the same position in a block, then their difference is $X \oplus X^* = 8000_x$. This difference propagates with certainty for keyed addition, because the difference 8000_x only affects the most significant bit. The problem to overcome is to propagate 8000_x across $\odot$ with high probability. This will be accomplished with weak (multiplicative) subkeys.

Notice that (1) is iterative, that is, the input and output differences are identical. This means that (1) can be concatenated with itself, covering an arbitrary number of rounds. This fact is motivated by criterium (i).

In order for (1) to hold for as many rounds as possible, it is necessary that the difference Δ can propagate across the KM half-rounds with high probability. This criterium is achieved by assuming the multiplicative subkeys $Z_1^{(i)}$, $Z_4^{(i)}$, $Z_6^{(i)}$ and $Z_7^{(i)}$ are weak. This condition is motivated by (iii).

If the weak-subkey condition is fulfilled, this characteristic effectively bypasses the MA-box, that is, the input (and output) differences to the MA-box are both equal to $(0,0,0,0)$. This way, we do not have to propagate differences across the MA-box. Notice that (1) is not a 'narrow' trail. It is actually a *wide trail*, but the fact that it avoids the MA-box altogether effectively improves the overall probability, which fulfills criterium (ii).

We have implemented a search algorithm to count how many user keys exist that lead to weak subkeys at the positions especified by $Z_1^{(i)}$, $Z_4^{(i)}$, $Z_6^{(i)}$ and $Z_7^{(i)}$, for as many rounds i as possible. It is a simple counting procedure. A weak subkey, being 0 or 1, means that its most significant fifteen bits have to be zero. We apply this simple reasoning to each of the four multiplicative subkeys listed above, and check how many of the 256 key bits are not affected, that is, can hold any value. For instance, for the first round, $4 \cdot 15 = 60$ bits of the user key K will have to be zero, while the remaining $256 - 60 = 196$ bits are free to hold any value. This means that for a subset (or a weak-key class) of 2^{196} keys K, (1) will hold with certainty.

Further, concatenating (1) with itself, that is, covering two rounds, will require zeroing 60 bits more, but due to overlapping bits which are already zero (from the first round), only 46 bits are set to zero. This leaves 150 free bits. It means a weak-key class of 2^{150} keys K satisfing the 2-round characteristic using (1), for which it holds with certainty.

This procedure, repeated for three rounds, leads to a weak-key class of 2^{124} keys K. For four rounds, the weak-key class has 2^{100} keys. For five rounds, 2^{76} weak keys. For six rounds, 2^{52} weak keys. For seven rounds, 2^{33} weak keys. For eight rounds, 2^{23} weak keys. For nine rounds, which covers all KM half-rounds, there are 2^{15} weak keys.

Therefore, there are 2^{15} keys for which (1) concatenated with itself nine times becomes a differential distinguisher holding across the full 8.5-round REESSE3+ with certainty. Such keys have free bits in the following fifteen positions: 20, 45, 121, 122, 123, 124, 125, 126, 127, 128, 151, 176, 201, 226, 251 (recall that K is numbered in left-to-right order from 0 to 255).

Even though REESSE3+ was defined with 8.5 rounds, one can extend it to more rounds, since the key schedule Sect. 3 can generate as many round subkeys as needed. See Sect. 5. Let us assume, hypothetically, that REESSE3+ had double the number of rounds instead. We can continue with the analysis above, up to 16 rounds (not including the output transformation). The result is a weak-key class containing 2 weak keys. Only the bit position 126 in K is free to hold any value.

This means that 16-round REESSE3+ still has weak keys (although its performance would be cut in half). The existence of a single weak-key value is enough to jeopardize its use in the construction of other cryptographic primitives, such as hash function, via a mode of operation such as Davies-Meyer [24] in which the key entry is part of the message to be hashed. Thus, an adversary could choose a weak key as message, and further manipulate the other entries to lead to a collision. This analysis has been performed for IDEA and PES ciphers in [25].

5.2 Linear Cryptanalysis Under Weak Keys

Similar to the attack described in Sect. 5.1, we started a linear cryptanalysis [22] of REESSE3+ looking for linear relations with low Hamming Weight, in the hope of exploiting a narrow linear trail for as many rounds of REESSE3+ as possible.

We found out that there are linear relations crossing the MA-box with the highest bias 2^{-1}, such as

$$(1,0,0,0) \overset{MA-box}{\rightarrow} (0,1,0,1). \tag{2}$$

See Fig. 1. This linear relation exploits the leftmost word input to the MA-box, which requires approximating the least significant bit of a single $\boxplus$ across the MA-box. Thus, this word seems to be the weakest entry concerning linear cryptanalysis. Other linear relations such as $(0,1,0,0) \overset{MA-box}{\rightarrow} (1,1,0,0)$ and $(0,0,0,1) \overset{MA-box}{\rightarrow} (1,0,1,0)$ requires crossing one $\odot$ in the MA-box, which costs bias $2^{-13.47}$.

The linear relation (2) leads to the following one-round linear approximation: $(1,1,1,0,0,0,0,1) \overset{1r}{\rightarrow} (0,0,1,0,0,0,0,1)$, requiring one weak-subkey condition: $Z_1^{(1)}$. The drawback is similar to the situation faced in Sect. 5.1: the round output mask is not the same as the input mask (it is not iterative), and building linear relations this way would not allow to bound the total bias to a small enough value. In the end, we concluded that the best strategy is to focus on iterative linear relations.

The use of weak (multiplicative) subkeys Z in linear masks M is similar to the reasoning using differences. Let $(X \odot Z) \cdot M$ denote a linear approximation to $\odot$ using mask M.

Under a weak subkey $(Z \in \{0, 1\})$, this approximation becomes either:

$- (X \odot 1) \cdot 1 = X \cdot 1,$
$- (X \odot 0) \cdot 1 = (X \odot (-1)) \cdot 1 = (-X) \cdot 1 = (X \oplus 8000_x) \cdot 1 = X \cdot 1.$

So, in both cases, the mask 1 only concerns the least significant bit of X. It means that the multiplication by a weak subkey does not affect linear approximations of the least significant bit of its operand.

The linear relation that achieves the best trade-off involving criteria (i), (ii) and (iii) in Sect. 5.1 is:

$$(1,1,1,1,1,1,1,1) \overset{1r}{\rightarrow} (1,1,1,1,1,1,1,1), \tag{3}$$

where '1' denotes a 16-bit mask which explores the least significant bit of a 16-bit word. This linear relation starts in the KM half-round. Across a keyed addition, this relation has bias 2^{-1}, since it approximates the least significant bit (lsb) of a word. Since there is no carry bit in the lsb, the bias is maximum. Across the four keyed multiplications, though, the propagation will depend on weak-subkey assumptions.

Note that the linear relation (3) is iterative, which means it can be concatenated with itself for an arbitrary number of rounds. The exact number of rounds is a trade-off between the number of weak-key conditions and the existence of keys that satisfy those conditions. Coincidently, the weak subkeys needed for (3) are in the exact same positions and in the exact same quantity as the ones needed for (1). Thus, the very same reasoning applies concerning the size of weak-key classes for increasing number of rounds as in Sect. 5.1.

Note that (3) is a wide linear trail, just like (1) is a wide differential trail.

Conclusion: differential and linear attacks under weak-key conditions are possible for the full 8.5-round REESSE3+ and even for 16-round REESSE3+. That is, at least one weak key exists.

6   Square Attacks

Recall that all internal operations in REESSE3+ are wordwise. This neat word-wise cipher framework makes it an ideal target to square attacks.

The original Square attack was developed by Knudsen *et al.* [11] as a dedicated technique against the Square block cipher, which is a byte-oriented design. This attack was later adapted to other word-oriented block ciphers. Since all internal operations in the REESSE3+ are on 16-bit words, Square attacks employ preferably this word size. Recall that Square (variant) attacks are among the best known attacks on reduced-round versions of the AES cipher [12].

Square attacks on REESSE3+ work with λ-sets, that is, multi-sets of 2^{16t} plaintexts in which different words (in a block) contain values with multiplicities ($t \geq 1$). We use the following terminology for different patterns of values: 'A' denotes an active 16-bit word (in which each value in the set $\{0, 1, 2, 3, \ldots, 2^{16} - 1\}$ appear exactly once in that word); 'P' denotes a passive word (in which each value in the word is the same); 'B' denotes a balanced word (in which the xor-sum of all values in the word is zero), and '?' an unbalanced word (there is no discernible pattern, e.g. the xor-sum of the values in a given word is unpredictable).

The design of the MA-box of REESSE3+ imply that the most effective λ-set that can propagate across the MA-box, with certainty, has the form (P, P, A, P) $\overset{MA-box}{\to} (A, A, A, A)$, that is, a single 'A' word in the third input position makes all four outputs also 'A'. Thus, the pattern (P, P, A, P) preserves the highest number of active words across the MA-box among all patterns 'A', 'P', 'B' and '?'.

Other patterns were not as successful across the MA-box: $(A, P, P, P) \overset{MA-box}{\to}$ $(?, ?, ?, A)$, $(P, P, P, A) \overset{MA-box}{\to} (A, A, ?, A)$ and, $(P, A, P, P) \overset{MA-box}{\to} (?, A, ?, A)$.

The pattern $(P, P, A, P) \overset{MA-box}{\to} (A, A, A, A)$ for the MA-box translates into the following one-round λ-sets: $(P, P, P, P, A, P, P, P) \overset{1r}{\to} (A, A, A, B, A, A, A, A)$ and $(P, P, P, P, P, P, A, P) \overset{1r}{\to} (A, A, A, A, A, B, A, A)$.

Further, across the MA-box

- the pattern $(B, ?, B, B) \overset{MA-box}{\to} (?, ?, ?, ?)$ mean that the previous one-round square distinguisher can be extended only across an additional KM half-round: $(P, P, P, P, A, P, P, P) \overset{1.5r}{\to} (A, A, A, ?, A, A, A, A)$,
- the pattern $(B, B, B, ?) \overset{MA-box}{\to} (?, ?, ?, ?)$ mean that the previous one-round square distinguisher can be extended only across an additional KM half-round: $(P, P, P, P, P, P, A, P) \overset{1.5r}{\to} (A, A, A, A, A, ?, A, A)$.

Both of them can be used as distinguishers of 2-round REESSE3+: from the output ciphertext λ-set pick pairs of words that reconstruct the input of the MA-boxes from the bottom up. For instance, let (a, b, c, d, e, f, g, h) denote the output λ-set. Then, the $(a \oplus c, b \oplus d, e \oplus g, f \oplus h)$ should be (B, B, B, B).

This 2-round distinguishing attack can be extended to a key-recovery attack on 2.5 rounds.

Without loss of generality, such an attack works as follows:

- choose 2^{16} plaintexts with word patterns according to the λ-set (P, P, P, P, A, P, P, P); the λ-set after 1.5 rounds is $(A, A, A, ?, A, A, A, A)$, but after two full rounds the λ-set contains only '?' in all eight word positions.
- guess the subkeys $Z_1^{(3)}$ and $Z_2^{(3)}$ (considering the word swap), and partially decrypt the last KM and MA half-round for all texts in the output λ-set;
- keep the subkey values for which the decrypted λ-set has the pattern 'B', the leftmost entry to the MA-box of the second round; each non-'?' word pattern is used as a distinguisher to test for the correct key guess;
- for a wrong subkey pair there is a chance of 2^{-16} that the word 'B' is satisfied: essentially the xor-sum should be 0 in all sixteen bits. Using two λ-sets, we get an error probability of 2^{-32}. Since there are $2^{32} - 1$ wrong subkeys, the expected number of wrong subkeys that survive this filtering is $(2^{32}-1) \cdot 2^{-32} < 1$, and only the correct subkey pair is expected to survive. The same procedure is repeated for the other four subkeys of the third KM half-round. The time complexities are the same, but the same data can be used for all subkeys, and at the same storage cost.

The attack complexity is $2 \cdot 2^{16}$ chosen plaintexts (CP), $4 \cdot 2^{32}$ KM half-round partial computations, which means $2^{34} \cdot 0.5/2.5 \approx 2^{32}$ 2.5-round computations, and 2^{17} blocks for storage.

Note that this attack applies equally well to three full rounds of REESSE3+. This fact follows from the absence of subkeys in the MA half-round (which can be easily removed by the adversary).

7 Impossible Differential Attack

The impossible-differential (ID) analysis technique was developed by Biham *et al.* in [1,2] and by Knudsen in [19]. For this analysis applied to REESSE3+, the difference operator used is exclusive-or.

The ID distinguishers for REESSE3+ are constructed using the Miss-in-the-Middle technique [2]: consider 2.5-round REESSE3+ starting from an MA half-round. The input difference $(\delta, 0, \delta, 0, 0, \delta, 0, \delta)$ cannot cause the output difference $(\delta, \delta, 0, 0, 0, 0, \delta, \delta)$ 2.5 rounds later, where δ is any nonzero (16-bit) word difference (the same difference value in every position). The reasoning is as follows:

- from an MA half-round, the input difference to the MA-box becomes $(0, 0, 0, 0)$ since all word pairs xored prior to the MA-box result in zero difference. The output of the MA half-round is $(\delta, \delta, 0, 0, 0, 0, \delta, \delta)$, already taking into account the word swap.
- after the KM half-round, the difference becomes $(\delta_1, \delta_2, 0, 0, 0, 0, \delta_3, \delta_4)$, where $\delta_i \neq 0$ for $i \in \{1, 2, 3, 4\}$, but not necessarily equal to δ.
- from the bottom-up direction, the difference $(\delta, \delta, 0, 0, 0, 0, \delta, \delta)$ leads to $(\delta, 0, \delta, 0, 0, \delta, 0, \delta)$ at the input to the last MA half-round since the pairs of entries that form the input to the MA-box have all difference zero.
- the input to the last KM half-round is $(\delta_5, 0, \delta_6, 0, 0, \delta_7, 0, \delta_8)$, where $\delta_j \neq 0$ for $j \in \{5, 6, 7, 8\}$, but they are not necessarily equal to δ.
- the output from the second MA half-round, before the word swap, becomes $(\delta_5, \delta_6, 0, 0, 0, 0, \delta_7, \delta_8)$.
- so, the leftmost word in a block have difference δ_1 from the top-down direction, before the xor, and difference δ_5 from the bottom-up direction, after the xor. Similarly, the second word from the left has difference δ_2 from the top-down, and difference δ_6 from the bottom-up.
- notice that the differences δ_1 and δ_5 are related by the leftmost output difference, say x, from the MA-box. Let us denote it by $x = \delta_1 \oplus \delta_5$. Likewise, δ_2 and δ_6 are related by the second leftmost output difference, say y, from the MA-box. Let us denote it by $y = \delta_2 \oplus \delta_6$.
- but, x is also xored to the third leftmost word in block, and y is xored to the fourth leftmost word in a block. Following the zero word difference from both the top and bottom of 2.5 rounds, we find out that $x = 0$ and $y = 0$. That means that the output difference of this MA-box is $(0, 0, 0, 0)$. But, the input difference to this MA-box is $(\delta_1, \delta_2, \delta_3, \delta_4)$, which is nonzero by construction. Moreover, the MA-box is a permutation mapping. Contradiction.
- therefore,

$$(\delta, 0, \delta, 0, 0, \delta, 0, \delta) \overset{2.5r}{\nrightarrow} (\delta, \delta, 0, 0, 0, 0, \delta, \delta), \tag{4}$$

is a 2.5-round ID distinguisher for REESSE3+, where the crossed arrow indicates the difference on the left-hand side can never cause the difference on the right-hand side of the arrow.

Analogously, by pairing other word differences both before and after 2.5 rounds of REESSE3+, the following ID distinguisher can be demonstrated:

$$(0, \delta, 0, \delta, \delta, 0, \delta, 0) \xrightarrow{2.5r} (0, 0, \delta, \delta, \delta, \delta, 0, 0), \tag{5}$$

Following the key schedule of REESSE3+, assume we recover the whole 128-bit round subkey from the i-th and $(i+3)$-th KM half-rounds surrounding a 2.5-round ID distinguisher. Let us denote them by K_i and K_{i+3}. They are related as $K_{i+3} = K_i \lll 75$. Since both K_i and K_{i+3} are derived from the same 256-bit user key, they share $128-75 = 53$ bits independent of i. Thus, without loss of generality, we attack the first 3.5 rounds.

The attack, using (4), works as follows: choose a structure of 2^{64} plaintexts of the form $(X, 0, Y, 0, 0, U, 0, V)$ where $X, Y, U, V \in \mathbb{Z}_2^{16}$ assume all possible 16-bit values, while the remaining four words are fixed constants. One such structure leads to $\frac{2^{64}(2^{64}-1)}{2} \approx 2^{127}$ text pairs with differences following the pattern $(\delta_1, 0, \delta_2, 0, 0, \delta_3, 0, \delta_4)$.

Let a plaintext block in the structure be denoted $(p_1, p_2, p_3, p_4, p_5, p_6, p_7, p_8)$, and a ciphertext block be denoted $(c_1, c_2, c_3, c_4, c_5, c_6, c_7, c_8)$.

At the ciphertext end, collect about $2^{127}/2^{64} = 2^{63}$ pairs which have zero difference in the four middle word positions, according to the difference pattern $(\delta_5, \delta_6, 0, 0, 0, 0, \delta_7, \delta_8)$.

For each remaining pair

- for all the 2^{32} possible values of $(Z_1^{(1)}, Z_3^{(1)})$ partially encrypt the topmost KM half-round

$$q_1 = (p_1 \odot Z_1^{(1)}) \oplus ((p_1 \oplus \delta_1) \odot Z_1^{(1)})$$

and

$$r_1 = (p_3 \boxplus Z_3^{(1)}) \oplus ((p_3 \oplus \delta_2) \boxplus Z_3^{(1)})$$

before the 2.5-round ID distinguisher and keep the subkey pairs whose differences are equal: $q_1 = r_1$. For all 2^{16} values of $Z_1^{(1)}$, compute and store q in 2^{16} words of memory (a table). Compute the other difference r for all 2^{16} values of $Z_3^{(1)}$ and look for a match in the table. This step costs 2^{16} words of memory and 2^{17} computations of q and r. About $2^{32}/2^{16} = 2^{16}$ shall match.

- similarly, for all the 2^{32} possible values of $(Z_6^{(1)}, Z_8^{(1)})$ partially encrypt the topmost KM half-round

$$q_2 = (p_6 \odot Z_6^{(1)}) \oplus ((p_6 \oplus \delta_3) \odot Z_6^{(1)})$$

and

$$r_2 = (p_8 \boxplus Z_8^{(1)}) \oplus ((p_8 \oplus \delta_4) \boxplus Z_8^{(1)})$$

before the 2.5-round ID distinguisher and keep the subkey pairs whose differences are equal: $q_2 = r_2$. This step costs the same as the previous step. About $2^{32}/2^{16} = 2^{16}$ shall match.

– try all the 2^{32} possible values of $(Z_1^{(4)}, Z_2^{(4)})$ and partially decrypt the bottommost KM half-round

$$q_3 = (c_1 \odot (Z_1^{(4)})^{-1}) \oplus ((c_1 \oplus \delta_5) \odot (Z_1^{(4)})^{-1})$$

and

$$r_3 = (c_2 \boxminus Z_2^{(4)}) \oplus ((c_2 \oplus \delta_6) \boxminus Z_2^{(4)})$$

and keep the subkey pairs whose differences are equal: $q_3 = r_3$. About $2^{32}/2^{16} = 2^{16}$ shall match.

– likewise, try all the 2^{32} possible values of $(Z_7^{(4)}, Z_8^{(4)})$ and partially decrypt the bottommost KM half-round

$$q_4 = (c_7 \odot (Z_7^{(4)})^{-1}) \oplus ((c_7 \oplus \delta_7) \odot (Z_7^{(4)})^{-1})$$

and

$$r_4 = (c_8 \boxminus Z_8^{(4)}) \oplus ((c_8 \oplus \delta_8) \boxminus Z_8^{(4)})$$

and keep the subkey pairs whose differences are equal: $q_4 = r_4$. About $2^{32}/2^{16} = 2^{16}$ shall match.

– make a list of all the 2^{64} 128-bit subkeys combining the previous steps. These subkeys cannot be the correct value because they satisfy the differences of an ID distinguisher.

Each pair defines a list of about 2^{64} wrong subkeys. Compute the union of these lists of wrong 128-bit subkeys. It is expected that after about 178 structures, the number of remaining wrong subkeys is $2^{128} \cdot (1 - 2^{-64})^{178 \cdot 2^{63}} \approx 2^{128} \cdot e^{-89} < 1$, and only the correct subkey value is expected to remain.

To mark the wrong subkey values, a bit vector of 2^{64} bits (2^{57} blocks) can be used. Initially, it is empty. Once a wrong subkey is found, it is marked in this vector. The only non-marked position indicates the correct subkey value.

This attack requires $178 \cdot 2^{64} \approx 2^{71}$ chosen plaintexts (CP) and about $4 \cdot 2^{63} \cdot 2^{17} \cdot 1/7 \approx 2^{79}$ 3.5-round computations (very rough estimate that a full KM half-round accounts for $1/7$ of 3.5 rounds).

This phase recovered subkeys $(Z_1^{(1)}, Z_3^{(1)}, Z_6^{(1)}, Z_8^{(1)}, Z_1^{(4)}, Z_2^{(4)}, Z_7^{(4)}, Z_8^{(4)})$.

Repeat the same procedure above using the ID distinguisher (5) to recover another set of 128-bit subkeys: $(Z_2^{(1)}, Z_4^{(1)}, Z_5^{(1)}, Z_7^{(1)}, Z_3^{(4)}, Z_4^{(4)}, Z_5^{(4)}, Z_6^{(4)})$.

In fact, recalling the key schedule of REESSE3+ (Sect. 3), the subkeys $(Z_6^{(1)}, Z_8^{(1)})$ share 16 key bits $(k_{112}, \ldots, k_{127})$, with the subkeys $(Z_3^{(4)}, Z_4^{(4)})$. Likewise, the subkeys $(Z_5^{(1)}, Z_7^{(1)})$ share 17 key bits, $(k_{25}, \ldots, k_{79}, k_{96}, \ldots, k_{106})$, with the subkeys $(Z_1^{(4)}, Z_2^{(4)})$.

Taking into account this bit overlap, there are 17 less key bits to recover with the ID distinguisher (5). Thus, the time and memory used is dominated by the attack using (4). The data complexity doubles. The same memory for the attack using (4) can be reused for (5).

These distinguishers allow a key-recovery attack on 3.5-round REESSE3+. In fact, since the MA half-round is key-independent, the attack reaches four rounds.

8   Even-Mansour Construction

The fact that the MA-box is a fixed permutation in all rounds of REESSE3+ makes it look like an r-round iterated Even-Mansour construction [15], that is, every single round of REESSE3+ consists of a fixed, public permutation (MA half-round) surrounded by keyed layers (KM half-rounds).

Previous results on analysis of Even-Mansour schemes include [6,8,9].

The result [8] is most relevant for REESSE3+:

- let us assume that the MA half-round of REESSE3+ can be modeled a 'random' permutation;
- REESSE3+ does not consist of a single 'random' permutation surrounded by keyed layers, as defined by Even and Mansour [15]. Rather, REESSE3+ consists of the same fixed permutation in every round, interleaved by keyed layers (KM half-rounds) which are not pairwise independent.

Nonetheless, [8] discusses a setting in which REESSE3+ could fit in: an iterated r-round Even-Mansour construction in which the round subkeys are not independent, and the fixed 'public' permutation is always the same mapping.

Assuming that the round subkeys of 8.5-round REESSE3+ are 'appropriate' according to the assumptions of [8], the 8.5-round REESSE3+ would be indistinguishable from a random permutation up to $O(2^{\frac{8n}{9}})$ queries of any adaptive adversary. Here, $n = 128$ bits, which means $O(2^{0.8 \cdot 128})$, that is, $O(2^{112.64})$ adaptive queries, which although far from practical: (i) is lower than the full codebook, (ii) is much lower than an exhaustive key search effort, (iii) is below the square-root bound for the key. Thus, the degree of indistinguishability-from-random of REESSE3+ is below half of that provided by a block cipher with a 256-bit key.

Table 1. Summary of attacks on REESSE3+ block cipher.

Type	#Rounds	Time	Data	Memory	Ref.
Differential	8.5–16	$O(1)$	$O(1)$ CP	$O(1)$	Sect. 5.1
Linear	8.5–16	$O(1)$	$O(1)$ KP	$O(1)$	Sect. 5.2
Square	2	2^{16}	2^{16} CP	$O(1)$	Sect. 6
Square	2.5–3	2^{32}	2^{17} CP	2^{17}	Sect. 6
ID	3.5–4	2^{79}	2^{72} CP	2^{57}	Sect. 7

CP: Chosen Plaintext; KP: Known Plaintext

9   Conclusions

This paper reported the first independent cryptanalyses of the REESSE3+ block cipher, with and without weak-key assumptions, both on the full 8.5-round and on reduced-round versions.

Some attacks, not yet finished, include the Biryukov-Demirci relation [14, 26], described in Appendix A.

A summary of the attack complexities in this paper are listed in Table 1.

Acknowledgements. I would like to thank the anonymous reviewers who provided detailed and valuable comments, which improved the readability and helped correct several mistakes in this paper.

A   Biryukov-Demirci Attack

The Biryukov-Demirci (BD) relation [14, 26] exploit a linear trail along two words in the middle of a block in the IDEA cipher framework, which are combined with intermediate cipher data only via $\oplus$ and $\boxplus$. For REESSE3+, the most promising linear trail involves the second and third words in the cipher framework (Fig. 1).

Let us denote the input plaintext block as $P = (p_1, p_2, p_3, p_4, p_5, p_6, p_7, p_8)$, the corresponding ciphertext block as $C = (c_1, c_2, c_3, c_4, c_5, c_6, c_7, c_8)$, and let the i-th MA-box output be denoted $(W_1^{(i)}, W_2^{(i)}, W_3^{(i)}, W_4^{(i)})$. Then, from Fig. 1, there is a linear relation involving the least significant bits (lsb) of $p_2 \oplus p_3$. As an example, for a single round (plus the output transformation):

$$\text{lsb}(p_2 \oplus p_3) = \text{lsb}(Z_2^{(i)} \oplus Z_3^{(i)} \oplus W_1^{(1)} \oplus W_2^{(1)} \oplus c_2 \oplus c_3), \qquad (6)$$

because there are only $\oplus$ and $\boxplus$ combined to p_2 and p_3 across 1.5 rounds, and the least significant bit of addition is not affected by carry bits. Likewise, similar relations can be derived for more rounds, leading to a key-dependent relation involving only two plaintext and two ciphertext bits, that hold with certainty.

The Biryukov-Demirci relation for r-round REESSE3+ is

$$\text{lsb}(p_2 \oplus p_3 \oplus c_2 \oplus c_3) = \text{lsb} \bigoplus_{i=1}^{r} (Z_2^{(i)} \oplus Z_3^{(i)} \oplus W_1^{(i)} \oplus W_2^{(i)}), \qquad (7)$$

This relation gives a 1-bit condition from the MA-box: $\text{lsb}(W_1^{(i)} \oplus W_2^{(i)})$, which changes from round to round, even though it is key-independent.

Unlike in IDEA, where this output bit from the MA-box could be traced back to the MA-box input (and a subkey), in REESSE3+, it is more involved, since (adapted from Sect. 2, with an indication to the i-th round): $W_1^{(i)} = (T_1^{(i)} \odot T_2^{(i)}) \boxplus ((T_1^{(i)} \odot T_2^{(i)}) \boxplus T_3^{(i)}) \odot T_4^{(i)}$ and $W_2^{(i)} = T_1^{(i)} \boxplus ((T_1^{(i)} \odot T_2^{(i)}) \boxplus T_3^{(i)}) \odot T_4^{(i)}$.

That is, $\text{lsb}(W_1^{(i)} \oplus W_2^{(i)}) = \text{lsb}((T_1^{(i)} \odot T_2^{(i)}) \oplus T_1^{(i)})$, where $T_1^{(i)}, T_2^{(i)}$ are the two leftmost inputs to the i-th MA-box.

The presence of the $\odot$ operator makes the relation (7) in REESSE3+ much more involved than the one in IDEA [14, 26], even though there are no subkeys in the MA-box.

The approach of [3] is an interesting topic for further research on the security of REESSE3+.

References

1. Biham, E., Biryukov, A., Shamir, A.: Cryptanalysis of skipjack reduced to 31 rounds using impossible differentials. In: Stern, J. (ed.) EUROCRYPT 1999. LNCS, vol. 1592, pp. 12–23. Springer, Heidelberg (1999)
2. Biham, E., Biryukov, A., Shamir, A.: Miss in the middle attacks on IDEA and Khufu. In: Knudsen, L.R. (ed.) FSE 1999. LNCS, vol. 1636, p. 124. Springer, Heidelberg (1999)
3. Biham, E., Dunkelman, O., Keller, N.: A new attack on 6-round IDEA. In: Biryukov, A. (ed.) FSE 2007. LNCS, vol. 4593, pp. 211–224. Springer, Heidelberg (2007)
4. Biham, E., Shamir, A.: Differential Cryptanalysis of the Data Encryption Standard. Springer, New York (1993)
5. Biryukov, A., Nakahara Jr., J., Preneel, B., Vandewalle, J.: New weak-key classes of IDEA. In: Deng, R.H., Qing, S., Bao, F., Zhou, J. (eds.) ICICS 2002. LNCS, vol. 2513, pp. 315–326. Springer, Heidelberg (2002)
6. Biryukov, A., Wagner, D.: Slide attacks. In: Knudsen, L.R. (ed.) FSE 1999. LNCS, vol. 1636, pp. 245–259. Springer, Heidelberg (1999)
7. Borst, J.: Differential-linear cryptanalysis of IDEA. Technical report, ESAT Department, COSIC group, pp. 96–102 (1996)
8. Chen, S., Lampe, R., Lee, J., Seurin, Y., Steinberger, J.: Minimizing the two-round even-mansour cipher. In: Garay, J.A., Gennaro, R. (eds.) CRYPTO 2014, Part I. LNCS, vol. 8616, pp. 39–56. Springer, Heidelberg (2014)
9. Daemen, J.: Limitations of the even-mansour construction. In: Matsumoto, T., Imai, H., Rivest, R.L. (eds.) ASIACRYPT 1991. LNCS, vol. 739, pp. 495–498. Springer, Heidelberg (1993)
10. Daemen, J., Govaerts, R., Vandewalle, J.: Weak keys for IDEA. In: Stinson, D.R. (ed.) CRYPTO 1993. LNCS, vol. 773, pp. 224–231. Springer, Heidelberg (1994)
11. Daemen, J., Knudsen, L.R., Rijmen, V.: The block cipher SQUARE. In: Biham, E. (ed.) FSE 1997. LNCS, vol. 1267, pp. 149–165. Springer, Heidelberg (1997)
12. Daemen, J., Rijmen, V.: AES Proposal: Rijndael. In: 1st AES Conference, California, USA (1998)
13. Demirci, H.: Square-like attacks on reduced rounds of IDEA. In: Nyberg, K., Heys, H. (eds.) SAC 2002. LNCS, vol. 2595, pp. 147–159. Springer, Heidelberg (2002)
14. Demirci, H., Ture, E., Selçuk, A.A.: A new meet-in-the-middle attack on the IDEA block cipher. In: Matsui, M., Zuccherato, R.J. (eds.) SAC 2003. LNCS, vol. 3006, pp. 117–129. Springer, Heidelberg (2004)
15. Even, S., Mansour, Y.: A construction of a cipher from a single pseudorandom permutation. J. Cryptol. **10**(3), 151–162 (1997)
16. Garfinkel, S.: PGP: Pretty Good Privacy. O'Reilly and Associates, Sebastopol (1994)
17. Hawkes, P.M.: Asymptotic bounds on differential probabilities and an analysis of the block cipher IDEA. The University of Queensland, St. Lucia, Australia (1998)
18. Khovratovich, D., Leurent, G., Rechberger, C.: Narrow-bicliques: cryptanalysis of full IDEA. In: Pointcheval, D., Johansson, T. (eds.) EUROCRYPT 2012. LNCS, vol. 7237, pp. 392–410. Springer, Heidelberg (2012)
19. Knudsen, L.R.: DEAL - a 128-bit block cipher. Technical report #151, University of Bergen, Department of Informatics, Norway (1998)
20. Lai, X.: On the design and security of block ciphers. In: Massey, J.L. (ed.) ETH Series in Information Processing, vol. 1. Hartung-Gorre Verlag, Konstanz (1995)

21. Lai, X., Massey, J.L.: Markov ciphers and differential cryptanalysis. In: Davies, D.W. (ed.) EUROCRYPT 1991. LNCS, vol. 547, pp. 17–38. Springer, Heidelberg (1991)
22. Matsui, M.: Linear cryptanalysis method for DES cipher. In: Helleseth, T. (ed.) EUROCRYPT 1993. LNCS, vol. 765, pp. 386–397. Springer, Heidelberg (1994)
23. Meier, W.: On the security of the IDEA block cipher. In: Helleseth, T. (ed.) EURO-CRYPT 1993. LNCS, vol. 765, pp. 371–385. Springer, Heidelberg (1994)
24. Menezes, A.J., van Oorschot, P.C., Vanstone, S.: Handbook of Applied Cryptography. CRC Press, Gary (1997)
25. Nakahara Jr., J., Preneel, B., Vandewalle, J.: A note on weak keys of PES, IDEA, and some extended variants. In: Boyd, C., Mao, W. (eds.) ISC 2003. LNCS, vol. 2851, pp. 267–279. Springer, Heidelberg (2003)
26. Nakahara Jr., J., Preneel, B., Vandewalle, J.: The Biryukov-Demirci attack on reduced-round versions of IDEA and MESH ciphers. In: Wang, H., Pieprzyk, J., Varadharajan, V. (eds.) ACISP 2004. LNCS, vol. 3108, pp. 98–109. Springer, Heidelberg (2004)
27. Nakahara Jr., J., Rijmen, V., Preneel, B., Vandewalle, J.: The MESH block ciphers. In: Chae, K.-J., Yung, M. (eds.) WISA 2003. LNCS, vol. 2908, pp. 458–473. Springer, Heidelberg (2004)
28. Su, S., Lu, S.: A 128-bit block cipher based on three group arithmetics. IACR ePrint archive, 2014/704 (2014)
29. Yıldırım, H.M.: Some linear relations for block cipher IDEA. The Middle East Technical University (2002)

Meet-in-the-Middle Attacks on Reduced-Round Hierocrypt-3

Ahmed Abdelkhalek, Riham AlTawy, Mohamed Tolba, and Amr M. Youssef[✉]

Concordia Institute for Information Systems Engineering,
Concordia University, Montréal, QC, Canada
youssef@ciise.concordia.ca

Abstract. Hierocrypt-3 is an SPN-based block cipher designed by Toshiba Corporation. It operates on 128-bit state using either 128, 192 or 256-bit key. In this paper, we present two meet-in-the-middle attacks in the single-key setting on the 4-round reduced Hierocrypt-3 with 256-bit key. The first attack is based on the differential enumeration approach where we propose a truncated differential characteristic in the first 2.5 rounds and match a multiset of state differences at its output. The other attack is based on the original meet-in-the-middle attack strategy proposed by Demirci and Selçuk at FSE 2008 to attack reduced versions of both AES-192 and AES-256. For our attack based on the differential enumeration, the master key is recovered with data complexity of 2^{113} chosen plaintexts, time complexity of 2^{238} 4-round reduced Hierocrypt-3 encryptions and memory complexity of 2^{218} 128-bit blocks. The data, time and memory complexities of our second attack are 2^{32}, 2^{245} and 2^{242}, respectively. To the best of our knowledge, these are the first attacks on 4-round reduced Hierocrypt-3.

Keywords: Cryptanalysis · Hierocrypt-3 · Meet-in-the-middle attack · Differential enumeration

1 Introduction

Hierocrypt-3 (HC-3) [11,24,28], designed by Toshiba Corporation in 2000, is a 128-bit block cipher that was submitted to the New European Schemes for Signatures, Integrity, and Encryption (NESSIE) project [23]. It is among the Japanese e-Government 2003 recommended ciphers list [10] and the 2013 candidate recommended ciphers list [9]. HC-3 employs a nested Substitution Permutation Network (SPN) structure as proposed in [24]. In nested SPN structure, each round has two substitution layers with distinct linear transformations and hence is equivalent to two rounds in normal SPN structure.

In the self-evaluation report done by Toshiba Corporation [28], HC-3 was concluded to be sufficiently secure against all well-known attacks at that time. Nevertheless, Barreto *et al.* presented an improved square attack against reduced round HC-3 [5]. They showed that HC-3 is vulnerable up to 3 rounds for 128-bit key, and up to 3.5 rounds for 192, 256-bit keys. These attacks are the best attacks

© Springer International Publishing Switzerland 2015
K. Lauter and F. Rodríguez-Henríquez (Eds.): LatinCrypt 2015, LNCS 9230, pp. 187–203, 2015.
DOI: 10.1007/978-3-319-22174-8_11

on HC-3 so far. Then, Cheon *et al.* presented a 2-round impossible differential that can be used to attack up to 3 rounds of HC-3 [20]. Furuya and Rijmen analyzed the key scheduling of HC-3 and showed many linear relations between the round key bits [27]. Finally, after the introduction of the biclique attack in 2011 [6], Rechberger evaluated the security of HC-3, among other 128-bit block ciphers, against the biclique attacks [25].

In 1977, Diffie and Hellman were the first to propose the meet-in-the-middle (MitM) attack for the cryptanalysis of Data Encryption Standard (DES) [16]. Since then, the attack has evolved to analyze many block ciphers such as PRINCE and PRESENT [8], KTANTAN [7], LBlock [4], mCrypton [18], and Kuznyechik [3]. In addition, the MitM technique was used on hash functions to present preimage or second preimage attacks exemplified by the work done on HAS-160 [19], Whirlpool [26], Streebog [1], and Whirlwind [2]. Demirci and Selçuk were the first to apply the MitM approach on AES [13] triggering a new line of research, this attack is hereafter referred to as the *plain* MitM attack. They have shown that if the input of 4 AES rounds has just one active byte then the value of each byte of the output can be described as a parameterized function of that active byte. The number of parameters was deduced to be 25 8-bit parameters in [13] and then reduced to 24 8-bit parameters in [14]. The reduction in the number of parameters was possible by noticing that the 25th parameter is a key byte that is constant for all functions. Therefore, by considering the differences of the functions rather than the mere values, only 24 parameters can be used. The main disadvantage of the *plain* MitM attack, even with the reduced number of parameters, is the high memory requirement to store a precomputation table of all the sequences resulting from all the possible combinations of these parameters. As such, the *plain* MitM attack only works for AES-256 and then a time/memory tradeoff is used to extend the attack to AES-192.

At ASIACRYPT 2010, Dunkelman, Keller, and Shamir [17] proposed a couple of new ideas to address the high memory requirement of the *plain* MitM attack. First, they showed that the precomputation table does not need to have the whole sequence, just its associated multiset, i.e., the unordered sequence with multiplicity rather than the ordered one. The introduction of the multiset concept reduced the size of the precomputation table by a factor of 4. However, the more significant reduction in the size of the precomputation table was due to the second and main idea which they called the differential enumeration. Differential enumeration reduced the number of parameters that describe the sequence or rather the multiset from 24 bytes to 16 bytes. This is achievable by relying on a low probability truncated differential characteristic where the generated sequence or multiset at its output can only take a restricted number of values. Consequently, the memory requirement has been reduced from 2^{192} to 2^{128}. However, the use of this truncated differential characteristic increases the data complexity as now we have to search through a large amount of input data pairs to find one pair that conforms to the used truncated differential characteristic.

Later on, at Eurocrypt 2013, Derbez *et al.* showed that it is possible, by borrowing ideas from the rebound attack [22], to enumerate the whole set of sequences more efficiently [15]. In particular, they showed that using their technique, which they called efficient enumeration, the whole set can take only 2^{80} values instead of 2^{128}. This means that the number of parameters is further reduced to 10 parameters only. The consequences of using the efficient enumeration technique were numerous. Firstly, the attack became feasible on AES-128 and in fact, their attack is considered the most efficient attack on 7-round reduced AES-128. Secondly, the use of a 5-round truncated differential characteristic is feasible which mounts to attacking 9-round reduced AES-256. Thirdly, the memory complexity is no longer the bottleneck of the attack.

Finally, Li *et al.* [21] employed a key-dependent sieve to further reduce the memory complexity of Derbez's attack. This technique helped them present an attack on 9-round reduced AES-192 using a 5-round truncated differential characteristic and an attack on 8-round reduced PRINCE.

In this work, we present two MitM attacks on HC-3; the first attack uses the idea of efficient differential enumeration while the second one utilizes the *plain* MitM attack in a data/memory tradeoff approach. Contrary to AES, HC-3 alternates between two distinct linear transformations; the first linear transformation, similar to the MixColumns transformation in AES, operates on 4 bytes while the other linear transformation acts on the whole 16-byte state. Nevertheless, we manage to construct a 2.5-round truncated differential characteristic that we use to mount an attack on 4-round reduced HC-3. In the second attack, we show that if the input has one active byte in a specific position then after 2 rounds of HC-3, certain bytes of the output can be described by a function of that active byte parameterized by 30 8-bit parameters. We use this 2-round distinguisher to attack 4-round reduced HC-3 as well with much less data complexity but higher memory and time complexities.

The rest of the paper is organized as follows. In Sect. 2, we describe the HC-3 block cipher and provide the notation used throughout the paper. Section 3 discusses the attack based on the efficient differential enumeration technique where we describe the chosen truncated differential characteristic and the adopted attack procedure. Afterwards in Sect. 4, we provide a brief description of our *plain* MitM attack on HC-3. Finally, we conclude the paper in Sect. 5.

2   Specification of Hierocrypt-3

HC-3 is an iterated nested SPN block cipher that operates on a 128-bit state with either 128, 192 or 256-bit key. In the nested SPN structure, the low level SPN structure is recursively used in the SPN of the higher level. That is, one large SBox is composed of two substitution layers with smaller SBoxes and one linear transformation in the middle of them. Therefore, one round in nested SPN structure has two substitution layers and hence corresponds to two rounds in normal SPN structure. The number of rounds in HC-3 varies with the cipher key size. For 128-bit keys, HC-3 has 6 rounds, for 192-bit keys there are 7 rounds, and for 256-bit keys, 8 rounds. In all cases, the last round is slightly different than the other rounds.

At the higher SPN level, an HC-3 round, as shown in Fig. 1, consists of three transformations:

- X: A subkey mixing layer consisting of xoring the 128-bit input with 16-byte subkey.
- XS: A layer of 4 parallel 32×32-bit keyed substitution boxes.
- H: A linear transformation consisting of xoring 8-bit subdata $x_{i(8)} (\in GF(2)^8; i = 0, 1, \cdots 15)$, which is represented by a matrix MDS_H. The matrix MDS_H and its inverse MDS_H^{-1} are given in Appendix A.

At the lower SPN level, each 32×32-bit XS-box consists of:

- S: A nonlinear layer composed of simultaneous application of 4 8×8-bit SBoxes.
- L: A bytewise linear transformation defined by a 4×4 matrix called mds_l. L has a branch number of 5, i.e., when the input (resp. output) has 1 active byte then the output (resp. input) must have 4 active bytes.
- X: A subkey mixing layer consisting of xoring the 32-bit input with 4-byte subkey.
- S: Another nonlinear layer composed of simultaneous application of 4 8×8-bit SBoxes. Hence, each round consists of two SBox layers and in our attacks below, the counting is done per SBox layer.

In the last round, the H transformation is replaced by a post-whitening key addition. Hence, the full encryption function of the 256-bit key version of HC-3 where the ciphertext C is evaluated from the plaintext P can be described as:

$$C = X[K_1^{(9)}] \circ ((S \circ X[K_2^{(8)}] \circ L \circ S) \circ X[K_1^{(8)}]) \circ \cdots$$
$$\circ (H \circ (S \circ X[K_2^{(1)}] \circ L \circ S) \circ X[K_1^{(1)}])(P)$$

For the convenience of describing our attacks, we use a different representation of HC-3, similar to the alternative expression used in [20]. As illustrated in Fig. 2, the state is represented as a 4×4 matrix where each element in the matrix represents a state byte. In this representation, mds_l operates column-wise, similar to the MixColumns operation in AES, and MDS_H acts on the whole matrix. In some cases, we are also interested in swapping the order of the linear transformations, either H or L, and the xor with the key X. These operations are linear and hence they can be interchanged, by first xoring the input with an equivalent key and then applying the linear transformation. The equivalent subkey at SBox layer i is denoted by U_i where $U_i = L^{-1}(K_i)$ when i is odd or $U_i = H^{-1}(K_i)$ when i is even. Additionally, we rely on the property of the SBox stated in Proposition 1 below:

Proposition 1 (Differential Property of S). *Given Δ_i and Δ_o two non-zero differences in $\mathbb{F}_{256}$, the equation: $S(x) + S(x + \Delta_i) = \Delta_o$ has one solution on average. This property also applies to S^{-1}.*

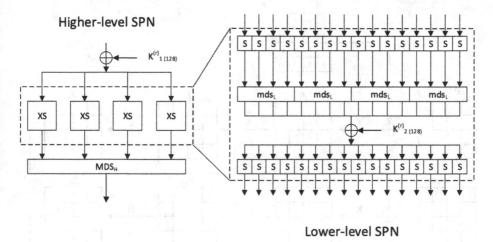

Fig. 1. One round of Hierocrypt-3.

Key Schedule. We now give a brief description of the 256-bit key schedule in HC-3.

The key state $Z_1 \| Z_2 \| Z_3 \| Z_4$ is 256-bit and undergoes 10 rounds to generate the 17 keys used in 8 rounds (2/1 keys per round/SBox layer) in addition to the final key. The first key state is denoted by $Z_1^{(-1)} \| Z_2^{(-1)} \| Z_3^{(-1)} \| Z_4^{(-1)}$ and instantiated with the master key. The first round is special as it omits a linear function and does not generate any round key. The other rounds can be split into 2 groups which we denote by $G1$ and $G2$. We focus our description on $G1$ as it is used to generate the round keys from round 1 up to round 5 which covers our attacked rounds. The key state words Z_3 and Z_4 are updated linearly every round, while Z_1 and Z_2 follow a Feistel structure with additional input from Z_3 and Z_4. Particularly, and as illustrated in Fig. 3, the key state words are updated as follows:

$$(Z_3^{(r)}, Z_4^{(r)}) = L_1(Z_3^{(r-1)}, Z_4^{(r-1)})$$
$$Z_1^{(r)} = Z_2^{(r-1)}$$
$$Z_2^{(r)} = Z_1^{(r-1)} \oplus F_\sigma(Z_2^{(r-1)} \oplus Z_3^{(r)})$$

where L_1 is a specific linear transformation and the function F_σ consists of a level of SBoxes followed by another, different than L_1, linear transformation. Then 64-bit keys $k_1^{(r)}, k_2^{(r)}, k_3^{(r)}$ and $k_4^{(r)}$ are generated as follows:

$$k_1^{(r)} = Z_1^{(r-1)} \oplus F_\sigma(Z_2^{(r-1)} \oplus Z_3^{(r)})$$
$$k_2^{(r)} = Z_3^{(r)} \oplus F_\sigma(Z_2^{(r-1)} \oplus Z_3^{(r)})$$
$$k_3^{(r)} = Z_4^{(r)} \oplus F_\sigma(Z_2^{(r-1)} \oplus Z_3^{(r)})$$
$$k_4^{(r)} = Z_4^{(r)} \oplus Z_2^{(r-1)}$$

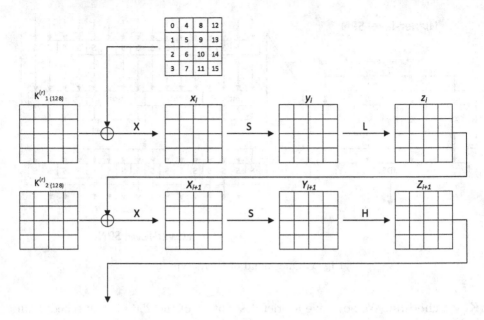

Fig. 2. Alternative representation of one round of Hierocrypt-3.

where $r = 1, 2, \cdots, 5$ then the 128-bit $K^{(r)}_{1(128)}$ is set to $k^{(r)}_1 \| k^{(r)}_2$ and $K^{(r)}_{2(128)}$ is set to $k^{(r)}_3 \| k^{(r)}_4$. It is to be noted that in our attacks, we number the keys sequentially from K_1 up to K_{17} (for the full cipher) where $K_i = K^{\lceil i/2 \rceil}_{1(128)}$ when i is odd and $K_i = K^{\lceil i/2 \rceil}_{2(128)}$ when i is even.

For further details regarding the SBox, the linear transformations or the key schedule, the reader is referred to [28].

2.1 Notations

The following notations are used throughout the paper:

- x_i, y_i: The 16-byte state after the X, S transformations at layer i, respectively.
- z_i: The 16-byte state after the linear transformation at layer i where the linear transformation is L (resp. H) when i is odd (resp. even).
- $x_i[j]$: The jth byte of the state x_i, where $j = 0, 1, \cdots, 15$, and the bytes are indexed as shown in Fig. 2.
- $x_i[j \cdots k]$: The bytes between the jth position and k^{th} position of the state x_i.
- $\Delta x_i, \Delta x_i[j]$: The difference at state x_i and byte $x_i[j]$, respectively.

We measure memory complexity of our attacks in number of 128-bit HC-3 blocks and time complexity in reduced-round HC-3 encryptions.

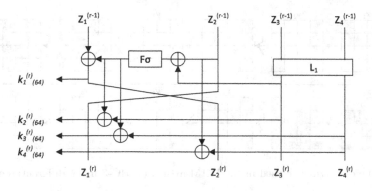

Fig. 3. 1 Round of Hierocrypt-3 key schedule.

3   A Differential Enumeration MitM Attack on HC-3

In a MitM attack, an r-round reduced block cipher is split into 3 consecutive parts of r_1, r_2 and r_3 rounds, $r = r_1 + r_2 + r_3$, such that a particular set of messages may verify a certain property in the middle r_2 rounds. In an offline phase, that particular property is evaluated independently of the keys used in the middle rounds. Then in an online phase, correct key candidates for the r_1 and r_3 rounds are checked whether they verify this distinguishing property or not. In our attacks, the chosen property is a truncated differential characteristic, such that when its input is a δ-set [12] captured by Definition 1, the set of each byte of the output states forms an ordered sequence or rather a multiset as in Dunkleman's attack and captured in Definition 2.

Definition 1. (δ-set of HC-3) *Let a δ-set be a set of 256 HC-3 states that are all different in one state byte (the active byte) and all equal in the other state bytes (the inactive bytes).*

Definition 2. (Multisets of bytes) *A multiset generalizes the set concept by allowing elements to appear more than once. In our case, a multiset of 256 bytes can take as many as* $\begin{pmatrix} 2^8 + 2^8 - 1 \\ 2^8 \end{pmatrix} \approx 2^{506.17}$ *different values.*

Our first proposed 8-layer (4-round) MitM attack relies on a 6-layer distinguisher. The distinguisher, as illustrated in Fig. 4, starts at x_1 and ends at the SBox transformation in layer 6, i.e., y_6. Proposition 2, below, is the core of our attack.

Proposition 2. *If a message m belongs to a pair of states conforming to the truncated differential characteristic of Fig. 4, then the multiset of differences $\Delta y_6[3]$ obtained from the δ-set constructed from m in $x_1[0]$ is fully determined by the following 27 bytes: $\Delta y_1[0]$, $x_2[0\cdots3]$, $x_3[0\cdots15]$, $\Delta y_6[3]$, $y_6[3]$ and $y_5[0\cdots3]$.*

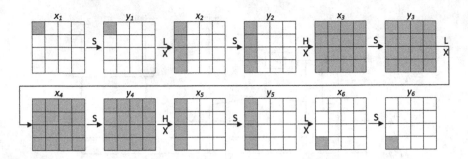

Fig. 4. The distinguisher used in the MitM attack on HC-3 using differential enumeration.

Proof. The proof uses rebound-like arguments borrowed from the hash function cryptanalysis [22] and used in [15]. Let (m, m') be a right pair that conforms to the truncated differential characteristic in Fig. 4. We show in the following how the knowledge of these 27 particular bytes is enough to compute the mulisets. The 27 bytes $\Delta y_1[0]$, $x_2[0\cdots 3]$, $x_3[0\cdots 15]$, $\Delta y_6[3]$, $y_6[3]$ and $y_5[0\cdots 3]$ can take as many as $2^{8\times 27} = 2^{216}$ possible values, and for each of them, we can determine the values of all the differences shown in Fig. 4. $\Delta y_1[0]$ can be propagated linearly through L to compute $\Delta x_2[0\cdots 3]$. With the knowledge of $x_2[0\cdots 3]$, we can bypass the SBox of layer 2 to reach y_2, then linearly through H to compute $\Delta x_3[0\cdots 15]$. With the knowledge of $x_3[0\cdots 15]$, we can bypass the SBox of layer 3 to reach y_3 and then linearly through L to compute $\Delta x_4[0\cdots 15]$. Similarly in the other direction, $\Delta y_6[3]$ and $y_6[3]$ enable us to bypass the SBox of layer 6 and compute $\Delta z_5[3]$, then linearly through L^{-1} we compute $\Delta y_5[0\cdots 3]$. With the knowledge of $y_5[0\cdots 3]$, we bypass the SBox of layer 5 to compute $\Delta z_4[0\cdots 3]$ and then linearly through H^{-1} we can compute $\Delta y_4[0\cdots 15]$. By the differential property of the HC-3 SBox (Proposition 1), there is, on average, one solution for each of the 16 bytes of the state x_4.

To construct the multiset for each of the 2^{216} possible values of the 27 bytes from Proposition 2, we consider all the 255 possible values for $\Delta y_1[0]$ and propagate them until y_6 with the help of the internal state solutions we have. This results in a multiset of 255 differences in $\Delta y_6[3]$. As the HC-3 SBox is a permutation over $\mathbb{F}_{256}$, the sequence in $\Delta y_1[0]$ allows to derive the sequence in $\Delta x_1[0]$.

Attack Procedure. Our attack recovers the 128-bit last round key K_9 and 13 bytes of $U_8 = L^{-1}(K_8)$ and is composed of a precomputation phase and an online phase.

Precomputation Phase. In the precomputation phase, we iterate over the 2^{216} possible values for the 27 bytes of Proposition 2 and for each of them, we deduce the possible values of the internal states as discussed in the above proof. These internal state values are then used to generate the multiset. We store all the multisets in a hash table.

Online Phase. The online phase is further divided into two steps; data collection and key recovery. First, we try to find pairs of messages that conform to the truncated differential characteristic in Fig. 5 in which the previous 6-layer characteristic is placed at the top. Next, the found pairs are used to create a δ-set and test them against the stored hash table to identify the correct key.

Data Collection. In this step, we query the encryption oracle with structures of chosen plaintexts to get enough pairs such that one conforms to the 8-layer truncated differential characteristic in Fig. 5. Each structure is composed of 2^8 plaintexts, where byte 0 takes all the 2^8 possible values while each of the other 15 bytes take any, possibly distinct, fixed value. Thus, each structure is expected to generate $2^8 \times (2^8 - 1)/2 \approx 2^{15}$ pairs. The probability that the whole truncated differential characteristic is verified is $2^{-7 \times 8 - 8 \times 8} = 2^{-120}$ because of the $16 \rightarrow 9$ transition over L^{-1} ($x_8 \rightarrow z'_7$) and the $9 \rightarrow 1$ transition over H^{-1} ($x_7 \rightarrow z'_6$), see Fig. 5. While the probability of the former transition over L^{-1} is trivial to deduce, the probability of the latter transition over H^{-1} was deduced by observing that if the input of H^{-1} consists of certain 9 active bytes having the same difference, then the output will have just one active byte in a specific position. This position can be either $3, 7, 11$ or 15 depending on the position of the 9 bytes. Hence, the probability of the $9 \rightarrow 1$ transition is equivalent to the probability of 9 active bytes having the same difference, i.e., $2^8/2^{8 \times 9} = 2^{-8 \times 8} = 2^{-64}$. This observation is also applicable to H, but the positions of the 9 active input bytes and the active output byte differs from the ones corresponding to H^{-1}. Since the probability of the whole truncated differential characteristic is 2^{-120}, then in order to find one pair that conforms to it, we need 2^{120} pairs which means 2^{105} structures of 2^8 messages. Therefore, we ask for the encryption of $2^{105+8} = 2^{113}$ messages to get the required 2^{120} pairs.

Key Recovery. For each of the 2^{120} pairs, we get $2^{8 \times (1+9)} = 2^{80}$ suggestions for the 25 key bytes: $K_9[0 \ldots 15]$ and $U_8[1, 3 \cdots 6, 8, 10, 13, 15]$. This is done as follows: we guess $\Delta y_7[1, 3 \cdots 6, 8, 10, 13, 15]$ and propagate it linearly through L to compute Δx_8. From the other side, Δx_9 is in fact the difference in the ciphertext pair and is also equal to Δy_8. So now, we have Δx_8 and Δy_8, so we use the differential property of the SBox and get a solution for each byte of x_8 and y_8 which enables us to deduce a key candidate for the whole K_9 by xoring the ciphertext with y_8, which in turn, helps compute z'_7. Then, we guess $\Delta y_6[3]$ and propagate it through H to get $\Delta x_7[1, 3 \cdots 6, 8, 10, 13, 15]$. Once again, we use the differential property of the SBox along with $\Delta y_7[1, 3 \cdots 6, 8, 10, 13, 15]$ which we guessed above to deduce a solution for the 9 bytes $y_7[1, 3 \cdots 6, 8, 10, 13, 15]$ which with z'_7 enables us to compute the corresponding 9 bytes in U_8. To summarize this part, we guess 10 bytes that help deduce 25 key bytes.

To compute the multiset at $\Delta y_6[3]$ which is equivalent to $\Delta z'_6[3]$, we noticed that extra key bytes are required. For example, if byte 3 is active in state y_i then, through H, it has an impact on 9 bytes in state z_i, these 9 bytes are $[1, 3 \cdots 6, 8, 10, 13, 15]$ (cf. 4th column in H). However, byte 3, through H^{-1}, is impacted by 11 bytes in state z_i, these 11 bytes are $[0, 2 \cdots 6, 8 \cdots 10, 12, 13]$ (cf. 4th row in H^{-1}). Hence, to compute the multiset, we need to guess the key bytes

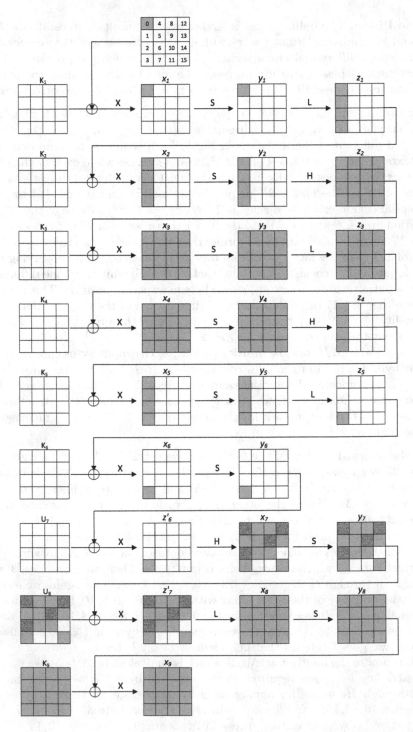

Fig. 5. Complete 8-layer truncated differential characteristic used in the attack using differential enumeration.

$[0, 2, 9, 12]$ on top of the above deduced keys. These key bytes are the gray cells in Fig. 5. This means that for each of the 2^{120} pairs, we have $2^{80+4\times 8} = 2^{112}$ key candidates. For each pair and for every key candidate, we build the plaintext δ-set, get the corresponding ciphertexts, compute the multiset and look for a match in the precomputation table. If a match is not found, we can discard that key candidate. The probability of a wrong key producing a valid multiset is given by $2^{120+112+216-467.6} = 2^{-19.6}$ which is negligible. Note that the probability of randomly having a match in the table is $2^{-467.6}$ (and not $2^{-506.7}$) because the number of ordered sequences associated to a multiset is not constant [15].

Now, our attack recovers the 16-byte K_9 and 13 bytes of U_8 which is the equivalent key of K_8 so, in order to recover the master key, we guess 3 bytes of U_8 and get 2^{24} candidates for K_8. This means that we have recovered the 64-bit keys $k_3^{(4)}, k_4^{(4)}, k_1^{(5)}$ and $k_2^{(5)}$. According to the key schedule, they are calculated as follows:

$$k_3^{(4)} = Z_4^{(4)} \oplus F_\sigma(Z_2^{(3)} \oplus Z_3^{(4)}) \tag{1}$$

$$k_4^{(4)} = Z_4^{(4)} \oplus Z_2^{(3)} \tag{2}$$

$$k_1^{(5)} = Z_1^{(4)} \oplus F_\sigma(Z_2^{(4)} \oplus Z_3^{(5)}) \tag{3}$$

$$k_2^{(5)} = Z_3^{(5)} \oplus F_\sigma(Z_2^{(4)} \oplus Z_3^{(5)}) \tag{4}$$

We start by guessing $Z_2^{(3)}$ which is of 64-bit length then from Eq. (2) we compute $Z_4^{(4)}$. From Eq. (1), we compute $Z_3^{(4)}$. From the key schedule, knowing $Z_3^{(4)}$ and $Z_4^{(4)}$ enables us to compute $Z_3^{(5)}$ and $Z_4^{(5)}$. Afterwards from Eq. (3) and considering that $Z_1^{(4)} = Z_2^{(3)}$, we compute $Z_2^{(4)}$. We use Eq. (4) as a 2^{-64} filter and we end up with one solution for $Z_2^{(3)}, Z_1^{(4)}, Z_2^{(4)}, Z_3^{(4)}, Z_4^{(4)}, Z_3^{(5)}$ and $Z_4^{(5)}$. Since we have $Z_1^{(4)}, Z_2^{(4)}, Z_3^{(4)}, Z_4^{(4)}$ we can recover the master key and get 2^{24} candidates for the master key corresponding to the 2^{24} candidates for K_8. We exhaustively search through these master key candidates to find the correct master key with no significant impact on the attack time complexity.

Attack Complexity. The memory requirement of the attack is due to the precomputation table needed to store 2^{216} multisets, each of 512 bits. Hence, the memory complexity of the attack is $2^{216} \times 512/128 = 2^{218}$ 128-bit blocks. The data complexity of the attack is attributed to the data collection step of the online phase where we query the encryption oracle with 2^{113} chosen plaintexts to generate 2^{120} pairs. The time complexity of the offline phase to construct the table is due to performing 2^{216} partial encryptions on 256 messages, which is equivalent to $2^{216+8} \times 2^{-2} = 2^{222}$ encryptions. The time complexity of the online phase to recover 29 key bytes (16-byte K_9 and 13 bytes from U_8) is the time required to partially decrypt the 2^8 values in a δ-set with all the 2^{112} key candidates for all the 2^{120} generated pairs, which is equivalent to $2^{120+112+8} \times 2^{-2} = 2^{238}$. The time complexity of the exhaustive search among the 2^{24} master key candidates using 2 plaintext/ciphertext pairs is $2 \times 2^{24} = 2^{25}$. Therefore, the time complexity of the attack is dominated by the time complexity of the online phase and is equivalent to $2^{238} + 2^{222} + 2^{25} \approx 2^{238}$.

4   A *plain* MitM Attack on HC-3

Our second proposed MitM attack is also launched on 8-layer of the 256-bit key version of HC-3 in a data/memory tradeoff approach. Although its memory and time complexities, as will be shown, are higher than the previous attack by a factor of 2^{24} and 2^7, respectively, it reduces the data complexity dramatically by a factor of 2^{81}. It follows the strategy used by Demirci and Selçuk, which we named above as the *plain* MitM. As illustrated in Fig. 6, the distinguisher in this attack covers 4 middle layers starting from x_2 to x_6. Proposition 3 is the base of our attack.

Proposition 3. *The ordered sequence of differences $\Delta x_6[2]$ obtained from the δ-set constructed by varying $x_2[3]$ is fully determined by the following 30 bytes: $x_2[3]$, $x_3[1, 3 \cdots 6, 8, 10, 13, 15]$, $x_4[0 \cdots 15]$ and $x_5[0 \cdots 3]$.*

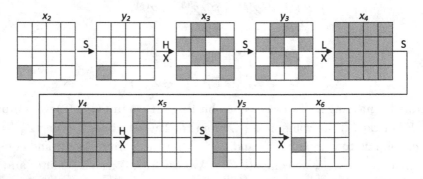

Fig. 6. The 4-layer distinguisher used in the MitM attack on HC-3 using Demirci and Selçuk approach.

Proof. The proof is similar to the proof of Proposition 2 without using the rebound-like arguments. We show in the following how the knowledge of these 30 particular bytes is enough to compute the ordered sequence of differences at $x_6[2]$. $\Delta x_2[3]$ and $x_2[3]$ help us bypass the SBox of layer 2 to compute $\Delta y_2[3]$ which is then propagated linearly through H to compute $\Delta x_3[1, 3 \cdots 6, 8, 10, 13, 15]$. With the knowledge of $x_3[1, 3 \cdots 6, 8, 10, 13, 15]$, we can bypass the SBox of layer 3 and then linearly through L, we compute $\Delta x_4[0 \cdots 15]$. With the knowledge of $x_4[0 \cdots 15]$, we can bypass the SBox of layer 4 and once again linearly through H, we compute $\Delta x_5[0 \cdots 15]$. With the knowledge of $x_5[0 \cdots 3]$, we can bypass the SBox of layer 5 in these 4 bytes to reach $y_5[0 \cdots 3]$ and then linearly through L to compute $\Delta x_6[2]$. The byte where the distinguisher starts was chosen such that it minimizes the number of parameters. $x_2[7, 11, 15]$ are other possible positions that would result in the same number of parameters. The byte where the ordered sequence is computed was chosen to minimize the number of key guesses in the online phase. $x_6[6, 10, 14]$ are other positions that would require the same number of key bytes.

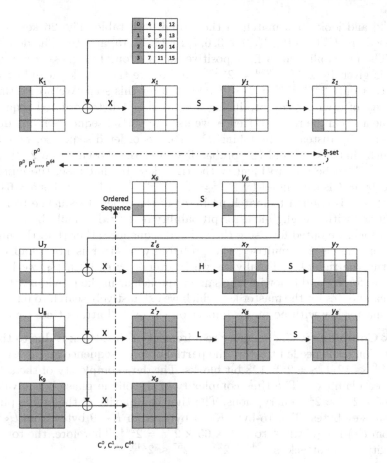

Fig. 7. The online phase of the 8-layer MitM attack on HC-3 using Demirci and Selçuk approach.

Attack Procedure. In this attack, we recover the 16-byte K_9, 9 bytes of $U_8 = L^{-1}(K_8)$, 1 byte of $U_7 = L^{-1}(K_7)$ and 4 bytes of K_1. Similar to the previous attack, this one is composed of a precomputation phase and an online phase. The precomputation phase in this attack is similar to the precomputation phase in the previous one, we build a hash table for the ordered sequence of $\Delta x_6[2]$ by iterating over all the possible values of the 30 parameters. The online phase is a bit different and thus explained in details below.

Online Phase. In the online phase, we choose an arbitrary plaintext as P^0, guess 4 key bytes $K_1[0 \cdots 3]$ and partially encrypt P^0 through the first layer to reach z_1. We create the δ-set at $z_1[3]$, which is equivalent to creating the δ-set at $x_2[3]$, and decrypt it backward to identify the plaintexts forming the δ-set containing P^0 and ask for their corresponding ciphertexts. From these ciphertexts and by guessing 26 key bytes, we compute the ordered sequence of $\Delta x_6[2]$ and look for a match in the stored hash table. The 26 key bytes to be guessed are $K_9[0\cdots15]$, $U_8[0,1,3,6,7,10,11,13,15]$ and $U_7[2]$ as depicted in Fig. 7. The probability for a false positive, i.e. a wrong key guess resulting in a match, is given by $2^{8\times30-2040} = 2^{-1800}$ and as we try 2^{240} key candidates, we expect that only $2^{240-1800} = 2^{-1560}$ remain after this step. We notice that the probability of a wrong key producing a valid $2^8 - 1$ bytes ordered sequence is negligible and can be relaxed. Hence, we use the partial sequence matching idea proposed in [3]. Instead of matching $2^8 - 1$ bytes ordered sequence, we match b bytes such that $b < 2^8$ and the probability of error, chosen to be 2^{-32}, is still small enough to be able to identity the right key. In that case, the number of required bytes b is calculated by $2^{-32} = 2^{8\times30+240-8b}$, which yields $b = 64$. This means that it is enough to match 64 bytes of the ordered sequence to identify the right key with a negligible error probability of 2^{-32}. It should be noted that in this attack, we opted for using the ordered sequence rather than the multiset because in the case of multiset, the probability of error is not small enough to identify the right key. Finally, as we recover 9 bytes from U_8, we have 2^{56} candidates for K_8 and following the same approach as in the previous attack, we get 2^{56} candidates for the master key which we exhaustively search to retrieve the correct master key with no major impact on the overall attack time complexity.

Attack Complexity. The memory requirement of the attack is due to the precomputation table needed to store the partial ordered sequence and is estimated to be $2^{240} \times 512/128 = 2^{242}$ 128-bit blocks. The data complexity of the attack is 2^{32} chosen plaintexts. The time complexity of the offline phase is equivalent to $2^{240} \times 65 \times 2^{-2} \approx 2^{244}$ encryptions. The time complexity of the online phase to recover 30 key bytes (The 16-byte K_9, 4 bytes from K_1, 9 bytes from U_8 and 1 byte from U_7) is equivalent to $2^{240} \times 65 \times 2^{-2} = 2^{244}$. Therefore, the total time complexity of the attack is $2^{244} + 2^{244} + 2^{57} \approx 2^{245}$.

5   Conclusion

In this work, we have presented two MitM attacks on the 256-bit key version of one of the Japanese e-Government candidate recommended block ciphers; Hierocrypt-3. The first attack employs the differential enumeration technique while the second one follows the strategy of Demirci and Selçuk. Both attacks are mounted against 8 SBox layers (4 rounds) and the complexities of the first attack are 2^{113} chosen plaintexts, 2^{238} 4-round reduced Hierocrypt-3 encryptions and 2^{218} 128-bit blocks. The other attack has much less data complexity of 2^{32} chosen plaintexts but higher time and memory complexities of 2^{245} encryptions and 2^{242} 128-bit block, respectively. We have noticed that the first attack based on the differential enumeration technique works in the chosen ciphertext context as well with exactly the same data, time and memory complexities. To the best of our knowledge, these are the best attacks on the 256-bit version of Hierocrypt-3 in the single-key setting.

A MDS_H and MDS_H^{-1}

MDS_H is represented by:

$$
\begin{pmatrix} y_{0(8)} \\ y_{1(8)} \\ y_{2(8)} \\ y_{3(8)} \\ y_{4(8)} \\ y_{5(8)} \\ y_{6(8)} \\ y_{7(8)} \\ y_{8(8)} \\ y_{9(8)} \\ y_{10(8)} \\ y_{11(8)} \\ y_{12(8)} \\ y_{13(8)} \\ y_{14(8)} \\ y_{15(8)} \end{pmatrix}
=
\begin{pmatrix}
1&0&1&0&1&0&1&0&1&1&0&1&1&1&1&1\\
1&1&0&1&1&1&0&1&1&1&1&0&0&1&1&1\\
1&1&1&0&1&1&1&0&1&1&1&1&0&0&1&1\\
0&1&0&1&0&1&0&1&1&0&1&0&1&1&1&0\\
1&1&1&1&0&1&0&1&0&1&0&1&1&1&0&1\\
0&1&1&1&1&1&0&1&1&1&0&1&1&1&1&0\\
0&0&1&1&1&1&1&0&1&1&1&0&1&1&1&1\\
1&1&1&0&0&1&0&1&0&1&0&1&1&0&1&0\\
1&1&0&1&1&1&1&1&1&0&1&0&1&0&1&0\\
1&1&1&0&0&1&1&1&1&1&0&1&1&1&0&1\\
1&1&1&1&0&0&1&1&1&1&1&0&1&1&1&0\\
1&0&1&0&1&1&1&0&0&1&0&1&0&1&0&1\\
1&0&1&0&1&1&0&1&1&1&1&1&1&0&1&0\\
1&1&0&1&1&1&1&0&0&1&1&1&1&1&0&1\\
1&1&1&0&1&1&1&1&0&0&1&1&1&1&1&0\\
0&1&0&1&1&0&1&0&1&1&1&0&0&1&0&1
\end{pmatrix}
\begin{pmatrix} x_{0(8)} \\ x_{1(8)} \\ x_{2(8)} \\ x_{3(8)} \\ x_{4(8)} \\ x_{5(8)} \\ x_{6(8)} \\ x_{7(8)} \\ x_{8(8)} \\ x_{9(8)} \\ x_{10(8)} \\ x_{11(8)} \\ x_{12(8)} \\ x_{13(8)} \\ x_{14(8)} \\ x_{15(8)} \end{pmatrix}
$$

The inverse matrix, MDS_H^{-1}, is given by:

$$
\begin{pmatrix} x_{0(8)} \\ x_{1(8)} \\ x_{2(8)} \\ x_{3(8)} \\ x_{4(8)} \\ x_{5(8)} \\ x_{6(8)} \\ x_{7(8)} \\ x_{8(8)} \\ x_{9(8)} \\ x_{10(8)} \\ x_{11(8)} \\ x_{12(8)} \\ x_{13(8)} \\ x_{14(8)} \\ x_{15(8)} \end{pmatrix}
=
\begin{pmatrix}
0&1&0&1&1&1&1&1&1&1&1&1&0&1&1&0\\
1&0&1&0&0&1&1&1&0&1&1&1&1&0&1&1\\
1&1&0&1&0&0&1&1&0&0&1&1&0&1&0&1\\
1&0&1&1&1&1&1&0&1&1&1&0&1&1&0&0\\
0&1&1&0&0&1&0&1&1&1&1&1&1&1&1&1\\
1&0&1&1&1&0&1&0&0&1&1&1&0&1&1&1\\
0&1&0&1&1&1&0&1&0&0&1&1&0&0&1&1\\
1&1&0&0&1&0&1&1&1&1&1&0&1&1&1&0\\
1&1&1&1&0&1&1&0&0&1&0&1&1&1&1&1\\
0&1&1&1&1&0&1&1&1&0&1&0&0&1&1&1\\
0&0&1&1&0&1&0&1&1&1&0&1&0&0&1&1\\
1&1&1&0&1&1&0&0&1&0&1&1&1&1&1&0\\
1&1&1&1&1&1&1&1&0&1&1&0&0&1&0&1\\
0&1&1&1&0&1&1&1&1&0&1&1&1&0&1&0\\
0&0&1&1&0&0&1&1&0&1&0&1&1&1&0&1\\
1&1&1&0&1&1&1&0&1&1&0&0&1&0&1&1
\end{pmatrix}
\begin{pmatrix} y_{0(8)} \\ y_{1(8)} \\ y_{2(8)} \\ y_{3(8)} \\ y_{4(8)} \\ y_{5(8)} \\ y_{6(8)} \\ y_{7(8)} \\ y_{8(8)} \\ y_{9(8)} \\ y_{10(8)} \\ y_{11(8)} \\ y_{12(8)} \\ y_{13(8)} \\ y_{14(8)} \\ y_{15(8)} \end{pmatrix}
$$

References

1. AlTawy, R., Youssef, A.M.: Preimage attacks on reduced-round Stribog. In: Pointcheval, D., Vergnaud, D. (eds.) AFRICACRYPT. LNCS, vol. 8469, pp. 109–125. Springer, Heidelberg (2014)

2. AlTawy, R., Youssef, A.M.: Second preimage analysis of Whirlwind. In: Lin, D., Yung, M., Zhou, J. (eds.) Inscrypt 2014. LNCS, vol. 8957, pp. 311–328. Springer, Heidelberg (2015)
3. AlTawy, R., Youssef, A.M.: Meet in the middle attacks on reduced round Kuznyechik. Cryptology ePrint Archive, Report 2015/096 (2015). http://eprint. iacr.org/
4. AlTawy, R., Youssef, A.M.: Differential sieving for 2-step matching meet-in-the-middle attack with application to Lblock. In: Eisenbarth, T., Öztürk, E. (eds.) LightSec 2014. LNCS, vol. 8898, pp. 126–139. Springer, Heidelberg (2015)
5. Barreto, P.S.L.M., Rijmen, V., Nakahara Jr, J., Preneel, B., Vandewalle, J., Kim, H.Y.: Improved SQUARE attacks against reduced-round HIEROCRYPT. In: Matsui, M. (ed.) FSE 2001. LNCS, vol. 2355, pp. 165–173. Springer, Heidelberg (2002)
6. Bogdanov, A., Khovratovich, D., Rechberger, C.: Biclique cryptanalysis of the full AES. In: Lee, D.H., Wang, X. (eds.) ASIACRYPT 2011. LNCS, vol. 7073, pp. 344–371. Springer, Heidelberg (2011)
7. Bogdanov, A., Rechberger, C.: A 3-subset meet-in-the-middle attack: cryptanalysis of the lightweight block cipher KTANTAN. In: Biryukov, A., Gong, G., Stinson, D.R. (eds.) SAC 2010. LNCS, vol. 6544, pp. 229–240. Springer, Heidelberg (2011)
8. Canteaut, A., Naya-Plasencia, M., Vayssière, B.: Sieve-in-the-middle: improved MITM attacks. In: Canetti, R., Garay, J.A. (eds.) CRYPTO 2013, Part I. LNCS, vol. 8042, pp. 222–240. Springer, Heidelberg (2013)
9. CRYPTEC: e-Government candidate recommended ciphers list (2013). http:// www.cryptrec.go.jp/english/method.html
10. CRYPTEC: e-Government recommended ciphers list (2003). http://www.cryptrec. go.jp/english/images/cryptrec_01en.pdf
11. CRYPTEC: Specification on a block cipher: Hierocrypt-3. http://www.cryptrec. go.jp/cryptrec_03_spec_cypherlist_files/PDF/08_02espec.pdf
12. Daemen, J., Knudsen, L.R., Rijmen, V.: The block cipher SQUARE. In: Biham, E. (ed.) FSE 1997. LNCS, vol. 1267, pp. 149–165. Springer, Heidelberg (1997)
13. Demirci, H., Selçuk, A.A.: A meet-in-the-middle attack on 8-round AES. In: Nyberg, K. (ed.) FSE 2008. LNCS, vol. 5086, pp. 116–126. Springer, Heidelberg (2008)
14. Demirci, H., Taşkın, I., Oban, M., Baysal, A.: Improved meet-in-the-middle attacks on AES. In: Roy, B., Sendrier, N. (eds.) INDOCRYPT 2009. LNCS, vol. 5922, pp. 144–156. Springer, Heidelberg (2009)
15. Derbez, P., Fouque, P.-A., Jean, J.: Improved key recovery attacks on reduced-round AES in the single-key setting. In: Johansson, T., Nguyen, P.Q. (eds.) EURO-CRYPT 2013. LNCS, vol. 7881, pp. 371–387. Springer, Heidelberg (2013)
16. Diffie, W., Hellman, M.E.: Special feature exhaustive cryptanalysis of the NBS Data Encryption Standard. Computer 10(6), 74–84 (1977)
17. Dunkelman, O., Keller, N., Shamir, A.: Improved single-key attacks on 8-round AES-192 and AES-256. In: Abe, M. (ed.) ASIACRYPT 2010. LNCS, vol. 6477, pp. 158–176. Springer, Heidelberg (2010)
18. Hao, Y., Bai, D., Li, L.: A meet-in-the-middle attack on round-reduced mCrypton using the differential enumeration technique. In: Au, M.H., Carminati, B., Kuo, C.-C.J. (eds.) NSS 2014. LNCS, vol. 8792, pp. 166–183. Springer, Heidelberg (2014)
19. Hong, D., Koo, B., Sasaki, Y.: Improved preimage attack for 68-step HAS-160. In: Lee, D., Hong, S. (eds.) ICISC 2009. LNCS, vol. 5984, pp. 332–348. Springer, Heidelberg (2010)
20. Cheon, J.H., Kim, M., Kim, K.: Impossible differential cryptanalysis of Hierocrypt-3 reduced to 3 rounds. NESSIE report (2002)

21. Li, L., Jia, K., Wang, X.: Improved meet-in-the-middle attacks on AES-192 and PRINCE. Cryptology ePrint Archive, Report 2013/573 (2013). http://eprint.iacr.org/
22. Mendel, F., Rechberger, C., Schläffer, M., Thomsen, S.S.: The rebound attack: cryptanalysis of reduced Whirlpool and Grøstl. In: Dunkelman, O. (ed.) FSE 2009. LNCS, vol. 5665, pp. 260–276. Springer, Heidelberg (2009)
23. New European Schemes for Signatures, Integrity, and Encryption. https://www.cosic.esat.kuleuven.be/nessie
24. Ohkuma, K., Muratani, H., Sano, F., Kawamura, S.: The block cipher Hierocrypt. In: Stinson, D.R., Tavares, S. (eds.) SAC 2000. LNCS, vol. 2012, p. 72. Springer, Heidelberg (2001)
25. Rechberger, C.: Security evaluation of 128-bit block ciphers AES, CIPHERUNICORN-A, and Hierocrypt-3 against biclique attacks. CRYPTREC (2012)
26. Sasaki, Y., Wang, L., Wu, S., Wu, W.: Investigating fundamental security requirements on Whirlpool: improved preimage and collision attacks. In: Wang, X., Sako, K. (eds.) ASIACRYPT 2012. LNCS, vol. 7658, pp. 562–579. Springer, Heidelberg (2012)
27. Furuya, S., Rijmen, V.: Observations on Hierocrypt-3/L1 key-scheduling algorithms. In: 2nd NESSIE Workshop (2001)
28. Toshiba Corporation: Block cipher family Hierocrypt. http://www.toshiba.co.jp/rdc/security/hierocrypt/index.htm

State-Recovery Analysis of Spritz

Ralph Ankele[1]([✉]), Stefan Kölbl[2], and Christian Rechberger[2]

[1] Graz University of Technology, Graz, Austria
ralph.ankele@alumni.tugraz.at
[2] DTU Compute, Technical University of Denmark, Lyngby, Denmark
{stek,crec}@dtu.dk

Abstract. RC4 suffered from a range of plaintext-recovery attacks using statistical biases, which use substantial, albeit close-to-practical, amounts of known keystream in applications such as TLS or WEP/WPA. Spritz was recently proposed at the rump session of CRYPTO 2014 as a slower redesign of RC4 by Rivest and Schuldt, aiming at reducing the statistical biases that lead to these attacks on RC4.

Even more devastating than those plaintext-recovery attacks from large amounts of keystream would be state- or key-recovery attacks from small amounts of known keystream. For RC4, there is unsubstantiated evidence that they may exist, the situation for Spritz is however not clear, as resistance against such attacks was not a design goal.

In this paper, we provide the first cryptanalytic results on Spritz and introduce three different state recovery algorithms. Our first algorithm recovers an internal state, requiring only a short segment of keystream, with an approximated complexity of 2^{1400}, which is much faster than exhaustive search through all possible states, but is still far away from a practical attack. Furthermore, we introduce a second algorithm that uses a pattern in the keystream to reduce the number of guessed values in our state recovery algorithm. Our third algorithm uses a probabilistic approach by considering the permutation table as probability distribution.

All in all, rather than showing a weakness, our analysis supports the conjecture that compared to RC4, Spritz may also provide higher resistance against potentially devastating state-recovery attacks.

Keywords: Spritz · RC4 · Stream cipher · State recovery · Cryptanalysis

1 Introduction

Spritz [1] is a stream cipher and hash function designed by Ronald L. Rivest and Jacob Schuldt. It is a complete redesign of the widely deployed stream cipher RC4. After the recent attacks on block ciphers in CBC mode in TLS, such as BEAST [2], Lucky 13 [3] and POODLE [4], the usage of RC4 was recommended. In 2013, AlFardan et. al. [5] published some new attacks on RC4 that require only 2^{24} TLS connections by exploiting new biases in RC4. RC4 has been well analyzed over the last years and some serious weaknesses have been found [6]. In practice it is often possible to mitigate such weaknesses by dropping the first

© Springer International Publishing Switzerland 2015
K. Lauter and F. Rodríguez-Henríquez (Eds.): LatinCrypt 2015, LNCS 9230, pp. 204–221, 2015.
DOI: 10.1007/978-3-319-22174-8_12

n bytes of the keystream. Nevertheless, the attacks on RC4 are getting very close to become practical and it is no longer recommended to use RC4. According to the community [7], some state organizations may have found a practical attack on RC4, which should further discourage the use of RC4 in any application. Recently, at CRYPTO 2014 (rump session), Ronald L. Rivest announced Spritz as a drop-in replacement of RC4.

The designers of Spritz claim their security bounds based on various statistical tests and extensive simulations, but did not publish a detailed security analysis in their proposal. In this paper, we want to amend this by providing the first cryptanalytic results on Spritz. If Spritz offers strong security bounds in various analyses, it could be used to replace RC4. Due to the big spectrum of applications and protocols in which RC4 is used, it is imperative though that a successor provides tight security bounds against all possible attack vectors.

Related Work. As Spritz was recently proposed as a drop-in replacement of RC4, no other cryptanalytic results are known to us.

There have been published some statistical weaknesses in the comparison between Spritz and VMPC-R [8]. Bartosz Zoltak [9], the designer of VMPC-R, published a statistical weakness in Spritz that shows a bias when observing the probability $\Pr(\text{output}(x) = \text{output}(x+2))$ for a simplified version of Spritz, that can distinguish the Spritz output from a random oracle after observing $2^{21.9}$ outputs.

Due to the different mode of operation and more complex structure of Spritz it is unclear to what extend previous state recovery attacks [10–12] on RC4 can be applied on Spritz. While those attacks typically have a very high complexity they might aid in comparing the security margins of RC4 and Spritz.

Contributions and Outline. In this paper, we provide the first results in the analysis of the recently announced stream cipher and hash function, Spritz. We first show some observations on Spritz like partial state rotations and that there exists a cycle in the keystream with length 6N, when no SWAP is done. Using these observations, we implemented three different state recovery attacks on Spritz and compared our results with the state recovery on RC4. Spritz is significantly slower than RC4 ([1], Sect. 9, Performance analysis), and as it turns out also provides higher security against state recovery attacks than RC4 (see Table 1).

Our first state recovery algorithm uses a recursive backtracking algorithm and has a lower complexity than exhaustive search through all possible states. Furthermore, we provide two different state recovery algorithms, where the first one searches for a pattern in the keystream, which allows an adversary to easily recover values in the permutation according to the known register values in a short window. Our third algorithm uses a probabilistic approach by considering the permutation table as a probability distribution similar to the ideas proposed in [10].

The source code of our attacks is public available for independent verification [13].

The remainder of this paper is structured as follows. Section 2 gives a detailed description of Spritz, where its structure and properties are discussed. In Sect. 3 we give some general observations on Spritz. Our state recovery attacks are introduced in Sect. 4. We conclude in Sect. 5 and give an outlook of further improvements.

Table 1. Comparison of state recovery attacks between RC4 and Spritz.

| Cipher | Permutation size | Complexity | | Reference |
		Time	Data	
RC4	32	2^{53}	2^{32}	[10]
Spritz	32	2^{99}	2^{5}	*This work*
RC4	64	2^{132}	2^{64}	[10]
RC4	64	2^{60}	2^{60}	[11]
Spritz	64	2^{249}	2^{6}	*This work*
RC4	128	2^{324}	2^{128}	[10]
RC4	128	2^{113}	2^{112}	[11]
Spritz	128	2^{599}	2^{7}	*This work*
RC4	256	2^{779}	2^{256}	[10]
RC4	256	2^{241a}	2^{240}	[11]
RC4	256	2^{262b}	2^{211}	[12]
Spritz	256	2^{1400}	2^{8}	*This work*

[a]Under some realistic assumptions. See [11] Sec. 6 for details.
[b] With N/10 pre-assigned entries to the state recovery table. See [12] for details.

2   Description of Spritz

Spritz [1] was proposed as an improved variant of the stream cipher RC4 and follows a similar design strategy. However, the update procedure is more complex and it uses a sponge-like construction.

Spritz builds upon a permutation and a few pointers (i.e. 8-bit registers). The UPDATE function of Spritz is the result of extensive simulations and statistical analysis. The designers created the UPDATE function by trying all possible candidates, constrained to six registers and a permutation of size N, and chose the one with the best security properties. The parameters were retrieved using a computing cluster with approximately five months of computation time, according to the designers [1].

Spritz uses a permutation $S = \{0, 1, \ldots, N - 1\}$ with N elements. Additionally, it consists of six registers: i, j, k, z, w and a. Registers i, j and k are used as pointers to the permutation S in the UPDATE function. The value of register w is always relatively prime to N and is used to update register i. Updating register i by addition of register w causes it to cycle through all values modulo N. Register a denotes the number of nibbles that have been absorbed. Register z stores the last generated value of the output keystream. The state of Spritz consists of the six registers and the permutation S. Spritz has a maximum of $N! \cdot N^6 \approx 2^{1730}$ different states, for $N = 256$. The key in Spritz is an arbitrary length byte array $K[0, \ldots, L - 1]$ where L denotes the length of the key. The default values for L are 16 and 32, resulting in key sizes of 128 and 256 bits, respectively.

The key schedule in Spritz is done by the ABSORB function, which takes an arbitrary length input of N-values as key. Each value is split into two half-nibbles, where register a counts the number of absorbed nibbles. For more details about the Spritz functions we refer to the Spritz proposal [1] and additionally give a short overview in Appendix B.

To produce an output word z_t the state of Spritz is updated as follows:

$$i_t = i_{t-1} + w \tag{1}$$

$$j_t = k_{t-1} + S_{t-1}[j_{t-1} + S_{t-1}[i_t]] \tag{2}$$

$$k_t = i_t + k_{t-1} + S_{t-1}[j_t] \tag{3}$$

$$S_t[i_t] = S_{t-1}[j_t], \ \ S_t[j_t] = S_{t-1}[i_t] \tag{4}$$

$$z_t = S_t[j_t + S_t[i_t + S_t[z_{t-1} + k_t]]] \tag{5}$$

where all operations are modulo N. In the next-state function, which is composed of Eqs. (1) to (5), which itself is a composition of the UPDATE and OUTPUT function, first the registers i, j and k are updated. Then the permutation S is swapped, at the index of registers i and j. In one update step, all words in the permutation remain the same except the swapped ones. In the end the output word z_i is generated.

3   Some Observations on Spritz

In our analysis of Spritz we made various observations, which might help to gain a better insight in Spritz. First, we show some partial state rotations that occur when Spritz is absorbing input of a specific form. Moreover, we analysed how Spritz works, if no swapping in the state update occurs and show that the output of Spritz quickly runs into a cycle, if the SWAP frequency is reduced to 0.

3.1   Partial State Rotations

We observed that the Spritz state partially rotates during ABSORB of particular messages. If we consider an internal state $S = \{s_0, s_1, \ldots, s_n\}$ and absorb a message block $M = \{0, 0, \ldots, 0\}$ where $|M| \leq N/4$ we can show that it holds that

$$S[i] = \begin{cases} s_{128}, & \text{if } i = 0 \\ s_{i-1}, & \text{if } 1 \leq i < 128 \\ s_0, & \text{if } i = 128 \\ s_i, & \text{if } 129 \leq i < 255 \end{cases}$$

Absorbing an input in Spritz increments the nibble counter a and the value of the permutation, at position of the nibble counter a is swapped with the value at $\lfloor N/2 \rfloor + x$, where x is the value of the absorbed nibble. If we absorb a message with the same content for each byte, Spritz results in a partial state rotation. These partial state rotations in Spritz are illustrated in Table 2.

Table 2. Partial state rotations in Spritz during ABSORB.

Step	$a\|\lfloor N/2 \rfloor + x\|$					$S^{(i)}$				
0	128	s_{128}	s_1	s_2	s_3	$\cdots$ s_{127}	s_0	$s_{129} \cdots s_{255}$		
1	128	s_{128}	s_0	s_2	s_3	$\cdots$ s_{127}	s_1	$s_{129} \cdots s_{255}$		
2	128	s_{128}	s_0	s_1	s_3	$\cdots$ s_{127}	s_2	$s_{129} \cdots s_{255}$		
3	128	s_{128}	s_0	s_1	s_2	$\cdots$ s_{127}	s_3	$s_{129} \cdots s_{255}$		
$\vdots$	$\vdots$	$\vdots$	$\vdots$	$\vdots$	$\vdots$	$\ddots$ $\vdots$	$\vdots$	$\vdots$ $\vdots$ $\vdots$		
127	128	s_{128}	s_0	s_1	s_1	$\cdots$ s_{126}	s_{127}	$s_{129} \cdots s_{255}$		

Since there is a call to SHUFFLE after absorbing $N/2$ nibbles, these state rotations only occur when absorbing less than $N/4$ bytes, and disappear after SHUFFLE gets called. Moreover, due to the addition of the nibble x in $\lfloor N/2 \rfloor + x$ in ABSORB, these state rotations are shifted by x if $x > 0$.

3.2 Cycle in the Keystream with No Swap

Using Spritz as a stream cipher generates a pseudorandom keystream that can be added to a plaintext stream to obtain a ciphertext stream. As the keystream is only pseudorandom, this implies that after some time it repeats and runs in a cycle. Fluhrer et al. [6] studied this behaviour on RC4 and showed that RC4 without the SWAP operation is useless as a keystream generator, as the keystream of RC4 becomes cyclic with a period of $2b$ (Note b in RC4 is 2^n, where n is the permutation's size). We observed that if no SWAP is done in the next-state function and no input is absorbed, the output of Spritz runs in a cycle of length $6N$. If no SWAP is done in RC4 and Spritz, the permutation S remains constant. The output z_n in RC4 depends on the still changing pointer registers i and j, where in Spritz we have an additional register k and the previous value of the keystream z_{n-1} that leads to the additional overhead, which increases the output cycle length of Spritz to $6N$, if no Swap is done.

4   State Recovery Attacks

Recovering the internal state of Spritz would enable an attacker to trivially predict any further output of the key stream. Furthermore, an adversary can try to recover the internal state, after the call to SHUFFLE in the output functions SQUEEZE or DRIP, by using an inverted next-state function and the knowledge of any internal state. Because of the call to SHUFFLE, it's not easily possible for an adversary to recover the initial state, since CRUSH in SHUFFLE aims to be non-invertible.

4.1   State Recovery with Backtracking

We introduce a recursive state recovery attack with backtracking and use a similar approach as Knudsen et al. [10] in their analysis of RC4. The state recovery attack needs only N output keystream bytes to successfully recover the internal state.

The idea of our recursive backtracking algorithm can be described as follows: We simulate the UPDATE and OUTPUT function of Spritz as long as all input to these functions is known. If a value is unknown we simply guess it and proceed. In the state recovery attack, we can start at any point, but we assume that the initial registers are correctly known (either from a previous different attack, through some heuristics or by simply guessing them). If we absorb no input and start at the beginning, we know the initial values of the registers, but after one application of SHUFFLE (e.g. in DRIP when any input was absorbed) we lose knowledge of the register values. In each step, we have to guess at most five unknown values (i.e. $S_{t-1}[i_t]$, $S_{t-1}[j_{t-1} + S_{t-1}[i_t]]$, $S_{t-1}[j_t]$, $S_t[z_{t-1} + k_t]$ and $S_t[i_t + S_t[z_{t-1} + k_t]]$) that we need to process to the next state.

For steps $t = 1 \ldots N$ we only proceed if $z'_t = z_t$, where z_t is the output word observed in the known output stream at step t, and z'_t is the output word from our simulation. In our tests we observed that the output word z'_t sometimes equals z_t even when we guessed the wrong entries in the partially recovered state. To avoid searching through dead branches in our search tree we impose several restrictions:

1. S is a permutation table, hence every element occurs only once. This reduces the number of possible values, which we have to guess, when a value is unknown.
2. If the known output word z'_t has been assigned to S during a previous guess, we can look if the index $j_t + S_t[i_t + S_t[z_{t-1} + k_t]]$ is equal to the position of the previous assigned value. If the indices are equal we can proceed to step $t + 1$. If not, we have a contradiction and can cut off the branch in the search tree.

3. If the known output word z'_t has not been assigned to S during a previous guess, we can again look if there is already a value at index $j_t + S_t[i_t + S_t[z_{t-1} + k_t]]$. If there is already a value, we again have a contradiction and can cut off the branch in the search tree. If not, we can set z'_t at position of index $j_t + S_t[i_t + S_t[z_{t-1} + k_t]]$ and then proceed with step $t + 1$.

Algorithm 1. State Recovery Algorithm with Backtracking

---

function RECOVERSTATE(t)

    $i_t \leftarrow i_{t-1} + w$

    if $S_{t-1}[i_t]$ is not assigned **then**

        Guess $S_{t-1}[i_t] \leftarrow v$ ▷ for $0 \le v < N$

    end if

    if $S_{t-1}[j_{t-1} + S_{t-1}[i_t]]$ is not assigned **then**

        Guess $S_{t-1}[j_{t-1} + S_{t-1}[i_t]] \leftarrow v$ ▷ for $0 \le v < N$

    end if

    $j_t \leftarrow k_{t-1} + S_{t-1}[j_{t-1} + S_{t-1}[i_t]]$

    if $S_{t-1}[j_t]$ is not assigned **then**

        Guess $S_{t-1}[j_t] \leftarrow v$ ▷ for $0 \le v < N$

    end if

    $k_t \leftarrow i_t + k_{t-1} + S_{t-1}[j_t]$

    $S_t[i_t] \leftarrow S_{t-1}[j_t]; S_t[j_t] \leftarrow S_{t-1}[i_t]$ ▷ Swap(S[i], S[j])

    if $S_t[z_{t-1} + k_t]$ is not assigned **then**

        Guess $S_t[z_{t-1} + k_t] \leftarrow v$ ▷ for $0 \le v < N$

    end if

    if $S_t[i_t + S_t[z_{t-1} + k_t]]$ is not assigned **then**

        Guess $S_t[i_t + S_t[z_{t-1} + k_t]] \leftarrow v$ ▷ for $0 \le v < N$

    end if

    $z'_t = S_t[j_t + S_t[i_t + S_t[z_{t-1} + k_t]]]$

    if z'_t equals any word in S **then**

        if $S_t[j_t + S_t[i_t + S_t[z_{t-1} + k_t]]] \ne$ position of any word in S **then**

            contradiction

        else

            RECOVERSTATEt + 1

        end if

    else

        if $j_t + S_t[i_t + S_t[z_{t-1} + k_t]] \ne$ position of any word in S **then**

            $S_t[j_t + S_t[i_t + S_t[z_{t-1} + k_t]]] = z'_t$

            RECOVERSTATEt + 1

        else

            contradiction

        end if

    end if

end function

We implemented the state recovery algorithm in a recursive function RECOV-ERSTATE() (see Algorithm 1), where most branches end up in contradictions.

If we achieve to fill up the internal state table in one branch (at maximum of N steps) we can verify the solution by calculating repeatingly output bytes and comparing it with the ciphertext stream. Afterwards, we calculate the initial state by inverting the next-state function. Furthermore, we can speed up the search if we pre-assign the state recovery table with a few previously known values.

Complexity. The complexity is given by the number of steps performed, until a solution is found. In case of our state recovery attack on Spritz, the complexity is measured in the total number of assignments made for all entries in the initial table S_0. We compute the complexity by splitting the algorithm in several cases $c_i(x)$ to which we assign probabilities according to the occurrence of each case. Afterwards, we compute the complexity based on the number of known bytes x in the permutation S and the assigned probabilities.

The complexity of the state recovery attack on Spritz is

$$\sum_{i=1}^{5} c_i(x) = \frac{x}{N} \cdot c_{i+1}(x) + (1 - \frac{x}{N}) \cdot (N - x) \cdot c_{i+1}(x+1) \tag{6}$$

$$c_6(x) = \frac{x}{N} \cdot ((1 - \frac{x}{N}) \cdot 1 + c_1(x)) + (1 - \frac{x}{N}) \cdot (\frac{x}{N} \cdot 1 + c_1(x+1)) \tag{7}$$

In our state recovery attack we have to guess at least five values. The first part of Eq. 6, which checks if the value is already in our partial recovered state table, will succeed on average with x/N. If it succeeds, we go to the next step without assigning a value. If not (second part of Eq. 6), we have to guess every possible value (which we do in this part of Eq. 6: $(1-x/N)\cdot(N-x)$), and proceed until we run into a contradiction (see Eq. 7) or succeed.

The complexity of the last step of our state recovery algorithm (Eq. 7) can be calculated in a similar way, but we additionally have to consider the probability that we run into a contradiction. Therefore, in $c_6(x)$ we first check if z'_n is already in our partly recovered permutation table, which again succeeds with x/N and a contradiction occurs with a probability of $1-x/N$. In the second part, the output value z'_n is not in our partly recovered permutation table, with a probability of $1 - x/N$. In this case, a contradiction can occur with probability x/N.

Another way of estimating the complexity is to experimentally observe it by counting how many assignments for all entries in the permutation S are made until we recover the initial state.

The results of our state recovery algorithm are shown in Table 3 where x gives the number of pre-assigned values in the state table. Furthermore, results for different N are given in Appendix C. The complexities are calculated using Eqs. 6 and 7. Additionally, we give the experimental complexity that we observed during testing our algorithm. The complexities are significantly lower than exhaustive search, but still far from practical for bigger N. Our algorithm becomes infeasible at $N = 32$, and is, therefore no threat to Spritz with $N = 256$. In our experiments we chose the number of pre-assigned values as low as possible to get a computable complexity. Moreover, we noted that there is a gap between the calculated values and our experimental observed values, which suggest that the attacks are even more powerful in practice than predicted by our complexity formulas.

Table 3. Calculated and experimental observed complexities for $N = 8, \ldots, 256$ using the backtracking algorithm, where x gives the number of pre-assigned values in the state table.

	calculated			experimental			
N	x	time comp.	$N!$	x	measured	time comp.	$(N - x)!$
8	0	$2^{13.7}$	$2^{15.2}$	0	$2^{11.7}$	$2^{13.7}$	$2^{15.2}$
16	0	$2^{44.3}$	$2^{44.2}$	5	$2^{18.5}$	$2^{22.7}$	$2^{25.3}$
32	0	$2^{99.8}$	$2^{117.6}$	19	$2^{21.3}$	$2^{28.5}$	$2^{32.5}$
64	0	$2^{249.0}$	2^{296}	47	$2^{20.8}$	$2^{42.2}$	$2^{48.3}$
128	0	$2^{599.4}$	$2^{716.1}$	109	$2^{18.7}$	2^{49}	$2^{56.8}$
256	0	2^{1400}	$2^{1683.9}$	-	-	-	-

4.2 Pattern Search and State Recovery

Our pattern search approach was inspired by the state recovery algorithm proposed by Maximov and Khovratovich [11]. In our algorithm, we applied a known plaintext attack, where we assumed that according to a pattern in the output keystream, all register values in a given window of length w_l are known. Therefore, we can easily derive a formula $S_{t-1}[j_t] = k_t - i_t - k_{t-1}$ from the update rule of register k. This allows us to compute some values of the permutation S without guessing additional values.

With the known entries in the permutation S and the knowledge of the register values during the window of length w_l, we can now apply our state recovery algorithm. It iteratively tries to recover an internal state. As long as we are inside our window, the register values are known and we can easily compute new values. If we lose knowledge of register j or k, outside of our window, we have to guess new permutation entries to proceed in our state recovery algorithm.

Assume at step t in a window, with length w_l, of the keystream z all the values $j_t, j_{t+1}, \ldots, j_{t+w_l}$ and $k_t, k_{t+1}, \ldots, k_{t+w_l}$ are known. Then we can according to

$$S_{t-1}[j_t] = k_t - i_t - k_{t-1}$$

calculate w_l entries for $S_{t-1}[j_t]$, which after SWAP become $S_t[i_t]$. Unfortunately, due to the uniform distribution of register j, the values are randomly distributed through our state recovery table. Nevertheless, we only have to guess three unknown values:

$$S_{t-1}[i_t], \quad S_{t-1}[j_{t-1} + S_{t-1}[i_t]], \quad S_{t-1}[j_t]$$

instead of five, as in our previous state recovery attack with backtracking (see Sect. 4.1). Our state recovery algorithm can be described as shown in Algorithm 2.

Algorithm 2. State Recovery Algorithm with Pattern Search

As long as registers i, j and k are known:

1. Calculate $S_{t-1}[j_t] = k_t - i_t - k_{t-1}$
2. Swap $S_t[i_t] \leftarrow S_{t-1}[j_t]; S_t[j_t] \leftarrow S_{t-1}[i_t]$
3. Check if $S_t[z_{t-1} + k_t]$ is already known
   3.1 If true $\rightarrow$ check if $S_t[i_t + s_t[z_{t-1} + k_t]]$ is already known
      3.1.1. If true $\rightarrow$ check if at index $j_t + S_t[i_t + s_t[z_{t-1} + k_t]]$ is already a value
         3.1.1.1. If true $\rightarrow$ compare if it is the same value as z_t
         3.1.1.2. If false $\rightarrow$ set z_t at index $j_t + S_t[i_t + s_t[z_{t-1} + k_t]]$

If registers j and k are no longer known:

4. Guess $S_{t-1}[i_t]$
5. Guess $S_{t-1}[j_{t-1} + S_{t-1}[i_t]]$
   5.1. Calculate $j_t = k_{t-1} + S_{t-1}[j_{t-1} + S_{t-1}[i_t]]$
6. Guess $S_{t-1}[j_t]$
   6.1. Calculate $k_t = i_t + k_{t-1} + S_{t-1}[j_t]$
7. Proceed with step 1

A contradiction can occur in steps $1 \ldots 3$ if the newly calculated value is already in another cell, in the state table. Moreover, a contradiction can occur, if there already exists a value, at the index of the current computed value, and the new computed value differs from the already existing value.

Our state recovery algorithm assumes that for a given window of length w_l all registers are known. This assumption is based on a pattern in the keystream that lets us determine the register values with high probability. Therefore, we need two definitions to describe these patterns in more detail (these definitions were defined by Maximov and Khovratovich [11] and are adjusted to Spritz):

Definition 1. *A* d-order pattern *is a 5-tuple defined as*

$$P_d = \{i, j, k, I, V\}, \qquad\qquad i, j, k \in \mathbb{Z}_N, I, V \in \mathbb{Z}_N^d$$

At step t the internal state of Spritz is compliant with pattern P_d if $i_t = i$, $j_t = j$, $k_t = k$ and d entries of permutation S contain their values in vector V and their corresponding indices in vector I.

Definition 2. *A pattern P_w is called* w-generative, *if for any internal state that is compliant to P_w, in the next w_l steps, all registers are known and let us derive w_l formulas, of the form $S_{t-1}[j_t] = k_t - i_t - k_{t-1}$.*

It is obvious that such patterns exist in the keystream of Spritz. However, for efficiency of our state recovery algorithm, we need to find patterns with a high d-order and patterns that are high w-generative, so that we initially know the values for the registers in a large window. We implemented a simple pattern search algorithm (described in Algorithm 3), that first fixes $i = 0$, and then tries all N values for each j and k before increasing i. If for a combination i, j and k the first index, value pair from the vectors I and V fits, we have found a pattern with d-order $= 1$. Our algorithm requires a keystream of length 2^N, where the search of such a pattern is a pre-computation stage of our attack.

Algorithm 3. Pattern Search

```
function SEARCHPATTERN(i, j, k, V, I)
    for n = 0 to 2^N do
        for {i, j, k} = 0 to N do
            if i_t = i and j_t = j and k_t = k then
                if keystream at position t equals V_t and I_t then
                    while keystream equals V and I do
                        P ← V and I
                        P ← i, j, k
                    end while
                end if
            end if
        end for
        STOREPATTERN(P)
    end for
end function
```

We implemented the state recovery algorithm and tested it for various sizes of N. If a contradiction occurs, we reset the responsible values in our state recovery table and continue with the recovery. In our tests we observed that even with some pre-assigned values, contradictions occurs too often, whereby we delete more information from our state recovery table, than we fill up with our state recovery attack. If we do not delete all responsible values when we reach a contradiction, we observed that either the complexity gets too high or we get too many wrong values in our state recovery table.

4.3 Probabilistic State Recovery

Knudsen et al. [10] proposed a probabilistic state recovery approach on RC4, that has been further improved by Golic and Morgari [12]. Our probabilistic state recovery algorithm follows a similar strategy and can be described as follows.

The initial state of the permutation S in Spritz depends on the secret key that is absorbed and is therefore, unknown to an attacker. We assume that all $N!$ possible states for the initial state are equally likely, i.e. the a priori probability distribution is uniform for the initial state. From the observation of the output keystream we gain information and can calculate an a posteriori probability distribution for the permutation S. After some steps the calculation for S should converge and we can recover the internal state.

In our probabilistic state recovery algorithm, we represent the information about the registers j, k and the permutation S by means of probability distributions. In each step, we calculate the a posteriori probability distribution of S_{dist}, j_{dist} and k_{dist}. We observe the keystream z and with the update rule for $z = S[j + S[i + S[z + k]]]$ and the Bayes rule we update the probability distributions. The distribution S_{dist} is represented as a $N \cdot N$ matrix (see Eq. 8), which represents the conditional probabilities of a given register and the associated entries in the permutation where the registers maps (e.g. $S[i][S_{t-1}[j_t]] = \Pr(S_{t-1}[i_t] \mid j)$).

$$S_{dist} = \begin{pmatrix} \frac{1}{N} & \frac{1}{N} & \frac{1}{N} & \frac{1}{N} \\ \frac{1}{N} & \frac{1}{N} & \frac{1}{N} & \frac{1}{N} \\ \frac{1}{N} & \frac{1}{N} & \frac{1}{N} & \frac{1}{N} \\ \frac{1}{N} & \frac{1}{N} & \frac{1}{N} & \frac{1}{N} \end{pmatrix} \tag{8}$$

We implemented this state recovery algorithm (see Algorithm 4) in a function PROBABILISTICRECOVERSTATE() that runs for a given amount of steps, where for each step it updates the probabilities of our S_{dist}, j_{dist} and k_{dist} distributions. If the algorithm converges, it stops and we can invert the next-state function of Spritz to recover the initial state. The convergence criteria in this case is that in the S_{dist} distribution each value is either zero or one. Based on that we can map our S_{dist} distribution to our state recovery table.

We tested our probabilistic state recovery algorithm with different sizes for N and with pre-assigned values (the distribution table was accordingly adjusted) for the state table, but even for small sizes of N our algorithm did not converge after a few steps and the complexity becomes infeasible.

Algorithm 4. Probabilistic State Recovery Algorithm

function PROBABILISTICRECOVERSTATE()
    INITIALIZESTATE()
    ABSORB(random key)
    $z \leftarrow$ DRIP   ▷ Store output keystream
    $\{S_{dist}, j_{dist}, k_{dist}\} \leftarrow$ INITIALIZEPROBABILITYDISTRIBUTIONS()
    for step i = 1 to steps **do**
        $\{S_{dist}, j_{dist}, k_{dist}\} \leftarrow$ UPDATEPROBABILITYDISTRIBUTIONS()
        if $\{S_{dist}, j_{dist}, k_{dist}\}$ converges **then**
            RECOVERINITIALSTATE()
        end if
    end for
end function

---

5 Conclusion

We have introduced three different state recovery algorithms for Spritz. The algorithms try to recover the initial state of Spritz by reconstructing any internal state. Our best state recovery algorithm has a complexity of 2^{1400} and requires only a small amount of known keystream. Nevertheless, the complexity is far from practical and is no real threat for Spritz using the default parameters. The consequences of a practical state recovery attack on any cipher would break the cipher. An adversary would be able to produce further output bytes, without knowing the secret key and therefore leverage the security provided by the secret key.

Furthermore, we implemented two additional state recovery algorithms based on pattern search and a probabilistic approach. However, the state recovery algorithm with the pattern search did not converge, because of too many contradictions and the probabilistic state recovery algorithm had a very high complexity, after very few steps.

As cryptanalytic results only get better over time we list here a few ideas for possible improvements. All three state recovery algorithms can be improved by heuristics that let an attacker pre-assign values into the state recovery table. Another possible improvement can be the combination of some of the state recovery algorithms, like combining the backtracking and probabilistic approach or only partially recover an internal state. It is our hope that our results will encourage others for further research on Spritz, to get a better understanding of its security margin.

A Pseudocode of the Spritz Functions

See Fig. 1.

INITIALIZESTATE(N)

```
1  i := j := k := z := a := 0
2  w := 1
3  for v = 0 to N - 1
4     S[v] := v
```

ABSORB(I)

```
1  for v = 0 to I.length - 1
2     ABSORBBYTE(I[v])
```

ABSORBBYTE(b)

```
1  ABSORBNIBBLE(LOW(b))
2  ABSORBNIBBLE(HIGH(b))
```

ABSORBNIBBLE(x)

```
1  if a = ⌊N/2⌋
2     SHUFFLE()
3  SWAP(S[a], S[⌊N/2⌋ + x])
4  a := a + 1
```

ABSORBSTOP()

```
1  if a = ⌊N/2⌋
2     SHUFFLE()
3  a := a + 1
```

SHUFFLE()

```
1  WHIP(2N)
2  CRUSH()
3  WHIP(2N)
4  CRUSH()
5  WHIP(2N)
6  a := 0
```

WHIP(r)

```
1  for v = 0 to r - 1
2     UPDATE()
3  do w := w + 1
4  until GCD(w, N) = 1
```

CRUSH()

```
1  for v = 0 to ⌊N/2⌋ - 1
2     if S[v] > S[N - 1 - v]
3        SWAP(S[v], S[N - 1 - v])
```

SQUEEZE(r)

```
1  if a > 0
2     SHUFFLE()
3  P := Array.New(r)
4  for v = 0 to r - 1
5     P[v] = DRIP()
6  return P
```

DRIP()

```
1  if a > 0
2     SHUFFLE()
3  UPDATE()
4  return OUTPUT()
```

UPDATE()

```
1  i := i + w
2  j := k + S[j + S[i]]
3  k := i + k + S[j]
4  SWAP(S[i], S[j])
```

OUTPUT()

```
1  z := S[j + S[i + S[z + k]]]
2  return z
```

Fig. 1. Pseudocode of Spritz. All additions are modulo N. When N is a power of 2, the last two lines in Whip are equivalent to $w = w + 2$.

B   Detailed Description of the Spritz Functions

InitializeState. This function initializes Spritz and sets the registers i, j, k, z and a to zero and register w to one. The permutation S is initialized with the identity permutation.

Absorb. The ABSORB function in Spritz absorbs arbitrary length input and updates the Spritz state accordingly. The input is split into bytes and absorbed with the ABSORBBYTE function. This function again splits the byte in two nibbles, while the lower nibble is absorbed first. For each nibble that is absorbed the register a is increased. If $a > \lfloor N/2 \rfloor$ is reached, Spritz is "full" (i.e. the capacity is reached) and Shuffle gets called. The ABSORBSTOP function is used in Spritz as padding function. ABSORBSTOP calls SHUFFLE if $a > \lfloor N/2 \rfloor$ and increments a by one. This is equivalent as absorbing a special stop symbol "▐" that is outside of the input alphabet.

Shuffle. SHUFFLE randomizes the Spritz state. To achieve good randomization three applications of WHIP and two applications of CRUSH are performed alternatively. WHIP calls UPDATE $2N$ times without producing any output, in order to randomly update the registers i, j and k as well as the permutation S. Additionally, WHIP updates register w every time it gets called to the next value relative prime to N. CRUSH maps $2^{N/2}$ states to one state, and looses information intentionally, which makes CRUSH a non-invertible transformation. In more detail, CRUSH compares iteratively the beginning to the end, and sorts the state ascending.

Squeeze, Drip. SQUEEZE and DRIP are the output functions of Spritz. First, register $a > 0$ is checked, and if necessary SHUFFLE is called which puts Spritz in "squeezing mode" (i.e. $a = 0$). Afterwards Squeeze, calls DRIP r times and returns the output in an array. DRIP uses UPDATE and OUTPUT to produce a new output byte.

Update, Output. UPDATE is the next-state function of Spritz. In UPDATE the registers i, j and k are updated and $S[i]$ and $S[j]$ are swapped. OUTPUT produces a single byte output by nested lookups in the permutation S mixed with the registers i, j, k and also feedback from the last produced output value.

C   Detailed Results of State Recovery Attacks

In this section, the results of our state recovery attacks are highlighted. We implemented three different state recovery algorithms. Our best one uses backtracking and cuts off branches that result in a contradiction. We measured the complexity experimentally, which is given in the next tables as *exp. complexity*. Additionally, we calculated the complexity which is given as *calc. complexity*. The parameter x denotes the number of pre-assigned values in our state recovery table. The last column shows the number of possible values in the permutation S given by $N!$.

C.1   Results of the State Recovery Attack with Backtracking

In Tables 4, 5 and 6 we applied our state recovery attack with random input, which leads to random initial states that we have to recover.

Table 4. Results for state recovery attack with backtracking for $N = 8$.

x	exp. complexity	calc. complexity	$(N - x)!$
07	2^0	2^0	2^0
06	$2^{1.7}$	2^1	2^1
05	$2^{2.7}$	$2^{2.6}$	$2^{2.6}$
04	$2^{4.6}$	$2^{4.6}$	$2^{4.6}$
03	$2^{6.2}$	$2^{6.9}$	$2^{6.9}$
02	$2^{8.1}$	$2^{9.5}$	$2^{9.5}$
01	$2^{9.0}$	$2^{11.3}$	$2^{12.3}$
00	$2^{11.7}$	$2^{13.7}$	$2^{15.3}$

Table 5. Results for state recovery attack with backtracking for $N = 16$.

x	exp. comp.	calc. comp.	$(N - x)!$
15	2^0	2^0	2^0
14	$2^{1.6}$	2^1	2^1
13	$2^{3.0}$	$2^{2.6}$	$2^{2.6}$
12	$2^{3.8}$	$2^{4.6}$	$2^{4.6}$
11	$2^{4.4}$	$2^{6.9}$	$2^{6.9}$
10	$2^{6.5}$	$2^{9.5}$	$2^{9.5}$
09	$2^{6.6}$	$2^{11.3}$	$2^{12.3}$
08	$2^{7.0}$	$2^{13.7}$	$2^{15.3}$
07	$2^{7.1}$	$2^{16.5}$	$2^{18.5}$
06	$2^{12.0}$	$2^{19.5}$	$2^{21.8}$
05	$2^{18.5}$	$2^{22.7}$	$2^{25.3}$
04	-	$2^{26.0}$	$2^{28.9}$
03	-	$2^{28.5}$	$2^{32.6}$
02	-	$2^{31.5}$	$2^{36.4}$
01	-	$2^{34.9}$	$2^{40.3}$
00	-	$2^{38.4}$	$2^{44.3}$

Table 6. Results for state recovery attack with backtracking for $N = 32$.

x	exp. comp.	calc. comp.	$(N - x)!$
31	2^0	2^0	2^0
30	$2^{1.5}$	2^1	2^1
29	$2^{2.3}$	$2^{2.6}$	$2^{2.6}$
28	$2^{3.5}$	$2^{4.6}$	$2^{4.6}$
27	$2^{5.4}$	$2^{6.9}$	$2^{6.9}$
26	$2^{9.4}$	$2^{9.5}$	$2^{9.5}$
25	$2^{15.3}$	$2^{11.3}$	$2^{12.3}$
24	$2^{15.9}$	$2^{13.7}$	$2^{15.3}$
23	$2^{18.9}$	$2^{16.5}$	$2^{18.5}$
22	$2^{19.3}$	$2^{19.5}$	$2^{21.8}$
21	$2^{19.7}$	$2^{22.7}$	$2^{25.3}$
20	$2^{20.1}$	$2^{26.0}$	$2^{28.9}$
19	$2^{21.3}$	$2^{28.5}$	$2^{32.6}$
18	-	$2^{31.5}$	$2^{36.4}$

C.2   Results of the State Recovery Attack with Backtracking and No Input

We can show that the performance of our state recovery attack is much higher if no input is absorbed (i.e. the initial state is the identity permutation) which is highlighted in Tables 7 and 8.

220   R. Ankele et al.

Table 7. Results for state recovery attack with backtracking for $N = 8$ and no input (i.e. identity permutation as initial state).

x	exp. comp.	calc. comp.	$(N-x)!$
07	2^0	2^0	2^0
06	2^1	2^1	2^1
05	$2^{1.6}$	$2^{2.6}$	$2^{2.6}$
04	2^2	$2^{4.6}$	$2^{4.6}$
03	$2^{2.4}$	$2^{6.9}$	$2^{6.9}$
02	$2^{2.6}$	$2^{9.5}$	$2^{9.5}$
01	$2^{2.9}$	$2^{11.3}$	$2^{12.3}$
00	-	$2^{13.7}$	$2^{15.3}$

Table 8. Results for state recovery attack with backtracking for $N = 16$ and no input (i.e. identity permutation as initial state).

x	exp. comp.	calc. comp.	$(N-x)!$
15	2^0	2^0	2^0
14	$2^{1.6}$	2^1	2^1
13	2^2	$2^{2.6}$	$2^{2.6}$
12	$2^{3.6}$	$2^{4.6}$	$2^{4.6}$
11	$2^{4.1}$	$2^{6.9}$	$2^{6.9}$
10	$2^{5.3}$	$2^{9.5}$	$2^{9.5}$
09	$2^{6.3}$	$2^{11.3}$	$2^{12.3}$
08	$2^{7.6}$	$2^{13.7}$	$2^{15.3}$
07	$2^{11.9}$	$2^{16.5}$	$2^{18.5}$
06	$2^{14.45}$	$2^{19.5}$	$2^{21.8}$
05	$2^{14.45}$	$2^{22.7}$	$2^{25.3}$
04	$2^{14.45}$	$2^{26.0}$	$2^{28.9}$
03	$2^{14.45}$	$2^{28.5}$	$2^{32.6}$
02	$2^{14.45}$	$2^{31.5}$	$2^{36.4}$
01	$2^{14.45}$	$2^{34.9}$	$2^{40.3}$
00	-	$2^{38.4}$	$2^{44.3}$

References

1. Rivest, R.L., Schuldt, J.C.N.: Spritz–a spongy RC4-like stream cipher and hash function (2014). http://people.csail.mit.edu/rivest/pubs/RS14.pdf
2. Duong, T., Rizzo, J.: Here come the ⊕ Ninjas. BEAST attack (2011)
3. Al Fardan, N., Paterson, K.: Lucky thirteen: breaking the TLS and DTLS record protocols. In: 2013 IEEE Symposium on Security and Privacy (SP), pp. 526–540 (2013)
4. Moeller, B., Duong, T., Kotowicz, K.: This POODLE Bites: Exploiting The SSL 3.0 Fallback (2014). https://www.openssl.org/bodo/ssl-poodle.pdf
5. AlFardan, N.J., Bernstein, D.J., Paterson, K.G., Poettering, B., Schuldt, J.C.N.: On the Security of RC4 in TLS and WPA (2013). http://www.isg.rhul.ac.uk/tls/RC4biases.pdf
6. Fluhrer, S.R., Mantin, I., Shamir, A.: Weaknesses in the key scheduling algorithm of RC4. In: Vaudenay, S., Youssef, A.M. (eds.) SAC 2001. LNCS, vol. 2259, p. 1. Springer, Heidelberg (2001)
7. Schneier, B.: The NSA Is Breaking Most Encryption on the Internet (2013). https://www.schneier.com/blog/archives/2013/09/the_nsa_is_brea.html
8. Bartosz, Z.: VMPC One-Way Function and Stream Cipher. In: Roy, B., Meier, W. (eds.) FSE 2004. LNCS, vol. 3017, pp. 210–225. Springer, Heidelberg (2004)
9. Bartosz, Z.: Statistical weakness in Spritz against VMPC-R: in search for the RC4 replacement. Cryptology ePrint Archive, Report 2014/985 (2014)
10. Knudsen, L.R., Meier, W., Preneel, B., Rijmen, V., Verdoolaege, S.: Analysis methods for (alleged) RC4. In: Ohta, K., Pei, D. (eds.) ASIACRYPT 1998. LNCS, vol. 1514, pp. 327–341. Springer, Heidelberg (1998)

11. Maximov, A., Khovratovich, D.: New state recovery attack on RC4. In: Wagner, D. (ed.) CRYPTO 2008. LNCS, vol. 5157, pp. 297–316. Springer, Heidelberg (2008)
12. Golic, J., Morgari, G.: Iterative Probabilistic Reconstruction of RC4 Internal States. Cryptology ePrint Archive, Report 2008/348 (2008). http://eprint.iacr.org/ 2008/348
13. Ankele, R.: (2015). https://github.com/ralphankele/Spritz

We Still Love Pairings

Computing Optimal 2-3 Chains for Pairings

Alex Capuñay[1] and Nicolas Thériault[2]([✉])

[1] Departamento de Matemáticas, Universidad de Chile, Santiago, Chile
alexmathe2012@hotmail.com
[2] Departamento de Matemática, Universidad Del Bío-Bío,
Avda. Collao 1202, Concepción, Chile
ntheriau@ubiobio.cl

Abstract. Using double-base chains to represent integers, in particular chains with bases 2 and 3, can be beneficial to the efficiency of scalar multiplication and the computation of bilinear pairings via (a variation of) Miller's algorithm. For one-time scalar multiplication, finding an optimal 2-3 chain could easily be more expensive than the scalar multiplication itself, and the associated risk of side-channel attacks based on the difference between doubling and tripling operations can produce serious complications to the use of 2-3 chains.

The situation changes when the scalar is fixed and public, as in the case of pairing computations. In such a situation, performing some extra work to obtain a chain that minimizes the cost associated to the scalar multiplication can be justified as the result may be re-used a large number of times. Even though this computation can be considered "attenuated" over several hundreds or thousands of scalar multiplications, it should still remain within the realm of "practical computations", and ideally be as efficient as possible.

An exhaustive search is clearly out of the question as its complexity grows exponentially in the size of the scalar. Up to now, the best practical approaches consisted in obtaining an approximation of the optimal chain via a greedy algorithm, or using the tree-based approach of Doche and Habsieger, but these offer no guarantee on how good the approximation will be. In this paper, we show how to find the optimal 2-3 chain in polynomial time, which leads to faster pairing computations. We also introduce the notion of "negative" 2-3 chains, where all the terms (except the leading one) are negative, which can provide near-optimal performance but reduces the types of operations used (reducing code size for the pairing implementation).

Keywords: Integer representations · Double-base chains · Tate pairings

1 Introduction

One of the most important operations in elliptic curve cryptosystems (ECC) is the scalar multiplication: given a point P in the group of the elliptic curve and an

This research was supported by FONDECYT grant 1151326 (Chile).

K. Lauter and F. Rodríguez-Henríquez (Eds.): LatinCrypt 2015, LNCS 9230, pp. 225–244, 2015.
DOI: 10.1007/978-3-319-22174-8_13

integer n, the scalar multiplication $[n]P$ is the sum of n copies of P. An important aspect for the efficiency of the scalar multiplication is the representation of the scalar (binary, NAF, w-NAF, ternary, etc.).

In 2005, Dimitrov et al. [9] proposed the use of double-base number systems (DBNS) for cryptographic application. A natural variation of this approach is the double-base chains, which are combined more naturally with scalar multiplication technique. The applicability of DBNS and double based chains are often impaired by the difficulty of finding a good representation (to reduce the cost of the scalar multiplication as much as possible) and by the risk associated to side-channel attacks (which could distinguish the different type of group operations performed and from this deduce the representation of the secret scalar).

Because of how Miller's algorithm computes the Tate pairing [12] (and similarly for the Ate and Eta pairings), scalar multiplication techniques can also have a direct impact on the cost of cryptographic pairings. After adapting Miller's algorithm, these representations can also be used in pairing computations [6]. In this situation, the scalar can be considered as fixed and is publicly known (so side-channel attacks on the representation of the scalar are irrelevant), and the representation can be computed once and used for a large number of scalar multiplications.

Although some techniques based on the greedy algorithm and on the tree-based approach of Doche and Habsieger [10] can produce "good" chains, these are only "approximations" (in terms of length) of the shortest chain. Up to now, the issue of finding the best possible chain remained an open problem.

In this paper, we focus on determining double-base chains that minimize either the Hamming weight (the number of terms) or the cost of the Tate pairing when they are used in Miller's algorithm. The algorithm we developed allows us to find optimal chains in both contexts (in general the chains that minimize the Hamming weight will not minimize the cost of pairing computation and vice-versa). The optimization for the cost of Miller's algorithm depend on the cost of the basic pairing operations (operations in the first point and evaluation in the second point), so these values should be considered parameters in the search for the optimal chain (entries of the algorithm along with the scalar). From this algorithm we can also study the effect of the different approaches for chain construction on pairing efficiency. We introduce the notion of *negative chains* which provide near-optimal pairing computation with simplified pairing implementations.

The paper is organized as follows: a background on double-base chains and pairings is given in Sect. 2. In Sect. 3 we present the definitions and notations used in the paper, the algorithm to obtain a minimal double-base chain (in terms of Hamming weight) and its analysis (with the proofs for some of the results in Appendix A). In Sect. 4 we explain how to adapt the algorithm to obtain a double-base chain that optimizes Tate pairing computations via Miller's algorithm (with finer details given in Appendix B). We study the impact of our algorithm on the efficiency of pairing computations in Sect. 5, illustrating with some numerical experiments. We conclude in Sect. 7.

2   Background

2.1   Double-Base Chains

Definition 1. *Given p, q two different prime numbers, a double-base number system (DBNS), is a representation scheme in which every positive integer n is represented as the sum or difference of numbers of the form $p^b q^t$, i.e.*

$$n = \sum_{i=1}^{m} s_i p^{b_i} q^{t_i}, \quad \text{with } s_i \in \{-1, 1\}, \text{ and } b_i, t_i \geq 0. \tag{1}$$

The Hamming weight of a DBNS is the number m of terms in (1).

In the following, we consider DBNS for $p = 2$ and $q = 3$. We note that the DBNS is quite redundant (and sparse) and a DBNS representation of n can easily be found with a *greedy-type approach*, namely find at each step the best approximation of a given integer in terms of an integer of the form $2^a 3^b$, then compute the difference and reapply the process until we reach zero.

In [9], Dimitrov *et al.* show that for any integer n, this greedy approach returns a DBNS expansion of n having at most $O\left(\frac{\log n}{\log \log n}\right)$ terms. However, in general a DBNS is not well suited for scalar multiplications since it often optimizes the number of sums without worrying on the number of doublings and triplings. For example, $2^{19} + 3^{12}$ is a minimal DBNS representation for 1055729, but requires more group operations (and more expensive ones) than the NAF representation $2^{20} + 2^{13} - 2^{10} - 2^4 + 1$.

In [8], Dimitrov *et al.* proposed the use of double-base chains to optimize the scalar multiplication in the group of an elliptic curve, minimizing the amount of multiplication-by-p and multiplication-by-q when using the DBNS in a scalar multiplication.

Definition 2. *A double-base chain (DBC) for n (or simply called a p-q chain for n) is an expansion of the form*

$$n = \sum_{i=1}^{m} s_i p^{a_i} q^{b_i}, \quad \text{with } s_i \in \{-1, 1\}, \tag{2}$$

such that $a_1 \geq a_2 \geq \cdots \geq a_m \geq 0$ and $b_1 \geq b_2 \geq \cdots \geq b_m \geq 0$.

The simplest case (in terms of group operations) comes from 2-3 chains. Because of its form, a 2-3 chain makes it possible to reach $[n]P$ using only doublings and triplings of the previous partial result and accumulating with additions and subtractions. One may be interested in choosing the best 2-3 chain for the efficiency of the scalar multiplications, but finding this chain is not a trivial problem.

To obtain a "reasonably good" double-base chain one can use a slightly modified version of the greedy algorithm for DBNS [9]. An alternative approach to obtain (better) 2-3 chains is to use the tree approach of Doche and Habsieger [10].

It consists of building a binary tree based on the sign of the terms in the chain (starting from the smallest one) and removing all factors of 2 and 3, until the scalar is reduced to ± 1. For efficiency reasons, only the "best" B chains are kept at each level (the other leaves are cut, to restrict the exponential growth of the tree). We discuss this technique in Sect. 3.4 after describing our own approach since the result of the full (unrestricted) binary tree can be obtained from our algorithm with some added restrictions.

2.2 Background on Tate Pairings

Let $E(\mathbb{F}_q)$ be the group of an elliptic curve over the finite field $\mathbb{F}_q$, and n a positive integer coprime to q (typically, n is also assumed to be prime). We denote by $E(\mathbb{F}_q)[n]$ the subgroup n-*torsion* points in $\mathbb{F}_q$. Let $P \in E(\mathbb{F}_q)[n]$ and $Q \in E(\mathbb{F}_{q^k})[n]$, then we denote by D_P and D_Q two divisors of E with disjoint support satisfying $D_P \sim (P) - (\infty)$ and $D_Q \sim (Q) - (\infty)$ respectively. Let f_P be a rational function on the elliptic curve whose divisor is $\mathrm{div}(f_P) = n(P) - n(\infty)$ (see [4] for a more detailed description).

The *reduced Tate pairing* τ_n is a well-defined, non-degenerate, bilinear mapping

$$\tau_n \,:\, E(\mathbb{F}_q)[n] \times E(\mathbb{F}_{q^k})[n] \longrightarrow \mathbb{F}_{q^k}^{*},$$

defined as

$$\tau_n(P, Q) = f_P(D_Q)^{(q^k - 1)/n}. \tag{3}$$

If $k > 1$ and Q is not $\mathbb{F}_q$-rational, one can avoid working with the divisor D_Q and simply work with the point Q, i.e. one can define the reduced Tate pairing as $\tau_n(P, Q) = f_P(Q)^{(q^k - 1)/n}$ (see [2]).

Miller's algorithm can be used to efficiently compute the Tate pairing. The main problem it addresses is how to build a rational function f_P such that $\mathrm{div}(f_P) = n(P) - n(\infty)$. Miller's idea is to combine the double-and-add algorithm for the scalar multiplication $[n]P$ with the evaluation of the straight lines passing through the points used in the addition/doubling process.

Let $[j]P$ and $[k]P$ be two distinct points in E (both multiples of P), with $j, k \in \mathbb{Z}$, and let $\ell_{jP, kP}$ be the (unique) straight line going through these two points, then

$$\mathrm{div}(\ell_{jP, kP}) = ([j]P) + ([k]P) + (-[j + k]P) - 3(\infty).$$

Furthermore, denote by ℓ_{jP} the vertical line through $[j]P$ and $-[j]P$, then

$$\mathrm{div}(\ell_{jP}) = ([j]P) + (-[j]P) - 2(\infty).$$

Lemma 1 (Miller). *Let $P \in E(\mathbb{F}_q)[n]$ and f_j be a function on the elliptic curve, such that its divisor $\mathrm{div}(f_j) = j(P) - ([j]P) - (j - 1)(\infty)$, $j \in \mathbb{Z}$. Then f_j can be chosen uniquely (up to constant multiple in $\mathbb{F}_q$) to satisfy the following conditions:*

1. $f_1 = 1$.

2. $f_{j+k} = f_j f_k \cdot \left(\dfrac{\ell_{jP,kP}}{\ell_{(j+k)P}} \right)$, for $j, k \in \mathbb{Z}$.

Miller's algorithm uses an addition chain for $[n]P$ to compute f_P, such that $\operatorname{div}(f_P) = n(P) - n(\infty)$. This algorithm is presented as an iteration through the binary expansion of n. From a representation of the scalar n in a *2-3 chain*, we can obtain a variation of Miller's algorithm based on scalar multiplication using the *2-3 chain* (see Algorithm 1).

Remark 1. In Algorithm 1, although addition/subtractions appear to be performed independently from doublings/triplings, this need not be the case in practice. Since at least one of u or v must be greater than zero, every addition and subtraction can be combined with either f_{2j} or f_{3j}.

Remark 2. In step 9 of Algorithm 1, it is possible to replace $(\ell_{Z,Z} \cdot \ell_{Z,2Z})/\ell_{2Z}$ by $\mathcal{P}_{Z,-3Z}$ which denote the equation of a parabola that passes through Z three times. This uses the technique K. Eisenträger *et al.* (for more details see [11]). In terms of evaluating the functions at the point Q, using the parabola allows us to reduce costs. Similar substitutions can be performed to reduce evaluation costs for the combined operations $2Z \pm P$ and $3Z \pm P$ (see [3] and [6] for more details).

Definition 3. *We write α_2 and α_3 (respectively) for the costs of doublings and triplings steps in Miller's algorithm. The total cost of a combined double-and-add step is written as $\alpha_2 + \beta_{2+}$ and combined double-and-subtraction as $\alpha_2 + \beta_{2-}$. The total cost of a combined triple-and-add step as $\alpha_3 + \beta_{3+}$ and combined triple-and-subtraction as $\alpha_3 + \beta_{3-}$.*

If the group operations $2P_i \pm P$ and $3P_i \pm P$ are performed as combined operations, then we expect $\beta_{2+} \neq \beta_{3+}$ and $\beta_{2-} \neq \beta_{3-}$, whereas $\beta_{2+} = \beta_{3+}$ and $\beta_{2-} = \beta_{3-}$ if the operations are performed separately (as in the original algorithm by Miller). The same notation can of course be used for "standard" (non-pairing) scalar multiplication operations, where α_2, α_3, β_{2+}, β_{2-}, β_{3+}, and β_{3-} should be interpreted as the cost of the associated group operation (without function evaluations).

Depending on the field arithmetic implementation, the type of processor used and the assumptions made on the curve equations (for example if the curve has automorphisms which permit to eliminate denominators in the pairing), the exact cost of each type of operation used in Miller's algorithm can vary greatly. For our purpose (optimizing the 2-3 chain), it is sufficient to know the costs of each type of operation relative to each other, so we can normalize to $\alpha_2 = 1$ and write the costs of the other operations as "multiples of the cost of doubling".

Giving a general proportion between the cost of the different operations is not practical. These proportions depend on the form of the curve equation (Weierstrass, Edwards, Jacobi, Doche-Icart-Kohel, to name a few), the type of

---

Algorithm 1. Miller's algorithm using 2-3 chain.

Input: $n = \sum_{i=1}^{m} s_i 2^{b_i} 3^{t_i}$, with $s_i \in \{-1, 1\}$, such that $b_1 \geq b_2 \geq \cdots \geq b_m \geq 0$,

and $t_1 \geq t_2 \geq \cdots \geq t_m \geq 0$. $P \in E(\mathbb{F}_q)[n]$, $Q \in E(\mathbb{F}_{q^k})[n]$.

Output: $f = f_n(Q)$.

1 **if** $s_1 = 1$ **then**
2 $\quad\lfloor\ f \leftarrow 1, \quad Z \leftarrow P$
3 **else**
4 $\quad\lfloor\ f \leftarrow f_{-1} = 1/\ell_P(Q), \quad Z \leftarrow -P$
5 **for** $i \leftarrow 1, 2, \ldots, m-1$ **do**
6 $\quad\quad u \leftarrow b_i - b_{i+1},$
7 $\quad\quad v \leftarrow t_i - t_{i+1},$
8 $\quad\quad$ **for** $j \leftarrow 1, \ldots, v$ **do**
9 $\quad\quad\quad\lfloor\ f \leftarrow f^3 \cdot \dfrac{\mathcal{P}_{Z,-3Z}(Q)}{\ell_{3Z}(Q)}, \quad Z \leftarrow 3Z.$
10 $\quad\quad$ **for** $j \leftarrow 1, \ldots, u$ **do**
11 $\quad\quad\quad\lfloor\ f \leftarrow f^2 \cdot \dfrac{\ell_{Z,Z}(Q)}{\ell_{2Z}(Q)}, \quad Z \leftarrow 2Z.$
12 $\quad\quad$ **if** $s_{i+1} = 1$ **then**
13 $\quad\quad\quad\lfloor\ f \leftarrow f \cdot \dfrac{\ell_{Z,P}(Q)}{\ell_{Z+P}(Q)}, \quad Z \leftarrow Z + P.$
14 $\quad\quad$ **else**
15 $\quad\quad\quad\lfloor\ f \leftarrow f \cdot \dfrac{\ell_{Z,-P}(Q)}{\ell_P(Q)\ell_{Z-P}(Q)}, \quad Z \leftarrow Z - P.$
16 **return** f

---

coordinates (affine, projective, Jacobian projective, Chudnovsky, etc.), if addition/subtraction operations are combined or not with doublings/triplings, in how the group operations are combined with function evaluations in pairing computations, etc.

Furthermore, the implementation of the underlying field arithmetic plays an important role in determining the proportions between the costs of distinct group operations, and this is heavily affected by the actual field used (in most field implementations, the relative costs of basic operations like multiplications, squares and inverses vary between field sizes). This means the relative costs of α_2, α_3, β_{2+}, β_{2-}, β_{3+}, and β_{3-} should be specific to the curve and field used. Examples of group operations that consider both doublings and triplings can be found in [5] and [7]. Combined group operations can also be found in [7] (including some combinations we did not consider in this work).

Examples of arithmetic operations for the field extensions used in the main pairing applications, along with their impact on the loop in Miller's algorithm can be found in [1]. For pairing computations using triplings (in characteristic three) see [2] and [11]. Combined operations for pairings in large characteristic are presented in [6].

For our numerical experiments in Sect. 6, we use the estimated operations costs of [6] since they cover the whole range of operations we are interested in. They should of course be replaced by case-specific proportions if an optimal chain is required for a specific implementation.

Remark 3. Since 2-3 chains rely on the ability to interchange doublings and triplings, they are more interesting if $\alpha_3/\alpha_2 \approx \log_2(3) = 1.58496\ldots$ (corresponding to the relative length of base-2 and base-3 expansions). If α_3/α_2 differs too much from $\log_2(3)$, our algorithm can still be used, but the output is very likely to be close to a single-base representation in the more advantageous operation.

Note that other types of pairings can also be used, for example the Ate and Eta pairings. One of the objectives of these pairings is to shorten the length of Miller's loop by using shorter scalars n (replacing the value of n in the Tate pairing by $q - n$ which has roughly half the bit-size). Obviously, our algorithm can still be applied to this new value of n.

3   Computing the Shortest 2-3 Chain

3.1   Definitions and Notations

Before presenting our algorithm, we introduce the notation used to describe and analyze it.

Definition 4. *Given a positive integer n, denote by $n_{i,j}$ the (unique) integer in $\{0, 1, \ldots, (2^i 3^j - 1)\}$ such that $n_{i,j} \equiv n \mod 2^i 3^j$, and by $\overline{n}_{i,j}$ the (unique) integer in $\{0, -1, \ldots, -(2^i 3^j - 1)\}$ such that $\overline{n}_{i,j} \equiv n \mod 2^i 3^j$.*

Definition 5. *We denote by $C_{i,j}$ any 2-3 chains for $n_{i,j}$ in which all terms are strict divisors of $2^i 3^j$ (that is to say, any term in the chain is of the form $\pm 2^a 3^b$ with at least $a < i$ or $b < j$), and by $\mathscr{C}_{i,j}$ a minimal 2-3 chain for $n_{i,j}$. Similarly, $\overline{C}_{i,j}$ is any 2-3 chains for $\overline{n}_{i,j}$ and $\overline{\mathscr{C}}_{i,j}$ is a minimal 2-3 chain for $\overline{n}_{i,j}$. If (and only if) no $C_{i,j}$ chain is possible for $n_{i,j}$, we write $C_{i,j} = \mathscr{C}_{i,j} = \emptyset$ (similarly $\overline{C}_{i,j} = \overline{\mathscr{C}}_{i,j} = \emptyset$ for $\overline{n}_{i,j}$).*

For Sect. 3, the minimality is in terms of the Hamming weight (the number of terms), whereas in Sect. 4 the minimality is in terms of the effect on Miller's algorithm. To simplify some arguments, we also consider all chains $C_{i,j}$ where either i or j is negative to be impossible. Since the process is inductive in both i and j, this avoids the need for writing special induction cases for $C_{0,j}$ and $C_{i,0}$ (which would depend on $C_{-1,j}$ and $C_{i,-1}$ respectively).

It should also be noted that not all values of $n_{i,j}$ (resp. $\overline{n}_{i,j}$) admit a chain $C_{i,j}$ (resp. $\overline{C}_{i,j}$) due to the bound on the exponents. For example, $n_{0,1} = 2$ does not admit a chain $C_{0,1}$ since the only chains possible where the largest term divides (but is not equal to) 3 are "0", "1" and "-1".

We also observe that except for $n_{i,j} = \overline{n}_{i,j} = 0$, the largest term (in absolute value) in $C_{i,j}$ (if such a chain exists) is positive, and the largest term in $\overline{C}_{i,j}$ (again if such a chain exists) is negative.

Lemma 2. *If $\mathscr{C}_{i,j}$ is a minimal chain for $n_{i,j}$ which contains a subchain $C_{a,b}$ for $n_{a,b}$ with $a \leq i$ and $b \leq j$, then $C_{a,b}$ must be minimal for $n_{a,b}$. Similarly, if $\mathscr{C}_{i,j}$ contains a subchain $\overline{C}_{a,b}$ for $\overline{n}_{a,b}$ with $a \leq i$ and $b \leq j$, then $\overline{C}_{a,b}$ must be minimal for $\overline{n}_{a,b}$.*

Proof. Suppose that $C_{a,b}$ is not minimal, then there would exist a chain $\mathscr{C}_{a,b}$ for $n_{a,b}$ with lower Hamming weight than $C_{a,b}$. Replacing $C_{a,b}$ by $\mathscr{C}_{a,b}$ in $\mathscr{C}_{i,j}$ produces a new chain for $n_{i,j}$ with lower Hamming weight than $\mathscr{C}_{i,j}$, contradicting the minimality. The argument for $\overline{C}_{a,b}$ is similar. □

In order to simplify the final step of our algorithm, we introduce one more notation:

Definition 6. *For every integer $0 \leq j \leq \lceil \log_3 n \rceil$, let i_j be the smallest non-negative integer such that $2^{i_j} 3^j > n$. If $\mathscr{C}_{i_j,j}$ exists, define $C_j = \mathscr{C}_{i_j,j}$, otherwise, $C_j = \emptyset$. Similarly, if $\overline{\mathscr{C}}_{i_j,j}$ exists, define $\overline{C}_j = 2^{i_j} 3^j + \overline{\mathscr{C}}_{i,j}$, otherwise, $\overline{C}_j = \emptyset$.*

3.2   The Algorithm

To understand the algorithm, it is essential to consider the relationship between a chain $C_{i,j}$ for $n_{i,j}$ and the (sub)chains for $n_{i-1,j}$, $\overline{n}_{i-1,j}$, $n_{i,j-1}$ and $\overline{n}_{i,j-1}$. First, we observe that for and $i, j \geq 0$,

$$\overline{n}_{i,j} = \begin{cases} 0 & \text{if } n_{i,j} = 0 \\ n_{i,j} - 2^i 3^j & \text{otherwise} \end{cases} \tag{4}$$

We now look at the evolution of the values of $n_{i,j}$ and $\overline{n}_{i,j}$. By definition of $n_{i,j}$ and $\overline{n}_{i,j}$, it is clear that if $i > 0$ then

$$\begin{aligned} n_{i,j} &\in \{n_{i-1,j}, n_{i-1,j} + 2^{i-1} 3^j\} \\ \overline{n}_{i,j} &\in \{\overline{n}_{i-1,j}, \overline{n}_{i-1,j} - 2^{i-1} 3^j\} \end{aligned} \tag{5}$$

and if $j > 0$ then

$$\begin{aligned} n_{i,j} &\in \{n_{i,j-1}, n_{i,j-1} + 2^i 3^{j-1}, n_{i,j-1} + 2 \cdot 2^i 3^{j-1}\} \\ \overline{n}_{i,j} &\in \{\overline{n}_{i,j-1}, \overline{n}_{i,j-1} - 2^i 3^{j-1}, \overline{n}_{i,j-1} - 2 \cdot 2^i 3^{j-1}\} \end{aligned} \tag{6}$$

However, terms of the form $\pm 2 \cdot 2^i 3^{j-1}$ are not allowed in a 2-3 chain (unless we write them as $\pm 2^{i+1} 3^{j-1}$), so they incompatible with our definition of $C_{i,j}$ and $\overline{C}_{i,j}$.

Given a chain for $n_{i,j}$ (where all terms are decreasing divisors $2^i 3^j$, all smaller than $2^i 3^j$ in absolute value), we can extract a chain for one of $n_{i-1,j}$, $\overline{n}_{i-1,j}$, $n_{i,j-1}$ or $\overline{n}_{i,j-1}$, and this gives us inductive relationships.

Lemma 3. *Let $C_{i,j}$ be any 2-3 chain for $n_{i,j}$, then*

- $C_{i,j} \neq \emptyset$ *if and only if one of the following cases occurs:*
  1. $C_{i,j} = C_{i-1,j}$ *(only if $n_{i,j} = n_{i-1,j}$);*
  2. $C_{i,j} = 2^{i-1}3^j + C_{i-1,j}$ *(only if $n_{i,j} = 2^{i-1}3^j + n_{i-1,j}$);*
  3. $C_{i,j} = 2^{i-1}3^j + \overline{C}_{i-1,j}$ *(only if $n_{i,j} = 2^{i-1}3^j + \overline{n}_{i-1,j}$);*
  4. $C_{i,j} = C_{i,j-1}$ *(only if $n_{i,j} = n_{i,j-1}$);*
  5. $C_{i,j} = 2^i3^{j-1} + C_{i,j-1}$ *(only if $n_{i,j} = 2^i3^{j-1} + n_{i,j-1}$);*
  6. $C_{i,j} = 2^i3^{j-1} + \overline{C}_{i,j-1}$ *(only if $n_{i,j} = 2^i3^{j-1} + \overline{n}_{i,j-1}$);*
- *For a given (fixed) chain $C_{i,j}$, the cases are mutually exclusive except for cases 1 and 4, and these occur at the same time if only if $n_{i,j} = n_{i-1,j-1}$.*
- *The possible cases and mutual exclusivity are similar for $\mathscr{C}_{i,j}$, $\overline{C}_{i,j}$ and $\overline{\mathscr{C}}_{i,j}$ (after the corresponding sign changes).*

Proof. If a $C_{i,j}$ chain exists, then we consider its largest term γ. If $\gamma = 2^{i-1}3^j$, then we have cases 2 or 3, depending on the sign of the second largest term (positive for case 2, negative for case 3). If $\gamma = 2^i3^{j-1}$, then we have cases 5 or 6, depending on the sign of the second largest term (positive for case 5, negative for case 6). If γ is of the form 2^a3^j with $a \leq i-2$, then we are (only) in case 1. If γ is of the form 2^i3^b with $b \leq j-2$, then we are (only) in case 4. If γ is of the form 2^a3^b with both $a \leq i-2$ and $b \leq j-2$, then the chains falls in both cases 1 and 4. The relations between $n_{i,j}$ and $n_{i-1,j}$ or $n_{i,j-1}$ follow directly. □

Lemma 3 gives "downward" relations between the chains $C_{i,j}$, that is, decreasing the values of i or j (or both), but it can easily be transformed into "upward" relations:

Corollary 1. *Given chains $C_{i-1,j}$, $\overline{C}_{i-1,j}$, $C_{i,j-1}$ and $\overline{C}_{i,j-1}$ (with value $= \emptyset$ only if no such chain is possible), then the possible "sources" for $C_{i,j}$ and $\overline{C}_{i,j}$ can be found in Tables 1 and 2.*

Table 1. Possible sources of $C_{i,j}$ and $\overline{C}_{i,j}$ when multiplying by 2

$n_{i-1,j}$	$n_{i,j}$	Possible $C_{i,j}$	Possible $\overline{C}_{i,j}$
$= 0$	0	0	0
$= 0$	$2^{i-1}3^j$	$2^{i-1}3^j$	$-2^{i-1}3^j$
> 0	$n_{i-1,j}$	$C_{i-1,j}$ and $2^{i-1}3^j + \overline{C}_{i-1,j}$	$-2^{i-1}3^j + \overline{C}_{i-1,j}$
> 0	$n_{i-1,j} + 2^{i-1}3^j$	$2^{i-1}3^j + C_{i-1,j}$	$\overline{C}_{i-1,j}$ and $-2^{i-1}3^j + C_{i-1,j}$

Proof. See Appendix A.

Proposition 1. *The largest term in chain C_j is either $2^{i_j-1}3^j$ or $2^{i_j}3^{j-1}$.*

Table 2. Possible sources of $C_{i,j}$ and $\overline{C}_{i,j}$ when multiplying by 3

$n_{i,j-1}$	$n_{i,j}$	Possible $C_{i,j}$	Possible $\overline{C}_{i,j}$
$= 0$	0	0	0
$= 0$	$2^i 3^{j-1}$	$2^i 3^{j-1}$	$\emptyset$
$= 0$	$2 \cdot 2^i 3^{j-1}$	$\emptyset$	$-2^i 3^{j-1}$
> 0	$n_{i,j-1}$	$C_{i,j-1}$ and $2^i 3^{j-1} + \overline{C}_{i,j-1}$	$\emptyset$
> 0	$n_{i,j-1} + 2^i 3^{j-1}$	$2^i 3^{j-1} + C_{i,j-1}$	$-2^i 3^{j-1} + \overline{C}_{i,j-1}$
> 0	$n_{i,j-1} + 2 \cdot 2^i 3^{j-1}$	$\emptyset$	$\overline{C}_{i,j-1}$ and $-2^i 3^{j-1} + C_{i,j-1}$

Proof. By definition of i_j, we have

$$2^{i_j} 3^{j-1} < 2^{i_j - 1} 3^j \leq n_{i_j, j} = n$$

so both $n_{i_j - 1, j} \neq n_{i_j, j}$ and $n_{i_j, j-1} \neq n_{i_j, j}$. From Lemma 3, $C_{i,j}$ must fall in cases 2, 3, 5 or 6, and the result follows directly.   □

Remark 4. By definition, the largest term in chain $\overline{C}_j$ is $2^{i_j} 3^j$.

Combining these results, we obtain Algorithm 2. The algorithm proceeds along a double induction: first along i (the powers of 2) and then along j (the powers of 3), both going up from 0.

To reduce memory requirements, the algorithm keeps only one copy at a time of both $C_{i,j}$ and $\overline{C}_{i,j}$ for any given i, which at the end of the algorithm will correspond to C_j and $\overline{C}_j$ in Definition 6. Note that Definition 6, Proposition 1, and Remark 4 are written in terms of $i = i_j = \lceil \log_2((n+1)/3^j) \rceil$, whereas the induction in the algorithm varies j for a "fixed" i. Since $\lceil \log_3((n+1)/2^i) \rceil \in \{\lceil \log_2((n+1)/3^j) \rceil, \lceil \log_2((n+1)/3^j) \rceil + 1\}$, i_m may be equal to $i-1$, in which case the proof of Theorem 1 will show that we de not need to consider the pair (i, m).

As mentioned in Sect. 3.1, we consider that $C_{-1,j} = \emptyset = C_{i,-1}$ for every i and j so the algorithm does not need special cases for the induction steps with $C_{0,j}$ (which calls $C_{-1,j}$) and $C_{i,0}$ (which calls $C_{i,-1}$). In the actual implementation it is slightly more efficient to deal with these cases separately and never use negative indices. In particular, from the definition one observes directly that $\mathscr{C}_{0,0} = 0 = \overline{\mathscr{C}}_{0,0}$.

3.3 Analysis of the Algorithm

Lemma 4. *For any pair $i, j \geq 0$, at least one of the two minimal chains $\mathscr{C}_{i,j}$ or $\overline{\mathscr{C}}_{i,j}$ exist (for $n_{i,j}$ and $\overline{n}_{i,j}$ respectively).*

Proof. See Appendix A.

Although Lemma 4 is not necessary to ensure the algorithm produces a 2-3 chain (for example it would be sufficient to show that the NAF representation is

---

Algorithm 2. Algorithm to compute a minimal 2-3 chain.

Input: Integer $n > 0$.
Output: Minimal 2-3 chain $\mathscr{C}$ for n.
1 $C_j \leftarrow \emptyset, \overline{C}_j \leftarrow \emptyset$ for every j
2 **for** $i \leftarrow 0$ *to* $\lceil \log_2(n+1) \rceil$ **do**
3      $m \leftarrow \lceil \log_3((n+1)/2^i) \rceil$
4      $i_m \leftarrow \lceil \log_2((n+1)/3^m) \rceil$
5   **if** $i_m < i$ **then**
6         $m \leftarrow m - 1$

7   **for** $j \leftarrow 0$ *to* m **do**
8   Using Table 1, $C_j \ (= \mathscr{C}_{i-1,j})$ and $\overline{C}_j$, find possibles $C_{i,j}$ and $\overline{C}_{i,j}$.
9   Using Table 2, $C_{j-1} \ (= \mathscr{C}_{i,j-1})$ and $\overline{C}_{j-1}$, find possibles $C_{i,j}$ and $\overline{C}_{i,j}$.
10       $C_j \leftarrow$ shortest chain among the $C_{i,j} \ (C_j = \mathscr{C}_{i,j})$.
11       $\overline{C}_j \leftarrow$ shortest chain among the $\overline{C}_{i,j} \ (\overline{C}_j = \mathscr{C}_{i,j})$.
12   **if** $i = i_j$ **then**
13         $\overline{C}_j \leftarrow 2^i 3^j + \overline{C}_j$.

14 $\mathscr{C} \leftarrow$ shortest chain among the C_j and $\overline{C}_j$
15 **return** $\mathscr{C}$

---

a valid output for either C_0 or $\overline{C}_0$), it indicates that we obtain potential minimal chains in $\{C_j, \overline{C}_j\}$ for every value of j (so no value of j can safely be ignored).

Theorem 1. *The set S of all C_j and $\overline{C}_j$ produced by Algorithm 2,*

$$S = \{C_j \mid 0 \le j \le \lceil \log_3(n+1) \rceil\} \cup \{\overline{C}_j \mid 0 \le j \le \lceil \log_3(n+1) \rceil\}$$

contains a minimal 2-3 chain for n.

Proof. See Appendix A.

Theorem 2. *Let n be a positive integer, then Algorithm 2 returns a minimal 2-3 chain in $O((\log n)^2)$ steps ($O((\log n)^4)$ bit operations), and requires $O((\log n)^3)$ bits of memory (it has $O(\log n)$ chains in memory).*

Proof. See Appendix A.

Remark 5. If the terms in the chains are recorded as their powers of 2 and 3 (e.g. 288 is recorded as the pair $(5, 2)$), then the memory requirements per chain decreases to $O(\log n \log \log n)$. If they are recorded as their difference (again in powers of 2 and 3) with the previous term, the memory requirement per chain is $O(\log n)$, so the memory requirements for Algorithm 2 can be decreased to $O((\log n)^2)$ bits.

Remark 6. If we assume fast integer arithmetic, the cost per step of algorithm decreases to $O^{\sim}(\log n)$ bit operations, and the total cost of the algorithm becomes $O^{\sim}((\log n)^3)$ bit operations.

3.4   Tree-Based Approach of Doche and Habsieger

The tree-based approach of Doche and Habsieger [10] can be seen as a special case of our approach where the maximal number of divisions by 2 and 3 are always applied to $n - \hat{n}$, where $\hat{n}$ is the current approximation of n. As a result, additions or subtractions to the chain are only allowed if neither a direct doubling or a direct tripling are possible at this step of the scalar multiplication. In other words, introducing terms of the form $\pm 2^i 3^j$ to the subchain $C_{i,j}$ is only permitted if $(n - n_{i,j})/(2^i 3^j)$ is coprime with 6. In our algorithm, this condition is not present, and neither is it necessary in the definition of 2-3 chains.

We can therefore run experiment for the output of an "unrestricted" version of the tree-based approach (where all leaves are followed to the end). The values computed by our modified algorithm are therefore *lower bounds* for the outputs of the Algorithm presented in [10].

Intuitively, this division condition may seem a reasonable assumption to obtain minimal chains, but in practice it often removes some essential subchains. For example, $3^{10} - 1$ is an optimal chain for 59048, but the best chains produced by the algorithm of Doche and Habsieger are: $2^{11} 3^3 + 2^7 3^3 + 2^5 3^2 + 2^3$ and $2^8 3^5 - 2^7 3^3 + 2^5 3^2 + 2^3$, which have twice the Hamming weight as the minimal chain.

4   Computing Optimal Chains for Pairings

In the previous section, the chains were optimized only for their Hamming weight (the number of terms). However, in practice this is usually not the best approach to obtain the fastest pairing computation. Depending on the underlying group arithmetic, some operations can be combined (for example a doubling followed by an addition becomes a "double-and-add"), and the sign of the term $2^i 3^j$ may change its impact (since point additions and point subtractions have a different effect in Miller's algorithm).

For this section, we consider the cost in terms of Miller's algorithm associated to each type of group operation coming from the 2-3 chain representation of n. Since our algorithm compute the scalar from the least significant term upward but Miller's algorithm works from the most significant term downward, we have to adjust the way we evaluate cost accordingly.

Furthermore, the way we choose to combine operations may depend on how many doublings and tripling are performed between two consecutive terms of the chain. We then need to separate the chains into sub-cases:

Definition 7. *The chains $C_{i,j}$ can be partitioned into three cases depending on the form of their largest term:*

- $C_{i,j}^{(3)}$ *if the largest term is of the form $2^i 3^b$ with $b < j$.*
- $C_{i,j}^{(2)}$ *if the largest term is of the form $2^a 3^j$ with $a < i$.*
- $C_{i,j}^{(6)}$ *if the largest term is of the form $2^a 3^b$ with $a < i$ and $b < j$.*

For arguments that apply to all three cases, we write $C_{i,j}^{(k)}$. The definition of cases for $\mathscr{C}_{i,j}$, $\overline{C}_{i,j}$ and $\overline{\mathscr{C}}_{i,j}$ are similar.

From this definition, we can refine Tables 1 and 2 to each cases, which gives Tables 3, 4, 5 and 6 in Appendix B.

To evaluate the cost of a subchain $C_{i,j}$ (respectively $\overline{C}_{i,j}$), we compute its cost as if the integer n is equal to $n_{i,j} + 2^i 3^j$ (respectively $\overline{n}_{i,j} + 2^i 3^j$). As a result, $C_{i,j}$ will have a "base count" of $i \cdot \alpha_2 + j \cdot \alpha_3$ to which we must add a number of β_{2+}, β_{2-}, β_{3+} and β_{3-} corresponding to the terms in the chain.

Table 7 details how the cost is updated as we increase the indices i and j. The first six cases (two "all" cases and four "new term" cases) should be self-explanatory. The "corrections" depending on the number of doublings and triplings since the last term of the chain was included. If we have both doublings and triplings since the last term, then the addition (respectively subtraction) should be combined with a doubling if $\beta_{2+} < \beta_{3+}$ (respectively $\beta_{2-} < \beta_{3-}$) and combined to a tripling otherwise. A correction may be necessary if the previous step had only doublings or only triplings (and therefore no choice on how to combine) but the current step as both.

This only leaves us with the question of what to do with the "extra term" $+2^i 3^j$ that we introduced to compute our costs. For most $C_{i,j}$ and $\overline{C}_{i,j}$ the "extra term" will not be a problem since it gets pushed along as i and j grow (it acts rather like a placeholder). For $\overline{C}_j$ (before introducing the final $+2^{i_j} 3^j$) this is in fact exactly the cost we want since the "extra term" is $+2^{i_j} 3^j$. The only problem comes with C_j, since we already added the final term and the "extra term" really overpasses n. However this problem is easily fixed since $(i,j) = (i_j, j) \iff n_{i,j} = n$, and in the steps where this occurs we simply have to keep the cost of $C_{i_j-1,j}$ or $C_{i_j,j-1}$ (depending on which one is used to obtain C_j). In Table 7, we refer to this as "remove $\alpha_2 + \beta_{2+}$ or $\alpha_3 + \beta_{3+}$".

It is now relatively straightforward to adapt Algorithm 2 to these cases, although writing out the detail case-by-case analysis becomes a rather long and repetitive process (due to the increased number of cases to deal with). Note that the relative costs of the operations (α_2, α_3, β_{2+}, β_{2-}, β_{3+}, and β_{3-}) are considered as entries for the algorithm, so the resulting optimal chain depends on the actual implementation of each operation.

5 Impact on Tate Pairings

5.1 Positive/Negative Chains

Under some conditions, the size of the code required for Miller's algorithm may be critical. In these situations, it may be interesting to limit how many specialized operations are used by Miller's algorithm. One approach is to keep doublings and triplings separated from additions and subtractions, but this loses the added efficiency of combining doubling with addition, etc.

Instead, we propose the notion of "positive chains" and "negative chains", where all the terms have the same sign (except the leading term in the negative

---

Algorithm 3. Algorithm to compute an optimal 2-3 chain for pairing computation.

Input: Integer $n > 0$, and operation costs α_2, α_3, β_{2+}, β_{2-}, β_{3+}, and β_{3-}.
Output: 2-3 chain $\mathscr{C}$ for n with minimal pairing cost.

1 $C_j^{(k)} \leftarrow \emptyset$, $\overline{C}_j^{(k)} \leftarrow \emptyset$ for every j
2 **for** $i \leftarrow 0$ ***to*** $\lceil \log_2(n+1) \rceil$ **do**
3 $\quad$ $m \leftarrow \lceil \log_3((n+1)/2^i) \rceil$
4 $\quad$ $i_m \leftarrow \lceil \log_2((n+1)/3^m) \rceil$
5 $\quad$ **if** $i_m < i$ **then**
6 $\quad\quad \lfloor \; m \leftarrow m - 1$
7 $\quad$ **for** $j \leftarrow 0$ ***to*** m **do**
8 $\quad\quad$ Using Table 3, $C_j^{(k)}$ $(= \mathscr{C}_{i-1,j}^{(k)})$ and $\overline{C}_j^{(k)}$, find possibles $C_{i,j}^{(k)}$.
9 $\quad\quad$ Using Table 4, $C_j^{(k)}$ and $\overline{C}_j^{(k)}$, find possibles $\overline{C}_{i,j}^{(k)}$.
10 $\quad\quad$ Using Table 5, $C_{j-1}^{(k)}$ $(= \mathscr{C}_{i,j-1}^{(k)})$ and $\overline{C}_{j-1}^{(k)}$, find possibles $C_{i,j}^{(k)}$.
11 $\quad\quad$ Using Table 6, $C_{j-1}^{(k)}$ and $\overline{C}_{j-1}^{(k)}$, find possibles $\overline{C}_{i,j}^{(k)}$.
12 $\quad\quad$ For each candidate, use Table 7 to compute its cost.
13 $\quad\quad$ $C_j^{(k)} \leftarrow$ chain of lowest among the $C_{i,j}^{(k)}$ $(C_j^{(k)} = \mathscr{C}_{i,j}^{(k)})$.
14 $\quad\quad$ $\overline{C}_j^{(k)} \leftarrow$ chain of lowest among the $\overline{C}_{i,j}^{(k)}$ $(\overline{C}_j^{(k)} = \overline{\mathscr{C}}_{i,j}^{(k)})$.
15 $\quad\quad$ **if** $i = i_j$ **then**
16 $\quad\quad\quad \lfloor \; \overline{C}_j^{(k)} \leftarrow 2^i 3^j + \overline{C}_j^{(k)}$.
17 $\quad\quad$ Correct the costs of $C_j^{(k)}$ and $\overline{C}_j^{(k)}$.

18 $\mathscr{C} \leftarrow$ chain of lowest cost among the $C_j^{(k)}$ and $\overline{C}_j^{(k)}$
19 **return** $\mathscr{C}$

---

chains). In this way, Miller's algorithm uses only four operations both for positive chains (*double, triple, double-and-add, triple-and-add*) and for negative chains (*double, triple, double-and-subtract, triple-and-subtract*).

Positive chains are built only from $\mathscr{C}_{i,j}$ (without allowing any $\overline{\mathscr{C}}_{i,j}$), all the way up to $\mathscr{C}_j$. Similarly, negative chains are built only from $\overline{\mathscr{C}}_{i,j}$ (without allowing any $\mathscr{C}_{i,j}$) all the way up to $\overline{\mathscr{C}}_j$ (where the final term is positive). Adapting Algorithm 2 to do this is straightforward.

For the operation costs presented in [6], negative chains are much more advantageous than positive chains (since both β_{2-} and β_{3-} are about three times smaller than β_{2+} and β_{3+}), and experimental results indicate that their cost is only slightly larger than the minimal chain.

6   Experimental Results

To better compare the efficiency of the different methods, we ran numerical experiments for a large number of primes over a range of bit-sizes (with 1000 randomly selected primes at each size). To simplify comparison at different bit-

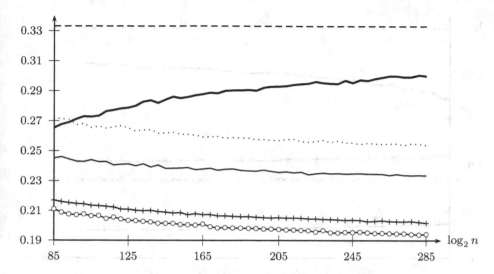

Fig. 1. Average Hamming weight per bit of n for 2-3 chains for the greedy algorithm (thick line), unrestricted version of Doche and Habsieger (+ symbols), minimized Hamming weight (o symbols), minimized Tate pairing cost (plain line), minimized negative chain (dotted line), and theoretical value for NAF (dashed line).

sizes, we considered the *Hamming weight per bit* (Hamming weight divided by $\log_2 n$) and the *cost per bit* (cost as multiples of α_2 divided by $\log_2 n$). The results are shown in Figs. 1 and 2. For the cost evaluations, we used the proportions between operations reported in [6]: $\alpha_2 = 1$, $\alpha_3 = 1.5705$, $\beta_{2+} = 0.3043$, $\beta_{2-} = 0.1107$, $\beta_{3+} = 0.3648$, and $\beta_{3-} = 0.1047$.

The average Hamming weight per bit of the "standard" binary representation (not included in the graph) is 0.50, and for the NAF it decreases to 0.33. From Fig. 1, the performance of the greedy algorithm appears to decrease as the size of n increases (slowly getting closer to that of a NAF), finishing around 0.30. On the other hand, the performance of the tree-based approach of Doche and Habsieger and our algorithm (all three cases considered) increases slightly as the size of n increases, but nowhere as near as would be expected for the sub-lineal DBNS.

From the graphs, it seems more reasonable to consider the 2-3 chains as having linear growth, with Hamming weight per bit around 0.25 for the minimized negative chains, 0.23 for the 2-3 chains with minimized cost and 0.19 for the 2-3 chains with minimal Hamming weight. For the tree-based approach, we observe a lower bound for the Hamming weight per bit of around 0.20, whereas Doche and Habsieger report around 0.215 (1/4.6419, see [10]). Recall that in [10] the number of live branches in the tree is limited, so the algorithm may in fact miss the most efficient branch of the tree (which is produced by our modified algorithm).

In terms of performance for pairing computations, we find that the 2-3 chains with minimal pairing cost offer an approximated 12.3 % gain over the original (standard binary) Miller algorithm ($1.152\alpha_2$ per bit), 4.22 % over a NAF-based pairing ($1.069\alpha_2$ per bit), 3.48 % over a greedy 2-3 chain, 1.40 % over the minimal

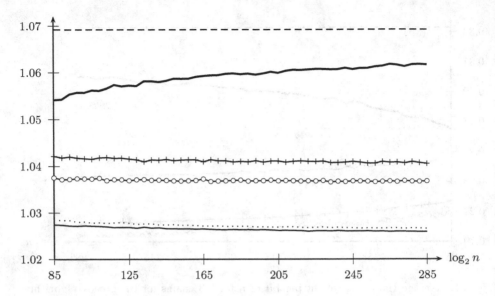

Fig. 2. Average pairing cost (multiples of α_2) per bit of n for 2-3 chains for the greedy algorithm (thick line), unrestricted version of Doche and Habsieger (+ symbols), minimized Hamming weight (o symbols), minimized Tate pairing cost (plain line), minimized negative chain (dotted line), and theoretical value for NAF (dashed line).

bound on Doche and Habsieger (roughly 1.74 % in practice, based on the values reported in [10]) and 1.06 % over a 2-3 chain with minimal Hamming weight. The difference between using negative chains and the truly minimized 2-3 chains appears to be less than 0.065 %, which could be compensated by the gains from having a simpler program (in particular due to reduced overheads). Note that at around $1.0259\alpha_2$ per bits, the optimized chain costs only 2.59 % more than a "squaring only" pairing (which can be considered an absolute lower bound on Miller-like algorithms).

7 Conclusion

We presented an algorithm to efficiently compute optimized 2-3 chain representations of an integer n. We described how to adapt our algorithm to minimize either the Hamming weight of the representation or to optimize the effect of the chain on a Tate pairing computation. These algorithms can easily be adapted to other situations (in particular if the cost of the various Tate pairing operations change).

Adapting to other multi-bases number systems is left as an open problem. Including other (more advanced) combined operations like the $2(3P_j) \pm P$ group operation of [7] may require only a simple adjustment of Table 7. Triple-based chains (for example 2-3-5 chains) are a natural extension, and would most likely require a three-dimensional version of our algorithm. One could also consider double-based chains with larger coefficient sets (see [5] for example). Adapting our algorithm to these representations would require new (larger) update tables, and most likely a larger set of possible chains at each position (i, j). Furthermore, we would also have to take into account the affect of the larger digit set on the costs in Miller's algorithm.

Because of the computational costs (compared to obtaining those required to compute a NAF), our algorithm may not be very attractive for scalar multiplication in a context of Diffie-Hellman key exchange (where the scalar is used only twice). However, in the context of bilinear pairings (Weil, Tate, Ate and Eta pairings) where the same (public) scalar can be used for a large number of operations (since the curve remains the same), it becomes a low-cost technique to produce significant savings (more than 3% over the previous best alternative). We also introduced the notion of negative chains, which offer near-optimal performance and allow to simplify the pairing implementation.

Acknowledgements. The authors would like to thanks the anonymous referees for their useful comments and suggestions.

A   Proofs of Lemmas and Theorems

Proof (Corollary 1). Since the tables process the updates from the values of $n_{i,j}$, $n_{i-1,j}$ and $n_{i,j-1}$, and not $\overline{n}_{i,j}$, $\overline{n}_{i-1,j}$ and $\overline{n}_{i,j-1}$, we must use Eq. 4, hence the distinctions between the cases $n_{i-1,j} = 0$ vs. $n_{i-1,j} > 0$ and $n_{i,j-1} = 0$ vs. $n_{i,j-1} > 0$.

Applying Lemma 3 to the cases $n_{i-1,j} = 0$ and $n_{i,j-1} = 0$ (with Eqs. 5 and 6) is straightforward.

To complete Table 1 we deal with two cases: If $n_{i,j} = n_{i-1,j}$, then $n_{i,j} = \overline{n}_{i-1,j} + 2^{i-1}3^j$ and $\overline{n}_{i,j} = n_{i-1,j} - 2 \cdot 2^{i-1}3^j = \overline{n}_{i-1,j} - 2^{i-1}3^j$. If $n_{i,j} = n_{i-1,j} + 2^{i-1}3^j$, then $n_{i,j} = \overline{n}_{i-1,j} + 2 \cdot 2^{i-1}3^j$ and $\overline{n}_{i,j} = n_{i-1,j} - 2^{i-1}3^j = \overline{n}_{i-1,j}$. We can then apply Lemma 3 to each case.

To complete Table 2 we deal with three cases: If $n_{i,j} = n_{i,j-1}$, then $n_{i,j} = \overline{n}_{i,j-1} + 2^i3^{j-1}$ and $\overline{n}_{i,j} = n_{i,j-1} - 3 \cdot 2^{i-1}3^j = \overline{n}_{i-1,j} - 2 \cdot 2^{i-1}3^j$. If $n_{i,j} = n_{i,j-1} + 2^i3^{j-1}$, then $n_{i,j} = \overline{n}_{i,j-1} + 2 \cdot 2^i3^{j-1}$ and $\overline{n}_{i,j} = n_{i,j-1} - 2 \cdot 2^{i-1}3^j = \overline{n}_{i-1,j} - 2^{i-1}3^j$. If $n_{i,j} = n_{i,j-1} + 2 \cdot 2^i3^{j-1}$, then $n_{i,j} = \overline{n}_{i,j-1} + 3 \cdot 2^i3^{j-1}$ and $\overline{n}_{i,j} = n_{i,j-1} - 2^{i-1}3^j = \overline{n}_{i-1,j}$. We can then apply Lemma 3 to each case.   □

Proof (Lemma 4). We observe that $\mathscr{C}_{0,0} = 0 = \overline{\mathscr{C}}_{0,0}$ (since $n_{0,0} = 0 = \overline{n}_{0,0}$), so at the "starting point" of the algorithm both chains exist. The result is then easily obtained by (double) induction. Table 1 ensures that at least one of the two chains exist when going from position $(i-1, j)$ to position (i, j), and Table 2 ensures that at least one of the two chains exist when going from position $(i, j-1)$ to position (i, j).   □

Proof (Theorem 1). From Definitions 4, 5, and 6, neither $C_{i,j}$ nor $2^i3^j + \overline{C}_{i,j}$ can be a chain for n if $i < i_j$. We now show that chains $C_{i,j}$ and $2^i3^j + \overline{C}_{i,j}$ need not be considered for $i > i_j$.

Let C be a chain with largest term $2^i 3^j$ with $i > i_j$, and let $\pm 2^a 3^b$ be its first term with $a > a_b$ (using the same definition as i_j). We define to subchains: $C_{low} = C_{a,b}$ or $\overline{C}_{a,b}$ (depending on the sign of its largest term), and C_{high} which consists of all the terms which are multiples of $2^a 3^b$. Due to the growth in a 2-3 chain (by factors of 2 and 3), any chain with terms bounded above by $2^r 3^s$ represents a number between $-(2^r 3^s - 1)$ and $2^r 3^s - 1$. Since none of the terms in C_{high} can affect the remainder modulo $2^a 3^b$, C_{low} must then represent either n or $n - 2^a 3^b$. If C_{low} represents n, then it is a better chain than C. If C_{low} represents $n - 2^a 3^b$, then $2^a 3^b + C_{low}$ is a better chain than C unless they are equal.

We can therefore restrict the search of a minimal chain for n to chains of the form $C_{i,j}$ and $2^i 3^j + \overline{C}_{i,j}$ where $i = i_j$. From Lemma 2, all subchains of a minimal chain must also be minimal, so we can restrict ourselves to the chains C_j and $\overline{C}_j$ produced by Algorithm 2.   □

Proof (Theorem 2). Since $0 \le i \le \lceil \log_2(n+1) \rceil$ and $0 \le j \le m \le \lceil \log_3(n+1) \rceil$, there are clearly $O((\log n)^2)$ steps in Algorithm 2. The operations performed at each step is easily bounded by $O((\log n)^2)$ bit operations. The algorithm produces $2(\lceil \log_3(n+1) \rceil + 1)$ chains (C_j and $\overline{C}_j$), each consisting in a sum of at most $\lceil \log_2(n+1) \rceil + 1$ terms of size bounded by $2n$, from which the result follows directly.   □

B   Update Tables for Optimal Chains

Table 3. Possible sources of $C_{i,j}^{(k)}$ when multiplying by 2 ($C_{i,j}^{(3)}$ not possible)

$n_{i-1,j}$	$n_{i,j}$	Possible $C_{i,j}^{(2)}$	Possible $C_{i,j}^{(6)}$
$= 0$	0	—	0
$= 0$	$2^{i-1}3^j$	$2^{i-1}3^j$	—
> 0	$n_{i-1,j}$	$C_{i-1,j}^{(2)}$ and $2^{i-1}3^j + \overline{C}_{i-1,j}^{(k)}$	$C_{i-1,j}^{(3)}$ and $C_{i-1,j}^{(6)}$
> 0	$n_{i-1,j} + 2^{i-1}3^j$	$2^{i-1}3^j + C_{i-1,j}^{(k)}$	—

Table 4. Possible sources of $\overline{C}_{i,j}^{(k)}$ when multiplying by 2 ($\overline{C}_{i,j}^{(3)}$ not possible)

$n_{i-1,j}$	$n_{i,j}$	Possible $\overline{C}_{i,j}^{(2)}$	Possible $\overline{C}_{i,j}^{(6)}$
$= 0$	0	—	0
$= 0$	$2^{i-1}3^j$	$-2^{i-1}3^j$	—
> 0	$n_{i-1,j}$	$-2^{i-1}3^j + \overline{C}_{i-1,j}^{(k)}$	—
> 0	$n_{i-1,j} + 2^{i-1}3^j$	$\overline{C}_{i-1,j}^{(2)}$ and $-2^{i-1}3^j + C_{i-1,j}^{(k)}$	$\overline{C}_{i-1,j}^{(3)}$ and $\overline{C}_{i-1,j}^{(6)}$

Table 5. Possible sources of $C_{i,j}^{(k)}$ when multiplying by 3 ($C_{i,j}^{(2)}$ not possible)

$n_{i-1,j}$	$n_{i,j}$	Possible $C_{i,j}^{(3)}$	Possible $C_{i,j}^{(6)}$
$=0$	0	—	0
$=0$	2^i3^{j-1}	2^i3^{j-1}	—
$=0$	$2\cdot2^i3^{j-1}$	—	—
>0	$n_{i,j-1}$	$C_{i,j-1}^{(3)}$ and $2^i3^{j-1}+\overline{C}_{i,j-1}^{(k)}$	$C_{i,j-1}^{(2)}$ and $C_{i,j-1}^{(6)}$
>0	$n_{i,j-1}+2^i3^{j-1}$	$2^i3^{j-1}+C_{i,j-1}^{(k)}$	—
>0	$n_{i,j-1}+2\cdot2^i3^{j-1}$	—	—

Table 6. Possible sources of $\overline{C}_{i,j}^{(k)}$ when multiplying by 3 ($\overline{C}_{i,j}^{(2)}$ not possible)

$n_{i-1,j}$	$n_{i,j}$	Possible $\overline{C}_{i,j}^{(3)}$	Possible $\overline{C}_{i,j}^{(6)}$
$=0$	0	—	0
$=0$	2^i3^{j-1}	—	—
$=0$	$2\cdot2^i3^{j-1}$	-2^i3^{j-1}	—
>0	$n_{i,j-1}$	—	—
>0	$n_{i,j-1}+2^i3^{j-1}$	$-2^i3^{j-1}+\overline{C}_{i,j-1}^{(k)}$	—
>0	$n_{i,j-1}+2\cdot2^i3^{j-1}$	$\overline{C}_{i,j-1}^{(3)}$ and $-2^i3^{j-1}+C_{i,j-1}^{(k)}$	$\overline{C}_{i,j-1}^{(2)}$ and $\overline{C}_{i,j-1}^{(6)}$

Table 7. Cost updates for $C_{i,j}^{(k)}$ and $\overline{C}_{i,j}^{(k)}$

Cases	Change		Effect on the cost
All	$C_{i-1,j}^{(k)}$ or $\overline{C}_{i-1,j}^{(k)} \rightarrow C_{i,j}^{(k)}$ or $\overline{C}_{i,j}^{(k)}$		$+\alpha_2$
	$C_{i,j-1}^{(k)}$ or $\overline{C}_{i,j-1}^{(k)} \rightarrow C_{i,j}^{(k)}$ or $\overline{C}_{i,j}^{(k)}$		$+\alpha_3$
New terms	$+2^{i-1}3^j \rightarrow C_{i,j}^{(2)}$		$+\beta_{2+}$
	$-2^{i-1}3^j \rightarrow \overline{C}_{i,j}^{(2)}$		$+\beta_{2-}$
	$+2^i3^{j-1} \rightarrow C_{i,j}^{(3)}$		$+\beta_{3+}$
	$-2^i3^{j-1} \rightarrow \overline{C}_{i,j}^{(3)}$		$+\beta_{3-}$
Corrections	$C_{i-1,j}^{(3)} \rightarrow C_{i,j}^{(6)}$	if $\beta_{2+} < \beta_{3+}$	$-\beta_{3+}+\beta_{2+}$
	$C_{i-1,j}^{(2)} \rightarrow C_{i,j}^{(6)}$	if $\beta_{2+} > \beta_{3+}$	$-\beta_{2+}+\beta_{3+}$
	$\overline{C}_{i-1,j}^{(3)} \rightarrow \overline{C}_{i,j}^{(6)}$	if $\beta_{2-} < \beta_{3-}$	$-\beta_{3-}+\beta_{2-}$
	$\overline{C}_{i-1,j}^{(2)} \rightarrow \overline{C}_{i,j}^{(6)}$	if $\beta_{2-} > \beta_{3-}$	$-\beta_{2-}+\beta_{3-}$
Final steps	$C_{i,j}^{(k)}$ with $i=i_j$		remove $\alpha_2+\beta_{2+}$ or $\beta_3+\beta_{3+}$
	$\overline{C}_{i,j}^{(k)}$ with $i=i_j$		–

References

1. Aranha, D.F., Karabina, K., Longa, P., Gebotys, C.H., López, J.: Faster explicit formulas for computing pairings over ordinary curves. In: Paterson, K.G. (ed.) EUROCRYPT 2011. LNCS, vol. 6632, pp. 48–68. Springer, Heidelberg (2011)
2. Barreto, P.S.L.M., Kim, H.Y., Lynn, B., Scott, M.: Efficient algorithms for pairing-based cryptosystems. In: Yung, M. (ed.) CRYPTO 2002. LNCS, vol. 2442, pp. 354–369. Springer, Heidelberg (2002)
3. Blake, I.F., Murty, V.K., Xu, G.: Refinements of Miller's algorithm for computing the weil/tate pairing. J. Algorithms **58**, 134–149 (2006)
4. Blake, I.F., Seroussi, G., Smart, N.P.: Advances in Elliptic Curve Cryptography. London Mathematical Society Lecture Note Series, vol. 317. Cambridge University Press, Cambridge (2005)
5. Bernstein, D.J., Birkner, P., Lange, T., Peters, C.: Optimizing double-base elliptic-curve single-scalar multiplication. In: Srinathan, K., Rangan, C.P., Yung, M. (eds.) INDOCRYPT 2007. LNCS, vol. 4859, pp. 167–182. Springer, Heidelberg (2007)
6. A. Capuñay. Multibase Scalar Multiplications in Cryptographic Pairings. preprint, 2015
7. Ciet, M., Joye, M., Lauter, K., Montgomery, P.L.: Trading inversions for multiplications in elliptic curve cryptography. Des. Codes Crypt. **39**(2), 189–206 (2006)
8. Dimitrov, V.S., Imbert, L., Mishra, P.K.: Efficient and secure elliptic curve point multiplication using double-base chains. In: Roy, B. (ed.) ASIACRYPT 2005. LNCS, vol. 3788, pp. 59–78. Springer, Heidelberg (2005)
9. Dimitrov, V.S., Jullien, G.A., Miller, W.C.: An algorithm for modular exponentiation. Inform. Process. Lett. **66**(3), 155–159 (1998)
10. Doche, C., Habsieger, L.: A tree-based approach for computing double-base chains. In: Mu, Y., Susilo, W., Seberry, J. (eds.) ACISP 2008. LNCS, vol. 5107, pp. 433–446. Springer, Heidelberg (2008)
11. Eisenträger, K., Lauter, K., Montgomery, P.L.: Fast elliptic curve arithmetic and improved weil pairing evaluation. In: Joye, M. (ed.) CT-RSA 2003. LNCS, vol. 2612, pp. 343–354. Springer, Heidelberg (2003)
12. Miller, V.S.: The Weil pairing, and its efficient calculation. J. Crypt. **17**(4), 235–261 (2004)

Subgroup Security in Pairing-Based Cryptography

Paulo S.L.M. Barreto[1], Craig Costello[2]([⊠]), Rafael Misoczki[1],
Michael Naehrig[2], Geovandro C.C.F. Pereira[1], and Gustavo Zanon[1]

[1] Escola Politécnica, University of São Paulo, São Paulo, Brazil
{pbarreto,rmisoczki,geovandro,gzanon}@larc.usp.br
[2] Microsoft Research, Redmond, WA, USA
{craigco,mnaehrig}@microsoft.com

Abstract. Pairings are typically implemented using ordinary pairing-friendly elliptic curves. The two input groups of the pairing function are groups of elliptic curve points, while the target group lies in the multiplicative group of a large finite field. At moderate levels of security, at least two of the three pairing groups are necessarily proper subgroups of a much larger composite-order group, which makes pairing implementations potentially susceptible to small-subgroup attacks.

To minimize the chances of such attacks, or the effort required to thwart them, we put forward a property for ordinary pairing-friendly curves called *subgroup security*. We point out that existing curves in the literature and in publicly available pairing libraries fail to achieve this notion, and propose a list of replacement curves that do offer subgroup security. These curves were chosen to drop into existing libraries with minimal code change, and to sustain state-of-the-art performance numbers. In fact, there are scenarios in which the replacement curves could facilitate faster implementations of protocols because they can remove the need for expensive group exponentiations that test subgroup membership.

Keywords: Pairing-based cryptography · Elliptic-curve cryptography · Pairing-friendly curves · Subgroup membership · Small-subgroup attacks

1 Introduction

In this paper we propose new instances of pairing-friendly elliptic curves that aim to provide stronger resistance against small-subgroup attacks [36]. A small-subgroup attack can be mounted on a discrete-logarithm-based cryptographic

Paulo S. L. M. Barreto, Rafael Misoczki, Geovandro C. C. F. Pereira—Supported by Intel Research grant "Energy-efficient Security for SoC Devices" 2012.
Paulo S. L. M. Barreto—Supported by CNPq research productivity grant 306935/2012-0.
Gustavo Zanon—Supported by the São Paulo Research Foundation (FAPESP) grant 2014/09200-5.

K. Lauter and F. Rodríguez-Henríquez (Eds.): LatinCrypt 2015, LNCS 9230, pp. 245–265, 2015.
DOI: 10.1007/978-3-319-22174-8_14

scheme that uses a prime-order group which is contained in a larger group of order divisible by small prime factors. By forcing a protocol participant to carry out an exponentiation of a non-prime-order group element with a secret exponent, an attacker could obtain information about that secret exponent. This is possible if the protocol implementation does not check that the group element being exponentiated belongs to the correct subgroup and thus has large prime order. In the worst case, the user's secret key could be fully revealed although the discrete logarithm problem (DLP) in the large prime-order subgroup is computationally infeasible. We start by illustrating the possibility of such attacks in the context of (pairing-based) *digital signature schemes*, many of which are based on the celebrated short signature scheme of Boneh, Lynn and Shacham (BLS) [10][1].

BLS Signatures. For both historical reasons and for ease of exposition, authors of pairing-based protocol papers commonly assume the existence of an efficient, *symmetric* bilinear map $e\colon \mathbb{G} \times \mathbb{G} \to \mathbb{G}_T$, where $\mathbb{G}$ and $\mathbb{G}_T$ are cryptographic groups of large prime order n. Let P be a public generator of $\mathbb{G}$ and let $\mathcal{H}\colon \{0,1\}^* \to \mathbb{G}$ be a suitably defined hash function. Boneh, Lynn and Shacham proposed a simple signature scheme [10] that works as follows. To sign a message $M \in \{0,1\}^*$ with her secret key $a \in \mathbb{Z}_n$, Alice computes $Q = \mathcal{H}(M) \in \mathbb{G}$ and sends the signature $\sigma = [a]Q$ to Bob. To verify this signature, Bob also computes $Q = \mathcal{H}(M)$ and then uses Alice's public key $[a]P$ to assert that $e([a]P, Q) = e(P, \sigma)$. It is shown in [10] that this scheme is a secure signature scheme if the *Gap Diffie-Hellman* (GDH) problem is hard.

Forged System Parameters. There are various threat models in which one might wish to think about the possible implications of small-subgroup attacks. In one of them, one assumes that it is possible for an attacker to even forge the public parameters used in the signature system. Such a possibility has for example been discussed for the Digital Signature Standard by Vaudenay in [52]. For BLS signatures, an attacker could forge the system parameters to use a base point P of non-prime order. Thus by means of a small-subgroup attack, Alice reveals information about her private key a to the attacker by simply publishing her public key $[a]P$.

In another example of public parameter manipulation, one might assume that the hash function $\mathcal{H}$ maps into the full composite order group, instead of into the prime order subgroup. Therefore, the hash of a message could be a group element of composite order and the BLS signature could leak information about Alice's private key. Such a faulty hash function might actually be the result of an implementation bug, for example the omission of cofactor exponentiations to move group elements to the right subgroup.

---

[1] We warn the reader that BLS is commonly used to abbreviate two different authorships in the context of pairing-based cryptography: BLS signatures [10] and BLS curves [4].

Valid System Parameters. Even if the system parameters are valid, there are scenarios in which small subgroup attacks might lead to a security breach. In this setting, we assume that, since P is a fixed public parameter (that is presumably asserted to be in $\mathbb{G}$) and Alice hashes elements into $\mathbb{G}$ herself, Alice's public key and signature are *guaranteed* to lie in $\mathbb{G}$, and are therefore protected by the hardness of the discrete logarithm problem (DLP) in $\mathbb{G}$. There is therefore no threat to Alice's secret key in context of BLS signatures, but this is not necessarily the case in the context of (pairing-based) *blind signatures*, as we discuss below.

Blind Signatures. Roughly speaking, blind signatures allow Alice to sign a message that is authored by a third party, Carol. The typical scenarios require that Carol and Alice interact with one another in such a way that Carol learns nothing about Alice's secret signing key and Alice learns nothing about Carol's message. In [9, Sect. 5], Boldyreva describes a simple blind signature scheme that follows naturally from BLS signatures. In order to "blindly" sign her message M, Carol computes $Q = \mathcal{H}(M)$ and sends Alice the blinded message $\tilde{Q} = Q + [r]P$, for some random $r \in \mathbb{Z}_n$ of Carol's choosing. Alice uses her secret key $a \in \mathbb{Z}_n$ to return the signed value $[a]\tilde{Q}$ to Carol, who then uses her random value r and Alice's public key $[a]P$ to compute $\sigma = [a]\tilde{Q} - [r]([a]P) = [a]Q$. Carol then sends σ to Bob who can assert that it is a valid BLS signature under Alice's key.

Unlike for the original BLS signatures, where Alice hashed the message into $\mathbb{G}$ herself before signing it, in the above scheme Alice signs the point that Carol sends her. If Carol maliciously sends Alice a point that belongs to a group in which the DLP is easy (e.g. via a small subgroup attack [36]), and if this goes undetected by Alice, then Carol can recover Alice's secret key.

Of course, in a well-designed version of the above protocol, Alice validates that the point she receives is in the correct group before using her secret key to create the signature. However, for the instantiations of bilinear pairings that are preferred in practice, this validation requires a full elliptic curve scalar multiplication. In addition, as is discussed in Remark 1 below, authors of pairing-based protocols often assume that certain group elements belong to the groups that they are supposed to. If these descriptions were translated into real-world implementations *unchanged*, then such instantiations could be susceptible to small subgroup attacks like the example above.

Asymmetric Pairings. The original papers that founded pairing-based cryptography [11,31,46] assumed the existence of a bilinear map of the form $e\colon \mathbb{G} \times \mathbb{G} \to \mathbb{G}_T$. Such *symmetric* pairings (named so because the two input groups are the same) only exist on supersingular curves, which places a heavy restriction on either or both of the underlying efficiency and security of the protocol; see [25] or [20] for further discussion. It was not long until practical instantiations of asymmetric pairings of the form $e\colon \mathbb{G}_1 \times \mathbb{G}_2 \to \mathbb{G}_T$ (with $\mathbb{G}_2 \neq \mathbb{G}_1$) were discovered [5,22], and were shown to be much more efficient than their symmetric counterparts, especially at high security levels. In recent times this performance gap has been stretched orders of magnitude further, both given the many advances in the asymmetric setting [2,6,29,51,53], and given that the fastest known instantiations of symmetric pairings are now considered broken [3]. Thus, all modern pairing libraries are built on ordinary elliptic curves in the asymmetric setting.

We note that transferring the above blind signature protocol to the asymmetric setting does not remove the susceptibility of the scheme to small subgroup attacks. Following the asymmetric version of BLS signatures [14,37], in Boldyreva's simple blind signature scheme Alice's public key could be $([a]P_1, [a]P_2)$ for fixed generators $(P_1, P_2) \in \mathbb{G}_1 \times \mathbb{G}_2$. If $\mathcal{H}\colon \{0,1\}^* \to \mathbb{G}_1$, then Carol can blind the message $Q = \mathcal{H}(M)$ by sending Alice $\tilde{Q} = Q + [r]P_1$. Upon receiving $[a]\tilde{Q}$ back from Alice, Carol removes the blinding factor as before by taking $[a]\tilde{Q} - [r]([a]P_1) = [a]Q$, and sends this result to Bob. Bob then uses Alice's public key to verify that $e([a]Q, P_2) = e(Q, [a]P_2)$. Here, if Alice does not pay the price of a scalar multiplication to assert that $\tilde{Q}$ is in fact in $\mathbb{G}_1$, then Carol could use small subgroup attacks to obtain Alice's private signing key.

In any case, subgroup attacks are inherent to (pairing-based) blind signatures, where signatures are performed *blindly* on points sent by third parties.

The Case of the Group $\mathbb{G}_2$. In the asymmetric pairing setting, protocols are often designed to perform the bulk of elliptic curve operations in the group $\mathbb{G}_1$, because $\mathbb{G}_1$ can be instantiated as the group of rational points on a pairing-friendly elliptic curve over the base field. Here group operations are more efficient than those in $\mathbb{G}_2$ and group elements have more compact representations. In some cases, the group of rational points even has prime order (i.e. it is equal to $\mathbb{G}_1$) and is thus resistant against subgroup attacks (assuming that valid system parameters are used), while the group $\mathbb{G}_2$ almost always lies in a much larger group with potentially many small prime factors dividing the cofactor. The above translation of the blind signature scheme would thus not be susceptible to the attack because the signed point is in $\mathbb{G}_1$. Only via forged parameters would Alice's public key $[a]P_2 \in \mathbb{G}_2$ leak information about her private key.

However, there are protocols that use the group $\mathbb{G}_2$ for the signing operation for efficiency reasons. This is indicated in the context of BLS signatures as credentials in the Boneh-Franklin identity-based encryption [11] in [43, Sect. 2]. An example for which a hash function $\mathcal{H} : \{0,1\}^* \to \mathbb{G}_2$ that is faulty and maps to a larger set of points containing non-prime order group elements, can lead to a subgroup attack is the oblivious signature-based envelope (OSBE) scheme by Li, Du, and Boneh [35]. The original OSBE scheme is described in the symmetric setting, but an asymmetric instantiation would be set up with signatures in $\mathbb{G}_2$. The scheme uses a trusted authority (TA) which hands out Boneh-Franklin credentials, which are BLS signatures on identities. Given an identity M and the master key $x \in \mathbb{Z}_n$, the TA computes and sends $[x]\mathcal{H}(M)$ to the receiver allowed to decrypt messages, which leaks information about x if $\mathcal{H}(M)$ does not have order n.

Subgroup Security. The main contribution of this paper is the definition of a new property for pairing-friendly curves, which we call *subgroup security*. A pairing-friendly curve is called *subgroup-secure* if the cofactors of all pairing groups, whenever they are of the same size as the prime group order n or larger, only contain prime factors larger than n. This is a realistic scenario because for curves targeting modern security levels, at least two of the pairing groups have very large cofactors. We slightly relax the condition to allow small inevitable cofactors that are imposed by the polynomial parametrizations in the popular constructions of pairing-friendly curves. This means that this property distinguishes those curves in a given family that provide as much resistance against small-subgroup attacks as possible.

We select subgroup-secure curves for four of the most efficient families of pairing-friendly curves that can replace existing curves in pairing libraries with minimal code change. For example, we find a low NAF-weight Barreto-Naehrig (BN) curve for which no (related) elliptic curve subgroup of order smaller than n exists. Replacing BN254 with this one could allow implementers to remove certain membership tests (via scalar multiplications). Returning to the blind signature scheme above, this would mean that Alice only needs to check that the point $\tilde{Q}$ is on the right curve before signing, and this is true whether the protocol is arranged such that $\tilde{Q}$ is intended to be in $\mathbb{G}_1$ or in $\mathbb{G}_2$. Even if Carol sends Alice a point that is not in the order n subgroup, Carol's job of recovering a from Q and $[a]Q$ is no easier since, by the application of Definition 1 to the BN family, the smallest prime factor dividing the order of Q will always be at least n.

While existing curves in the literature are not subgroup-secure and may therefore require expensive operations to guarantee discrete-log security in the pairing groups, the curves we propose can, wherever possible, maintain their discete-log security even in the absence of some of the subgroup membership checks. Our performance benchmarks show that replacing existing curves with subgroup-secure curves incurs only a minor performance penalty in the pairing computation; on the other hand, all group operations remain unaffected by this stronger security notion and retain their efficiency.

Related Work. The comments made by Chen, Cheng and Smart [16] are central to the theme of this work. We occasionally refer back to the following remark, which quotes [16, Sect. 2] verbatim.

Remark 1 ([16]). "An assumption is [often] made that all values passed from one party to another lie in the correct groups. Such assumptions are often implicit within security proofs. However, one needs to actually:

(i) check that given message flows lie in the group,
(ii) force the messages to lie in the group via additional computation, or
(iii) choose parameters carefully so as the problem does not arise.

Indeed, some attacks on key agreement schemes, such as the small-subgroup attack [36], are possible because implementors do not test for subgroup membership. For pairing-based systems one needs to be careful whether and how one implements these subgroup membership tests as it is not as clear as for standard discrete logarithm based protocols."

The overall aim of this paper is to explore and optimize option (iii) above.

In the paper introducing small-subgroup attacks [36], Lim and Lee suggest that a strong countermeasure is to ensure that the intended cryptographic subgroup is the smallest subgroup within *the* large group. In the context of pairing-based cryptography, Scott [49] showed a scenario in which a small-subgroup attack could be possible on elements of the third pairing group $\mathbb{G}_T$, the target group, and subsequently he adapted the Lim-Lee solution to put forward the notion of "$\mathbb{G}_T$-strong" curves. Our definition of subgroup security (see Definition 1) applies this solution to all three of the pairing groups, the two elliptic curve input groups $\mathbb{G}_1$, $\mathbb{G}_2$ as well as the target group $\mathbb{G}_T$, and therefore this paper can be seen as an extension and generalization of Scott's idea: while he gave an example of a BN curve that is $\mathbb{G}_T$-strong, we give replacement curves from several families that are both $\mathbb{G}_T$-strong *and* $\mathbb{G}_2$-strong – this is the optimal situation for the families used in practice[2].

2   Pairing Groups and Pairing-Friendly Curves

For modern security levels, the most practical pairings make use of an ordinary elliptic curve E defined over a large prime field $\mathbb{F}_p$ whose *embedding degree* (with respect to a large prime divisor n of $\#E(\mathbb{F}_p)$) is k, i.e. k is the smallest positive integer such that $n \mid p^k - 1$. In this case, there exists a pairing $e : \mathbb{G}_1 \times \mathbb{G}_2 \to \mathbb{G}_T$, where $\mathbb{G}_1$ is the subgroup $E(\mathbb{F}_p)[n]$ of order n of $E(\mathbb{F}_p)$, $\mathbb{G}_2$ is a specific subgroup of order n of $E(\mathbb{F}_{p^k})$ contained in $E(\mathbb{F}_{p^k}) \setminus E(\mathbb{F}_p)$, and $\mathbb{G}_T$ is the subgroup of n-th roots of unity, in other words, the subgroup of order n of the multiplicative group $\mathbb{F}_{p^k}^\times$.

Let $t \in \mathbb{Z}$ be the trace of the Frobenius endomorphism on $E/\mathbb{F}_p$, so that $\#E(\mathbb{F}_p) = p + 1 - t$ and $|t| \le 2\sqrt{p}$. Write $t^2 - 4p = Dv^2$ for some square-free negative integer D and some integer v. All of the pairing-friendly curves in this paper have $D = -3$ and $6 \mid k$; together, these two properties ensure that we can always write the curves as $E/\mathbb{F}_p\colon y^2 = x^3 + b$, and that we can make use of a *sextic twist* $E'/\mathbb{F}_{p^{k/6}}\colon y^2 = x^3 + b'$ of $E(\mathbb{F}_{p^k})$ to instead represent $\mathbb{G}_2$ by the isomorphic group $\mathbb{G}_2' = E'(\mathbb{F}_{p^{k/6}})[n]$, such that coordinates of points in $\mathbb{G}_2'$ lie in the much smaller subfield $\mathbb{F}_{p^{k/6}}$ of $\mathbb{F}_{p^k}$. Henceforth we abuse notation and rewrite $\mathbb{G}_2$ as $\mathbb{G}_2 = E'(\mathbb{F}_{p^{k/d}})[n]$.

For a particular curve $E/\mathbb{F}_q\colon y^2 = x^3 + b$ with $D = -3$, where $q = p^e$ for some $e \ge 1$, there are six *twists* of E defined over $\mathbb{F}_q$, including E itself. These twists are isomorphic to E when considered over $\mathbb{F}_{q^6}$. The following lemma (cf. [30, Sect. A.14.2.3]) determines the group orders of these twists over $\mathbb{F}_q$, and is used several times in this work.

---

[2] Our definition of subgroup security incorporates $\mathbb{G}_1$ for completeness, since for curves from the most popular families of pairing-friendly curves the index of $\mathbb{G}_1$ in $E(\mathbb{F}_p)$ is both greater than one and much less than the size of $\mathbb{G}_1$, thereby necessarily containing small subgroups. The only exceptions are the prime order families like the MNT [38], Freeman [19], and BN [6] curve families, for which this index is 1.

Lemma 1. *Let t be the trace of Frobenius of the elliptic curve $E/\mathbb{F}_q\colon y^2 = x^3 + b$, and let $v \in \mathbb{Z}$ such that $t^2 - 4q = -3v^2$. Up to isomorphism, there are at most six curves (including E) defined over $\mathbb{F}_q$ with trace t' such that $t'^2 - 4q = -3v'^2$ for some square-free $v' \in \mathbb{Z}$. The six possibilities for t' are $t, -t, (t+3v)/2, -(t+3v)/2, (t-3v)/2,$ and $-(t-3v)/2$.*

In this work, we focus on four of the most popular families of ordinary pairing-friendly curves: the Barreto-Naehrig (BN) family [6] with $k = 12$ which is favorable at the 128-bit security level; the Barreto-Lynn-Scott (BLS) cyclotomic family [4] with $k = 12$ and the Kachisa-Schaefer-Scott (KSS) family [32] with $k = 18$, both of which are suitable at the 192-bit security level; and the cyclotomic BLS family with $k = 24$, which is well suited for use at the 256-bit security level.

The above examples of pairing-friendly curves are all *parameterized families*. This means that the parameters p, t and n of a specific curve from each family are computed via the evaluation of univariate polynomials $p(u)$, $t(u)$ and $n(u)$ in $\mathbb{Q}[u]$ at some $u_0 \in \mathbb{Z}$. The typical way to find a good curve instance is to search over integer values u_0 of low NAF-weight (i.e. with as few non-zero entries in signed-binary, non-adjacent form (NAF) representation as possible) and of a suitable size, until $p(u_0)$ and $n(u_0)$ are simultaneously prime. Since our curves are all of the form $E/\mathbb{F}_p\colon y^2 = x^3 + b$, and since Lemma 1 states that there are at most 6 isomorphism classes over $\mathbb{F}_p$, the correct curve is quickly found by iterating through small values of b and testing non-zero points $P \neq \mathcal{O}$ on E for the correct order, i.e. testing whether $[p(u_0) + 1 - t(u_0)]P = \mathcal{O}$.

3   Subgroup-Secure Pairing-Friendly Curves

In this section we recall small-subgroup attacks and define the notion of *subgroup security*, a property that is simple to achieve in practice and that strengthens the resistance of pairing-friendly curves against *subgroup attacks*. After that, we discuss the four most popular choices of pairing-friendly curve families, BN ($k = 12$), KSS ($k = 18$) and BLS ($k = 12$ and $k = 24$) curves and provide examples of subgroup-secure curves suitable for efficient implementation of optimal pairings at the 128-, 192-, and 256-bit security levels.

3.1   Small-Subgroup Attacks

Small-subgroup attacks against cryptographic schemes based on the discrete logarithm problem (DLP) were introduced by Lim and Lee [36]. The following is a brief description of the basic idea in a general group setting.

Suppose that $\mathbb{G}$ is a group of prime order n (written additively), which is contained in a larger, finite abelian group $\mathcal{G}$, and let h be the index of $\mathbb{G}$ in $\mathcal{G}$, $|\mathcal{G}| = h \cdot n$. Suppose that the DLP is hard in any subgroup of $\mathcal{G}$ of large enough prime order. In particular, assume that the prime n is large enough such that the DLP is infeasible in $\mathbb{G}$. If the index h has a small prime factor r, then there exists a group element P of order a multiple of r, and if r is small enough, the DLP in $\langle P \rangle$ can be easily solved modulo r. If an attacker manages to force a protocol participant to use P for a group exponentiation involving a secret exponent, instead of using a valid element from $\mathbb{G}$, solving the DLP in $\langle P \rangle$ provides partial information on the secret exponent. If h has several small prime factors, the Pohlig-Hellman attack [45] may be able to recover the full secret exponent.

Such small-subgroup attacks can be avoided by *membership testing*, i.e. by checking that any point P received during a protocol actually belongs to the group $\mathbb{G}$ and cannot have a smaller order (see point (i) in Remark 1). Another way to thwart these attacks is a *cofactor exponentiation* or *cofactor multiplication* (which is a solution to achieve point (ii) in Remark 1). If every received element P is multiplied by the index h, which also means that the protocol needs to be adjusted to work with the point $[h]P$ instead of P, then points of small order are mapped to $\mathcal{O}$ and any small-order component of P is cleared by this exponentiation.

3.2   Subgroup Security

If $h > 1$ and it does not contain any prime factors smaller than n, then $\mathbb{G}$ is one of the subgroups in $\mathcal{G}$ with the weakest DLP security. In other words, for any randomly chosen element $P \in \mathcal{G}$, the DLP in the group $\langle P \rangle$ is guaranteed to be at least as hard as the DLP in $\mathbb{G}$, since even if $|\langle P \rangle| = |\mathcal{G}|$, the Pohlig-Hellman reduction [45] requires the solution to a DLP in a subgroup of prime order at least n. Depending on the protocol design, it might be possible to omit membership testing and cofactor multiplication if parameters are chosen such that h does not have prime factors smaller than n: this is one possibility that addresses point (iii) in Remark 1.

One might consider omitting the test as to whether an element belongs to the group $\mathbb{G}$ if this is a costly operation. For example, if testing membership for $\mathbb{G}$ requires a relatively expensive group exponentiation, and testing membership for $\mathcal{G}$ is relatively cheap (i.e. costs no more than a few group operations), one can replace the costly check by the cheaper one given that the index h does not have any small factors. When the group $\mathcal{G}$ is the group of $\mathbb{F}_q$-rational points on an elliptic curve E, and $\mathbb{G}$ is a prime order subgroup, then testing whether a point P belongs to $\mathcal{G}$ is relatively cheap, because it only requires to check validity of the curve equation, while testing whether a point belongs to $\mathbb{G}$ additionally requires either a scalar multiplication $[n]P$ to check whether P has the right order, or a cofactor multiplication $[h]P$ to force the resulting point to have the right order. If the cofactor is small, the latter cost is low, but for large cofactors, it might be more efficient to refrain from carrying out any of the exponentiations when working with suitable parameters.

An attempt to define the notion of subgroup security could be to demand that the index h (if it is not equal to 1) only contains prime factors of size n or larger, in which case both exponentiations are very costly. However, in the case of elliptic curve cryptography (ECC), such a definition does not make sense, since curves are chosen such that the cofactor is equal to 1 or a very small power of 2 (such as 4 or 8) depending on the curve model that is selected for efficiency and security reasons. Although there are good reasons to require cofactor $h = 1$, it would unnecessarily exclude curve models which allow performance gains by having a small cofactor (such as Montgomery [39] or Edwards [18] curve models). Therefore, demanding only large prime factors in h only makes sense if the group inherently has large, unavoidable cofactors by construction. This is the case for some of the groups that arise from pairing-friendly curves.

For the three pairing (sub)groups $\mathbb{G}_1$, $\mathbb{G}_2$ and $\mathbb{G}_T$ defined in Sect. 2, there are very natural choices of three associated groups $\mathcal{G}_1$, $\mathcal{G}_2$ and $\mathcal{G}_T$ for which testing membership is easy. Namely, we define $\mathcal{G}_1$, $\mathcal{G}_2$, and $\mathcal{G}_T$ as follows:

$$\mathbb{G}_1 \subseteq \mathcal{G}_1 = E(\mathbb{F}_p), \qquad \mathbb{G}_2 \subseteq \mathcal{G}_2 = E'(\mathbb{F}_{p^{k/d}}), \qquad \mathbb{G}_T \subseteq \mathcal{G}_T = G_{\Phi_k(p)},$$

where $G_{\Phi_k(p)}$ is the cyclotomic subgroup[3] of order $\Phi_k(p)$ in $\mathbb{F}_{p^k}^\times$. Scott also chose $\mathcal{G}_T$ that way when proposing $\mathbb{G}_T$-strong curves [49]. Note that testing membership in $\mathcal{G}_1$ or $\mathcal{G}_2$ simply amounts to checking the curve equation for $E(\mathbb{F}_p)$ or $E'(\mathbb{F}_{p^{k/d}})$, respectively, and that testing whether an element is in $\mathcal{G}_T$ can also "be done at almost no cost using the Frobenius" [49, Sect. 8.3]. We give more details on this check in Sect. 5.2, where we also discuss why $\mathcal{G}_T$ is chosen as the cyclotomic subgroup of order $\Phi_k(p)$, rather than the full multiplicative group $\mathbb{F}_{p^k}^\times$.

Since $|\mathbb{G}_1| = |\mathbb{G}_2| = |\mathbb{G}_T| = n$, the relevant indices $h_1, h_2, h_T \in \mathbb{Z}$ are defined as

$$h_1 = \frac{|\mathcal{G}_1|}{n}, \qquad h_2 = \frac{|\mathcal{G}_2|}{n}, \qquad h_T = \frac{|\mathcal{G}_T|}{n}.$$

The sizes of these cofactors are determined by the properties of the pairing-friendly curve. For all of the curves in this paper, both $\mathbb{G}_2$ and $\mathbb{G}_T$ are groups of order n in the much larger groups $\mathcal{G}_2$ and $\mathcal{G}_T$ and the cofactors h_2 and h_T are at least of a similar size as n. The group $\mathcal{G}_1$ is typically not that large, and comes closer to the case of a group used in plain ECC. Therefore the cofactor h_1 is smaller than n, and in almost all cases larger than 1.

The next attempt at a definition of subgroup security could demand that for any of the three pairing groups for which the cofactor is of size similar to n or larger, it must not have prime factors significantly smaller than n. This is a more useful definition since it focuses on the case in which large cofactors exist. However, most pairing-friendly curves are instances of parameterized families and their parameters are derived as the evaluation of rational polynomials at an integer value. And for certain families, these polynomials may also necessarily produce small factors in the indices (cf. Remark 2 below).

The following definition of subgroup security accounts for this fact in capturing – for a given polynomial family of pairing-friendly curves – the best that can be achieved within that family. We make use of the fact that, for the parameterized families of interest in this work, the three cofactors above are also parameterized as $h_1(u), h_2(u), h_T(u) \in \mathbb{Q}[u]$.

---

[3] Here Φ_k denotes the k-th cyclotomic polynomial.

Definition 1 *(Subgroup security)*. *Let $p(u), t(u), n(u) \in \mathbb{Q}[u]$ parameterize a family of ordinary pairing-friendly elliptic curves, and for any particular $u_0 \in \mathbb{Z}$ such that $p = p(u_0)$ and $n = n(u_0)$ are prime, let E be the resulting pairing-friendly elliptic curve over $\mathbb{F}_p$ of order divisible by n. We say that E is subgroup-secure if all $\mathbb{Q}[u]$-irreducible factors of $h_1(u)$, $h_2(u)$ and $h_T(u)$ that can represent primes and that have degree at least that of $n(u)$, contain no prime factors smaller than $n(u_0) \in \mathbb{Z}$ when evaluated at $u = u_0$.*

It should be pointed out immediately that the wording of "smaller" in Definition 1 can be relaxed in cases where the difference is relatively close. Put simply, Definition 1 aims to prohibit the existence of any unnecessary subgroups of size smaller than n inside the larger groups for which validation is easy. We note that, for simplicity, Definition 1 says that subgroup security is dependent on the pairing-friendly curve E. However, given that the property is dependent on the three groups $\mathbb{G}_1$, $\mathbb{G}_2$ and $\mathbb{G}_T$, it would be more precise to say that the property is based on the *pairing* that is induced by E and n.

In Table 1, we have collected popular pairing-friendly curves that have been used in the literature and in pairing implementations because of their efficiency. We have evaluated all such curves according to their subgroup security. This means that we had to (partially) factor the indices h_1, h_2 and h_T. Note that h_1 is quite small in all cases, and since it is smaller than n, there is no need to find its factorization in order to test for subgroup security. To find the (partial) factorizations of h_2 and h_T, we used the implementation of the ECM method[4] [34] in Magma [13]. To illustrate the factorizations, we use p_m and c_m to denote some m-bit prime and some m-bit composite number respectively.

It is important to note that the curves chosen from the literature in Table 1 were not chosen strategically; none of them are subgroup secure, but the chances of a curve miraculously achieving this property (without being constructed to) is extremely small. Thus, these curves are a fair representation of all ordinary pairing-friendly curves proposed in previous works, since we could not find any prior curve that is subgroup secure according to Definition 1 (the closest example being the BN curve given by Scott [49, Sect. 9], which is $\mathbb{G}_T$-strong but not $\mathbb{G}_2$-strong).

In Sects. 3.3–3.6, we focus on achieving subgroup security for the four popular parameterized families of pairing-friendly curves mentioned in Sect. 2. Our treatment of each family follows the same recipe: the polynomial parameterizations of p, n and t immediately give us parameterizations for h_T as $h_T(u) = \Phi_k(p(u))/n(u)$, but in each case it takes some more work to determine the parameterization of the cofactor h_2; this is done in Propositions 1-4. To find a subgroup-secure curve instance from each family, we searched through $u = u_0$ values of a fixed length and of low NAF-weight, increasing the NAF-weight (and exhausting all possibilities each time) until a curve was found with $p(u_0)$, $n(u_0)$, $h_2(u_0)$ and $h_T(u_0)$ all prime. In theory we could have relaxed the search condition of $h_2(u_0)$ and $h_T(u_0)$ being prime to instead having no prime factors smaller

---

[4] We tweaked the parameters according to http://www.loria.fr/~zimmerma/records/ecm/params.html, until enough factors were found.

Table 1. Subgroup security for pairing-friendly curves previously used in the literature, considering curves from the Barreto-Naehrig (BN) family [6] with $k = 12$; the Barreto-Lynn-Scott (BLS) cyclotomic families [4] with $k = 12$ and $k = 24$ and the Kachisa-Schaefer-Scott (KSS) family [32] with $k = 18$. The columns for p and n give the bitsizes of these primes. The column marked "where?" provides reference to the literature in which the specific curves have been used in implementations. The column $\text{wt}(u_0)$ displays the NAF-weight of the parameter u_0. The symbols p_m and c_m in the columns that display factors of the indices h_1, h_2, and h_T are used to denote an unspecified prime of size m bits or a composite number of size m bits, respectively.

sec level	family k	p (bits)	n (bits)	Curve choices					sub sec.?
				where?	$\text{wt}(u_0)$	h_1	h_2	h_T	
		256	256	[41]	23	1	$c_{17}p_{239}$	$c_{74}c_{692}$	no
128	BN	254	254	[2,42,44,48,54]	3	1	$c_{96}p_{158}$	$c_{79}c_{681}$	no
	12	254	254	Example 1	6	1	p_{254}	p_{762}	yes
		638	427	[1]	4	c_{212}	$c_{48}c_{802}$	$c_{58}c_{2068}$	no
192	BLS	635	424	[12]	4	c_{211}	$c_{15}c_{831}$	$c_{33}c_{2082}$	no
	12	635	425	Example 2	6	c_{211}	p_{845}	p_{2114}	yes
		511	378	[48]	8	c_{133}	$c_{50}c_{1106}$	$c_{26}c_{2660}$	no
192	KSS	508	376	[1]	4	c_{133}	$c_{85}c_{1063}$	$c_{15}c_{2656}$	no
	18	508	376	Example 3	9	c_{133}	$3p_{1146}$	p_{2671}	yes
		639	513	[17]	4	c_{127}	2^2c_{2040}	$c_{41}c_{4556}$	no
256	BLS	629	505	[48]	4	c_{125}	$2^2p_{69}c_{1940}$	$c_{132}c_{4392}$	no
	24	629	504	Example 4	8	c_{125}	p_{2010}	p_{4524}	yes

than n, but finding or proving such factorizations requires an effort beyond the efforts of current factorization records. The fixed length of u_0 was chosen so that the parameter sizes closely match the sizes of curves already in the literature and in online libraries; we also aimed to make sure the parameters matched in terms of efficient constructions of the extension field towerings. In order to compare to previous curves, we have included the subgroup-secure curves found in each family in Table 1.

3.3   BN Curves with $k = 12$

The Barreto-Naehrig (BN) family [6] of curves is particularly well-suited to the 128-bit security level. BN curves are found via the parameterizations $p(u) = 36u^4 + 36u^3 + 24u^2 + 6u + 1$, $t(u) = 6u^2 + 1$, and $n(u) = 36u^4 + 36u^3 + 18u^2 + 6u + 1$.

In this case $\#E(\mathbb{F}_p) = n(u)$, so $\mathbb{G}_1 = \mathcal{G}_1 = E(\mathbb{F}_p)$, meaning $h_1(u) = 1$. The cofactor in $\mathcal{G}_T = G_{\Phi_{12}(p)}$ is parameterized as $h_T(u) = (p(u)^4 - p(u)^2 + 1)/(n(u))$. The following proposition gives the cofactor $h_2(u)$.

Proposition 1. *With parameters as above, the correct sextic twist $E'/\mathbb{F}_{p^2}$ for a BN curve has group order $\#E'(\mathbb{F}_{p^2}) = h_2(u) \cdot n(u)$, where*

$$h_2(u) = 36u^4 + 36u^3 + 30u^2 + 6u + 1.$$

Proof. [40, Rem.2.13] says that BN curves always have $h_2(u) = p(u) - 1 + t(u)$. $\square$

Example 1. The BN curve $E/\mathbb{F}_p$: $y^2 = x^3 + 5$ with $u_0 = 2^{62} + 2^{59} + 2^{55} + 2^{15} + 2^{10} - 1$ has both $p = p(u_0)$ and $n = n(u_0) = \#E(\mathbb{F}_p)$ as 254-bit primes. A model for the correct sextic twist over $\mathbb{F}_{p^2} = \mathbb{F}_p[i]/(i^2+1)$ is $E'/\mathbb{F}_{p^2}$: $y^2 = x^3 + 5(i+1)$, and its group order is $\#E'(\mathbb{F}_{p^2}) = h_2 \cdot n$, where $h_2 = h_2(u_0)$ is also a 254-bit prime. Thus, once points are validated to be in $\mathcal{G}_1 = E(\mathbb{F}_p)$ or $\mathcal{G}_2 = E'(\mathbb{F}_{p^2})$, no cofactor multiplications are required to avoid subgroup attacks on this curve, i.e. there are no points of order less than n in $E(\mathbb{F}_p)$ or $E'(\mathbb{F}_{p^2})$. Furthermore, the group $\mathcal{G}_T$ has order $|\mathcal{G}_T| = h_T \cdot n$, where $h_T = h_T(u_0)$ is a 762-bit prime, so once $\mathbb{F}_{p^{12}}$ elements are validated to be in $\mathcal{G}_T = G_{\Phi_{12}(p)}$, no further cofactor multiplications are necessary for discrete log security here either. For completeness, we note that $\mathbb{F}_{p^{12}}$ can be constructed as $\mathbb{F}_{p^6} = \mathbb{F}_{p^2}[v]/(v^3 - (i+1))$ and $\mathbb{F}_{p^{12}} = \mathbb{F}_{p^6}[w]/(w^2 - v)$; E and E' are then isomorphic over $\mathbb{F}_{p^{12}}$ via $\Psi: E' \to E$, $(x', y') \mapsto (x'/v, y'/(vw))$.

3.4   BLS Curves with $k = 12$

The Barreto-Lynn-Scott (BLS) family [4] with $k = 12$ was shown to facilitate efficient pairings at the 192-bit security level [1]. This family has the parameterizations $p(u) = (u-1)^2 \cdot (u^4 - u^2 + 1)/3 + u$, $t(u) = u + 1$, and $n(u) = u^4 - u^2 + 1$.

Here $\#E(\mathbb{F}_p) = h_1(u) \cdot n(u)$ with $h_1(u) = (u-1)^2/3$, so there is always a cofactor that is much smaller than n in $\mathcal{G}_1$. Again, the cofactor in $\mathcal{G}_T = G_{\Phi_{12}(p)}$ is $h_T(u) = (p(u)^4 - p(u)^2 + 1)/(n(u))$. The following proposition gives the cofactor $h_2(u)$.

Proposition 2. *With parameters as above, the correct sextic twist $E'/\mathbb{F}_{p^2}$ for a $k = 12$ BLS curve has group order $\#E'(\mathbb{F}_{p^2}) = h_2(u) \cdot n(u)$, where*

$$h_2(u) = (u^8 - 4u^7 + 5u^6 - 4u^4 + 6u^3 - 4u^2 - 4u + 13)/9.$$

Proof. Write $\#E(\mathbb{F}_{p^2}) = p_2 + 1 - t_2$, where $p_2 = p^2$ and $t_2 = t^2 - 2p$ [8, CorollaryVI.2]. The CM equation for $E(\mathbb{F}_{p^2})$ is $t_2^2 - 4p_2 = -3v_2^2$, which gives $v_2 = (x-1)(x+1)(2x^2-1)/3$. Lemma 1 reveals that $t' = (t_2 - 3v_2)/2$ gives rise to the correct sextic twist $E'/\mathbb{F}_{p^2}$ with $n \mid \#E'(\mathbb{F}_{p^2}) = p^2 + 1 - t'$, and the cofactor follows as $h_2 = (p^2 + 1 - t')/n$. $\square$

Example 2. The $k = 12$ BLS curve $E/\mathbb{F}_p$: $y^2 = x^3 - 2$ with $u_0 = -2^{106} - 2^{92} - 2^{60} - 2^{34} + 2^{12} - 2^9$ has $p = p(u_0)$ as a 635-bit prime and $\#E(\mathbb{F}_p) = h_1 \cdot n$, where $n = n(u_0)$ is a 425-bit prime and the composite cofactor $h_1 = h_1(u_0)$ is 211 bits. A model for the correct sextic twist over $\mathbb{F}_{p^2} = \mathbb{F}_p[i]/(i^2 + 1)$ is $E'/\mathbb{F}_{p^2}$: $y^2 = x^3 - 2/(i+1)$, and its group order is $\#E'(\mathbb{F}_{p^2}) = h_2 \cdot n$, where $h_2 = h_2(u_0)$

is an 845-bit prime. Furthermore, the group $\mathcal{G}_T$ has order $|\mathcal{G}_T| = h_T \cdot n$, where $h_T = h_T(u_0)$ is a 2114-bit prime. Thus, once elements are validated to be in either $\mathcal{G}_2 = E'(\mathbb{F}_{p^2})$ or $\mathcal{G}_T = G_{\Phi_{12}(p)}$, no cofactor multiplications are required to avoid subgroup attacks. On the other hand, a scalar multiplication (by either h_1 or n) may be necessary to ensure that points in $E(\mathbb{F}_p)$ have the requisite discrete log security, and this is unavoidable across the $k = 12$ BLS family. For completeness, we note that $\mathbb{F}_{p^{12}}$ can be constructed as $\mathbb{F}_{p^6} = \mathbb{F}_{p^2}[v]/(v^3 - (i+1))$ and $\mathbb{F}_{p^{12}} = \mathbb{F}_{p^6}[w]/(w^2 - v)$; E and E' are then isomorphic over $\mathbb{F}_{p^{12}}$ via $\Psi \colon E' \to E$, $(x', y') \mapsto (x' \cdot v, y' \cdot vw)$.

3.5   KSS Curves with $k = 18$

The Kachisa-Schaefer-Scott (KSS) family [32] with $k = 18$ is another family that is suitable at the 192-bit security level. This family has the parameterizations $p(u) = u^8 + 5u^7 + 7u^6 + 37u^5 + 188u^4 + 259u^3 + 343u^2 + 1763u + 2401$, $t(u) = (u^4 + 16u + 7)/7$, and $n(u) = (u^6 + 37u^3 + 343)/7^3$.

Here $\#E(\mathbb{F}_p) = h_1(u) \cdot n(u)$ with $h_1(u) = (49u^2 + 245u + 343)/3$, so again there is always a cofactor much smaller than n in $\mathcal{G}_1$. The cofactor in $\mathcal{G}_T = G_{\Phi_{18}(p)}$ is $h_T(u) = (p(u)^6 - p(u)^3 + 1)/(n(u))$. The proposition below gives the cofactor $h_2(u)$.

Proposition 3. *With parameters as above, the correct sextic twist $E'/\mathbb{F}_{p^3}$ for a $k = 18$ KSS curve has group order $\#E'(\mathbb{F}_{p^3}) = h_2(u) \cdot n(u)$, where*

$$h_2(u) = (u^{18} + 15u^{17} + 96u^{16} + 409u^{15} + 1791u^{14} + 7929u^{13} + 27539u^{12} + 81660u^{11} +$$
$$256908u^{10} + 757927u^9 + 1803684u^8 + 4055484u^7 + 9658007u^6 + 19465362u^5 +$$
$$30860595u^4 + 50075833u^3 + 82554234u^2 + 88845918u + 40301641)/27.$$

Proof. Write $\#E(\mathbb{F}_{p^3}) = p_3 + 1 - t_3$, where $p_3 = p^3$ and $t_3 = t^3 - 3pt$ [8, CorollaryVI.2]. The CM equation for $E(\mathbb{F}_{p^3})$ is $t_3^2 - 4p_3 = -3v_3^2$, which gives $v_3 = (x^4 + 7x^3 + 23x + 119)(5x^4 + 14x^3 + 94x + 259)(4x^4 + 7x^3 + 71x + 140)/3087$. Lemma 1 reveals that $t' = (t_3 + 3v_3)/2$ gives rise to the correct sextic twist $E'/\mathbb{F}_{p^3}$ with $n \mid \#E'(\mathbb{F}_{p^3}) = p^3 + 1 - t'$, and the cofactor follows as $h_2 = (p^3 + 1 - t')/n$. $\square$

Remark 2. The KSS parameterization requires $u \equiv 14 \mod 42$. Under this condition, it is straightforward to see that $h_2(u) \equiv 0 \mod 3$. Thus, there is always a factor of 3 in the cofactor of $\mathcal{G}_2$ in this family.

Example 3. The $k = 18$ KSS curve $E/\mathbb{F}_p \colon y^2 = x^3 + 2$ with $u_0 = 2^{64} + 2^{47} + 2^{43} + 2^{37} + 2^{26} + 2^{25} + 2^{19} - 2^{13} - 2^7$ has $p = p(u_0)$ as a 508-bit prime and $\#E(\mathbb{F}_p) = h_1 \cdot n$, where $n = n(u_0)$ is a 376-bit prime and the composite cofactor $h_1 = h_1(u_0)$ is 133 bits. A model for the correct sextic twist over $\mathbb{F}_{p^3} = \mathbb{F}_p[v]/(v^3 - 2)$ is $E'/\mathbb{F}_{p^3} \colon y^2 = x^3 + 2/v$, and its group order is $\#E'(\mathbb{F}_{p^3}) = 3 \cdot h_2 \cdot n$ (see Remark 2), where $h_2 = h_2(u_0)$ is a 1146-bit prime. Thus, once points are validated to be in $E'(\mathbb{F}_{p^3})$, it may be necessary to multiply points by 3 to clear this cofactor. Furthermore, a scalar multiplication by h_1 or n may be necessary to ensure that random points in $E(\mathbb{F}_p)$ are in $\mathbb{G}_1 = E(\mathbb{F}_p)[n]$ before any secret scalar multiplications take place. On the other hand, once points are validated to be in $\mathcal{G}_T = G_{\Phi_{18}(p)}$, no cofactor multiplications are required to avoid subgroup attacks since $h_T = h_T(u_0)$ is a 2671-bit prime in this case. For completeness, we note that $\mathbb{F}_{p^{18}}$ can be constructed as $\mathbb{F}_{p^9} = \mathbb{F}_{p^3}[v]/(w^3 - v)$ and $\mathbb{F}_{p^{18}} = \mathbb{F}_{p^9}[z]/(z^3 - w)$; E and E' are then isomorphic over $\mathbb{F}_{p^{18}}$ via $\Psi \colon E' \to E$, $(x', y') \mapsto (x' \cdot w, y' \cdot wz)$.

3.6   BLS Curves with $k = 24$

The Barreto-Lynn-Scott (BLS) family [4] with $k = 24$ is well suited to the 256-bit security level. This family has the parameterizations $p(u) = (u - 1)^2 \cdot (u^8 - u^4 + 1)/3 + u$, $t(u) = u + 1$, and $n(u) = u^8 - u^4 + 1$.

Here $\#E(\mathbb{F}_p) = h_1(u) \cdot n(u)$ with $h_1(u) = (u - 1)^2/3$, so once more there is always a cofactor which is much smaller than n in $\#\mathcal{G}_1$. Here the cofactor for $\mathcal{G}_T = G_{\Phi_{24}(p)}$ is $h_T(u) = (p(u)^8 - p(u)^4 + 1)/(n(u))$. The following proposition gives the cofactor $h_2(u)$.

Proposition 4. *With parameters as above, the correct sextic twist $E'/\mathbb{F}_{p^4}$ for a $k = 24$ BLS curve has group order $\#E'(\mathbb{F}_{p^4}) = h(u) \cdot n(u)$, where $h_2(u) = (u^{32} - 8u^{31} + 28u^{30} - 56u^{29} + 67u^{28} - 32u^{27} - 56u^{26} + 160u^{25} - 203u^{24} + 132u^{23} + 12u^{22} - 132u^{21} + 170u^{20} - 124u^{19} + 44u^{18} - 4u^{17} + 2u^{16} + 20u^{15} - 46u^{14} + 20u^{13} + 5u^{12} + 24u^{11} - 42u^{10} + 48u^9 - 101u^8 + 100u^7 + 70u^6 - 128u^5 + 70u^4 - 56u^3 - 44u^2 + 40u + 100)/81$.*

Proof. Write $\#E(\mathbb{F}_{p^4}) = p_4 + 1 - t_4$, where $p_4 = p^4$ and $t_4 = t^4 - 4pt^2 + 2p^2$ [8, CorollaryVI.2]. The CM equation for $E(\mathbb{F}_{p^4})$ is $t_4^2 - 4p_4 = -3v_4^2$, which gives $v_4 = (x - 1)(x + 1)(2x^4 - 1)(2x^{10} - 4x^9 + 2x^8 - 2x^6 + 4x^5 - 2x^4 - x^2 - 4x - 1)/9$. Lemma 1 reveals that $t' = (t_4 + 3v_4)/2$ gives rise to the correct sextic twist $E'/\mathbb{F}_{p^4}$ with $n \mid \#E'(\mathbb{F}_{p^3}) = p^4 + 1 - t'$, and the cofactor follows as $h_2 = (p^4 + 1 - t')/n$. $\square$

Example 4. The $k = 24$ BLS curve $E/\mathbb{F}_p\colon y^2 = x^3 + 1$ with $u_0 = -(2^{63} - 2^{47} - 2^{31} - 2^{26} - 2^{24} + 2^8 - 2^5 + 1)$ has $p = p(u_0)$ as a 629-bit prime and $\#E(\mathbb{F}_p) = h_1 \cdot n$, where $n = n(u_0)$ is a 504-bit prime and the composite cofactor h_1 is 125 bits. If $\mathbb{F}_{p^4}$ is constructed by taking $\mathbb{F}_{p^2} = \mathbb{F}_p[i]/(i^2 + 1)$ and $\mathbb{F}_{p^4} = \mathbb{F}_{p^2}[v]/(v^2 - (i + 1))$, then a model for the correct sextic twist is $E'/\mathbb{F}_{p^4}\colon y^2 = x^3 + 1/v$, and its group order is $\#E'(\mathbb{F}_{p^4}) = h_2 \cdot n$, where $h_2 = h_2(u_0)$ is a 2010-bit prime. Furthermore, the group $\mathcal{G}_T$ has order $|\mathcal{G}_T| = h_T \cdot n$, where $h_T = h_T(u_0)$ is a 4524-bit prime. Thus, once elements are validated to be in either $\mathcal{G}_2 = E'(\mathbb{F}_{p^4})$ or $\mathcal{G}_T = G_{\Phi_{24}(p)}$, no cofactor multiplications are required to avoid subgroup attacks. On the other hand, once random points are validated to be on $E(\mathbb{F}_p)$, a scalar multiplication by h_1 or n is required to ensure points are in $\mathbb{G}_1$. In this case we note that $\mathbb{F}_{p^{24}}$ can be constructed as $\mathbb{F}_{p^{12}} = \mathbb{F}_{p^4}[w]/(w^3 - v)$ and $\mathbb{F}_{p^{24}} = \mathbb{F}_{p^{12}}[z]/(z^2 - w)$; E and E' are then isomorphic over $\mathbb{F}_{p^{24}}$ via $\Psi \colon E' \to E$, $(x', y') \mapsto (x' \cdot w, y' \cdot wz)$.

4   Performance Comparisons: The Price of Subgroup Security

As we saw in Table 1, subgroup-secure curves are generally found with a search parameter of larger NAF-weight than non-subgroup-secure curves because of the additional (primality) restrictions imposed in the former case[5]. Thus, pairings computed with the subgroup-secure curves will naturally be more expensive. In this section, we give performance numbers that provide a concrete comparison between our subgroup-secure curves and the speed-record curves that have appeared elsewhere in the literature. Table 2 shows the approximate factor slowdowns incurred by choosing subgroup-secure curves. We note that these slowdowns are only reported in the computation of the pairing; optimal methods for group exponentiations are unrelated to the search parameter and will therefore remain unchanged when using a subgroup-secure curve of the same size in the same curve family. On the other hand, operations like hashing to $\mathbb{G}_2$ [21,50] do benefit from a low hamming-weight and will also experience a slowdown when using subgroup secure curves.

Table 2. Benchmarks of the optimal ate pairing on non-subgroup-secure pairing-friendly curves used previously compared to subgroup-secure curves (according to Definition 1) from the same family. Timings show the rounded average over 10000 measurements for which Turbo Boost and Hyperthreading were disabled. The experiments only reflect the difference in the NAF-weight of the parameter u_0 leading to an increased number of Miller steps in the Miller loop and multiplications in the final exponentiation. All other parameters are kept the same.

sec level	family k	p (bits)	u (bits)	NAF-weight of param. u_0	where?	optimal ate ($\times 10^6$ clock cycles)	subgroup secure
128	BN	254	63	3	[2,42,44,48,54]	7.68	no
	12	254	63	6	Example 1	8.20	yes
		approximate slowdown factor				**1.07**	
192	BLS	635	106	4	[12]	51.00	no
	12	635	107	6	Example 2	51.98	yes
		approximate slowdown factor				**1.02**	
192	KSS	508	65	4	[1]	85.10	no
	18	508	65	9	Example 3	94.06	yes
		approximate slowdown factor				**1.11**	
256	BLS	629	63	3	[48]	123.79	no
	24	629	63	8	Example 4	139.37	yes
		approximate slowdown factor				**1.13**	

---

[5] We note that Scott [49, Sect. 9] hinted at this "negative impact" when discussing a $\mathbb{G}_T$-strong curve.

We used a pairing library written in C to obtain the performance numbers in Table 2, benchmarked on an Intel Xeon E5-2620 clocked at 2.0 GHz. We note that our library does not perform as fast as some other pairing libraries as it was written entirely in C without using any assembly-level optimizations. Nevertheless, it uses all of the state-of-the-art high-level optimizations such as the optimal ate pairing [53] with a fast final exponentiation [51], as well as taking advantage of efficient extension field towerings [7] and enhanced operations in the cyclotomic subgroups [28]. Moreoever, comparison to speed-record implementations in the literature is immaterial; our point here is to compare the price of a pairing on a subgroup-secure curve to the price of a pairing on one of the popular curves used in the literature, using the same implementation in order to get a fair performance ratio. The pairing functions use the NAF representation of the loop parameter u_0 for the Miller loop as well as the final exponentiation. The implementation computes the runtime for pairings on the subgroup-secure curves by only changing the value for u_0 in the Miller loop and final exponentiation in the implementation of the original curves, all other parameters remain the same. We note that the fastest implementations of field arithmetic for ordinary pairing-friendly curves, e.g. [2], do not take advantage of the NAF-weight of the prime p. The results therefore provide a gauge as to the relative slowdown one can expect in the pairing when employing a subgroup-secure curve, indicating that, in the worst case, a slowdown factor of 1.13 can be expected.

5   How to Use Subgroup-Secure Curves

In this section we discuss the implications of working with a subgroup-secure pairing-friendly curve and point out possible efficiency improvements. As we have seen in Sect. 4, there is a small performance penalty in the pairing algorithm when switching from the currently used "speed-record curves" to subgroup-secure curves, which is incurred by the increase in the NAF-weight of the parameter u_0. Note that this penalty only affects the pairing computation; it does not have any consequences for elliptic curve or finite field arithmetic in the groups themselves.

As we discussed earlier, an important subtlety that is rarely[6] factored into pairing-based protocol papers is the notion of (testing) subgroup membership [16, Sect. 2.2]. Naturally then, the cost for performing these checks is often not reflected in the pairing literature. When using a subgroup-secure curve, there is the potential to reduce the cost for these checks as hinted to in Sect. 3.2, and possibly to mitigate the performance penalty.

5.1   Reducing the Cost of Subgroup Membership Checks

We emphasize that we do not recommend skipping subgroup membership checks. What we do recommend, though, is that if such checks are in place to guarantee

---

[6] Menezes and Chatterjee recently pointed out another interesting example of this [15].

DLP security, then protocols should be examined to see if these checks can be replaced by less expensive measures, such as omitting costly scalar multiplications in the presence of a subgroup-secure curve. Next, we discuss the different possibilities.

For the group $\mathcal{G}_1$, the index h_1 of $\mathbb{G}_1$ is typically much smaller than n, which means that we cannot select the parameters to avoid prime factors smaller than n in $|\mathcal{G}_1|$. Therefore, one must carry out either a scalar multiplication by n to check for the correct order or by the cofactor h_1 to force points to have the right order. Let $\#E(\mathbb{F}_p) = h_1 \cdot n$ for some *cofactor* h_1 and recall that $\log_2 h_1 \ll \log_2 n$ for all of the above curve families. Thus, for a random point $P \in E(\mathbb{F}_p)$, it is faster to compute $R = [h_1]P$ to guarantee that $R \in \mathbb{G}_1$ than it is to check whether $[n]P = \mathcal{O}$ and was in $\mathbb{G}_1$ to begin with. However, this solution requires the protocol to allow the point R to replace the orginal point P, and this might require slight changes to the protocol; for example, it may require more than one party to perform the same scalar multiplication by h_1 such that it would have been less expensive (overall) for a single party to check that $[n]P = \mathcal{O}$.

In the group $\mathcal{G}_2$, the picture is different. Let $\#E'(\mathbb{F}_{p^{k/6}}) = h_2 \cdot n$ for some cofactor h_2, and recall from Sect. 3.2 that $h_2 > n$ for the families in this paper. In this case, guaranteeing that a point is in the order n subgroup $\mathbb{G}_2$ through a naive cofactor multiplication by h_2 seems to be at least as costly as checking that a point was in $\mathbb{G}_2$ to begin with; in particular, for the $k = 18$ and $k = 24$ families above, the bit length of h_2 is around 3 and 4 times that of n, respectively. However, the work by Scott et al. [50] and improvements by Fuentes-Castañeda et al. [21] show that cofactor multiplication with h_2 in $\mathcal{G}_2$ is significantly faster than multiplication by the group order n. As in $\mathcal{G}_1$, for a curve that is not subgroup-secure and if the protocol allows, it is thus cheaper to move a point $Q' \in E'(\mathbb{F}_{p^{k/6}})$ to $\mathbb{G}_2$ by computing $[h_2]Q'$ than it is to check the condition $[n]Q' = \mathcal{O}$. On the other hand, if the curve is subgroup-secure, one could check the curve equation to ensure that $Q' \in E'$ and omit either check at no risk of compromising the DLP security. However, not every protocol might allow working with points of order different than n. Thus the application of this optimization needs to be evaluated in every specific protocol. For example, the pairing implementation might not be bilinear when applied to points of order other than n, or the subgroup membership check may be in place for a reason different than discrete log security.

In the context of the Tate pairing, Scott [47, Sect. 4.4] pointed out that during the pairing computation, one can check whether the first input $P \in \mathcal{G}_1$ to the pairing function actually has order n, i.e. whether it is in $\mathbb{G}_1$. This is possible because the Miller loop in the Tate pairing inherently computes $[n]P$ alongside the pairing value, so there is no additional effort required to assert that $[n]P = \mathcal{O}$. However, when using optimal pairings [53], this is not true anymore. Due to the shortening of the Miller loop and the swapping of the input groups, optimal pairings only compute $[\lambda]Q'$ for λ much smaller than n, and for $Q' \in \mathcal{G}_2$. The trick outlined by Scott can therefore only help to save part of the exponentiation $[n]Q'$.

Elliptic curve scalar multiplications in both $\mathcal{G}_1$ and $\mathcal{G}_2$ can benefit from GLV/GLS decompositions [23,26,27]. In $\mathcal{G}_1$, one can use precomputed 2-dimensional GLV decompositions to speed up the scalar multiplications by h_1 and n. In $\mathcal{G}_2$, one can use even higher-dimensional GLV+GLS decompositions of the scalar n. In both cases, since n and h_1 are fixed system parameters, their decomposition can be computed offline. Moreover, these fixed multiplications are not by secret scalars and therefore need not be implemented in *constant time*.

Finally, the index of $\mathbb{G}_T$ in $\mathcal{G}_T = G_{\Phi_k(p)}$ is $h_T = \Phi_k(p)/n$, which is at least three times larger than n for the families in this paper. Thus, for a subgroup-secure curve, h_T is prime (up to possibly small factors given by the polynomial parameterization) and a subgroup membership test for $\mathbb{G}_T$ may be replaceable by a cheap membership test for $\mathcal{G}_T$ (see Sect. 5.2 below). Again, this is contingent on the ability of the protocol to allow $\mathcal{G}_T$-elements of order other than n. If membership tests can not be avoided, then the fixed exponentiation by n can take advantage of several techniques that accelerate arithmetic in the cyclotomic subgroup $\mathcal{G}_T = G_{\Phi_k(p)}$; these include cyclotomic squarings [28], exponent decompositions [26], and trace-based methods (cf. [49, Sect. 8.1]).

5.2   Checking Membership in $\mathcal{G}_T$

We elaborate on Scott's observation [49, Sect. 8.3] concerning the ease of checking membership in $\mathcal{G}_T = G_{\Phi_k(p)}$. For the $k = 12$ BN and BLS families, checking that $g \in \mathcal{G}_T$ amounts to asserting $g^{p^4 - p^2 + 1} = 1$, i.e. asserting that $g^{p^4} \cdot g = g^{p^2}$. Here the required Frobenius operations are a small cost compared to the multiplication that is needed, so this check essentially costs one multiplication in $\mathbb{F}_{p^{12}}$. Similarly, the tests for the $k = 18$ KSS and $k = 24$ BLS families check that $g^{p^6} \cdot g = g^{p^3}$ and $g^{p^8} \cdot g = g^{p^4}$ respectively, which also cost around one extension field multiplication.

The reason we take $\mathcal{G}_T$ to be the subgroup of order $\Phi_k(p)$, rather than the full multiplicative group $\mathbb{F}_{p^k}^{\times}$, is because it is extremely difficult to achieve subgroup security in $\mathbb{F}_{p^k}^{\times}$. As Scott points out when $k = 12$, the number of elements in $\mathbb{F}_{p^{12}}^{\times}$ factors as $p^{12} - 1 = (p-1) \cdot (p^2+1) \cdot (p^2+p+1) \cdot (p^2-p+1) \cdot (p+1) \cdot ((p^4-p^2+1)/n) \cdot n$, so here there are 6 factors (excluding n) that we would need to be almost prime if we were to deem $\mathbb{F}_{p^k}^{\times}$ as subgroup-secure. Even if it were possible to find a u_0 value such that these 6 factors were almost prime, it would certainly no longer have a sparse NAF representation, and the resulting loss in pairing efficiency would be drastic. On the other hand, taking $\mathcal{G}_T = G_{\Phi_k(p)}$ means that we can search for only one additional factor (i.e. $(p^4 - p^2 + 1)/n)$) being almost prime, meaning that sparse u_0 values (and therefore state-of-the-art performance numbers) are still possible and the cost of asserting membership in $\mathcal{G}_T$ remains negligible.

Acknowledgements. We are grateful to Melissa Chase and Greg Zaverucha for their interest in this work, and for their help pointing out implications for pairing-based protocols. We also thank Francisco Rodríguez-Henríquez for his suggestions to improve the paper.

References

1. Aranha, D.F., Fuentes-Castañeda, L., Knapp, E., Menezes, A., Rodríguez-Henríquez, F.: Implementing pairings at the 192-bit security level. In: Abdalla, M., Lange, T. (eds.) Pairing 2012. LNCS, vol. 7708, pp. 177–195. Springer, Heidelberg (2013)
2. Aranha, D.F., Karabina, K., Longa, P., Gebotys, C.H., López, J.: Faster explicit formulas for computing pairings over ordinary curves. In: Paterson, K.G. (ed.) EUROCRYPT 2011. LNCS, vol. 6632, pp. 48–68. Springer, Heidelberg (2011)
3. Barbulescu, R., Gaudry, P., Joux, A., Thomé, E.: A heuristic quasi-polynomial algorithm for discrete logarithm in finite fields of small characteristic. In: Nguyen, P.Q., Oswald, E. (eds.) EUROCRYPT 2014. LNCS, vol. 8441, pp. 1–16. Springer, Heidelberg (2014)
4. Barreto, P.S.L.M., Lynn, B., Scott, M.: Constructing elliptic curves with prescribed embedding degrees. In: Cimato, S., Galdi, C., Persiano, G. (eds.) SCN 2002. LNCS, vol. 2576, pp. 257–267. Springer, Heidelberg (2003)
5. Barreto, P.S.L.M., Lynn, B., Scott, M.: Efficient implementation of pairing-based cryptosystems. J. Cryptol. 17(4), 321–334 (2004)
6. Barreto, P.S.L.M., Naehrig, M.: Pairing-Friendly Elliptic Curves of Prime Order. In: Preneel, B., Tavares, S. (eds.) SAC 2005. LNCS, vol. 3897, pp. 319–331. Springer, Heidelberg (2006)
7. Benger, N., Scott, M.: Constructing tower extensions of finite fields for implementation of pairing-based cryptography. In: Hasan, M.A., Helleseth, T. (eds.) WAIFI 2010. LNCS, vol. 6087, pp. 180–195. Springer, Heidelberg (2010)
8. Blake, I.F., Seroussi, G., Smart, N.: Elliptic Curves in Cryptography, vol. 265. Cambridge University Press, Cambridge (1999)
9. Boldyreva, A.: Threshold signatures, multisignatures and blind signatures based on the Gap-Diffie-Hellman-group signature scheme. In: Desmedt, Y. (ed.) PKC 2003. Lecture Notes in Computer Science, vol. 2567, pp. 31–46. Springer, Heidelberg (2003)
10. Boneh, D., Lynn, B., Shacham, H.: Short signatures from the weil pairing. In: Boyd, C. (ed.) ASIACRYPT 2001. LNCS, vol. 2248, p. 514. Springer, Heidelberg (2001)
11. Boneh, D., Franklin, M.: Identity-based encryption from the weil pairing. In: Kilian, J. (ed.) CRYPTO 2001. LNCS, vol. 2139, pp. 213–229. Springer, Heidelberg (2001)
12. Bos, J.W., Costello, C., Naehrig, M.: Exponentiating in pairing groups. In: Lange, T., Lauter, K., Lisoněk, P. (eds.) SAC 2013. LNCS, vol. 8282, pp. 438–455. Springer, Heidelberg (2014)
13. Bosma, W., Cannon, J., Playoust, C.: The magma algebra system I: the user language. J. Symbolic Comput. 24(3–4), 235–265 (1997). Computational algebra and number theory (London, 1993)
14. Chatterjee, S., Hankerson, D., Knapp, E., Menezes, A.: Comparing two pairing-based aggregate signature schemes. Des. Codes Crypt. 55(2–3), 141–167 (2010)
15. Chatterjee, S., Menezes, A.: Type 2 structure-preserving signature schemes revisited. Cryptology ePrint Archive, Report 2014/635 (2014). http://eprint.iacr.org/
16. Chen, L., Cheng, Z., Smart, N.P.: Identity-based key agreement protocols from pairings. Int. J. Inf. Sec. 6(4), 213–241 (2007)
17. Costello, C., Lauter, K., Naehrig, M.: Attractive subfamilies of BLS curves for implementing high-security pairings. In: Bernstein, D.J., Chatterjee, S. (eds.) INDOCRYPT 2011. LNCS, vol. 7107, pp. 320–342. Springer, Heidelberg (2011)

18. Edwards, H.M.: A normal form for elliptic curves. Bull. Am. Math. Soc. **44**(3), 393–422 (2007)

19. Freeman, D.: Constructing pairing-friendly elliptic curves with embedding degree 10. In: Hess, F., Pauli, S., Pohst, M. (eds.) ANTS 2006. LNCS, vol. 4076, pp. 452–465. Springer, Heidelberg (2006)

20. Freeman, D., Scott, M., Teske, E.: A taxonomy of pairing-friendly elliptic curves. J. Crypt. **23**(2), 224–280 (2010)

21. Fuentes-Castañeda, L., Knapp, E., Rodríguez-Henríquez, F.: Faster hashing to G_2. In: Miri, A., Vaudenay, S. (eds.) SAC 2011. LNCS, vol. 7118, pp. 412–430. Springer, Heidelberg (2012)

22. Galbraith, S.D., Harrison, K., Soldera, D.: Implementing the tate pairing. In: Fieker, C., Kohel, D.R. (eds.) ANTS 2002. LNCS, vol. 2369, p. 324. Springer, Heidelberg (2002)

23. Galbraith, S.D., Lin, X., Scott, M.: Endomorphisms for faster elliptic curve cryptography on a large class of curves. J. Crypt. **24**(3), 446–469 (2011)

24. Galbraith, S.D., Paterson, K.G. (eds.): Pairing 2008. Lecture Notes in Computer Science, vol. 5209. Springer, Heidelberg (2008)

25. Galbraith, S.D., Paterson, K.G., Smart, N.P.: Pairings for cryptographers. Discrete Appl. Math. **156**(16), 3113–3121 (2008)

26. Galbraith, S.D., Scott, M.: Exponentiation in pairing-friendly groups using homomorphisms. In: Galbraith, S.D., Paterson, K.G. (eds.) Pairing 2008. LNCS, vol. 5209, pp. 211–224. Springer, Heidelberg (2008)

27. Gallant, R.P., Lambert, R.J., Vanstone, S.A.: Faster point multiplication on elliptic curves with efficient endomorphisms. In: Kilian, J. (ed.) CRYPTO 2001. LNCS, vol. 2139, pp. 190–200. Springer, Heidelberg (2001)

28. Granger, R., Scott, M.: Faster squaring in the cyclotomic subgroup of sixth degree extensions. In: Nguyen, P.Q., Pointcheval, D. (eds.) PKC 2010. LNCS, vol. 6056, pp. 209–223. Springer, Heidelberg (2010)

29. Hess, F., Smart, N.P., Vercauteren, F.: The eta pairing revisited. IEEE Trans. Inf. Theo. **52**(10), 4595–4602 (2006)

30. IEEE P1363 Working Group. Standard Specifications for Public-Key Cryptography - IEEE Std 1363–2000 (2000)

31. Joux, A.: A one round protocol for tripartite Diffie-Hellman. J. Crypt. **17**(4), 263–276 (2004)

32. Kachisa, E.J., Schaefer, E.F., Scott, M.: Constructing Brezing-Weng pairing-friendly elliptic curves using elements in the cyclotomic field. In: Galbraith, S.D., Paterson, K.G. (eds.) Pairing 2008. LNCS, vol. 5209, pp. 126–135. Springer, Heidelberg (2008)

33. Kilian, J. (ed.): CRYPTO 2001. Lecture Notes in Computer Science, vol. 2139. Springer, Heidelberg (2001)

34. Lenstra Jr., H.W.: Factoring integers with elliptic curves. Ann. Math. **126**, 649–673 (1987)

35. Li, N., Du, W., Boneh, D.: Oblivious signature-based envelope. In: Borowsky, E., Rajsbaum, S. (eds.) PODC 2003, pp. 182–189. ACM, New York (2003)

36. Lim, C.H., Lee, P.J.: A key recovery attack on discrete log-based schemes using a prime order subgroup. In: Kaliski Jr., B.S. (ed.) CRYPTO 1997. LNCS, vol. 1294, pp. 249–263. Springer, Heidelberg (1997)

37. Menezes, A.: Asymmetric pairings. Talk at ECC 2009. Slides at http://math.ucalgary.ca/ecc/files/ecc/u5/Menezes_ECC2009.pdf

38. Miyaji, A., Nakabayashi, M., Takano, S.: New explicit conditions of elliptic curve traces for FR-reduction. IEICE Trans. Fundam. Electron. Commun. Comput. Sci. **84**(5), 1234–1243 (2001)
39. Montgomery, P.L.: Speeding the pollard and elliptic curve methods of factorization. Math. Comput. **48**(177), 243–264 (1987)
40. Naehrig, M.: Constructive and computational aspects of cryptographic pairings. Ph.D. thesis, Eindhoven University of Technology, May 2009
41. Naehrig, M., Niederhagen, R., Schwabe, P.: New software speed records for cryptographic pairings. In: Abdalla, M., Barreto, P.S.L.M. (eds.) LATINCRYPT 2010. LNCS, vol. 6212, pp. 109–123. Springer, Heidelberg (2010)
42. Nogami, Y., Akane, M., Sakemi, Y., Katou, H., Morikawa, Y.: Integer variable chi-based ate pairing. In: Galbraith and Paterson [24], pp. 178–191
43. Page, D., Smart, N.P., Vercauteren, F.: A comparison of MNT curves and supersingular curves. IACR Cryptology ePrint Archive, vol. 2004, p. 165 (2004)
44. Pereira, G.C.C.F., Simplício Jr., M.A., Naehrig, M., Barreto, P.S.L.M.: A family of implementation-friendly BN elliptic curves. J. Syst. Softw. **84**(8), 1319–1326 (2011)
45. Pohlig, S.C., Hellman, M.E.: An improved algorithm for computing logarithms over GF(p) and its cryptographic significance. IEEE Trans. Inf. Theo. **24**(1), 106–110 (1978)
46. Sakai, R., Ohgishi, K., Kasahara, M.: Cryptosystems based on pairing. In: The 2000 Symposium on Cryptography and Information Security, Okinawa, Japan, pp. 135–148 (2000)
47. Scott, M.: Computing the tate pairing. In: Menezes, A. (ed.) CT-RSA 2005. LNCS, vol. 3376, pp. 293–304. Springer, Heidelberg (2005)
48. Scott, M.: On the efficient implementation of pairing-based protocols. In: Chen, L. (ed.) IMACC 2011. LNCS, vol. 7089, pp. 296–308. Springer, Heidelberg (2011)
49. Scott, M.: Unbalancing pairing-based key exchange protocols. Cryptology ePrint Archive, Report 2013/688 (2013). http://eprint.iacr.org/2013/688
50. Scott, M., Benger, N., Charlemagne, M., Dominguez Perez, L.J., Kachisa, E.J.: Fast hashing to G_2 on pairing-friendly curves. In: Shacham, H., Waters, B. (eds.) Pairing 2009. LNCS, vol. 5671, pp. 102–113. Springer, Heidelberg (2009)
51. Scott, M., Benger, N., Charlemagne, M., Dominguez Perez, L.J., Kachisa, E.J.: On the final exponentiation for calculating pairings on ordinary elliptic curves. In: Shacham, H., Waters, B. (eds.) Pairing 2009. LNCS, vol. 5671, pp. 78–88. Springer, Heidelberg (2009)
52. Vaudenay, S.: Hidden collisions on DSS. In: Koblitz, N. (ed.) CRYPTO 1996. LNCS, vol. 1109, pp. 83–88. Springer, Heidelberg (1996)
53. Vercauteren, F.: Optimal pairings. IEEE Trans. Inf. Theo. **56**(1), 455–461 (2010)
54. Zavattoni, aE., Dominguez Perez, L.J., Mitsunari, S., Sánchez-Ramírez, A.H., Teruya, T., Rodríguez-Henríquez, F.: Software implementation of an attribute-based encryption scheme (2015)

Curves in Cryptography

Twisted Hessian Curves

Daniel J. Bernstein[1,2]($\boxtimes$), Chitchanok Chuengsatiansup[1]($\boxtimes$), David Kohel[3]($\boxtimes$), and Tanja Lange[1]($\boxtimes$)

[1] Department of Mathematics and Computer Science, Technische Universiteit Eindhoven, P.O. Box 513, 5600 MB Eindhoven, The Netherlands
c.chuengsatiansup@tue.nl, tanja@hyperelliptic.org
[2] Department of Computer Science, University of Illinois at Chicago, Chicago 60607–7045, USA
djb@cr.yp.to
[3] Institut de Mathématiques de Marseille, Aix-Marseille Université, 163, Avenue de Luminy, Case 907, 13288 Marseille Cedex 09, France
David.Kohel@univ-amu.fr

Abstract. This paper presents new speed records for arithmetic on a large family of elliptic curves with cofactor 3: specifically, 8.77M per bit for 256-bit variable-base single-scalar multiplication when curve parameters are chosen properly. This is faster than the best results known for cofactor 1, showing for the first time that points of order 3 are useful for performance and narrowing the gap to the speeds of curves with cofactor 4.

Keywords: Efficiency · Elliptic-curve arithmetic · Double-base chains · Fast arithmetic · Hessian curves · Complete addition laws

1 Introduction

For efficiency reasons, it is desirable to take the cofactor to be as small as possible. — "Recommended elliptic curves for federal government use", National Institute of Standards and Technology, 1999 [47]

All of NIST's standard prime-field elliptic curves have cofactor 1. However, by now there is overwhelming evidence that cofactor 1 does not provide the best performance/security tradeoff for elliptic-curve cryptography. All of the latest speed records for ECC are set by curves with cofactor divisible by 2, with base

This work was supported by the U.S. National Science Foundation under grants 0716498 and 1018836; by the Agence Nationale de la Recherche grant ANR-12-BS01-0010-01; by the Netherlands Organisation for Scientific Research (NWO) under grants 639.073.005 and 613.001.011; and by the European Commission under Contract ICT-645421 ECRYPT-CSA. This work was started during the ESF exploratory workshop "Curves, Coding Theory, and Cryptography" in March 2009; the first and fourth author would like to thank the ESF for financial support. Permanent ID of this document: 1ad9e9d82a9e27e390be46e1fe7b895f. Date: 2015.06.03.

© Springer International Publishing Switzerland 2015
K. Lauter and F. Rodríguez-Henríquez (Eds.): LatinCrypt 2015, LNCS 9230, pp. 269–294, 2015.
DOI: 10.1007/978-3-319-22174-8_15

fields $\mathbf{F}_q$ where q is a square, and with extra endomorphisms: Faz-Hernández–Longa–Sánchez [27] use a twisted Edwards GLS curve with cofactor 8 over $\mathbf{F}_q$ where $q = (2^{127} - 5997)^2$; Oliveira–López–Aranha–Rodríguez-Henríquez [49] use a GLV+GLS curve with cofactor 2 over $\mathbf{F}_q$ where $q = 2^{254}$; and Costello–Hisil–Smith [20] use a Montgomery $\mathbf{Q}$-curve with cofactor 4 (and twist cofactor 8) over $\mathbf{F}_q$ where $q = (2^{127} - 1)^2$. Similarly, for "conservative" ECC over prime fields without extra endomorphisms, Bernstein [5] uses a Montgomery curve with cofactor 8 (and twist cofactor 4), and Bernstein–Duif–Lange–Schwabe–Yang [7] use an equivalent twisted Edwards curve.

The very fast Montgomery ladder for Montgomery curves [42] was published at the dawn of ECC, and its speed always relied on a cofactor divisible by 4. However, for many years the benefit of such cofactors seemed limited to ladders for variable-base single-scalar multiplication. Cofactor 1 seemed slightly faster than cofactor 4 for signature generation and signature verification; NIST's curves were published in the context of a signature standard. Many years of investigations of addition formulas for a wide range of curve shapes (see, e.g., [17], [19], [34], [41], and [13]) failed to produce stronger arguments for cofactors above 1 — until the advent [24] and performance analysis [9] of Edwards curves.

Cofactor 3. Several papers have tried to exploit a different cofactor, namely 3, as follows. Hessian curves $x^3 + y^3 + 1 = dxy$, which always have points of order 3 over finite fields, have a very simple and symmetric addition law due to Sylvester. Chudnovsky–Chudnovsky in [17] already observed that this law requires just 12M in projective coordinates. However, Hessian doublings were much slower than Jacobian-coordinate Weierstrass doublings, and this slowdown outweighed the addition speedup, since (in most applications) doublings are much more frequent than additions. The best way to handle a curve with cofactor 3 was to forget about the points of order 3 and simply use the same formulas used for curves with cofactor 1.

What we show in this paper, for the first time, is how to use cofactor 3 to beat the best available results for cofactor 1. We do not claim to have beaten cofactor 4, but we have significantly narrowed the gap.

We now review previous speeds and compare them to our speeds. We adopt the following rules to maximize comparability:

- For individual elliptic-curve operations we count multiplications and squarings. $\mathbf{M}$ is the cost of a multiplication, and $\mathbf{S}$ is the cost of a squaring. We do not count additions or subtractions. (Computer-verified operation counts for our formulas, including counts of additions and subtractions, appear in the latest update of EFD [8].)
- In summaries of scalar-multiplication performance we take $\mathbf{S} = 0.8\mathbf{M}$. Of course, squarings are much faster than multiplications in characteristic 2, but we emphasize the case of large characteristic.
- We also count multiplications by curve parameters: e.g., $\mathbf{M}_d$ is the cost of multiplying by d. We assume that curves are sensibly chosen with small d. In summaries we take $\mathbf{M}_d = 0$.

- We do not include the cost of final conversion to affine coordinates. We also assume that inversion is not fast enough to make intermediate inversions useful. Consequently the exact cost of inversion does not appear.
- We focus on the traditional case of variable-base single-scalar multiplication, in particular for average 256-bit scalars. Beware that this is only loosely correlated with other scalar-multiplication tasks. (Other tasks tend to rely more on additions, so the fast complete addition law for twisted Hessian curves should provide an even larger benefit compared to Weierstrass curves.)

Bernstein–Lange in [10] analyzed scalar-multiplication performance on several curve shapes and concluded, under these assumptions, that Weierstrass curves $y^2 = x^3 - 3x + a_6$ in Jacobian coordinates used 9.34M per bit on average, and that Hessian curves were slower. Bernstein–Birkner–Lange–Peters in [6] used double-base chains (doublings, triplings, and additions) to considerably speed up Hessian curves to 9.65M per bit and to slightly speed up Weierstrass curves to 9.29M per bit. Hisil in [32, Table 6.4], without double-base chains, reported more than 10M per bit for Hessian curves.

Our new results are just 8.77M per bit. This means that one actually gains something by taking advantage of a point of order 3. The new speeds require a base field with $6 \neq 0$ and with fast multiplication by a primitive cube root of 1, such as a field of the form $\mathbf{F}_p[\omega]/(\omega^2 + \omega + 1)$ where $p \in 2 + 3\mathbf{Z}$. This quadratic field structure might seem to constrain the applicability of the results, but (1) GLS-curve and $\mathbf{Q}$-curve results already show that a quadratic field structure is desirable for performance; (2) there is also a fast primitive cube root of 1 in, e.g., the prime field $\mathbf{F}_p$ where $7p = 2^{298} + 2^{149} + 1$; (3) we do not lose much speed from more general fields (the cost of a tripling increases by 0.4M). Note that the 8.77M per bit does not use the speedups in (1).

Completeness, Side Channels, and Precomputation. For a large fraction of curves, the formulas we use have a further benefit not reflected in the multiplication counts stated above: namely, the formulas are *complete*. This means that the formulas work for all curve points. The implementor does not have to waste any time checking for exceptional cases, and does not have to worry that an attacker can generate inputs that trigger exceptional cases: there are no exceptional cases. (For comparison, a *strongly unified* but incomplete addition law works for most additions and works for most doublings, but still has exceptional cases. The traditional addition law for Weierstrass curves is not even strongly unified: it consistently fails for doublings.)

Often completeness is used as part of a side-channel defense; see, e.g., [9, Section 8]. In this paper we focus purely on speed: we do not limit attention to scalar-multiplication techniques that are safe inside applications that expose secret scalars to side-channel attacks. Note that scalars are public in many cryptographic protocols, such as signature verification, and also in many other elliptic-curve computations, such as the elliptic-curve method of integer factorization.

We also allow scalar-multiplication techniques that rely on scalar-dependent precomputation. This is reasonable for applications that reuse a single scalar many times. For example, in the context of signatures, the *signer* can carry out the precomputation and compress the results into the signature. The signer can also choose different techniques for different scalars: in particular, there are some scalars where our cofactor-3 techniques are even faster than cofactor 4. One can easily find, and we suggest choosing, curves of cofactor 12 that simultaneously allow the current cofactor-3 and cofactor-4 methods; these curves are also likely to be able to take advantage of any future improvements in cofactor-3 and cofactor-4 methods.

Tools and Techniques. At a high level, we use a tree search for double-base chains, allowing windows and taking account of the costs of doublings, triplings, and additions. At a lower level, we use tripling formulas that take $6\mathbf{M} + 6\mathbf{S}$, doubling formulas that take $6\mathbf{M} + 2\mathbf{S}$, and addition formulas that take $11\mathbf{M}$; in this overview we ignore multiplications by constants. These formulas work in projective coordinates for Hessian curves.

Completeness relies on two further tools. First, we use a rotated addition law. Unlike the standard (Sylvester) addition law, the rotated addition law is strongly unified. In fact, the rotated addition law works in every case where the standard addition law fails; i.e., the two laws together form a complete system of addition laws. Second, we work more generally with twisted Hessian curves $ax^3 + y^3 + 1 = dxy$. If a is not a cube then the rotated addition law by itself is complete. The doubling formulas and tripling formulas are also complete, meaning that they have no exceptional cases. The generalization also provides more flexibility in finding curves with small parameters.

For comparison, Jacobian coordinates for Weierstrass curves $y^2 = x^3 - 3x + a_6$ use $7\mathbf{M} + 7\mathbf{S}$ for tripling, $3\mathbf{M} + 5\mathbf{S}$ for doubling, and $11\mathbf{M} + 5\mathbf{S}$ for addition. This saves $3(\mathbf{M} - \mathbf{S})$ in doubling but loses $\mathbf{M} + \mathbf{S}$ in tripling and loses $5\mathbf{S}$ in addition. Given these operation counts it is not a surprise that we beat Weierstrass curves.

$6\mathbf{M} + 6\mathbf{S}$ triplings were achieved once before, namely by tripling-oriented Doche–Icart–Kohel curves [22]. Those curves also offer $2\mathbf{M} + 7\mathbf{S}$ doublings, competitive with our $6\mathbf{M} + 2\mathbf{S}$. However, the best addition formulas known for those curves take $11\mathbf{M} + 6\mathbf{S}$, even slower than Weierstrass curves.

As noted earlier, Edwards curves are still faster for average scalars, thanks to their particularly fast doublings and additions. However, we do beat Edwards curves for scalars that involve many triplings.

Credits and Priority Dates. Hessian curves and the standard addition law are classical material. The rotated addition law, the fact that the rotated addition law is strongly unified, the concept of twisted Hessian curves, the generalization of the addition laws to twisted Hessian curves, the complete system of addition laws, and the completeness of the rotated addition law for non-cube a are all due to this paper. We announced the essential details online in July 2009 (e.g., stating the completeness result in [4, p. 40], and contributing a "twisted

Table 1.1. Costs of various formulas for Hessian curves in projective coordinates. Costs are sorted using the assumption $S \approx 0.8M$; note that S/M is normally much smaller in characteristic 2. "T" means that the formula was stated for twisted Hessian curves, not just Hessian curves; all of the "T" formulas are complete for suitable curves. "S" means "strongly unified": an addition formula that also works for doubling. "2" means that the formula works in characteristic 2. "3" means that the formula works in characteristic 3. ">" means that the formula works in characteristic above 3.

Operation	T	S	2	3	>	Cost	Source
doubling			✓	✓	✓	$6M + 3S \approx 8.4M$	1986 Chudnovsky–Chudnovsky [17]
doubling	✓		✓	✓	✓	$6M + 3S \approx 8.4M$	our 2009 announcement
doubling			✓	✓		$3M + 6S \approx 7.8M$	2007 Hisil–Carter–Dawson [30]
doubling			✓	✓	✓	$7M + 1S \approx 7.8M$	2007 Hisil–Carter–Dawson [30]
doubling	✓			✓	✓	$6M + 2S \approx 7.6M$	this paper
addition		✓		✓	✓	$9M + 6S \approx 13.8M$	2009 Hisil–Wong–Carter–Dawson [31]
addition			✓	✓	✓	$12M \quad = 12.0M$	1986 Chudnovsky–Chudnovsky [17]
addition	✓	✓	✓	✓	✓	$12M \quad = 12.0M$	our 2009 announcement
addition	✓	✓		✓	✓	$11M \quad = 11.0M$	2010 Hisil [32]
tripling			✓	✓		$8M + 6S \approx 12.8M$	2007 Hisil–Carter–Dawson [30]
tripling	✓			✓	✓	$8M + 6S \approx 12.8M$	our 2009 announcement
tripling	✓	✓				$7M + 6S \approx 11.8M$	2010 Farashahi–Joye [25]
tripling	✓			✓		$8M + 4S \approx 11.2M$	2013 Farashahi–Wu–Zhao [26]
tripling	✓		✓	✓		$8M + 4S \approx 11.2M$	2015 Kohel [39]
tripling	✓		✓	✓	✓	$8M + 4S \approx 11.2M$	this paper
tripling	✓	✓			✓	$6M + 6S \approx 10.8M$	this paper, assuming fast primitive $\sqrt[3]{1}$

Hessian" section to EFD), but this paper is our first formal publication of these results.

The speeds that we announced at that time for twisted Hessian curves were no better than known speeds for standard formulas for Hessian curves: $8M + 6S$ for tripling, $6M + 3S$ for doubling, and $12M$ for addition. Followup work found better formulas for all of these operations. Almost all of those formulas are superseded by formulas that we now announce; the only exception is that we use $11M$ addition formulas [32] from Hisil. See Table 1.1 for an overview.

Tripling: One of the followup papers [25], by Farashahi–Joye, reported $7M + 6S$ for twisted Hessian tripling, but only for characteristic 2. Another followup paper [26], by Farashahi–Wu–Zhao, reported 4 multiplications and 4 cubings, overall $8M + 4S$, for Hessian tripling, but only for characteristic 3. Further followup work [39], by Kohel, reported 4 multiplications and 4 cubings for twisted Hessian tripling in any odd characteristic. In Sect. 6 we generalize the approach of [39] and show how a better specialization reduces cost to just 6 cubings, assuming that the field has a fast primitive cube root of 1.

Doubling: In Sect. 6 we present four doubling formulas, starting with $6\mathbf{M}+3\mathbf{S}$ and culminating with $6\mathbf{M}+2\mathbf{S}$. In the case $a = 1$, the first formula was already well known before our work. Hisil, Carter and Dawson in [30] had already introduced doubling formulas using $3\mathbf{M}+6\mathbf{S}$, and also introduced doubling formulas using $7\mathbf{M}+1\mathbf{S}$, using techniques that seem to be specific to small cube a such as $a = 1$; see also [32]. Our $6\mathbf{M}+2\mathbf{S}$ is better than $7\mathbf{M}+1\mathbf{S}$ if $\mathbf{S} < \mathbf{M}$, and is better than $3\mathbf{M}+6\mathbf{S}$ if $\mathbf{S} > 0.75\mathbf{M}$.

At a higher level, double-base chains have been explored in several papers. The idea of a tree search for double-base chains was introduced by Doche and Habsieger in [21]. The tree search in [21] tries to minimize the number of additions used in a double-base chain, ignoring the cost of doublings and triplings; we do better by using the cost of doublings and triplings to adjust the weights of nodes in the tree.

2   Twisted Hessian Curves

Let k be a field. A **projective twisted Hessian curve over** k is a curve of the form $aX^3 + Y^3 + Z^3 = dXYZ$ in $\mathbf{P}^2$ with specified point $(0 : -1 : 1)$, where a, d are elements of k with $a(27a - d^3) \neq 0$. Theorem 2.1 below states that any projective twisted Hessian curve is an elliptic curve. The correponding affine curve $ax^3 + y^3 + 1 = dxy$ with specified point $(0, -1)$ is an **affine twisted Hessian curve**.

We state theorems for the projective curve, and allow the reader to deduce corresponding theorems for the affine curve. When we say "Let H be the twisted Hessian curve $aX^3 + Y^3 + Z^3 = dXYZ$ over k" we mean that a, d are elements of k, that $a(27a - d^3) \neq 0$, and that H is the projective twisted Hessian curve $aX^3 + Y^3 + Z^3 = dXYZ$ in $\mathbf{P}^2$ with specified point $(0 : -1 : 1)$. Some theorems need, and state, further assumptions such as $d \neq 0$.

The special case $a = 1$ of a twisted Hessian curve is simply a **Hessian curve**. The twisted Hessian curve $aX^3 + Y^3 + Z^3 = dXYZ$ is isomorphic to the Hessian curve $\bar{X}^3 + Y^3 + Z^3 = (d/a^{1/3})\bar{X}YZ$ over any extension of k containing a cube root $a^{1/3}$ of a: simply take $\bar{X} = a^{1/3}X$. Similarly, taking $\bar{X} = dX$ when $d \neq 0$ shows that the twisted Hessian curve for (a, d) is isomorphic to the twisted Hessian curve for $(a/d^3, 1)$; but we retain a and d as separate parameters to allow more curves with small parameters and thus with fast arithmetic.

Hessian curves have a long history, but twisted Hessian curves do not. The importance of twisted Hessian curves, beyond their extra generality, is that they have a complete addition law when a is not a cube. See Theorem 4.5 below.

Proof Strategy: Twisted Hessian Curves as Foundations. One can use the first isomorphism stated above to derive many features of twisted Hessian curves from corresponding well-known features of Hessian curves. We instead give direct proofs in the general case, meant as replacements for the older proofs in the special case: in other words, we propose starting with the theory of twisted Hessian curves rather than starting with the theory of Hessian curves.

This reduces the total proof length: the extra cost of tracking a through the proofs is smaller than the extra cost of applying the isomorphism.

We do not claim that this tracking involves any particular difficulty. In one case the tracking has been done before: specifically, some of the nonsingularity computations in Theorem 2.1 are special cases of classical discriminant computations for ternary cubics $aX^3 + bY^3 + cZ^3 = dXYZ$. See, e.g., [2] and [16]. However, the classical computations were carried out in characteristic 0, and the range of validity of the computations is not always obvious. Many of the computations fail in characteristic 3, even though Theorem 2.1 is valid in characteristic 3. Since the complete proofs are straightforward we simply include them here.

Similarly, one can derive many features of twisted Hessian curves from corresponding well-known features of Weierstrass curves, but we instead give direct proofs. We do use Weierstrass curves inside Theorem 5.2, which proves a property of all elliptic curves having points of order 3.

Notes on Definitions: Hessian Curves. There are various superficial differences among the definitions of Hessian curves in the literature. First, often characteristic 3 is prohibited. For example, [50] considers only base fields $\mathbf{F}_q$ with $q \in 2 + 3\mathbf{Z}$, and [34] considers only characteristics larger than 3. Our main interest is in the case $q \in 1 + 3\mathbf{Z}$, and in any event we see no reason to restrict the characteristic in the definition.

Second, often constants are introduced into the parameter d. For example, [34] defines a Hessian curve as $X^3 + Y^3 + Z^3 = 3dXYZ$, and the curve actually considered by Hesse in [29, p. 90, formula 54] was $X^3 + Y^3 + Z^3 + 6dXYZ = 0$.

Third, the specified point is often taken as a point at infinity, specifically $(-1 : 1 : 0)$; see, e.g., [17]. We use an affine point $(0 : -1 : 1)$ to allow completeness of the *affine* twisted Hessian curve rather than merely completeness of the *projective* twisted Hessian curve; if a is not a cube then there are no points at infinity for implementors to worry about. Converting addition laws (and twists and so on) between these two choices of neutral element is a trivial matter of permuting X, Y, Z.

Notes on Definitions: Elliptic Curves. There are also various differences among the definitions of elliptic curves in the literature.

The most specific definitions would say that Hessian curves are not elliptic curves: for example, Koblitz in [36, p. 117] defines elliptic curves to have long Weierstrass form. Obviously we do not use such restrictive definitions.

Two classical definitions that allow Hessian curves are as follows: (1) an elliptic curve is a nonsingular cubic curve in $\mathbf{P}^2$ with a specified point; (2) an elliptic curve is a nonsingular cubic curve in $\mathbf{P}^2$ with a specified *inflection* point. The importance of the inflection-point condition is that it allows the traditional geometric addition law: three distinct curve points on a line have sum 0; more generally, all curve points on a line, counted with multiplicity, have sum 0. If the specified point were not an inflection point then the addition law would be more complicated. See, e.g., [33, Chapter 3, Theorem 1.2].

We take the first of these two definitions. The statement that any twisted Hessian curve H is elliptic (Theorem 2.1) thus means that H is a nonsingular cubic curve with a specified point. We prove separately (Theorem 2.2) that the specified point $(0 : -1 : 1)$ is an inflection point.

Theorem 2.1. *Let H be the twisted Hessian curve $aX^3 + Y^3 + Z^3 = dXYZ$ over a field k. Then H is an elliptic curve.*

Proof. $aX^3 + Y^3 + Z^3 = dXYZ$ is a cubic curve in $\mathbf{P}^2$, and $(0 : -1 : 1)$ is a point on the curve. What remains is to prove that this curve is nonsingular.

A singularity $(X : Y : Z) \in \mathbf{P}^2$ of $aX^3 + Y^3 + Z^3 = dXYZ$ satisfies $3aX^2 = dYZ$, $3Y^2 = dXZ$, and $3Z^2 = dXY$. We will deduce $X = Y = Z = 0$, contradicting $(X : Y : Z) \in \mathbf{P}^2$.

Case 1: $3 \neq 0$ in k. Multiply to obtain $27aX^2Y^2Z^2 = d^3X^2Y^2Z^2$, i.e., $(27a - d^3)X^2Y^2Z^2 = 0$. By hypothesis $27a - d^3 \neq 0$, so $X^2Y^2Z^2 = 0$, so $X = 0$ or $Y = 0$ or $Z = 0$.

Case 1.1: $X = 0$. Then $3Y^2 = 0$ and $3Z^2 = 0$ so $Y = 0$ and $Z = 0$ as claimed.

Case 1.2: $Y = 0$. Then $3aX^2 = 0$ and $3Z^2 = 0$, and $a \neq 0$ by hypothesis, so $X = 0$ and $Z = 0$ as claimed.

Case 1.3: $Z = 0$. Then $3aX^2 = 0$ and $3Y^2 = 0$, and again $a \neq 0$, so $X = 0$ and $Y = 0$ as claimed.

Case 2: $3 = 0$ in k. Then $dYZ = 0$ and $dXZ = 0$ and $dXY = 0$. By hypothesis $a(-d^3) \neq 0$, so $d \neq 0$, so at least two of the coordinates X, Y, Z are 0.

Case 2.1: $X = Y = 0$. Then the curve equation $aX^3 + Y^3 + Z^3 = dXYZ$ forces $Z^3 = 0$ so $Z = 0$ as claimed.

Case 2.2: $X = Z = 0$. Then the curve equation forces $Y^3 = 0$ so $Y = 0$ as claimed.

Case 2.3: $Y = Z = 0$. Then the curve equation forces $aX^3 = 0$, and $a \neq 0$ by hypothesis, so $X = 0$ as claimed. □

Theorem 2.2. *Let H be the twisted Hessian curve $aX^3 + Y^3 + Z^3 = dXYZ$ over a field k. Then $(0 : -1 : 1)$ is an inflection point on H.*

Proof. We claim that $(0 : -1 : 1)$ is the only point of intersection of the line $-3(Y + Z) = dX$ with the curve $aX^3 + Y^3 + Z^3 = dXYZ$ over any extension of k. Consequently, by Bézout's theorem, this point has intersection multiplicity 3. (An alternative proof, involving essentially the same calculation, computes the multiplicity directly from its definition.)

To prove the claim, assume that $-3(Y + Z) = dX$ and $aX^3 + Y^3 + Z^3 = dXYZ$. Then $(27a - d^3)X^3 = 27aX^3 - (-3(Y + Z))^3 = 27(aX^3 + (Y + Z)^3) = 27(aX^3 + Y^3 + Z^3 + 3(Y + Z)YZ) = 27(dXYZ - dXYZ) = 0$ so $X^3 = 0$ so $X = 0$. Now $Y + Z = 0$: this follows from $-3(Y + Z) = dX = 0$ if $3 \neq 0$ in k, and it follows from $Y^3 + Z^3 = 0$ if $3 = 0$ in k. Thus $(X : Y : Z) = (0 : -1 : 1)$. □

3   The Standard Addition Law

Theorem 3.2 states an addition law for twisted Hessian curves. We originally derived this addition law as follows:

- Start from Sylvester's addition law for $X^3 + Y^3 + Z^3 = dXYZ$. See, e.g., [17, p. 425,equation 4.21i].
- Observe, as noted in [17], that the addition law is independent of d.
- Conclude that the addition law also works for $X^3 + Y^3 + Z^3 = (d/c)XYZ$, where c is a cube root of a.
- Permute X, Y, Z to our choice of neutral element.
- Replace X with cX.
- Rescale the outputs X_3, Y_3, Z_3 by a factor c.

The resulting polynomials X_3, Y_3, Z_3 are identical to Sylvester's addition law: they are independent of curve parameters, and in particular are independent of a. We refer to this addition law as the **standard addition law**. For reasons explained in Sect. 2, we prove Theorem 3.2 here by giving a direct proof of the standard addition law for the general case, rather than deriving the general case from the special case $a = 1$.

The standard addition law is never complete: it fails whenever $(X_2 : Y_2 : Z_2) = (X_1 : Y_1 : Z_1)$. More generally, it fails if and only if $(X_2 : Y_2 : Z_2) - (X_1 : Y_1 : Z_1)$ has the form $(0 : -\omega : 1)$ where $\omega^3 = 1$, or equivalently $(X_2 : Y_2 : Z_2) = (\omega^2 X_1 : \omega Y_1 : Z_1)$. See Theorem 4.6 for the equivalence, and Theorem 3.3 for the failure analysis.

A different way to analyze the failure cases, with somewhat less calculation, is as follows. First prove that $(X_2 : Y_2 : Z_2)$ has the form $(0 : -\omega : 1)$ if and only if the addition law fails to add the neutral element $(0 : -1 : 1)$ to $(X_2 : Y_2 : Z_2)$. Then use a theorem of Bosma and Lenstra [14, Theorem 2] stating that the set of failure cases of a degree-$(2, 2)$ addition law for a cubic elliptic curve in $\mathbf{P}^2$ is a union of shifted diagonals $\Delta_S = \{(P_1, P_1 + S)\}$. The theorems in [14] are stated only for Weierstrass curves, but they are invariant under linear equivalence and thus also apply to twisted Hessian curves. See [38] for a generalization to elliptic curves embedded in projective space of any dimension.

Theorems 4.2 and 4.5 below introduce a new addition law that (1) works for all doublings on any twisted Hessian curve and (2) is complete for any twisted Hessian curve with non-cube a.

Theorem 3.1. *Let H be the twisted Hessian curve $aX^3 + Y^3 + Z^3 = dXYZ$ over a field k. Let X_1, Y_1, Z_1 be elements of k such that $(X_1 : Y_1 : Z_1) \in H(k)$. Then $-(X_1 : Y_1 : Z_1) = (X_1 : Z_1 : Y_1)$.*

Proof. Recall that the specified neutral element of the curve is $(0 : -1 : 1)$.

Case 1: $(X_1 : Y_1 : Z_1) \neq (X_1 : Z_1 : Y_1)$. Then $X_1(Y + Z) = X(Y_1 + Z_1)$ is a line in $\mathbf{P}^2$: if all its coefficients $-Y_1 - Z_1, X_1, X_1$ are 0 then $(X_1 : Y_1 : Z_1) = (0 : -1 : 1) = (X_1 : Z_1 : Y_1)$, contradiction. This line intersects the curve at the distinct points $(0 : -1 : 1)$, $(X_1 : Y_1 : Z_1)$, and $(X_1 : Z_1 : Y_1)$. Hence $-(X_1 : Y_1 : Z_1) = (X_1 : Z_1 : Y_1)$.

Case 2: $(X_1 : Y_1 : Z_1) = (X_1 : Z_1 : Y_1)$ and $X_1 \neq 0$. Again $(X_1 : Y_1 : Z_1) \neq (0 : -1 : 1)$, and again $X_1(Y + Z) = X(Y_1 + Z_1)$ is a line. This line intersects the curve at both $(0 : -1 : 1)$ and $(X_1 : Y_1 : Z_1)$, and we show in a moment that it is tangent to the curve at $(X_1 : Y_1 : Z_1)$. Hence $-(X_1 : Y_1 : Z_1) = (X_1 : Y_1 : Z_1) = (X_1 : Z_1 : Y_1)$.

For the tangent calculation we take coordinates $y = Y/X$ and $z = Z/X$. The curve is then $a + y^3 + z^3 = dyz$; the point P_1 is $(y_1, z_1) = (Y_1/X_1, Z_1/X_1)$, which by hypothesis satisfies $y_1 = z_1$; and the line is $y + z = y_1 + z_1$. The curve is symmetric between y and z, so its slope at $(y_1, z_1) = (z_1, y_1)$ must be -1, which is the same as the slope of the line.

Case 3: $(X_1 : Y_1 : Z_1) = (X_1 : Z_1 : Y_1)$ and $X_1 = 0$. Then $Y_1^3 + Z_1^3 = 0$ by the curve equation so $Y_1 = \lambda Z_1$ for some λ with $\lambda^3 = -1$; but $(Y_1 : Z_1) = (Z_1 : Y_1)$ implies $\lambda = 1/\lambda$, so $\lambda = -1$, so $(X_1 : Y_1 : Z_1) = (0 : -1 : 1)$. Hence $-(X_1 : Y_1 : Z_1) = (0 : -1 : 1) = (0 : 1 : -1) = (X_1 : Z_1 : Y_1)$.   □

Theorem 3.2. *Let H be the twisted Hessian curve $aX^3 + Y^3 + Z^3 = dXYZ$ over a field k. Let $X_1, Y_1, Z_1, X_2, Y_2, Z_2$ be elements of k such that $(X_1 : Y_1 : Z_1), (X_2 : Y_2 : Z_2) \in H(k)$. Define*

$$X_3 = X_1^2 Y_2 Z_2 - X_2^2 Y_1 Z_1,$$
$$Y_3 = Z_1^2 X_2 Y_2 - Z_2^2 X_1 Y_1,$$
$$Z_3 = Y_1^2 X_2 Z_2 - Y_2^2 X_1 Z_1.$$

If $(X_3, Y_3, Z_3) \neq (0, 0, 0)$ then $(X_1 : Y_1 : Z_1) + (X_2 : Y_2 : Z_2) = (X_3 : Y_3 : Z_3)$.

Proof. The polynomial identity

$aX_3^3 + Y_3^3 + Z_3^3 - dX_3 Y_3 Z_3$
$= (X_1^3 Y_2^3 Z_2^3 + Y_1^3 X_2^3 Z_2^3 + Z_1^3 X_2^3 Y_2^3 - 3X_1 Y_1 Z_1 X_2^2 Y_2^2 Z_2^2)(aX_1^3 + Y_1^3 + Z_1^3 - dX_1 Y_1 Z_1)$
$- (X_2^3 Y_1^3 Z_1^3 + Y_2^3 X_1^3 Z_1^3 + Z_2^3 X_1^3 Y_1^3 - 3X_2 Y_2 Z_2 X_1^2 Y_1^2 Z_1^2)(aX_2^3 + Y_2^3 + Z_2^3 - dX_2 Y_2 Z_2)$

implies that $(X_3 : Y_3 : Z_3) \in H(k)$. The rest of the proof uses the chord-and-tangent definition of addition to show that $(X_1 : Y_1 : Z_1) + (X_2 : Y_2 : Z_2) = (X_3 : Y_3 : Z_3)$.

If $(X_1 : Y_1 : Z_1) = (X_2 : Y_2 : Z_2)$ then $(X_3, Y_3, Z_3) = (0, 0, 0)$, contradiction. Assume from now on that $(X_1 : Y_1 : Z_1) \neq (X_2 : Y_2 : Z_2)$.

The line through $(X_1 : Y_1 : Z_1)$ and $(X_2 : Y_2 : Z_2)$ is $(Z_1 Y_2 - Z_2 Y_1)X + (X_1 Z_2 - X_2 Z_1)Y + (X_2 Y_1 - X_1 Y_2)Z = 0$. The polynomial identity

$$(Z_1 Y_2 - Z_2 Y_1)X_3 + (X_1 Z_2 - X_2 Z_1)Z_3 + (X_2 Y_1 - X_1 Y_2)Y_3 = 0$$

shows that $(X_3 : Z_3 : Y_3)$ is also on this line.

One would now like to conclude that $(X_1 : Y_1 : Z_1) + (X_2 : Y_2 : Z_2) = -(X_3 : Z_3 : Y_3)$, so $(X_1 : Y_1 : Z_1) + (X_2 : Y_2 : Z_2) = (X_3 : Y_3 : Z_3)$ by Theorem 3.1. The only difficulty is that $(X_3 : Z_3 : Y_3)$ might be the same as $(X_1 : Y_1 : Z_1)$ or $(X_2 : Y_2 : Z_2)$; the rest of the proof consists of verifying that, in these two cases, the line is tangent to the curve at $(X_3 : Z_3 : Y_3)$.

We use two more polynomial identities. First, $X_1Y_2Y_3 + Y_1Z_2X_3 + Z_1X_2Z_3 = 0$. Second, $aX_1X_2X_3 + Z_1Z_2Y_3 + Y_1Y_2Z_3 = (aX_1^3 + Y_1^3 + Z_1^3)X_2Y_2Z_2 - (aX_2^3 + Y_2^3 + Z_2^3)X_1Y_1Z_1$; the curve equations for $(X_1 : Y_1 : Z_1)$ and $(X_2 : Y_2 : Z_2)$ then imply $aX_1X_2X_3 + Z_1Z_2Y_3 + Y_1Y_2Z_3 = 0$.

Case 1: $(X_3 : Z_3 : Y_3) = (X_1 : Y_1 : Z_1)$. The two identities above then imply $X_1Y_2Z_1 + Y_1Z_2X_1 + Z_1X_2Y_1 = 0$ and $aX_1^2X_2 + Z_1^2Z_2 + Y_1^2Y_2 = 0$ respectively. Our line is $(Z_1Y_2 - Z_2Y_1)X + (X_1Z_2 - X_2Z_1)Y + (X_2Y_1 - X_1Y_2)Z = 0$, while the tangent to the curve at $(X_1 : Y_1 : Z_1)$ is $(3aX_1^2 - dY_1Z_1)X + (3Y_1^2 - dX_1Z_1)Y + (3Z_1^2 - dX_1Y_1)Z = 0$. To see that these lines are the same, observe that the cross product

$$\begin{pmatrix} (3Y_1^2 - dX_1Z_1)(X_2Y_1 - X_1Y_2) - (3Z_1^2 - dX_1Y_1)(X_1Z_2 - X_2Z_1) \\ (3Z_1^2 - dX_1Y_1)(Z_1Y_2 - Z_2Y_1) - (3aX_1^2 - dY_1Z_1)(X_2Y_1 - X_1Y_2) \\ (3aX_1^2 - dY_1Z_1)(X_1Z_2 - X_2Z_1) - (3Y_1^2 - dX_1Z_1)(Z_1Y_2 - Z_2Y_1) \end{pmatrix}$$

is exactly

$$\begin{pmatrix} 3X_2 & -3X_1 & dX_1 \\ 3Y_2 & -3Y_1 & dY_1 \\ 3Z_2 & -3Z_1 & dZ_1 \end{pmatrix} \begin{pmatrix} aX_1^3 + Y_1^3 + Z_1^3 - dX_1Y_1Z_1 \\ aX_1^2X_2 + Z_1^2Z_2 + Y_1^2Y_2 \\ X_1Y_2Z_1 + Y_1Z_2X_1 + Z_1X_2Y_1 \end{pmatrix} = \begin{pmatrix} 0 \\ 0 \\ 0 \end{pmatrix}.$$

Case 2: $(X_3 : Z_3 : Y_3) = (X_2 : Y_2 : Z_2)$. Exchanging $(X_1 : Y_1 : Z_1)$ with $(X_2 : Y_2 : Z_2)$ replaces (X_3, Y_3, Z_3) with $(-X_3, -Y_3, -Z_3)$ and moves to case 1. □

Theorem 3.3. *In the situation of Theorem 3.2, $(X_3, Y_3, Z_3) = (0, 0, 0)$ if and only if $(X_2 : Y_2 : Z_2) = (\omega^2 X_1 : \omega Y_1 : Z_1)$ for some $\omega \in k$ with $\omega^3 = 1$.*

Proof. If $(X_2 : Y_2 : Z_2) = (\omega^2 X_1 : \omega Y_1 : Z_1)$ and $\omega^3 = 1$ then (X_3, Y_3, Z_3) is proportional to $(X_1^2 \omega Y_1 Z_1 - \omega^4 X_1^2 Y_1 Z_1, Z_1^2 \omega^2 X_1 \omega Y_1 - Z_1^2 X_1 Y_1, Y_1^2 \omega^2 X_1 Z_1 - \omega^2 Y_1^2 X_1 Z_1) = (0, 0, 0)$.

Conversely, assume that $(X_3, Y_3, Z_3) = (0, 0, 0)$. Then $X_1^2 Y_2 Z_2 = X_2^2 Y_1 Z_1$, $Z_1^2 X_2 Y_2 = Z_2^2 X_1 Y_1$, and $Y_1^2 X_2 Z_2 = Y_2^2 X_1 Z_1$.

If $X_1 = 0$ then $Y_1^3 + Z_1^3 = 0$ by the curve equation, so $Y_1 \neq 0$ and $Z_1 \neq 0$. Write $\lambda_1 = Y_1/Z_1$; then $(X_1 : Y_1 : Z_1) = (0 : \lambda_1 : 1)$ and $\lambda_1^3 = -1$. Furthermore $X_2^2 Y_1 Z_1 = 0$ so $X_2 = 0$ so $(X_2 : Y_2 : Z_2) = (0 : \lambda_2 : 1)$ where $\lambda_2^3 = -1$. Define $\omega = \lambda_2/\lambda_1$; then $\omega^3 = \lambda_2^3/\lambda_1^3 = 1$ and $(X_2 : Y_2 : Z_2) = (0 : \lambda_2 : 1) = (0 : \omega\lambda_1 : 1) = (\omega^2 X_1 : \omega Y_1 : Z_1)$.

If $X_2 = 0$ then similarly $X_1 = 0$. Assume from now on that $X_1 \neq 0$ and $X_2 \neq 0$. Write $y_1 = Y_1/X_1$, $z_1 = Z_1/X_1$, $y_2 = Y_2/X_2$, and $z_2 = Z_2/X_2$. Rewrite the three equations $X_3 = 0$, $Y_3 = 0$, and $Z_3 = 0$ as $y_2z_2 = y_1z_1$, $z_1^2y_2 = z_2^2y_1$, and $y_1^2z_2 = y_2^2z_1$. The first two equations imply $z_1^3y_1 = z_1^2y_2z_2 = z_2^3y_1$, so $(z_1^3 - z_2^3)y_1 = 0$; the first and third equations imply $y_1^3z_1 = y_1^2y_2z_2 = y_2^3z_1$, so $(y_1^3 - y_2^3)z_1 = 0$.

If $y_1 = 0$ then $z_1^2y_2 = 0$ by the second equation. The curve equation $a + y_1^3 + z_1^3 = dy_1z_1$ forces $a + z_1^3 = 0$ so $z_1 \neq 0$; hence $y_2 = 0$. The curve equation $a + y_2^3 + z_2^3 = dy_2z_2$ similarly forces $a + z_2^3 = 0$ so $z_2^3 = z_1^3$. Write $\omega = z_2/z_1$; then

$\omega^3 = 1$ and $(X_2 : Y_2 : Z_2) = (1 : y_2 : z_2) = (1 : 0 : z_2) = (1 : 0 : \omega z_1) = (\omega^2 : \omega y_1 : z_1) = (\omega^2 X_1 : \omega Y_1 : Z_1)$.

Assume from now on that $y_1 \neq 0$. Similarly assume that $z_1 \neq 0$. Then $z_1^3 = z_2^3$ and $y_1^3 = y_2^3$. Write $\omega = y_1/y_2$; then $\omega^3 = 1$. The equation $X_3 = 0$ forces $\omega = z_2/z_1$. Hence $(X_2 : Y_2 : Z_2) = (1 : y_2 : z_2) = (1 : \omega^{-1}y_1 : \omega z_1) = (\omega^2 X_1 : \omega Y_1 : Z_1)$. □

4   The Rotated Addition Law

Theorem 4.2 states a new addition law for twisted Hessian curves. This addition law is obtained as follows:

- Subtract $(1 : -c : 0)$ from one input, using Theorem 4.1, where c is a cube root of a.
- Use the standard addition law in Theorem 3.2.
- Add $(1 : -c : 0)$ to the output, using Theorem 4.1 again.

The formulas in Theorem 4.1 are linear, so the resulting addition law has the same bidegree as the standard addition law. This is an example of what Bernstein and Lange in [11, Section 8] call **rotation** of an addition law.

This rotated addition law is new, even in the case $a = 1$. Unlike the standard addition law, the rotated addition law works for doublings. Specializing the rotated addition law to doublings, and further to $a = 1$, produces exactly the Joye–Quisquater doubling formula from [34, Proposition 2]. Even better, the rotated addition law is complete when a is not a cube; see Theorem 4.5 below.

Theorem 4.7 states that the standard addition law and the rotated addition law form a complete system of addition laws for any twisted Hessian curve: any pair of input points can be added by at least one of the two laws. This system is vastly simpler than the Bosma–Lenstra complete system [14] of addition laws for Weierstrass curves, and arguably even simpler than the Bernstein–Lange complete system [11] of addition laws for twisted Edwards curves: each output coordinate here is a difference of just two degree-$(2, 2)$ monomials, as in [11], but here there are just three output coordinates while in [11] there were four.

One can easily rotate the addition law again (or, equivalently, exchange the two inputs) to obtain a third addition law with the same features as the second addition law. One can also prove that these three addition laws are a basis for the space of degree-$(2, 2)$ addition laws for H: it is easy to see that the laws are linearly independent, and Bosma and Lenstra showed in [14, Section 4] that the whole space has dimension 3.

Theorem 4.1. *Let H be the twisted Hessian curve $aX^3 + Y^3 + Z^3 = dXYZ$ over a field k. Assume that $c \in k$ satisfies $c^3 = a$. Then $(1 : -c : 0) \in H(k)$. Furthermore, if X_1, Y_1, Z_1 are elements of k such that $(X_1 : Y_1 : Z_1) \in H(k)$, then $(X_1 : Y_1 : Z_1) + (1 : -c : 0) = (Y_1 : cZ_1 : c^2 X_1)$.*

Proof. First $a(1)^3 + (-c)^3 + (0)^3 = 0$ so $(1 : -c : 0) \in H(k)$.

Case 1: $Z_1 \neq 0$. Write $(X_2, Y_2, Z_2) = (1, -c, 0)$, and define (X_3, Y_3, Z_3) as in Theorem 3.2. Then $X_3 = -Y_1 Z_1$, $Y_3 = -cZ_1^2$, and $Z_3 = -c^2 X_1 Z_1$, so $(X_3 : Y_3 : Z_3) = (Y_1 : cZ_1 : c^2 X_1)$, so $(X_1 : Y_1 : Z_1) + (1 : -c : 0) = (Y_1 : cZ_1 : c^2 X_1)$ by Theorem 3.2.

Case 2: $Z_1 = 0$. Then $aX_1^3 + Y_1^3 = 0$ by the curve equation. Write $\omega = Y_1/(-cX_1)$; then $\omega^3 = Y_1^3/(-aX_1^3) = 1$, and $(X_1 : Y_1 : Z_1) = (1 : -\omega c : 0)$.

Case 2.1: $\omega \neq 1$. The line $Z = 0$ intersects the curve at the three distinct points $(1 : -c : 0)$, $(1 : -\omega c : 0)$, and $(1 : -\omega^{-1} c : 0)$, so $(1 : -c : 0) + (1 : -\omega c : 0) = -(1 : -\omega^{-1} c : 0) = (1 : 0 : -\omega^{-1} c) = (-\omega c : 0 : c^2) = (Y_1 : cZ_1 : c^2 X_1)$ by Theorem 3.1.

Case 2.2: $\omega = 1$, i.e., $(X_1 : Y_1 : Z_1) = (1 : -c : 0)$. The line $3c^2 X + 3cY + dZ = 0$ intersects the curve at $(1 : -c : 0)$. We will see in a moment that it has no other intersection points. Consequently $3(1 : -c : 0) = 0$; i.e., $(X_1 : Y_1 : Z_1) + (1 : -c : 0) = 2(1 : -c : 0) = -(1 : -c : 0) = (1 : 0 : -c) = (-c : 0 : c^2) = (Y_1 : cZ_1 : c^2 X_1)$ by Theorem 3.1.

We finish by showing that the only intersection is $(1 : -c : 0)$. Assume that $3c^2 X + 3cY + dZ = 0$ and $aX^3 + Y^3 + Z^3 = dXYZ$. Then $-dZ = 3c(cX + Y)$, but also $(cX + Y)^3 = aX^3 + Y^3 + 3c^2 X^2 Y + 3cXY^2 = -Z^3$, so $-d^3 Z^3 = 27a(cX + Y)^3 = -27aZ^3$. By hypothesis $27a \neq d^3$, so $Z^3 = 0$, so $Z = 0$, so $cX + Y = 0$, so $(X : Y : Z) = (1 : -c : 0)$. $\square$

Theorem 4.2. *Let H be the twisted Hessian curve $aX^3 + Y^3 + Z^3 = dXYZ$ over a field k. Let $X_1, Y_1, Z_1, X_2, Y_2, Z_2$ be elements of k such that $(X_1 : Y_1 : Z_1), (X_2 : Y_2 : Z_2) \in H(k)$. Define*

$$X_3' = Z_2^2 X_1 Z_1 - Y_1^2 X_2 Y_2,$$
$$Y_3' = Y_2^2 Y_1 Z_1 - aX_1^2 X_2 Z_2,$$
$$Z_3' = aX_2^2 X_1 Y_1 - Z_1^2 Y_2 Z_2.$$

If $(X_3', Y_3', Z_3') \neq (0, 0, 0)$ then $(X_1 : Y_1 : Z_1) + (X_2 : Y_2 : Z_2) = (X_3' : Y_3' : Z_3')$.

Proof. Fix a field extension K of k containing a cube root c of a. Replace k, X_1, Y_1, Z_1 with $K, Z_1, c^2 X_1, cY_1$ respectively throughout Theorem 3.2. This replaces X_3, Y_3, Z_3 with $-Z_3', -c^2 X_3', -cY_3'$ respectively. Hence $(Z_1 : c^2 X_1 : cY_1) + (X_2 : Y_2 : Z_2) = (Z_3' : c^2 X_3' : cY_3')$ if $(X_3', Y_3', Z_3') \neq (0, 0, 0)$.

Now add $(1 : -c : 0)$ to both sides. Theorem 4.1 implies $(1 : -c : 0) + (Z_1 : c^2 X_1 : cY_1) = (c^2 X_1 : c^2 Y_1 : c^2 Z_1) = (X_1 : Y_1 : Z_1)$, so $(X_1 : Y_1 : Z_1) + (X_2 : Y_2 : Z_2) = (X_3' : Y_3' : Z_3')$ if $(X_3', Y_3', Z_3') \neq (0, 0, 0)$. $\square$

Theorem 4.3. *In the situation of Theorem 4.2, $(X_3', Y_3', Z_3') = (0, 0, 0)$ if and only if $(X_2 : Y_2 : Z_2) = (Z_1 : \gamma^2 X_1 : \gamma Y_1)$ for some $\gamma \in k$ with $\gamma^3 = a$.*

Proof. Fix a field extension K of k containing a cube root c of a. Replace k, X_1, Y_1, Z_1 with $K, Z_1, c^2 X_1, cY_1$ respectively throughout Theorem 3.2 and Theorem 3.3 to see that $(-Z_3', -c^2 X_3', -cY_3') = (0, 0, 0)$ if and only if $(X_2 : Y_2 : Z_2) = (\omega^2 Z_1 : \omega c^2 X_1 : cY_1)$ for some $\omega \in K$ with $\omega^3 = 1$.

If $(X_2 : Y_2 : Z_2) = (Z_1 : \gamma^2 X_1 : \gamma Y_1)$ for some $\gamma \in k$ with $\gamma^3 = a$ then this condition is satisfied by the ratio $\omega = \gamma/c \in K$ so $(X_3', Y_3', Z_3') = (0,0,0)$.

Conversely, if $(X_3', Y_3', Z_3') = (0,0,0)$ then $(X_2 : Y_2 : Z_2) = (\omega^2 Z_1 : \omega c^2 X_1 : cY_1)$ for some $\omega \in K$ with $\omega^3 = 1$, so $(X_2 : Y_2 : Z_2) = (Z_1 : \gamma^2 X_1 : \gamma Y_1)$ where $\gamma = c\omega$. To see that $\gamma \in k$, note that at least two of X_1, Y_1, Z_1 are nonzero. If X_1, Y_1 are nonzero then Y_2, Z_2 are nonzero and $(\gamma^2 X_1)/(\gamma Y_1) = Y_2/Z_2$ so $\gamma = (Y_2/Z_2)(Y_1/X_1) \in k$. If Y_1, Z_1 are nonzero then X_2, Z_2 are nonzero and $(\gamma Y_1)/Z_1 = Z_2/X_2$ so $\gamma = (Z_2/X_2)(Z_1/Y_1) \in k$. If X_1, Z_1 are nonzero then X_2, Y_2 are nonzero and $(\gamma^2 X_1)/Z_1 = Y_2/X_2$ so $\gamma^2 = (Y_2/X_2)(Z_1/X_1) \in k$; but also $\gamma^3 = c^3 = a \in k$, so $\gamma = a/\gamma^2 \in k$. $\qquad\square$

Theorem 4.4. *In the situation of Theorem 4.2,* $(X_3', Y_3', Z_3') \neq (0,0,0)$ *if* $(X_2 : Y_2 : Z_2) = (X_1 : Y_1 : Z_1)$.

Proof. Suppose $(X_3', Y_3', Z_3') = (0,0,0)$. Then $(X_2, Y_2, Z_2) = (Z_1, \gamma^2 X_1, \gamma Y_1)$ for some $\gamma \in k$ with $\gamma^3 = a$ by Theorem 4.3, so $(X_2 : Y_2 : Z_2) + (1 : -\gamma : 0) = (\gamma^2 X_1 : \gamma^2 Y_1 : \gamma^2 Z_1) = (X_1 : Y_1 : Z_1)$ by Theorem 4.1. Subtract $(X_2 : Y_2 : Z_2) = (X_1 : Y_1 : Z_1)$ to obtain $(1 : -\gamma : 0) = (0 : -1 : 1)$, contradiction.

Alternative proof, showing more directly that $Y_3' \neq 0$ or $Z_3' \neq 0$: Write (X_2, Y_2, Z_2) as $(\lambda X_1, \lambda Y_1, \lambda Z_1)$ for some $\lambda \neq 0$. Then $Y_3' = \lambda^2 Z_1(Y_1^3 - aX_1^3)$ and $Z_3' = \lambda^2 Y_1(aX_1^3 - Z_1^3)$.

Case 1: $Y_1 = 0$. Then $aX_1^3 = -Z_1^3$ by the curve equation, so $Y_3' = -\lambda^2 Z_1^4$. If $Y_3' = 0$ then $Z_1 = 0$ so $aX_1^3 = 0$ so $X_1 = 0$ so $(X_1, Y_1, Z_1) = (0,0,0)$, contradiction. Hence $Y_3' \neq 0$.

Case 2: $Z_1 = 0$. Then $aX_1^3 = -Y_1^3$ by the curve equation, so $Z_3' = -\lambda^2 Y_1^4$. If $Z_3' = 0$ then $Y_1 = 0$ so $aX_1^3 = 0$ so $X_1 = 0$ so $(X_1, Y_1, Z_1) = (0,0,0)$, contradiction. Hence $Z_3' \neq 0$.

Case 3: $Y_1 \neq 0$ and $Z_1 \neq 0$. If $Y_3' = 0$ and $Z_3' = 0$ then $aX_1^3 = Y_1^3$ and $aX_1^3 = Z_1^3$; in particular $X_1 \neq 0$. so $3aX_1^3 = dX_1 Y_1 Z_1$ by the curve equation, so $27a^3 X_1^9 = dX_1^3 Y_1^3 Z_1^3 = da^2 X_1^9$, so $27a = d^3$, contradiction. Hence $Y_3' \neq 0$ or $Z_3' \neq 0$. $\qquad\square$

Theorem 4.5. *In the situation of Theorem 4.2, assume that a is not a cube in k. Then* $(X_3', Y_3', Z_3') \neq (0,0,0)$ *and* $(X_1 : Y_1 : Z_1) + (X_2 : Y_2 : Z_2) = (X_3' : Y_3' : Z_3')$.

Proof. By hypothesis no $\gamma \in k$ satisfies $\gamma^3 = a$. By Theorem 4.3, $(X_3', Y_3', Z_3') \neq (0,0,0)$. By Theorem 4.2, $(X_1 : Y_1 : Z_1) + (X_2 : Y_2 : Z_2) = (X_3' : Y_3' : Z_3')$.

We also give a second, more direct, proof that $Z_3' \neq 0$. The curve equation forces $Z_1 \neq 0$ and $Z_2 \neq 0$. Write $x_1 = X_1/Z_1$, $y_1 = Y_1/Z_1$, $x_2 = X_2/Z_2$, and $y_2 = Y_2/Z_2$. Suppose that $Z_3' = 0$, i.e., $y_2 = ax_1 y_1 x_2^2$. Eliminate y_2 in the curve equation $ax_2^3 + y_2^3 + 1 = dx_2 y_2$ to obtain $ax_2^3 + (ax_1 y_1 x_2^2)^3 + 1 = dax_1 y_1 x_2^3$. Use the curve equation at (x_1, y_1) to eliminate d and rewrite $(ax_1 y_1 x_2^2)^3 = -ax_2^3 - 1 + ax_2^3(ax_1^3 + y_1^3 + 1) = ax_2^3(ax_1^3 + y_1^3) - 1$ which factors as $(a^2 x_1^3 x_2^3 - 1)(ax_2^3 y_1^3 - 1) = 0$, implying that a is a cube in k. $\qquad\square$

Theorem 4.6. *Let H be the twisted Hessian curve $aX^3 + Y^3 + Z^3 = dXYZ$ over a field k. Assume that $\omega \in k$ satisfies $\omega^3 = 1$. Then $(0 : -\omega : 1) \in H(k)$.*

Furthermore, if X_1, Y_1, Z_1 *are elements of* k *such that* $(X_1 : Y_1 : Z_1) \in H(k)$, *then* $(X_1 : Y_1 : Z_1) + (0 : -\omega : 1) = (\omega^2 X_1 : \omega Y_1 : Z_1)$.

Proof. Take $(X_2, Y_2, Z_2) = (0, -\omega, 1)$ in Theorem 3.2 to obtain $(X_3, Y_3, Z_3) = (-\omega X_1^2, -X_1 Y_1, -\omega^2 X_1 Z_1)$. If $X_1 \neq 0$ then $(X_3, Y_3, Z_3) \neq (0, 0, 0)$ and $(X_1 : Y_1 : Z_1) + (0 : -\omega : 1) = (X_3 : Y_3 : Z_3) = (\omega^2 X_1 : \omega Y_1 : Z_1)$.

Also take $(X_2, Y_2, Z_2) = (0, -\omega, 1)$ in Theorem 4.2 to obtain $(X_3', Y_3', Z_3') = (X_1 Z_1, \omega^2 Y_1 Z_1, \omega Z_1^2)$. If $Z_1 \neq 0$ then $(X_3', Y_3', Z_3') \neq (0, 0, 0)$ and $(X_1 : Y_1 : Z_1) + (0 : -\omega : 1) = (X_3' : Y_3' : Z_3') = (\omega^2 X_1 : \omega Y_1 : Z_1)$.

At least one of X_1, Z_1 must be nonzero, so at least one of these cases applies.
$\square$

Theorem 4.7. *Let* H *be the twisted Hessian curve* $aX^3 + Y^3 + Z^3 = dXYZ$ *over a field* k. *Let* $X_1, Y_1, Z_1, X_2, Y_2, Z_2$ *be elements of* k *such that* $(X_1 : Y_1 : Z_1), (X_2 : Y_2 : Z_2) \in H(k)$. *Define* (X_3, Y_3, Z_3) *as in Theorem 3.2, and* (X_3', Y_3', Z_3') *as in Theorem 4.2. Then* $(X_3, Y_3, Z_3) \neq (0, 0, 0)$ *or* $(X_3', Y_3', Z_3') \neq (0, 0, 0)$.

Proof. Suppose that $(X_3, Y_3, Z_3) = (0, 0, 0)$ and $(X_3', Y_3', Z_3') = (0, 0, 0)$. Then $(X_2 : Y_2 : Z_2) = (\omega^2 X_1 : \omega Y_1 : Z_1)$ for some $\omega \in k$ with $\omega^3 = 1$ by Theorem 3.3, so $(X_2 : Y_2 : Z_2) = (X_1 : Y_1 : Z_1) + (0 : -\omega : 1)$ by Theorem 4.6. Furthermore $(X_2 : Y_2 : Z_2) = (Z_1 : \gamma^2 X_1 : \gamma Y_1)$ for some $\gamma \in k$ with $\gamma^3 = a$ by Theorem 4.3, so $(X_2 : Y_2 : Z_2) = (X_1 : Y_1 : Z_1) - (1 : -\gamma : 0)$ by Theorem 4.1. Hence $(0 : -\omega : 1) = -(1 : -\gamma : 0) = (1 : 0 : -\gamma)$, contradiction.
$\square$

5 Points of Order 3

Each projective twisted Hessian curve over $\mathbf{F}_q$ has a rational point of order 3. See Theorem 5.1. In particular, for $q \in 1 + 3\mathbf{Z}$, the point $(0 : -\omega : 1)$ is a rational point of order 3, where ω is a primitive cube root of 1 in $\mathbf{F}_q$.

Conversely, if $q \in 1 + 3\mathbf{Z}$, then each elliptic curve over $\mathbf{F}_q$ with a point P_3 of order 3 is isomorphic to a twisted Hessian curve via an isomorphism that takes P_3 to $(0 : -\omega : 1)$. We prove this converse in two steps:

- Over any field, each elliptic curve with a point P_3 of order 3 is isomorphic to a curve of the form $y^2 + dxy + ay = x^3$, where $a(27a - d^3) \neq 0$, via an isomorphism taking P_3 to $(0, 0)$. This is a standard fact; see, e.g., [23, Section 13.1.5.b]. To keep this paper self-contained we include a proof as Theorem 5.2. We refer to $y^2 + dxy + ay = x^3$ as a **triangular curve** because its Newton polygon is a triangle of minimum area (equivalently, minimum number of boundary lattice points) among all Newton polygons of Weierstrass curves.
- Over a field with a primitive cube root ω of 1, this triangular curve is isomorphic to the twisted Hessian curve $(d^3 - 27a)X^3 + Y^3 + Z^3 = 3dXYZ$ via an isomorphism that takes $(0, 0)$ to $(0 : -\omega : 1)$. See Theorem 5.3.

Furthermore, over any field, this triangular curve is 3-isogenous to the twisted Hessian curve $aX^3 + Y^3 + Z^3 = dXYZ$, provided that $d \neq 0$. See Theorem 5.4. This gives an alternate proof, for $d \neq 0$, that $aX^3 + Y^3 + Z^3 = dXYZ$ has a point of order 3 over $\mathbf{F}_q$: the triangular curve $y^2 + dxy + ay = x^3$ has a point of order 3, namely $(0,0)$, so its group order over $\mathbf{F}_q$ is a multiple of 3; the isogenous twisted Hessian curve $aX^3 + Y^3 + Z^3 = dXYZ$ has the same group order, and therefore also a point of order 3. This isogeny also leads to extremely fast tripling formulas; see Sect. 6.

For comparison: Over a field where all elements are cubes, such as a field $\mathbf{F}_q$ with $q \in 2 + 3\mathbf{Z}$, Smart in [50, Section 3] states an isomorphism from the triangular curve to a Hessian curve, taking $(0,0)$ to the point $(-1 : 0 : 1)$ of order 3 (modulo permutation of coordinates to put the neutral element at infinity). We instead emphasize the case $q \in 1 + 3\mathbf{Z}$ since this is the case that allows completeness.

Theorem 5.1. *Let H be the twisted Hessian curve $aX^3 + Y^3 + Z^3 = dXYZ$ over a finite field k. Then $H(k)$ has a point of order 3.*

Proof. Case 1: $\#k \in 1 + 3\mathbf{Z}$. There is a primitive cube root ω of 1 in k. The point $(0 : -\omega : 1)$ is in $H(k)$ by Theorem 4.6, is nonzero since $\omega \neq 1$, satisfies $2(0 : -\omega : 1) = (0 : -\omega^2 : 1) = (0 : 1 : -\omega)$ by Theorem 4.6, and satisfies $-(0 : -\omega : 1) = (0 : 1 : -\omega)$ by Theorem 3.1, so it is a point of order 3.

Case 2: $\#k \notin 1 + 3\mathbf{Z}$. There is a cube root c of a in k. The point $(1 : -c : 0)$ is in $H(k)$ by Theorem 4.1, is visibly nonzero, satisfies $2(1 : -c : 0) = (-c : 0 : c^2) = (1 : 0 : -c)$ by Theorem 4.1, and satisfies $-(1 : -c : 0) = (1 : 0 : -c)$ by Theorem 3.1, so it is a point of order 3. $\qquad\square$

Theorem 5.2. *Let E be an elliptic curve over a field k. Assume that $E(k)$ has a point P_3 of order 3. Then there exist a, d, ϕ such that $a, d \in k$; $a(27a - d^3) \neq 0$; ϕ is an isomorphism from E to the triangular curve $y^2 + dxy + ay = x^3$; and $\phi(P_3) = (0,0)$.*

Proof. Write E in long Weierstrass form $v^2 + e_1 uv + e_3 v = u^3 + e_2 u^2 + e_4 u + e_6$. The point P_3 is nonzero so it is affine, say (u_3, v_3).

Substitute $u = x + u_3$ and $v = t + v_3$ to obtain an isomorphic curve C in long Weierstrass form $t^2 + c_1 xt + c_3 t = x^3 + c_2 x^2 + c_4 x + c_6$. This isomorphism takes P_3 to the point $(0,0)$. This point has order 3, so the tangent line to C at $(0,0)$ intersects the curve at that point with multiplicity 3, so it does not intersect the point at infinity, so it is not vertical; i.e., it has the form $t = \lambda x$ for some $\lambda \in k$.

Substitute $y = t - \lambda x$ to obtain an isomorphic curve A in long Weierstrass form $y^2 + a_1 xy + a_3 y = x^3 + a_2 x^2 + a_4 x + a_6$. This isomorphism preserves $(0,0)$, and now the line $y = 0$ intersects A at $(0,0)$ with multiplicity 3. Hence $a_2 = a_4 = a_6 = 0$; i.e., the curve is $y^2 + a_1 xy + a_3 y = x^3$. Write $d = a_1$ and $a = a_3$.

The discriminant of this curve is $a^3(d^3 - 27a)$ so $a \neq 0$ and $27a - d^3 \neq 0$. More explicitly, if $a = 0$ then $(0,0)$ is singular; if $d^3 = 27a$ and $3 = 0$ in k then $(-(a^2/4)^{1/3}, -a/2)$ is singular; if $d^3 = 27a$ and $3 \neq 0$ in k then $(-d^2/9, a)$ is singular. $\qquad\square$

Theorem 5.3. *Let a, d be elements of a field k such that $a(27a - d^3) \neq 0$. Let ω be an element of k with $\omega^3 = 1$ and $\omega \neq 1$. Let E be the triangular* curve $VW(V + dU + aW) = U^3$. Then there is an isomorphism ϕ from E to the twisted Hessian curve $(d^3 - 27a)X^3 + Y^3 + Z^3 = 3dXYZ$, defined by $\phi(U : V : W) = (X : Y : Z)$ where $X = U$, $Y = \omega(V + dU + aW) - \omega^2 V - aW$, $Z = \omega^2(V + dU + aW) - \omega V - aW$. Furthermore $\phi(0 : 0 : 1) = (0 : -\omega : 1)$.

Proof. Note that $3 \neq 0$ in k: otherwise $(\omega - 1)^3 = \omega^3 - 1 = 0$ so $\omega - 1 = 0$, contradiction.

Write H for the curve $a'X^3 + Y^3 + Z^3 = d'XYZ$, where $a' = d^3 - 27a$ and $d' = 3d$. Then $a'(27a' - (d')^3) = (d^3 - 27a)(27(d^3 - 27a) - 27d^3) = 27^2 a(27a - d^3) \neq 0$, so H is a twisted Hessian curve over k.

The identity $a'X^3 + Y^3 + Z^3 - d'XYZ = 27a(VW(V + dU + aW) - U^3)$ in the ring $\mathbf{Z}[a, d, U, V, W, \omega]/(\omega^2 + \omega + 1)$ shows that ϕ maps E to H.

The map ϕ is invertible on $\mathbf{P}^2$: specifically, $\phi^{-1}(X : Y : Z) = (U : V : W)$ where $U = X$, $V = -(dX + \omega Y + \omega^2 Z)/3$, and $W = -(dX + Y + Z)/(3a)$. The same identity shows that ϕ^{-1} maps H to E.

Hence ϕ is an isomorphism of curves from H to E. To see that it is an isomorphism of elliptic curves, observe that it maps the neutral element of E to the neutral element of H: specifically, $\phi(0 : 1 : 0) = (0 : \omega - \omega^2 : \omega^2 - \omega) = (0 : -1 : 1)$.

Finally $\phi(0 : 0 : 1) = (0 : \omega a - a : \omega^2 a - a) = (0 : \omega - 1 : \omega^2 - 1) = (0 : -\omega : 1)$.
□

Theorem 5.4. *Let H be the twisted Hessian curve $aX^3 + Y^3 + Z^3 = dXYZ$ over a field k. Assume that $d \neq 0$. Let E be the triangular curve $VW(V + dU + aW) = U^3$. Then there is an isogeny ι from H to E defined by $\iota(X : Y : Z) = (-XYZ : Y^3 : X^3)$; there is an isogeny ι' from E to H defined by*

$$\iota'(U : V : W)$$
$$= \left(\frac{R^3 + S^3 + V^3 - 3RSV}{d} : RS^2 + SV^2 + VR^2 - 3RSV : RV^2 + SR^2 + VS^2 - 3RSV \right)$$

where $Q = dU$, $R = aW$, and $S = -(V + Q + R)$; and $\iota'(\iota(P)) = 3P$ for each point P on H.

Proof. If $U = -XYZ$, $V = Y^3$, and $W = X^3$ then $VW(V + dU + aW) - U^3 = X^3 Y^3 (aX^3 + Y^3 + Z^3 - dXYZ)$. Hence ι is a rational map from H to E. The neutral element $(0 : -1 : 1)$ of H maps to the neutral element $(0 : 1 : 0)$ of E, so ι is an isogeny from H to E. Note that ι is defined everywhere on H: each point $(X : Y : Z)$ on H has $X \neq 0$ or $Y \neq 0$, so $(-XYZ, Y^3, X^3) \neq (0, 0, 0)$.

If $Q = dU$, $R = aW$, $S = -(V + Q + R)$, $X = (R^3 + S^3 + V^3 - 3RSV)/d$, $Y = RS^2 + SV^2 + VR^2 - 3RSV$, and $Z = RV^2 + SR^2 + VS^2 - 3RSV$ then the following identities hold:

$$aX^3 + Y^3 + Z^3 - dXYZ$$
$$= a(Q^2 + 3QR + 3R^2 + 3QV + 3VR + 3V^2)^3(VW(V + dU + aW) - U^3);$$
$$a(R + S + V)^3 - d^3 RSV = ad^3(VW(V + dU + aW) - U^3);$$
$$dX + 3Y + 3Z = (R + S + V)^3 - 27RSV.$$

The first identity implies that ι' is a rational map from E to H. The neutral element $(0 : 1 : 0)$ of E maps to the neutral element $(0 : -1 : 1)$ of H, so ι' is an isogeny from E to H. The remaining identities imply that ι' is defined everywhere on E. Indeed, if $(X, Y, Z) = (0, 0, 0)$ then $a(R+S+V)^3 - d^3RSV = 0$ and $(R+S+V)^3 - 27RSV = dX + 3Y + 3Z = 0$ so $(d^3 - 27a)RSV = 0$, implying $R = 0$ or $S = 0$ or $V = 0$. If $R = 0$ then $0 = Y = SV^2$ so $S = 0$ or $V = 0$; if $S = 0$ then $0 = Y = VR^2$ so $V = 0$ or $R = 0$; if $V = 0$ then $0 = Y = RS^2$ so $R = 0$ or $S = 0$. In all cases at least two of R, S, V are 0, but also $R+S+V = 0$, so all three are 0. Thus implies $W = 0$, $Q = 0$, and $U = 0$, contradicting $(U : V : W) \in \mathbf{P}^2$.

What remains is to prove that $\iota' \circ \iota$ is tripling on H. Take a point $(X_1 : Y_1 : Z_1)$ on H. Define $(X_2, Y_2, Z_2) = ((Z_1^3 - Y_1^3)X_1, (Y_1^3 - aX_1^3)Z_1, (aX_1^3 - Z_1^3)Y_1)$; then $(X_2 : Y_2 : Z_2) = 2(X_1 : Y_1 : Z_1)$ by Theorem 4.2 and Theorem 4.3. Define (X_3, Y_3, Z_3) and (X_3', Y_3', Z_3') as in Theorem 3.2 and Theorem 4.2 respectively. Define $(U, V, W) = (-X_1Y_1Z_1, Y_1^3, X_1^3)$; then $(U : V : W) = \iota(X_1 : Y_1 : Z_1)$. Define $Q = dU$, $R = aW$, $S = -(V + Q + R)$, $X = (R^3 + S^3 + V^3 - 3RSV)/d$, $Y = RS^2 + SV^2 + VR^2 - 3RSV$, and $Z = RV^2 + SR^2 + VS^2 - 3RSV$; then $\iota'(\iota(X_1 : Y_1 : Z_1)) = (X : Y : Z)$. Write C for the polynomial $aX_1^3 + Y_1^3 + Z_1^3 - dX_1Y_1Z_1$.

Case 1: $X_1 \neq 0$. The identities

$$X_3 = X_1(-X + C(2aX_1^3 + 2Y_1^3 - Z_1^3 - dX_1Y_1Z_1)X_1Y_1Z_1),$$
$$Y_3 = X_1(-Y + C(a^2X_1^6 - adX_1^4Y_1Z_1 - aX_1^3Z_1^3 + 4aX_1^3Y_1^3 - Y_1^6)),$$
$$Z_3 = X_1(-Z + C(-a^2X_1^6 - dX_1Y_1^4Z_1 - Y_1^3Z_1^3 + 4aX_1^3Y_1^3 + Y_1^6))$$

show that $(X_3 : Y_3 : Z_3) = (X : Y : Z)$. In particular, $(X_3, Y_3, Z_3) \neq (0, 0, 0)$, so $3(X_1 : Y_1 : Z_1) = (X_3 : Y_3 : Z_3)$ by Theorem 3.2, so $3(X_1 : Y_1 : Z_1) = (X : Y : Z)$.

Case 2: $Y_1 \neq 0$. The identities

$$X_3' = Y_1(X - C(2aX_1^3 + 2Y_1^3 - Z_1^3 - dX_1Y_1Z_1)X_1Y_1Z_1),$$
$$Y_3' = Y_1(Y - C(a^2X_1^6 - adX_1^4Y_1Z_1 - aX_1^3Z_1^3 + 4aX_1^3Y_1^3 - Y_1^6)),$$
$$Z_3' = Y_1(Z - C(-a^2X_1^6 - dX_1Y_1^4Z_1 - Y_1^3Z_1^3 + 4aX_1^3Y_1^3 + Y_1^6))$$

show that $(X_3' : Y_3' : Z_3') = (X : Y : Z)$. In particular, $(X_3', Y_3', Z_3') \neq (0, 0, 0)$, so $3(X_1 : Y_1 : Z_1) = (X_3' : Y_3' : Z_3')$ by Theorem 4.2, so $3(X_1 : Y_1 : Z_1) = (X : Y : Z)$.

At least one of X_1 and Y_1 must be nonzero, so at least one of these cases applies. $\square$

6   Cost of Additions, Doublings, and Triplings

This section analyzes the cost of various formulas for arithmetic on twisted Hessian curves. Input and output points are assumed to be represented in projective coordinates $(X : Y : Z)$.

All of the formulas in this section are complete when a is not a cube. In particular, the addition formulas use the rotated addition law rather than the standard addition law. Switching back to the standard addition law is a straightforward rotation exercise and saves $1\mathbf{M}_a$ in addition, at the expense of completeness. If incomplete formulas are acceptable then one can achieve the same savings in the rotated addition law by taking $a = 1$, although this would force somewhat larger constants in doublings and triplings.

Addition. The following formulas compute addition $(X_3 : Y_3 : Z_3) = (X_1 : Y_1 : Z_1) + (X_2 : Y_2 : Z_2)$ in $12\mathbf{M} + 1\mathbf{M}_a$.

$$A = X_1 \cdot Z_2;\ B = Z_1 \cdot Z_2;\ C = Y_1 \cdot X_2;\ D = Y_1 \cdot Y_2;\ E = Z_1 \cdot Y_2;$$
$$F = aX_1 \cdot X_2;\ X_3 = A \cdot B - C \cdot D;\ Y_3 = D \cdot E - F \cdot A;\ Z_3 = F \cdot C - B \cdot E.$$

Mixed addition, computing $(X_3 : Y_3 : Z_3) = (X_1 : Y_1 : Z_1) + (X_2 : Y_2 : 1)$, takes only $10\mathbf{M} + 1\mathbf{M}_a$: eliminate the two multiplications by Z_2 in the above formulas.

In followup work, Hisil has saved $1\mathbf{M}$ as follows, achieving $11\mathbf{M} + 1\mathbf{M}_a$ for addition (and $9\mathbf{M} + 1\mathbf{M}_a$ for mixed addition), assuming $2 \neq 0$ in the field:

$$A = X_1 \cdot Z_2;\ B = Z_1 \cdot Z_2;\ C = Y_1 \cdot X_2;\ D = Y_1 \cdot Y_2;\ E = Z_1 \cdot Y_2;$$
$$F = aX_1 \cdot X_2;\ G = (D + B) \cdot (A - C);\ H = (D - B) \cdot (A + C);$$
$$J = (D + F) \cdot (A - E);\ K = (D - F) \cdot (A + E);$$
$$X_3 = G - H;\ Y_3 = K - J;\ Z_3 = J + K - G - H - 2(B - F) \cdot (C + E).$$

Theorem 4.5 shows that all of these formulas are complete if a is not a cube. In particular, these formulas can be used to compute doublings. This is one way to reduce side-channel leakage in twisted Hessian coordinates. However, faster doublings are feasible as we show below.

Doubling. Each of the following formulas is a complete doubling formula, i.e., correctly doubles all curve points, whether or not a is a cube. To see this, substitute $(X_2, Y_2, Z_2) = (X_1, Y_1, Z_1)$ in Theorem 4.2, and observe that the resulting vector (X_3', Y_3', Z_3') is, up to sign (and scaling by a power of 2 for the formulas labeled as requiring $2 \neq 0$), the same as the vector (X_3, Y_3, Z_3) computed here. Recall that Theorem 4.2 is always usable for doublings by Theorem 4.4.

The first doubling formulas use $6\mathbf{M} + 3\mathbf{S} + 1\mathbf{M}_a$. Note that the formulas compute the squares of all input values as a step towards cubing them. They are not used individually, so the formulas would benefit from dedicated cubings.

$$A = X_1^2;\ B = Y_1^2;\ C = Z_1^2;\ D = A \cdot X_1;\ E = B \cdot Y_1;\ F = C \cdot Z_1;\ G = aD;$$
$$X_3 = X_1 \cdot (E - F);\ Y_3 = Z_1 \cdot (G - E);\ Z_3 = Y_1 \cdot (F - G).$$

The second doubling formulas require $2 \neq 0$ in the field and require the field to contain an element i with $i^2 = -1$. These formulas use $8\mathbf{M} + 1\mathbf{M}_i + 1\mathbf{M}_d$.

$$J = iZ_1;\ A = (Y_1 - J) \cdot (Y_1 + J);\ P = Y_1 \cdot Z_1;$$
$$C = (A - P) \cdot (Y_1 + Z_1);\ D = (A + P) \cdot (Z_1 - Y_1);\ E = 3C - 2dX_1 \cdot P;$$
$$X_3 = -2X_1 \cdot D;\ Y_3 = (D - E) \cdot Z_1;\ Z_3 = (D + E) \cdot Y_1.$$

The third doubling formulas eliminate the multiplication by i, further improve cost to $7\mathbf{M} + 1\mathbf{S} + 1\mathbf{M}_d$, and eliminate the requirement for the field to contain i, although they still require $2 \neq 0$ in the field.

$$P = Y_1 \cdot Z_1; \ Q = 2P; \ R = Y_1 + Z_1;$$
$$A = R^2 - P; \ C = (A - Q) \cdot R; \ D = A \cdot (Z_1 - Y_1); \ E = 3C - dX_1 \cdot Q;$$
$$X_3 = -2X_1 \cdot D; \ Y_3 = (D - E) \cdot Z_1; \ Z_3 = (D + E) \cdot Y_1.$$

The fourth doubling formulas, also requiring $2 \neq 0$ in the field, improve cost even more, to $6\mathbf{M} + 2\mathbf{S} + 1\mathbf{M}_d$.

$$R = Y_1 + Z_1; \ S = Y_1 - Z_1; \ T = R^2; \ U = S^2; \ V = T + 3U; \ W = 3T + U;$$
$$C = R \cdot V; \ D = S \cdot W; \ E = 3C - dX_1 \cdot (W - V);$$
$$X_3 = -2X_1 \cdot D; \ Y_3 = (D + E) \cdot Z_1; \ Z_3 = (D - E) \cdot Y_1.$$

In most situations the fastest approach is to choose small d and use the fourth doubling formulas. Characteristic 3 typically has fast cubings, making the first doubling formulas faster. Characteristic 2 allows only the first doubling formulas.

Tripling. Assume that $d \neq 0$. The 3-isogenies in Theorem 5.4 then lead to efficient tripling formulas that compute $(X_3 : Y_3 : Z_3) = 3(X_1 : Y_1 : Z_1)$ significantly faster than a doubling followed by an addition. This is useful in, e.g., scalar multiplications using double-base chains; see Sect. 7.

Specifically, define

$$U = -X_1 Y_1 Z_1; \qquad V = Y_1^3; \qquad W = X_1^3; \qquad Q = dU = -dX_1 Y_1 Z_1;$$
$$R = aW = aX_1^3; \qquad S = -(V + Q + R) = -(Y_1^3 - dX_1 Y_1 Z_1 + aX_1^3) = Z_1^3;$$
$$X_3 = (R^3 + S^3 + V^3 - 3RSV)/d;$$
$$Y_3 = RS^2 + SV^2 + VR^2 - 3RSV; \qquad Z_3 = RV^2 + SR^2 + VS^2 - 3RSV.$$

Then the isogenies ι and ι' in Theorem 5.4 satisfy $\iota(X_1 : Y_1 : Z_1) = (U : V : W)$ and $3(X_1 : Y_1 : Z_1) = \iota'(U : V : W) = (X_3 : Y_3 : Z_3)$. All tripling formulas that we consider begin by computing $R = aX_1^3$, $V = Y_1^3$, and $S = Z_1^3$ with three cubings (normally $3\mathbf{M} + 3\mathbf{S}$, except for fields supporting faster cubing) and then compute X_3, Y_3, Z_3 from R, S, V. Note that computing S as Z_1^3 is faster than computing U as $-X_1 Y_1 Z_1$, and there does not seem to be any benefit in computing U or $Q = dU$.

The following straightforward formulas compute X_3, Y_3, Z_3 from R, S, V in $5\mathbf{M} + 3\mathbf{S} + \mathbf{M}_{1/d}$, assuming $2 \neq 0$ in the field, where $\mathbf{M}_{1/d}$ means the cost of multiplying by the curve parameter $1/d$:

$$A = (R - V)^2; \ B = (R - S)^2; \ C = (V - S)^2; \ D = A + C; \ E = A + B;$$
$$X_3 = (1/d)(R + V + S) \cdot (B + D); \ Y_3 = 2RC - V \cdot (C - E);$$
$$Z_3 = 2VB - R \cdot (B - D).$$

The total cost for tripling this way is $8\mathbf{M} + 6\mathbf{S} + \mathbf{M}_a + \mathbf{M}_{1/d}$. For the case $a = 1$ the same cost had been achieved by Hisil, Carter, and Dawson in [30]. One can of course scale X_3, Y_3, Z_3 by a factor of d, replacing $\mathbf{M}_{1/d}$ with $2\mathbf{M}_d$.

Here is a technique to produce faster formulas, building upon the structure used in the proofs in Sect. 5. Start with the polynomial identity

$$(\alpha R + \beta S + \gamma V)(\alpha S + \beta V + \gamma R)(\alpha V + \beta R + \gamma S)$$
$$= \alpha\beta\gamma dX_3 + (\alpha\beta^2 + \beta\gamma^2 + \gamma\alpha^2)Y_3 + (\beta\alpha^2 + \gamma\beta^2 + \alpha\gamma^2)Z_3 + (\alpha + \beta + \gamma)^3 RSV.$$

Specialize this identity to three choices of constants (α, β, γ), and use the curve equation $d^3 RSV = a(R + S + V)^3$ appearing in the proof of Theorem 5.4, to obtain four linear equations for dX_3, Y_3, Z_3, RSV. If the constants are sensibly chosen then the equations are independent.

We now give three examples of this technique. First: Taking $(\alpha, \beta, \gamma) = (1, 1, 1)$ gives $(R + S + V)^3 = dX_3 + 3Y_3 + 3Z_3 + 27RSV$, as already used in the proof of Theorem 5.4. Taking $(\alpha, \beta, \gamma) = (1, -1, 0)$ gives $(R - S)(S - V)(V - R) = Y_3 - Z_3$, and taking $(\alpha, \beta, \gamma) = (1, 1, 0)$ gives $(R + S)(S + V)(V + R) = Y_3 + Z_3 + 8RSV$. These equations, together with $a(R + S + V)^3 = d^3 RSV$, are linearly independent except in characteristic 2: we have

$$dX_3 = (1 - 3a/d^3)(R+S+V)^3 - 3(R+S)(S+V)(V+R),$$
$$2Y_3 = (R+S)(S+V)(V+R) + (R-S)(S-V)(V-R) - 8(a/d^3)(R+S+V)^3,$$
$$2Z_3 = (R+S)(S+V)(V+R) - (R-S)(S-V)(V-R) - 8(a/d^3)(R+S+V)^3.$$

Computing $(2X_3, 2Y_3, 2Z_3)$ from these formulas takes one cubing for $(R+S+V)^3$, $2\mathbf{M}$ for $(R+S)(S+V)(V+R)$, $2\mathbf{M}$ for $(R-S)(S-V)(V-R)$, one multiplication by a/d^3 (or, alternatively, a multiplication of $R + S + V$ by $1/d$ and a subsequent multiplication by a), one multiplication by $1/d$, and several additions, for a total cost of $8\mathbf{M} + 4\mathbf{S} + \mathbf{M}_a + \mathbf{M}_{a/d^3} + \mathbf{M}_{1/d}$; i.e., $8\mathbf{M} + 4\mathbf{S}$ when both a and $1/d$ are chosen to be small. As noted in the introduction, this result is due to Kohel [39], as a followup to our preliminary announcements of results in this paper.

Second example: For characteristic 2 one must take at least one vector (α, β, γ) outside $\mathbf{F}_2^3$, creating more multiplications by constants. The overall cost is still $8\mathbf{M} + 4\mathbf{S}$ if all constants are chosen to be small and $(1, 1, 1)$ is used as an (α, β, γ).

Third example: Assume that the base field k is $\mathbf{F}_p[\omega]/(\omega^2 + \omega + 1)$ where $p \in 2 + 3\mathbf{Z}$, or more generally has any primitive cube root ω of 1 for which multiplications by ω are fast. Now take the vectors $(\alpha, \beta, \gamma) = (1, \omega^i, \omega^{2i})$ and observe that the left side of the above identity is always a cube:

$$(R + \omega S + \omega^2 V)^3 = dX_3 + 3\omega^2 Y_3 + 3\omega Z_3,$$
$$(R + \omega^2 S + \omega V)^3 = dX_3 + 3\omega Y_3 + 3\omega^2 Z_3.$$

These equations and $(1 - 27a/d^3)(R + S + V)^3 = dX_3 + 3Y_3 + 3Z_3$ are linearly independent; the matrix of coefficients of $dX_3, 3Y_3, 3Z_3$ is a Fourier matrix. We apply the inverse Fourier matrix to obtain $dX_3, 3Y_3, 3Z_3$ with a few more multiplications by ω. Overall this tripling algorithm costs just 6 cubings, i.e., $6\mathbf{M} + 6\mathbf{S}$.

One way to understand the appearance of the Fourier matrix here is to observe that the polynomial $dX_3 + 3Y_3t + 3Z_3t^2 + 9(1 + t + t^2)RSV$ is the cube of $V + St + Rt^2$ modulo $t^3 - 1$. We compute the cube of $V + St + Rt^2$ separately modulo $t - 1$, $t - \omega$, and $t - \omega^2$.

7   Cost of Scalar Multiplication

This section analyzes the cost of scalar multiplication using twisted Hessian curves. In particular, this section explains how we obtained a cost of just 8.77M per bit for average 256-bit scalars.

Since our new twisted-Hessian formulas provide very fast tripling and reasonably fast doubling, the results of [6] suggest that it will be fastest to represent scalars using $\{2, 3\}$-double-base chains. Scalar multiplication then involves not only doubling and addition but also tripling. A well-known advantage of double-base representations is that the number of additions is smaller than in the binary representation.

We use a newer algorithm to generate double-base chains, shown in Fig. 7.1. This algorithm is an improved version of the basic "tree-based" algorithm proposed and analyzed by Doche and Habsieger in [21].

In the basic algorithm, n is computed recursively from either $(n-1)/(2^{\cdots}3^{\cdots})$ or $(n+1)/(2^{\cdots}3^{\cdots})$, where the exponents of 2 and 3 are chosen to be as large as possible. The algorithm explores the branching tree of possibilities in breadth-first fashion until it reaches $n = 1$. To limit time and memory usage, the algorithm keeps only the smallest B nodes at each level. We chose $B = 200$.

We use an extension to this algorithm mentioned but not analyzed in [21]. The extension uses not just $n - 1$ and $n + 1$, but all $n - c$ where c is in a precomputed set (including both positive and negative values). We include the cost of precomputing this set. We chose 21 different possibilities for the precomputed set, namely the 21 sets listed in [6].

We change the way to add new nodes as follows:

– n has *one* child node $n/2$ if n is divisible by 2;
– otherwise, n has *one* child node $n/3$ if n is divisible by 3;
– otherwise, n has *several* child nodes $n - c$, one for each $c \in S$.

We improve the algorithm by continuing to search the tree until we have found C chains, rather than stopping with the first chain; we then take the lowest-cost chain. We chose $C = 200$.

We further improve the algorithm by taking the lowest-weight B nodes at each level instead of the smallest B nodes at each level; here "weight" takes account not just of smallness but also of the cost of operations used to reach the node. More precisely, we define "weight" as $\text{cost} + 8 \cdot \log_2(n)$.

We ran this algorithm for 10000 random 256-bit scalars, i.e., integers between 2^{255} and $2^{256} - 1$, using as input the costs of twisted Hessian operations. The average cost of the resulting chain was 8.77M per bit.

Input: An integer n, precomputation set S, and bounds B and C
Output: A double-base chain computing n

> **for** each precomputation set S **do**
>   counter $\leftarrow 0$
>   Initialize a tree T with root node n
>   **while** (counter $< C$) **do**
>   **for** each leaf node m in T **do**
>   **if** m divisible by 2 **then**
>   Insert child $\leftarrow f_2(m)$ ⊳ $f_2(m) = m/2^{v_2(m)}$
>   **if** $f_2(m)$ equals 1 **then**
>   counter $\leftarrow$ counter $+1$
>   **else if** m divisible by 3 **then**
>   Insert child $\leftarrow f_3(m)$ ⊳ $f_3(m) = m/3^{v_3(m)}$
>   **if** $f_3(m)$ equals 1 **then**
>   counter $\leftarrow$ counter $+1$
>   **else**
>   **for** each element c in precomputation set S **do**
>   **if** $m - c > 0$ **then**
>   Insert child $\leftarrow f(m - c)$
>   **if** $m - c$ equals 1 **then**
>   counter $\leftarrow$ counter $+ 1$
>   Discard all but the B smallest weight leaf nodes
> **return** The smallest cost chain

Fig. 7.1. The algorithm we used to generate double-base chains. "End" statements are implied by indentation, as in Python.

To more precisely assess the advantage of cofactor 3 over cofactor 1, we carried out a larger series of experiments for smaller scalars, comparing the cost of twisted Hessian curves to the cost of short Weierstrass curves $y^2 = x^3 - 3x + a_6$ in Jacobian coordinates. Specifically, for each b from 2 through 16, we constructed double-base chains for all b-bit integers; for each b from 17 through 64, we constructed double-base chains for 1000 randomly chosen b-bit integers. The top of Fig. 7.2 plots pairs (x, y) where x is the cost to multiply by n on a twisted Hessian curve and $x + y$ is the cost to multiply by the same integer n on a Weierstrass curve; i.e., switching from Weierstrass to twisted Hessian saves $y\mathbf{M}$. We reduced the number of dots plotted in this figure to avoid excessive PDF file sizes and display times, but a full plot is similar. Dots along the x-axis represent integers with the same cost for both curve shapes. Different colors are used for different bit-sizes b.

We have generated similar plots for some other pairs of curve shapes. For example, the bottom of Fig. 7.2 shows that Edwards is faster than Hessian for most values of n. In some cases, such as Hessian vs. tripling-oriented Doche–Icart–Kohel curves, the plots are concentrated much more narrowly around a line, since these curve shapes favor similar integers that use many triplings.

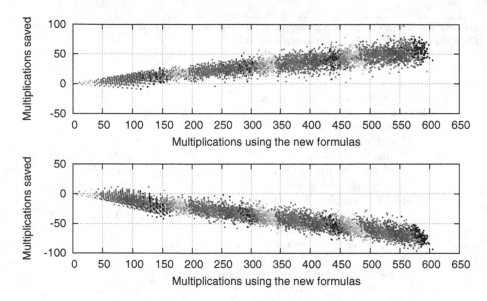

Fig. 7.2. Top: Points (x, y) for 100 randomly sampled b-bit integers n for each $b \in \{2, 3, \ldots, 64\}$. Here $x\mathbf{M}$ are used to compute $P \mapsto nP$ on a twisted Hessian curve in projective coordinates; $(x + y)\mathbf{M}$ are used to compute $P \mapsto nP$ on a Weierstrass curve $y^2 = x^3 - 3x + a_6$ in Jacobian coordinates; and the color is a function of b. Bottom: Similar, but using a twisted Edwards curve rather than a Weierstrass curve.

References

1. Aréne, C., Lange, T., Naehrig, M., Ritzenthaler, C.: Faster computation of the Tate pairing. J. Number Theor. **131**, 842–857 (2011)
2. Aronhold, S.H.: Zur Theorie der homogenen Functionen dritten Grades von drei Variabeln. Crelles J. für die reine und angewandte Mathematik **1850**(39), 140–159 (1850)
3. Benaloh, J. (ed.): CT-RSA 2014. LNCS, vol. 8366. Springer, Heidelberg (2014)
4. Bernstein, D.J.: Complete addition laws for all elliptic curves over finite fields (talk slides) (2009). http://cr.yp.to/talks/2009.07.17/slides.pdf
5. Bernstein, D.J.: Curve25519: new Diffie-Hellman speed records. In: PKC 2006 [52], pp. 207–228 (2006)
6. Bernstein, D.J., Birkner, P., Lange, T., Peters, C.: Optimizing double-base ellipticcurve single-scalar multiplication. In: Indocrypt 2007 [51], pp. 167–182 (2007)
7. Bernstein, D.J., Duif, N., Lange, T., Schwabe, P., Yang, B.-Y.: High-speed high-security signatures. J. Cryptographic Eng. **2**, 77–89 (2012)
8. Bernstein, D.J., Lange, T.: Explicit-formulas database (2007). https://hyperelliptic.org/EFD
9. Bernstein, D.J., Lange, T.: Faster addition and doubling on elliptic curves. In: Asiacrypt 2007 [40], pp. 29–50 (2007)
10. Bernstein, D.J., Lange, T.: Analysis and optimization of elliptic-curve single-scalar multiplication, In: Fq8 [44], pp. 1–19 (2008)
11. Bernstein, D.J., Lange, T.: A complete set of addition laws for incomplete Edwards curves. J. Number Theor. **131**, 858–872 (2011)

12. Bertoni, G., Coron, J.-S. (eds.): CHES 2013. LNCS, vol. 8086, pp. 142–158. Springer, Heidelberg (2013)
13. Billet, O., Joye, M.: The Jacobi model of an elliptic curve and side-channel analysis. In: AAECC 2003 [28], pp. 34–42 (2003)
14. Bosma, W., Lenstra Jr., H.W.: Complete systems of two addition laws for elliptic curves. J. Number Theor. **53**, 229–240 (1995)
15. Brankovic, L., Susilo, W. (eds.): Australasian information security conference (AISC 2009), Wellington, New Zealand, January 2009. In: Conferences in Research and Practice in Information Technology (CRPIT), 1998. Australian Computer Society Inc. (2009)
16. Cayley, A.: On the 34 concomitants of the ternary cubic. Am. J. Math. **4**, 1–15 (1881)
17. Chudnovsky, D.V., Chudnovsky, G.V.: Sequences of numbers generated by addition in formal groups and new primality and factorization tests. Adv. Appl. Math. **7**, 385–434 (1986)
18. Cohen, H., Frey, G. (eds.): Handbook of elliptic and hyperelliptic curve cryptography. CRC Press (2005)
19. Cohen, H., Miyaji, A., Ono, T.: Efficient elliptic curve exponentiation using mixed coordinates. In: Asiacrypt 1998 [48], pp. 51–65 (1998)
20. Costello, C., Hisil, H., Smith, B.: Faster compact Diffie–Hellman: endomorphism-son the x-line. In: Eurocrypt 2014 [45], pp. 183–200 (2014)
21. Doche, C., Habsieger, L.: A tree-based approach for computing double-base chains. In: ACISP 2008 [43], pp. 433–446 (2008)
22. Doche, C., Icart, T., Kohel, D.R.: Efficient scalar multiplication by isogeny decompositions. In: PKC 2006 [52], pp. 191–206 (2006)
23. Doche, C., Lange, T.: Arithmetic of elliptic curves. In: HEHCC [18], pp. 267–302 (2005)
24. Edwards, H.M.: A normal form for elliptic curves. Bull. Am. Math. Soc. **44**, 393–422 (2007)
25. Farashahi, R.R., Joye, M.: Efficient arithmetic on Hessian curves. In: PKC 2010 [46], pp. 243–260 (2010)
26. Farashahi, R.R., Wu, H., Zhao, C.-A.: Efficient Arithmetic on Elliptic Curves over Fields of Characteristic Three. In: SAC 2012 [35], pp. 135–148 (2013)
27. Faz-Hernández, A., Longa, P., Sánchez, A.H.: Efficient and Secure Algorithms for GLV-Based Scalar Multiplication and Their Implementation on GLV-GLS Curves. In: CT-RSA 2014 [3], pp. 1–27 (2014)
28. Fossorier, M.P.C., Høholdt, T., Poli, A. (eds.): AAECC 2003. LNCS, vol. 2643. Springer, Heidelberg (2003)
29. Hesse, O.: Über die Elimination der Variabeln aus drei algebraischen Gleichungen vom zweiten Grade mit zwei Variabeln. J. für die Reine und Angewandte Mathematik **28**, 68–96 (1844)
30. Hisil, H., Carter, G., Dawson, E.: New formulae for efficient elliptic curve arithmetic. In: Indocrypt 2007 [51], pp. 138–151 (2007)
31. Hisil, H., Wong, K.K-H., Carter, G., Dawson, E.: Faster group operations on elliptic curve. In: AISC 2009 [15], pp. 7–19 (2009)
32. Hisil, H.: Elliptic curves, group law, and efficient computation, Ph.D. thesis, Queensland University of Technology (2010)
33. Husemöller, D.: Elliptic Curves. Graduate Texts in Mathematics, vol. 111, 2nd edn. Springer, New York (2003)
34. Joye, M., Quisquater, J.-J.: Hessian Elliptic Curves and Side-Channel Attacks. In: CHES 2001 [37], pp. 402–410 (2001)

35. Knudsen, L.R., Wu, H. (eds.): SAC 2012. LNCS, vol. 7707. Springer, Heidelberg (2013)
36. Koblitz, N.: Algebraic Aspects of Cryptography. Algorithms and Computation in Mathematics, vol. 3. Springer, Heidelberg (1998)
37. Koç, Ç.K., Naccache, D., Paar, C. (eds.): CHES 2001. LNCS, vol. 2162. Springer, Heidelberg (2001)
38. Kohel, D.: Addition law structure of elliptic curves. J. Number Theor. **131**, 894–919 (2011)
39. Kohel, D.: The geometry of efficient arithmetic on elliptic curves. In: Arithmetic, Geometry, Coding Theory and Cryptography, vol. 637. pp. 95–109 (2015)
40. Kurosawa, K. (ed.): ASIACRYPT 2007. LNCS, vol. 4833. Springer, Heidelberg (2007)
41. Liardet, P.-Y., Smart, N.P.: Preventing SPA/DPA in ECC Systems Using the Jacobi Form. In: CHES 2001 [37], pp. 391–401 (2001)
42. Montgomery, P.L.: Speeding the Pollard and elliptic curve methods of factorization. Math. Comput. **48**, 243–264 (1987)
43. Mu, Y., Susilo, W., Seberry, J. (eds.): ACISP 2008. LNCS, vol. 5107. Springer, Heidelberg (2008)
44. Mullen, G.L., Panario, D., Shparlinski, I.E. (eds.): Finite Fields and applications. In: papers from the 8th international conference held in Melbourne, July 9–13, 2007, Contemporary Mathematics, 461, American Mathematical Society (2008)
45. Nguyen, P.Q., Oswald, E. (eds.): EUROCRYPT 2014. LNCS, vol. 8441. Springer, Heidelberg (2014)
46. Nguyen, P.Q., Pointcheval, D. (eds.): PKC 2010. LNCS, vol. 6056. Springer, Heidelberg (2010)
47. National Institute of Standards and Technology: Recommended elliptic curves for federal government use (1999). http://csrc.nist.gov/groups/ST/toolkit/documents/dss/NISTReCur.pdf
48. Ohta, K., Pei, D. (eds.): ASIACRYPT 1998. LNCS, vol. 1514. Springer, Heidelberg (1998)
49. Oliveira, T., López, J., Aranha, D.F., Rodríguez-Henríquez, F.: Lambda Coordinates for Binary Elliptic Curves. In: CHES 2013 [12], pp. 311–330 (2013)
50. Smart, N.P.: The Hessian Form of an Elliptic Curve. In: CHES 2001 [37], pp. 118–125(2001)
51. Srinathan, K., Rangan, C.P., Yung, M. (eds.): INDOCRYPT 2007. LNCS, vol. 4859. Springer, Heidelberg (2007)
52. Yung, M., Dodis, Y., Kiayias, A., Malkin, T. (eds.): PKC 2006. LNCS, vol. 3958. Springer, Heidelberg (2006)

Improved Sieving on Algebraic Curves

Vanessa Vitse[1] and Alexandre Wallet[2,3]($\boxtimes$)

[1] Institut Fourier, UJF-CNRS, UMR 5582, 38402 Saint-martin d'hères, France
`vanessa.vitse@ujf-grenoble.fr`
[2] Sorbonnes Universités, UPMC Univ Paris 06, CNRS, INRIA, LIP6 UMR 7606,
4 Place Jussieu, 75005 Paris, France
[3] Projet POLSYS, INRIA Rocquencourt, 78153 Le Chesnay Cedex, France
`alexandre.wallet@lip6.fr`

Abstract. The best algorithms for discrete logarithms in Jacobians of algebraic curves of small genus are based on index calculus methods coupled with large prime variations. For hyperelliptic curves, relations are obtained by looking for reduced divisors with smooth Mumford representation (Gaudry); for non-hyperelliptic curves it is faster to obtain relations using special linear systems of divisors (Diem, Kochinke). Recently, Sarkar and Singh have proposed a sieving technique, inspired by an earlier work of Joux and Vitse, to speed up the relation search in the hyperelliptic case. We give a new description of this technique, and show that this new formulation applies naturally to the non-hyperelliptic case with or without large prime variations. In particular, we obtain a speed-up by a factor approximately 3 for the relation search in Diem and Kochinke's methods.

Keywords: Discrete logarithm · Index calculus · Algebraic curves · Curve-based cryptography

1 Introduction

Given a commutative group $(G, +)$, and two elements g, h of G, the discrete logarithm problem (DLP) consists in finding, if it exists, an integer x such that $x \cdot g = h$. It is considered as a major computational challenge and much work has been done on the principal families of groups in the last decades. Jacobians of algebraic curves defined over finite fields are of particular interest, largely because of their link with elliptic curves (which, as genus 1 curves, are their own Jacobians). Indeed, transfer attacks as introduced in [5] can reduce certain discrete logarithm instances on elliptic curves to instances on higher genus curves, see for instance the record computations of [7]. Consequently, even though for applications in cryptography only genus 1 (i.e. elliptic curves) or genus 2 curves are currently considered, assessing the exact difficulty of the DLP in higher genus is still very important from both the practical and the number-theoretical points of view.

A popular approach to the discrete logarithm problem is given by the index calculus family of algorithms, that rely heavily on the structures surrounding

© Springer International Publishing Switzerland 2015
K. Lauter and F. Rodríguez-Henríquez (Eds.): LatinCrypt 2015, LNCS 9230, pp. 295–307, 2015.
DOI: 10.1007/978-3-319-22174-8_16

the group. The general picture is always the same. We start by choosing a so-called "factor base" $\mathcal{B}$, i.e. a subset of the elements of the group. Then in a first phase we search for relations involving elements of the factor base, yielding linear equations between their discrete logarithms; we call this first phase "harvesting" throughout this paper. In a second phase we basically solve the linear system of equations given by the relation matrix and deduce the wanted discrete logarithm. This linear algebra phase does not depend on the structure of the group, but its complexity clearly depends on the cardinality N of the factor base, since it is roughly the size of the relation matrix.

It is possible to simplify the linear algebra at the expense of the harvesting phase by decreasing the size of the factor base. This must be done with added care since less elements in the factor base means lower probability of obtaining a relation. In the so-called "large prime" variants[1] (see [2,6,8,11]), we choose or construct a subset $\mathcal{B}_s$ of the factor base $\mathcal{B}$ called the set of "small primes", of size $O(N^\alpha)$ with $0 < \alpha < 1$. The set $\mathcal{B} \backslash \mathcal{B}_s$ is the set of large primes. Then we accept only relations involving elements in $\mathcal{B}_s$ and at most two large primes. This decreases the probability of a relation, but it is shown in [2,6] that finding $O(N)$ relations involving at most two large primes is enough to form $O(N^\alpha)$ relations involving only small primes. The linear algebra phase over $\mathcal{B}_s$ then runs in $O(N^{2\alpha})$ and it remains to choose α to balance both main phases of index calculus. More details are given at the end of Sect. 2.1.

In this work we focus on improving the known harvesting methods, dedicating to Jacobian varieties of algebraic curves defined over finite fields. We emphasize that all the other aspects of the index calculus method (such as the choice of the facto r base, the processing of large prime relations and the linear algebra phase) are not modified. Recently, Sarkar and Singh proposed in [10] to use a sieving technique for harvesting relations in the hyperelliptic case, instead of the standard approach of Gaudry [4] based on smooth reduced divisors. A very similar sieve had actually been used before by Joux and Vitse in [7], but in the different context of curves defined over extension fields and Weil restrictions. It turns out that Sarkar and Singh's sieve has a simpler interpretation, which allows to generalize it to the index calculus introduced by Diem [2] for non-hyperelliptic curves, or more exactly small degree planar curves. In our experiments, the new non-hyperelliptic sieve improves Diem's original method, as well as its development by Diem and Kochinke [3], by a factor approximately 3.

The presentation follows these steps. We begin by the case of hyperelliptic curves, recalling the classical approach of Gaudry based on smoothness check. We present and analyze the sieving variant of Sarkar and Singh before introducing its simpler reformulation. The next section deals with algebraic curves of genus g admitting a plane model of degree $d \le g+1$. Again we start by the classical ideas of using principal divisors associated to equations of lines to generate relations (Diem). We then give the adaptation of our sieve to small degree curves, and compare it to Diem's method. We also briefly present the singularity-based

---

[1] The terminology of index calculus stems from the context of integer factorization. In our setting, "large primes" are arbitrary elements, and involve no notion of size.

technique of Diem and Kochinke and show that our sieve adapts again to this setting. Experiments and timings are reported in the last section.

2   Sieving for Hyperelliptic Curves

2.1   Gaudry's Relation Search

Index calculus on hyperelliptic curves relies on the search of smooth Mumford representations of divisors. Let $\mathcal{H}$ be a genus g imaginary hyperelliptic curve defined over $\mathbb{F}_q$, with equation $y^2 = h(x)$ (so that $\deg h = 2g + 1$) and point at infinity P_∞. For simplicity we assume that we are in the odd characteristic case, but everything can be easily adapted to the characteristic two case. We recall that any element $[D]$ in the Jacobian variety $\mathrm{Jac}_\mathcal{H}(\mathbb{F}_q)$ can be uniquely represented by a couple $[u(x), v(x)]$ of polynomials such that:

- u is monic and $\deg u \leq g$;
- $\deg v < \deg u$;
- $u|(h - v^2)$.

More precisely, this is the Mumford representation of the unique reduced divisor $D = (P_1) + \cdots + (P_{\deg u}) - \deg u\,(P_\infty)$ in the class $[D]$. The roots of u are exactly the x-coordinates of the points $P_i \in \mathcal{H}(\overline{\mathbb{F}_q})$, and v satisfies $v(x_{P_i}) = y_{P_i}$. The Mumford representation is actually defined for every *semi*-reduced divisor, i.e. of the form $(P_1) + \cdots + (P_w) - w(P_\infty)$ with $P_i \neq \imath P_j$ if $i \neq j$, where $\imath$ stands for the hyperelliptic involution $(x, y) \mapsto (x, -y)$. The integer $w = \deg u$ is called the weight of the divisor, which is reduced if and only if $w \leq g$.

With this setting we see that if the polynomial u splits over $\mathbb{F}_q$, then for each i the point P_i is $\mathbb{F}_q$-rational so that the class of $(P_i) - (P_\infty)$ defines an element of $\mathrm{Jac}_\mathcal{H}(\mathbb{F}_q)$. Furthermore, we have $D = \sum_i ((P_i) - (P_\infty))$ (i.e. D is 1-smooth) and the same equality holds when taking linear equivalence classes. Gaudry's algorithm [4] stems from this observation, and its harvesting phase can be summarized as follows.

- The factor base $\mathcal{B}$ is the set $\{(P) - (P_\infty) : P \in \mathcal{H}(\mathbb{F}_q)\}$, or rather a set of representatives of its quotient by the hyperelliptic involution, accounting for the trivial relations $(\imath P) - (P_\infty) \sim -((P) - (P_\infty))$. It contains $\Theta(q)$ elements.
- At each step, we compute (using a semi-random walk) the Mumford representation $[u, v]$ of a reduced divisor $D \sim aD_0 + bD_1$, where D_0 and D_1 are the entries of the DLP challenge.
- If u splits over $\mathbb{F}_q$ as $\prod_i (x - x_i)$ then a relation is found: $aD_0 + bD_1 \sim \sum_i ((P_i) - (P_\infty))$ where $P_i = (x_i, v(x_i))$.

We see that each step of the harvesting phase requires a few operations in $\mathrm{Jac}_\mathcal{H}$ followed by the factorization of u, which is generically a degree g polynomial. The probability that u actually splits over $\mathbb{F}_q$ is about $1/g!$, so we need about $g!$ trials before finding a relation.

A precise analysis of the complexity is given by the author in [4]. However, asymptotically when g is fixed and q grows to $+\infty$, the cost of finding one relation is in $\tilde{O}(1)$, so the complexity of the harvesting phase is in $\tilde{O}(q)$. By contrast, the linear algebra phase costs $\tilde{O}(q^2)$ and dominates the complexity. Balancing the two phases has first been proposed by Harley and improved by Thériault [11], who introduced the large prime variants in this context. Asymptotically, the best result[2] is obtained by Gaudry, Thériault, Thomé and Diem [6] using the double large prime variation with a factor base $\mathcal{B}_s$ of size $\approx q^{1-1/g}$, yielding a complexity of $\tilde{O}(q^{2-2/g})$ for both the harvesting and the linear algebra phase. Giving a detailed description of the double large variant is outside the scope of this paper, but we mention that there are two distinct approaches regarding the construction of the small prime factor base. A first possibility is to define directly $\mathcal{B}_s$ as a subset of $\mathcal{B}$ whose elements satisfy an easy to check condition, so that membership testing and enumeration are very fast; alternately, the set $\mathcal{B}_S$ can be constructed progressively from the first relations in order to simplify the future elimination of large primes (as done in by Laine and Lauter [8] in the non-hyperelliptic case).

2.2 Sarkar and Singh's Sieve

A recent result of Sarkar and Singh [10] proposes a sieving approach to the relation search for hyperelliptic curves. In this method, with the same factor base $\mathcal{B}$ as in Sect. 2.1, we start from a weight g reduced divisor $D = [u, v] = \sum_{i=1}^{g}(P_i) - g(P_\infty)$, usually related to the challenge. We then consider all the weight $g + 1$ semi-reduced divisors $D' = [u', v']$ that are linearly equivalent to $-D$; a relation is obtained each time u' is split (the factor base $\mathcal{B}$ is the same as in the previous version). The set of all the decompositions of $-D$ as

$$-D \sim \sum_{i=1}^{g+1}(Q_i) - (g+1)(P_\infty),$$

i.e. the set of all weight $g+1$ semi-reduced divisors linearly equivalent to $-D$, is in one-to-one correspondence with the set of divisors of functions in the Riemann-Roch space $\mathcal{L}(-D + (g+1)(P_\infty)) = \mathcal{L}(-\sum_{i=1}^{g}(P_i) + (2g+1)(P_\infty))$. This space is equal to $Span(u(x), y - v(x))$ (since functions in this space have poles at P_∞ only, of order at most $2g + 1$, and vanish at the support of D), and thus the decompositions of $-D$ can be parametrized by an element $\lambda \in \mathbb{F}_q$.

We begin with the non-large-prime, non-sieving version of the algorithm. The relation search consists of two main loops, the outer one being simply a semi-random walk iterating through reduced divisors $D = [u, v] \sim aD_0 + bD_1$. The inner loop iterates over the value of the parameter $\lambda \in \mathbb{F}_q$. For each λ, we consider the function $f_\lambda = y - v(x) + \lambda u(x)$ and the corresponding semi-reduced

---

[2] This is only true asymptotically. For actual instances of the DLP many other factors have to be taken into account, and large prime variations are not always appropriate.

divisor $D_\lambda = -D + \text{div}(f_\lambda)$. The Mumford representation $[u_\lambda, v_\lambda]$ of D_λ is given by the formulae

$$\begin{cases} u_\lambda = c\frac{(\lambda u - v)^2 - h}{u} = c(\lambda^2 u - 2\lambda v + \frac{v^2 - h}{u}) \\ v_\lambda = v - \lambda u \mod u_\lambda \end{cases},$$

where $c \in \mathbb{F}_q$ is the constant that makes u_λ monic. We obtain a relation each time D_λ is 1-smooth, i.e. when u_λ is split over $\mathbb{F}_q$; this happens heuristically with probability $1/(g+1)!$.

The main advantage of this relation search is that it admits a sieving version, in the spirit of [7]. The idea is to replace the inner loop in λ by an inner loop in $x \in \mathbb{F}_q$.

For each value of x, we compute the expression

$$S(x, \lambda) = \lambda^2 u(x) - 2\lambda v(x) + \frac{v(x)^2 - h(x)}{u(x)},$$

which becomes a quadratic polynomial in λ, and find the corresponding roots (for simplicity we can skip the values of x for which $u(x) = 0$). There are two distinct roots λ_0 and λ_1 if and only if $h(x)$ is a square in $\mathbb{F}_q$, and those roots are given by:

$$\lambda_0 = \frac{v(x) + h(x)^{1/2}}{u(x)}, \quad \lambda_1 = \frac{v(x) - h(x)^{1/2}}{u(x)}.$$

As explained by the authors, this step is very fast if a table containing a square root of $h(x)$ (if it exists) for each $x \in \mathbb{F}_q$ has been precomputed. We then store the corresponding couples (λ_0, x) and (λ_1, x). At the end of the inner loop, we look for the values of λ that have appeared $g + 1$ times: this means that the corresponding polynomial u_λ has $g + 1$ distinct roots, so that D_λ yields a relation, i.e. a decomposition of $-D$. In practice, we can either store each value of x in an array L of lists indexed by λ; each time a value of λ is obtained as a root of the quadratic expression, we append x to $L[\lambda]$. When $\#L[\lambda] = g + 1$, we directly have the x-coordinates of the points in the support of $\text{div}(f_\lambda) - D$, and a last step is then to compute back the y-coordinates using f_λ. Alternatively, we can simply maintain a counter array $\texttt{Ctr}$ indexed by λ and increment $\texttt{Ctr}[\lambda]$ each time λ is obtained as a root. When this counter reaches $g + 1$, we factorize the corresponding split polynomial u_λ. This variant has the merit of saving memory at the expense of some duplicate computations, but is more interesting when g increases since the proportion of λ's yielding a relation becomes small.

The main speed-up is provided by the fact that at each iteration of the inner loop, we replace the splitting test and the eventual factorization of either the degree g polynomial u (in Gaudry's version) or the degree $g + 1$ polynomial $S(\lambda, x)$ evaluated in λ, by the resolution of the much simpler equation $S(\lambda, x) = 0$, evaluated in x. This comes at the expense of a slightly lower decomposition probability, namely $1/(g+1)!$ instead of $1/g!$, and higher memory requirement.

As already noticed in [7], a second advantage of this sieve is its compatibility with the double large prime variation. Indeed, once the "small prime" factor base $\mathcal{B}_s$ is constructed, it is sufficient to sieve among the values of $x \in \mathbb{F}_q$ corresponding to abscissae of its elements (the full sieving as described above can still be used in the construction steps of $\mathcal{B}_s$ if necessary). Since the cardinality of $\mathcal{B}_s$ is in $\Theta(q^\alpha)$ with $\alpha = 1 - 1/g$, this shortened sieving only costs $\tilde{O}(q^\alpha)$ instead of $\tilde{O}(q)$. We then look for the values of λ that have been obtained at least $g - 1$ times. The corresponding polynomials u_λ have at least $g-1$ roots corresponding to small primes, and it just remains to test if it is indeed split, which happens with heuristic probability $1/2$ (in the case where λ has been obtained exactly $g - 1$ times). Note that we cannot simply scan the array L or $\mathtt{Ctr}$, as it would cost $\tilde{O}(q)$ (even with a very small hidden constant) and defeat our purpose. So additional care must be taken in the implementation in order to recover the interesting values of λ in only $\tilde{O}(q^\alpha)$, for instance using associative arrays, see [10] for details. Although it is not specified in the original paper, one can show that the asymptotic complexity of this variant is still in $\tilde{O}(q^{2-2/g})$ for fixed g, as in the work of Gaudry, Thomé, Thériault and Diem [6], but it is more efficient in practice, and the authors report a significant speed-up.

2.3   Sarkar and Singh's Sieve Revisited

As mentioned above, precomputing a table containing an eventual square root of $h(x)$ for each $x \in \mathbb{F}_q$ can significantly speed up the sieving phase (for a $\tilde{O}(q)$ overhead). But this table is actually nothing more than a list of the rational points of $\mathcal{H}$. Indeed, if y is a square root of $h(x)$ then (x, y) and $(x, -y)$ are exactly the two points in $\mathcal{H}(\mathbb{F}_q)$ with abscissa x, and this precomputation is actually performed when the factor base $\mathcal{B} = \{(P) - (P_\infty) : P \in \mathcal{H}(\mathbb{F}_q)\}$ is enumerated.

This means that we can modify Sarkar and Singh's sieve as follows. Recall that we are looking for functions $f_\lambda = y - v(x) - \lambda u(x)$ such that $-D + \mathrm{div}(f_\lambda)$ is 1-smooth. Instead of sieving over the value of $x \in \mathbb{F}_q$, or in a small subset corresponding to small primes, we directly sieve over $P = (x_P, y_P) \in \mathcal{B}$ or $\mathcal{B}_s$, and the corresponding value of λ is simply recovered as $\frac{y_P - v(x_P)}{u(x_P)}$. We give a pseudo-code of this sieve in Algorithm 1.

This pseudo-code corresponds to the non-large-prime version. Details like the management of the list or associative array L and the update of M are omitted. As mentioned above, a simple counter array $\mathtt{Ctr}$ can be used instead of L, requiring the factorization of $S(x, \lambda)$ for the update of M. If the double large prime variation is used, then the first inner loop iterates only over the elements of the small factor base $\mathcal{B}_s$, and in the second we test if $\#L[\lambda] \geq g - 1$ and subsequently if the remaining factor of $S(x, \lambda)$ splits.

An easy improvement, not included in the pseudo-code for the sake of clarity, is to use the action of the hyperelliptic involution to divide by two the size of the factor base. We can then compute simultaneously the values of λ corresponding to $P = (x_P, y_P)$ and $\iota P = (x_P, -y_P)$. This saves one evaluation of u and of v at x_P, and one inversion of $u(x_P)$, although it is also possible to precompute all inverses. It is clear that this is basically a rewriting of Sarkar and Singh's

---

Algorithm 1. Sieving in the hyperelliptic case

Input: the set of rational points $\mathcal{B}$ of $\mathcal{H}$.
Output: the relation matrix M.
  $n_{rel} = 0$.
  repeat
    Choose a random reduced divisor $D = [u, v] \sim aD_0 + bD_1$.
    Initialize an array of lists L.
    for $P = (x_P, y_P) \in \mathcal{B}$ **do**
      Compute $u(x_P)$ and $v(x_P)$.
      if $u(x_P) \neq 0$ **then**
        Compute $\lambda = (y_P - v(x_P))/u(x_P)$.
        Append P to $L[\lambda]$.
      end if
    end for
    for $\lambda \in \mathbb{F}_q$ **do**
      if $\#(L[\lambda]) = g + 1$ **then**
        Update M.
        Increment n_{rel}.
      end if
    end for
  until $n_{rel} > \#\mathcal{B}$
  return the matrix M.

---

original sieve, so that the performances of both should be similar. However, we will now see that it is easier to adapt to the non-hyperelliptic case.

3 Sieving for Small Degree Curves

3.1 Diem's Relation Search

Gaudry's algorithm can be adapted to non-hyperelliptic curves. Most divisors can still be represented by Mumford coordinates, but computations in the Jacobian variety are not as tractable and checking for 1-smoothness is less obvious. However, all these operations are in $\tilde{O}(1)$ when g is fixed, so that the asymptotic complexity is still in $\tilde{O}(q^{2-2/g})$, albeit with a larger hidden constant than in the hyperelliptic case.

In [2], Diem devised a different harvesting technique for plane curves whose degree is really close to the genus. More precisely, if a curve of genus $g \geq 3$ is general enough (which rules out hyperelliptic curves), we can find in polynomial time a plane model of expected degree $d \leq g + 1$ using a probabilistic algorithm. This means that Diem's algorithm applies to almost all non-hyperelliptic curves, with the (then unexpected) consequence that the DLP is easier on non-hyperelliptic curves than on hyperelliptic ones.

Harvesting is done by considering relations coming from principal divisors corresponding to equations of lines. For any couple of rational points P_1 and P_2 on the curve $\mathcal{C}$, we consider the affine function $f = ax + by + c \in \mathbb{F}_q(\mathcal{C})$ such that the equation of the line L passing through P_1 and P_2 is $f = 0$. The intersection of L and the affine part of $\mathcal{C}$ contains up to d rational points, of which we already know two; determining the remaining $d - 2$ amounts to finding the roots of a degree $d - 2$ polynomial. If there are exactly $d - 2$ other rational intersection points $P_3, \ldots, P_d$ then we obtain a relation of the form

$$\mathrm{div}(f) = (P_1) + \cdots + (P_d) - D_\infty \sim 0,$$

where D_∞ is the divisor corresponding to the intersection of $\mathcal{C}$ with the line at infinity. This happens with probability $1/(d-2)!$, which is better than the $1/g!$ probability for hyperelliptic curves as soon as $d \leq g + 1$.

We can summarize Diem's harvesting technique as follows. The curve $\mathcal{C}$ is defined by the equation $F(x, y) = 0$, and we denote by $\mathcal{C}_0$ its affine, non-singular part. We no longer consider only (classes of) degree 0 divisors, so technically we are working in the full divisor class group and not only its degree 0 part, but in practice it makes no difference.

- The factor base is $\mathcal{B} = \{(P) : P \in \mathcal{C}_0(\mathbb{F}_q)\} \cup \{D_\infty\}$.
- We choose two points $P_1 = (x_1, y_1)$ and $P_2 = (x_2, y_2)$ in $\mathcal{B}$ such that $x_1 \neq x_2$ (for simplicity) and compute $\lambda = (y_2 - y_1)/(x_2 - x_1)$ and $\mu = y_1 - \lambda x_1$.
- We test if the degree $d - 2$ polynomial $\frac{F(x, \lambda x + \mu)}{(x - x_1)(x - x_2)}$ splits over $\mathbb{F}_q$. If it is the case, we compute its roots $x_3, \ldots, x_d$ and the associated y-coordinates $y_3 = \lambda x_3 + \mu, \ldots$, and we store the relation

$$(P_1) + (P_2) + (P_3) + \cdots + (P_d) - D_\infty \sim 0$$

where $P_i = (x_i, y_i)$, provided these points are non-singular.
- We go back to the second step until enough relations are found.

Note that a descent phase is needed to express the entries of the DLP challenge in terms of elements of the factor base, see [2].

The whole routine is particularly well-suited to a two large prime variation (see also [8] for another version differing mainly on the construction of the factor base and the large prime graph). Instead of selecting two points in $\mathcal{B}$, we pick them in the small factor base $\mathcal{B}_s$ and keep only relations involving at most two large primes. The main advantage (as compared to the hyperelliptic case) is that we ensure in this way that each potential relation contains already two small primes; this greatly increases the probability of finding relations with only two large primes. In particular if $d = 4$, every relation found by the above method automatically satisfies the two large prime condition.

A precise complexity analysis is given in [2] and, if $\mathcal{C}$ admits a plane model of degree $g + 1$, gives an asymptotic running time of $\tilde{O}(q^{2-2/(d-2)})$, for a small factor base $\mathcal{B}_s$ of size $\Theta(q^{1-1/(d-2)})$. If $d = g + 1$ this is $\tilde{O}(q^{2-2/(g-1)})$, which improves over the $\tilde{O}(q^{2-2/g})$ complexity of the hyperelliptic case. Note however that the size of the small factor base is such that in order to find enough relations, almost all lines going through pairs of points of $\mathcal{B}_s$ have to be considered. This is troublesome because each line, and thus each relation, can be obtained several times, namely $n(n-1)/2$ times if it contains n small factor base points. This is not really an issue if $d = 4$, but for higher d some extra care has to be taken in order to prevent duplicate relations.

3.2   The Sieving Technique

We can easily adapt our sieving formulation to Diem's setting. The factor base remains the same set of points. Basically, in a first loop we iterate over $P_1 = (x_1, y_1) \in \mathcal{C}_0(\mathbb{F}_q)$; the equation of a non-vertical line passing through P_1 is $(y - y_1) - \lambda(x - x_1) = 0$. The task is now to find the values of λ such that the line has d rational points of intersection with $\mathcal{C}$ without checking for smoothness. For this we then loop in $P_2 = (x_2, y_2) \in \mathcal{C}_0(\mathbb{F}_q)$ and compute the corresponding $\lambda = (y_2 - y_1)/(x_2 - x_1)$. But instead of looking for the intersection of the line with $\mathcal{C}$, we just append P_2 to the list $L[\lambda]$, where L is an array of lists; alternatively, we can simply increment a counter $\mathtt{Ctr}[\lambda]$. If this counter reaches $d - 1$, or if $L[\lambda]$ contains $d - 1$ elements, we know that the line contains enough points and yields a relation. This is made precise in the pseudo-code of Algorithm 2.

Algorithm 2 . Sieving for small degree curves

Input: the list of rational non-singular affine points $\mathcal{B} = \mathcal{C}_0(\mathbb{F}_q)$.
Output: the relation matrix M.

  $n_{rel} = 0$.
  for $i = 1$ to $\#\mathcal{B}$ **do**
    $(x_1, y_1) \leftarrow \mathcal{B}[i]$
    Initialize an array of lists L.
    for $j = i + 1$ to $\#\mathcal{B}$ **do**
      $(x_2, y_2) \leftarrow \mathcal{B}[j]$.
      if $x_2 \neq x_1$ **then**
        Compute $\lambda = (y_2 - y_1)/(x_2 - x_1)$.
        Append (x_2, y_2) to $L[\lambda]$.
      end if
    end for
    for $\lambda \in \mathbb{F}_q$ **do**
      if $\#L[\lambda] = d - 1$ **then**
        Update M.
        Increment n_{rel}.
      end if
      if $n_{rel} > \#\mathcal{B}$ **then**
        return the matrix M.
      end if
    end for
  end for

Note that in the inner loop we do not iterate over the elements of $\mathcal{B}$ that have already been considered in the outer loop. Indeed, after an iteration of the outer loop all the lines passing through the given point $P_1 = \mathcal{B}[i]$ have been surveyed, so there is no reason to scan this point again. In this way no line can be considered twice, and we avoid completely having to check for duplicate relations.

In Diem's version, each step requires the computation and factorization of $\frac{F(x,\lambda x+\mu)}{(x-x_1)(x-x_2)}$. The probability of finding a relation is $1/(d-2)!$, so that after q steps about $q/(d-2)!$ relations are harvested. By comparison, in our sieving each step requires a single division (or multiplication if the inverses are pretabulated). The inner loop ends after about $\#\mathcal{B} \approx q$ steps, and yields approximately $q/(d-1)!$ relations: all the lines through $P_1 = (x_1, y_1)$ have been explored, and contain $d-1$ other points with probability $1/(d-1)!$. Thus we need $d-1$ times as many steps to obtain the same number of relations, but each step is much simpler, and the experiments of the next section confirm the important speed-up.

This sieve can be adapted straightforwardly to the double large prime variation : we just have to restrict both loops to the small factor base $\mathcal{B}_s$ (once it is constructed, if the version of [8] is followed), then we recover the values of λ such that $\#L[\lambda] \geq d-3$. When $\#L[\lambda] = d-3$, we still have to check if the remaining two points on the line are rational, which amounts to factorizing a degree 2 polynomial. If $d = 4$, in Diem's version there are at most two remaining points on any line anyway; our new sieve is thus basically equivalent and does not provide a significant speed-up when using double large primes. However as soon as $d \geq 5$ it outperforms Diem's version, but the asymptotic complexity remains in $\tilde{O}(q^{2-2/(d-2)})$.

3.3   Sieving with Singularities

An article of Diem and Kochinke [3] tries to improve on the asymptotic complexity of the above method. The basic idea is to consider singular small degree plane models, and use a singular point as one of the points defining the lines cutting out the curve. Indeed, a singular point appears with a multiplicity greater than one in any line passing through it, so that there are fewer remaining points of intersection with $\mathcal{C}$, and the degree of the polynomial to test for smoothness is less than when two regular points are used. Unfortunately in general there are not enough singular points on a given planar curve to obtain sufficiently many relations. Thus an important part of Diem and Kochinke's work is to find a way to compute new singular plane models of degree $d \leq g + 1$ for a given genus g curve, but this is outside of the scope of the present article; furthermore, the computation of the maps between the different models is not asymptotically relevant. Using Brill-Noether theory and considerations on special linear systems, they show that this method works for "general enough" non-hyperelliptic curves, of genus $g \geq 5$.

So we assume that we are given a degree d curve $\mathcal{C}$, of equation $F(x,y) = 0$, with a rational singular point P_1 of multiplicity $m \geq 2$ (in most cases $m = 2$). The factor base is given by the rational points of the desingularization $\tilde{\mathcal{C}}$ of $\mathcal{C}$, i.e. $\mathcal{B} = \{(P) : P \in \tilde{\mathcal{C}}(\mathbb{F}_q)\}$. In the original version, for each other point P_2 in $\mathcal{B}$ or in the small factor base $\mathcal{B}_s$, the intersection of $\mathcal{C}$ with the line passing through P_1 and P_2 is computed as before. This amounts to finding the roots of the polynomial

$$\frac{F(x, \lambda x + \mu)}{(x - x_1)^m (x - x_2)},$$

which has degree $d - m - 1$. If it splits, which happens with probability $\approx 1/(d - m - 1)!$, we compute the intersection points $P_3, \ldots, P_{d-m-1}$ and obtain a relation that we can write as

$$D \sim (P_2) + (P_3) + \cdots + (P_{d-m-1}),$$

where D involves the singularity and the points at infinity. In the double large prime variation we keep this relation only if it involves no more than two large primes. To get rid of the divisor D on the left-hand side we would like to subtract one such relation from all the other ones. But in order to do this (using large primes) we need one relation involving only small primes ; if it does not exist a solution is then to add some points to $\mathcal{B}_s$. Since there are less points on the right-hand side than in Diem's first algorithm, the probability of finding a relation increases, and one can show that the overall complexity becomes $\tilde{O}(q^{2-2/(g-2)})$. Note that here again, some care must be taken to avoid duplicate relations, and in particular not all points P_2 but only a fraction of the factor base should be considered.

Now it is clear that our sieve can be naturally adapted to this new setting. Indeed, we can keep the inner loops of Algorithm 2 ; the point (x_1, y_1) is now the singular point P_1, and we look for the values of λ that have been obtained $d - m$ times, or $d - m - 2$ times in the double large prime variation. Once again, this replaces the factorization of a degree $d - m - 1$ polynomial by a single division, and avoids checking for duplicate relations.

4   Experiments

We have experimented the harvesting techniques presented in this article for several curves of different genera, defined over different finite fields. All computations have been done using the computer algebra system Magma [1] on an AMD Opteron™ 6176 SE@2.3 GHz processor. We only implemented the non-large-prime version of the algorithms, the main reason being that we wanted the tests to be as simple as possible[3]. The curves have been generated with the command RandomCurveByGenus, which always returned a degree g curve (instead of $g + 1$) for $g \geq 6$; for this reason the results in genus 6 are very close to those in genus 5 and we did not report them. For the non-sieving versions, we used associative arrays and sets to automate the check for duplicate relations, but this is more and more costly as the number of relations grows.

---

[3] More fundamentally, large prime variations are interesting for the asymptotic complexity analysis, but are not always well-suited in practice ; other methods such as the Gaussian structured elimination [9] can be more efficient.

We give in Table 1 the comparison between Diem's method and our sieve; the values are the timings in seconds (on an Intel© Core i5@2.00 Ghz processor) to obtain $p \approx \#\mathcal{F}$ relations, averaged over several curves.

Table 1. Comparisons of the new sieve with Diem's classical method

p		78137	177167	823547	1594331
Genus 3, degree 4	Diem	11.57	27.54	135.1	266.1
	Diem + sieving	3.65	9.38	46.96	94.60
	Ratio	3.16	2.95	2.88	2.81
Genus 4, degree 5	Diem	51.85	122.4	595.8	1174
	Diem + sieving	15.58	40.01	195.1	387.6
	Ratio	3.33	3.06	3.05	3.03
Genus 5, degree 6	Diem	229.4	535.8	2581	5062
	Diem + sieving	75.66	199.0	969.3	1909
	Ratio	3.03	2.69	2.66	2.65
Genus 7, degree 7	Diem	1382	3173	14990	29280
	Diem + sieving	458.5	1199	5859	11510
	Ratio	3.02	2.65	2.56	2.54

In Table 2 we give timings comparing the new sieve with Diem and Kochinke's method. We did not implement the change of plane models; instead, we simply chose random curves possessing rational singular points, and used one of them as the base point for the relation search. In the sieving version all the relations involving lines passing through the singularity were computed, whereas in the non-sieving case we only iterated through half of the basis, as suggested in [3]. For this reason the values correspond to the timings in seconds needed to obtain 1000 relations, again averaged over several curves.

Table 2. Comparisons of the new sieve with Diem and Kochinke's method

p		78137	177167	823547	1594331
Genus 5, degree 6	Diem and Kochinke	1.58	1.60	1.69	1.76
	DK + sieving	0.43	0.45	0.52	0.61
	Ratio	3.67	3.60	3.23	2.90
Genus 7, degree 7	Diem &Kochinke	8.59	8.68	8.97	9.20
	DK + sieving	1.21	1.25	1.56	1.93
	Ratio	7.13	6.96	5.74	4.77

5   Conclusion

We have shown in this work that a reformulation of Sarkar and Singh's sieve [10], namely sieving over points instead of x-coordinates, gives a simpler presentation of the harvesting phase of the index calculus algorithm on hyperelliptic curves. More importantly, it can be naturally adapted to Diem and Kochinke's index calculus for non-hyperelliptic curves [2,3]. Our experiments show that the new sieve clearly outperforms the relation search of the other methods in all circumstances and should always be preferred.

Acknowlegdements. We would like to thank the anonymous referees for their useful comments during the elaboration of the article.

References

1. Bosma, W., Cannon, J., Playoust, C.: The Magma algebra system. I. The user language. J. Symbolic Comput. **24**(3–4), 235–265 (1997). Computational algebra and number theory (London, 1993)
2. Diem, C.: An index calculus algorithm for plane curves of small degree. In: Hess, F., Pauli, S., Pohst, M. (eds.) ANTS 2006. LNCS, vol. 4076, pp. 543–557. Springer, Heidelberg (2006)
3. Diem, C., Kochinke, S.: Computing discrete logarithms with special linear systems (2013). http://www.math.uni-leipzig.de/diem/preprints/dlp-linear-systems.pdf
4. Gaudry, P.: An algorithm for solving the discrete log problem on hyperelliptic curves. In: Preneel, B. (ed.) EUROCRYPT 2000. LNCS, vol. 1807, pp. 19–34. Springer, Heidelberg (2000)
5. Gaudry, P., Hess, F., Smart, N.P.: Constructive and destructive facets of Weil descent on elliptic curves. J. Cryptol. **15**(1), 19–46 (2002)
6. Gaudry, P., Thomé, E., Thériault, N., Diem, C.: A double large prime variation for small genus hyperelliptic index calculus. Math. Comput. **76**(257), 475–492 (2007)
7. Joux, A., Vitse, V.: Cover and decomposition index calculus on elliptic curves made practical. In: Pointcheval, D., Johansson, T. (eds.) EUROCRYPT 2012. LNCS, vol. 7237, pp. 9–26. Springer, Heidelberg (2012)
8. Laine, K., Lauter, K.: Time-memory trade-offs for index calculus in genus 3. J. Math. Cryptol. **9**(2), 95–114 (2015)
9. LaMacchia, B.A., Odlyzko, A.M.: Computation of discrete logarithms in prime fields. Des. Codes Crypt. **1**(1), 47–62 (1991)
10. Sarkar, P., Singh, S.: A new method for decomposition in the Jacobian of small genus hyperelliptic curves. Cryptology ePrint Archive, Report 2014/815 (2014)
11. Thériault, N.: Index calculus attack for hyperelliptic curves of small genus. In: Laih, C.-S. (ed.) ASIACRYPT 2003. LNCS, vol. 2894, pp. 75–92. Springer, Heidelberg (2003)

Attacking a Binary GLS Elliptic Curve with Magma

Jesús-Javier Chi and Thomaz Oliveira[✉]

Computer Science Department, CINVESTAV-IPN, Mexico City, Mexico
thomaz.figueiredo@gmail.com

Abstract. In this paper we present a complete Magma implementation for solving the discrete logarithm problem (DLP) on a binary GLS curve defined over the field $\mathbb{F}_{2^{62}}$. For this purpose, we constructed a curve vulnerable against the gGHS Weil descent attack and adapted the algorithm proposed by Enge and Gaudry to solve the DLP on the Jacobian of a genus-32 hyperelliptic curve. Furthermore, we describe a mechanism to check whether a randomly selected binary GLS curve is vulnerable against the gGHS attack. Such method works with all curves defined over binary fields and can be applied to each element of the isogeny class.

1 Introduction

In the last two decades, the elliptic curve cryptosystems introduced by Koblitz and Miller [23,28] have been increasingly employed to instantiate public-key standards [30] and protocols [5,36]. The main reason for that is their reduced key size, which accommodate fast and lightweight implementations.

In 2011, Galbraith, Lin and Scott (GLS) [12] introduced efficient computable endomorphisms for a large class of elliptic curves defined over $\mathbb{F}_{p^2}$, where p is a prime number. Later, Hankerson, Karabina and Menezes [19] analyzed the GLS curves defined over characteristic two fields $\mathbb{F}_{2^{2n}}$, with prime n.

Since then, many authors combined the GLS efficient endomorphisms with the Gallant-Lambert-Vanstone decomposition method [14] to present high-performance scalar multiplication software implementations over binary [19,31] and prime [2,8,21,24] fields.

The theoretical security of an elliptic curve is given by the complexity for solving the discrete logarithm problem (DLP) on its group of points. Given an elliptic curve E defined over a field $\mathbb{F}_q$, a generator point $P \in E(\mathbb{F}_q)$ of order r and a challenge point $Q \in \langle P \rangle$, the DLP on E consists in computing the integer $\lambda \in \mathbb{Z}_r$ such that $Q = \lambda P$.

Among the classical methods for solving the DLP on $E(\mathbb{F}_q)$ we can cite the Baby Step Giant Step [3] and Pollard Rho [32] algorithms. Both of them run in time $O(\sqrt{q})$. In 1993, Menezes, Okamoto and Vanstone [27] presented a method that uses the Weil pairing to reduce the DLP on $E(\mathbb{F}_q)$ to the same problem on $\mathbb{F}_{q^m}^*$. In curves such that m is small, the attack is highly effective, because there

© Springer International Publishing Switzerland 2015
K. Lauter and F. Rodríguez-Henríquez (Eds.): LatinCrypt 2015, LNCS 9230, pp. 308–326, 2015.
DOI: 10.1007/978-3-319-22174-8_17

exist quasi-polynomial algorithms for solving the discrete logarithm on finite fields [1]. For binary curves, we also have algorithms based on the index-calculus approach [7] which run in time $O(\sqrt{q})$.

In 2000, Gaudry, Hess and Smart (GHS) [17] applied the ideas in [9,13] to reduce any instance of the DLP on a binary curve $E/\mathbb{F}_{2^{ln}}$ to one on the Jacobian of a hyperelliptic curve defined over a subfield $\mathbb{F}_{2^l}$. Afterwards, Galbraith, Hess and Smart [11] extended the attack by using isogenies. Next, Hess [20] generalized the attack (gGHS) to arbitrary Artin-Schreier extensions.

The analysis of the practical implications of the GHS Weil descent method were made by Menezes and Qu [25] who demonstrated that the attack is infeasible for elliptic curves over $\mathbb{F}_{2^n}$ with primes $n \in [160, \ldots, 600]$ and by Menezes, Teske and Weng [26] who showed that the attack can be applied to curves defined over composite extensions of a binary field. Finally, the authors in [19] analyzed the application of the gGHS attack over GLS binary curves $E/\mathbb{F}_{2^{2n}}$ and concluded that for $n \in [80, \ldots, 256]$, the degree-127 extension is the only one that contains vulnerable curve isogeny classes.

Our Contributions. In this work, we wanted to get a practical perspective of the gGHS Weil descent attack. In order to achieve this goal, we implemented the attack against a binary GLS elliptic curve on the Magma computer algebra system. The implementation included the construction of vulnerable curves, the search for susceptible isogenous curves and the adaptation of the Enge-Gaudry algorithm [6] to solve the discrete logarithm problem on the generated hyperelliptic curve.

Moreover, we proposed a mechanism to check for unsafe binary curve parameters against the gGHS attack. The Magma source code for the algorithms presented in this document is available at http://computacion.cs.cinvestav.mx/~thomaz/gls.tar.gz. Our program can easily adapted for any extension field and is able to execute on single and multi-core architectures.

2   Hyperelliptic Curves

Let $\mathbb{F}_{2^l}$ be a finite field of 2^l elements, for some positive integer l, and let $\mathbb{F}_{2^{ln}}$ be a degree-n extension field of $\mathbb{F}_{2^l}$. A hyperelliptic curve $H/\mathbb{F}_{2^l}$ of genus g is given by the following non-singular equation,

$$H/\mathbb{F}_{2^l} : y^2 + h(x)y = f(x), \tag{1}$$

where $f, h \in \mathbb{F}_{2^l}[x]$, $\deg(f) = 2g + 1$ and $\deg(h) \leq g$. The set of $\mathbb{F}_{2^{ln}}$-rational points on H is $H(\mathbb{F}_{2^{ln}}) = \{(x, y) : x, y \in \mathbb{F}_{2^{ln}}, y^2 + h(x)y = f(x)\} \cup \{\mathcal{O}\}$. The opposite of a point $P = (x, y) \in H(\mathbb{F}_{2^{ln}})$ is denoted as $\overline{P} = (x, y + h(x))$ and $\overline{\mathcal{O}} = \mathcal{O}$.

The group law is not defined over the curve itself but on the Jacobian of H, denoted by $J_H(\mathbb{F}_{2^l})$, which is defined in terms of the set of divisors on H. A divisor is a finite formal sum of points on the curve and the set of all divisors on H yield an abelian group denoted by $\mathrm{Div}(H)$. Let c_i be an integer, then for each divisor $D = \sum_{P_i \in H} c_i(P_i)$, $\deg(D) = \sum c_i$ is the degree of D. The set $\mathrm{Div}^0(H)$ of degree-zero divisors forms a subgroup of $\mathrm{Div}(H)$.

The function field $\mathbb{F}_{2^{ln}}(H)$ of H is the set of rational functions on H. For each non-zero function $\varphi \in \mathbb{F}_{2^{ln}}(H)$, we can associate a divisor $\mathrm{div}(\varphi) = \sum_{P_i \in H} \nu_{P_i}(\varphi)(P_i)$, where $\nu_{P_i}(\varphi)$ is an integer defined as follows:

$$\nu_{P_i}(\varphi) = \begin{cases} \text{the multiplicity of } P_i \text{ with respect to } \varphi & \text{if } \varphi \text{ has a zero at } P_i \\ \text{the negative of the multiplicity of } P_i & \\ \text{with respect to } \varphi & \text{if } \varphi \text{ has a pole at } P_i \\ 0 & \text{otherwise.} \end{cases}$$

A non-zero rational function has only finitely many zeroes and poles. In addition, the number of poles equals the number of zeroes (with multiplicity). Therefore, $\nu_{P_i}(\varphi)$ is equal to zero for almost all P_i and $\mathrm{div}(\varphi)$ is consequently well defined.

The divisor $\mathrm{div}(\varphi)$ is called principal. Given two functions φ_0 and $\varphi_1 \in \mathbb{F}_{2^{ln}}(H)$, the difference of two principal divisors $\mathrm{div}(\varphi_0)$ and $\mathrm{div}(\varphi_1)$ is also a principal divisor, corresponding to the fraction of the two functions. The set $\mathcal{P}(H)$ of principal divisors contains 0 as $\mathrm{div}(1)$ and it is a subgroup of $\mathrm{Div}^0(H)$. The Jacobian of the curve H is then given by the quotient group $J_H(\mathbb{F}_{2^l}) = \mathrm{Div}^0(H)/\mathcal{P}(H)$.

A consequence of the Riemann-Roch theorem [3] is that every element of the Jacobian can be represented by a divisor of the form

$$D = (P_1) + (P_2) \cdots + (P_r) - r(\mathcal{O}) \tag{2}$$

where $P_i \in H$ for $i = 1, \ldots, r$ and $r \leq g$. Furthermore, if $P_i \neq \overline{P_j}$ for all $i \neq j$, then D is called a reduced divisor. A reduced divisor can be uniquely represented by a pair of polynomials $U, V \in \mathbb{F}_{2^l}[x]$ such that (i) $\deg(V) < \deg(U) \leq g$; (ii) U is monic; and (iii) $U | (V^2 + Vh - f)$.

If U and V are two polynomials that satisfy the above conditions, we denote by $\mathrm{div}(U, V)$ the corresponding element of $J_H(\mathbb{F}_{2^l})$. When U is irreducible in $\mathbb{F}_{2^l}[x]$ we say that $\mathrm{div}(U, V)$ is a prime divisor. Let $D = \mathrm{div}(U, V) \in J_H(\mathbb{F}_{2^l})$ and $U = \prod U_i$, where each U_i is an irreducible polynomial in $\mathbb{F}_{2^l}[x]$, and let $V_i = V \bmod U_i$. Then $D_i = \mathrm{div}(U_i, V_i)$ is a prime divisor and $D = \sum D_i$.

3   The Hyperelliptic Curve Discrete Logarithm Problem

Let $q = 2^l$. The discrete logarithm problem on $J_H(\mathbb{F}_q)$ is defined as follows: given $D_1 \in J_H(\mathbb{F}_q)$ of order r and $D_2 \in \langle D_1 \rangle$, find $\lambda \in \mathbb{Z}_r$ such that $D_2 = \lambda D_1$.

Besides the Pollard Rho algorithm, whose expected running time is in $O(\sqrt{\frac{\pi q^g}{2}})$, the methods proposed in the literature for solving the DLP on H are index-calculus-based algorithms:

1. Gaudry in [15] proposed an algorithm whose expected running time is in $O(g^3 q^2 \log^2 q + g^2 g! q \log^2 q)$. If one considers a fixed genus g, the algorithm executes in time $O(q^{(2+\epsilon)})$. In [17], the algorithm is modified to perform in time $O(q^{(\frac{2g}{g+1}+\epsilon)})$.

2. The Enge-Gaudry algorithm [6] has an expected running time of $L_{q^g}[\sqrt{2}]$ when $g/\log q \to \infty$. Here, $L_x[c]$ denotes the expression $e^{((c+o(1))\mathit{sqrt}\log x \sqrt{\log \log x})}$.

3. In [18], Gaudry et al. propose a double large prime variation in order to improve the relation collection phase. For curves with fixed genus $g \geq 3$ the algorithm runs in time $\tilde{O}(q^{2-\frac{2}{g}})$.

4. The approach from Sarkar and Singh [33,34], based on the Nagao's work [29], avoids the requirement of solving a multi-variate system and combines a sieving method proposed by Joux and Vitse [22]. They showed that it is possible to obtain a single relation in about $(2g+3)!$ trials.

4   The Gaudry-Hess-Smart (GHS) Weil Descent Attack

Let $\mathbb{F}_{2^{ln}}$ be a degree-n extension of $\mathbb{F}_{2^l}$ and let E be an elliptic curve defined over $\mathbb{F}_{2^{ln}}$ given by the equation

$$E/\mathbb{F}_{2^{ln}} : y^2 + xy = x^3 + ax^2 + b \qquad a \in \mathbb{F}_{2^{ln}}, \, b \in \mathbb{F}_{2^{ln}}^*. \qquad (3)$$

The GHS Weil descent attack [17] consists of the following steps,

1. The Weil descent:
   (a) Construct the Weil restriction $W_{E/\mathbb{F}_{2^l}}$ of scalars of E, which is an n-dimensional abelian variety over $\mathbb{F}_{2^l}$. One can construct this variety as follows. Let $\beta = \{\phi_1, \dots, \phi_n\}$ be a basis of $\mathbb{F}_{2^{ln}}$ viewed as a vector space over $\mathbb{F}_{2^l}$. Then write a, b, x and y in terms of β,

   $$a = \sum_{i=1}^{n} a_i \phi_i, \quad b = \sum_{i=1}^{n} b_i \phi_i, \quad x = \sum_{i=1}^{n} x_i \phi_i \text{ and } y = \sum_{i=1}^{n} y_i \phi_i. \qquad (4)$$

   Given that β is a linearly independent set, by substituting the equations (4) into the equation (3) we obtain an n-dimensional abelian variety A defined over $\mathbb{F}_{2^l}$. Moreover, the group law of A is similar to the elliptic curve E group law.
   (b) Intersect A with $n-1$ hyperplanes (e.g. $x_1 = x_2 = \cdots = x_n = x$) to obtain a subvariety of A, and then use its linear independence property to obtain a curve H over $\mathbb{F}_{2^l}$.
2. Reduce the DLP on $E(\mathbb{F}_{2^{ln}})$ to the DLP on $J_H(\mathbb{F}_{2^l})$.
3. Solve the DLP on $J_H(\mathbb{F}_{2^l})$.

Let $\gamma \in \mathbb{F}_{2^{ln}}$, $\sigma \colon \mathbb{F}_{2^{ln}} \to \mathbb{F}_{2^{ln}}$ be the Frobenius automorphism defined as $\sigma(\gamma) = \gamma^{2^l}$, $\gamma_i = \sigma^i(\gamma)$ for all $i \in \{0, \dots, n-1\}$ and $m = m(\gamma) = \dim(\mathrm{Span}_{\mathbb{F}_2}\{(1, \gamma_0^{1/2}), \dots, (1, \gamma_{n-1}^{1/2})\})$. Finally, let us assume that either n is odd or $m(b) = n$ or $\mathrm{Tr}_{\mathbb{F}_{2^{ln}}/\mathbb{F}_2}(a) = 0$. 　　　　(5)

Then the GHS Weil descent attack constructs an explicit group homomorphism $\chi\colon E(\mathbb{F}_{2^{ln}}) \to J_H(\mathbb{F}_{2^l})$, where H is a hyperelliptic curve defined over $\mathbb{F}_{2^l}$ of genus $g = 2^{m-1}$ or $g = 2^{m-1} - 1$.

4.1　The Generalized GHS (gGHS) Weil Descent Attack

In [20] Hess generalized the GHS restrictions (5) as follows. Let $\wp(x) = x^2 + x$ and $F = \mathbb{F}_{2^{ln}}(x)$, $\Delta = f\mathbb{F}_2[\sigma] + \wp(F)$ where $f = \gamma_1/x + \gamma_3 + x\gamma_2$ for $\gamma_1, \gamma_2, \gamma_3 \in \mathbb{F}_{2^{ln}}$ such that $\gamma_1\gamma_2 \neq 0$.

Given a polynomial $p = \sum_{i=0}^{d} p_i x^i \in \mathbb{F}_2[x]$ of degree d we write $p(\sigma)(x) = \sum_{i=0}^{d} p_i x^{2^{li}}$. For each element $\gamma \in \mathbb{F}_{2^{ln}}$, $\mathrm{Ord}_\gamma(x)$ is the unique monic polynomial $p \in \mathbb{F}_2[x]$ of least degree such that $p(\sigma)(\gamma) = 0$. Furthermore, we define the m-degree $\mathrm{Ord}_{\gamma_1,\gamma_2,\gamma_3}$ as,

$$\mathrm{Ord}_{\gamma_1,\gamma_2,\gamma_3} = \begin{cases} \mathrm{lcm}(\mathrm{Ord}_{\gamma_1}, \mathrm{Ord}_{\gamma_2}) & \text{if } \mathrm{Tr}_{\mathbb{F}_{2^{ln}}/\mathbb{F}_2}(\gamma_3) = 0 \\ \mathrm{lcm}(\mathrm{Ord}_{\gamma_1}, \mathrm{Ord}_{\gamma_2}, x+1) & \text{otherwise.} \end{cases}$$

Then $\Delta/\wp(F) \cong \mathbb{F}_2[x]/\mathrm{Ord}_{\gamma_1,\gamma_2,\gamma_3}$ and the Frobenius automorphism σ of F with respect to $\mathbb{F}_{2^{ln}}$ extends to a Frobenius automorphism of a function field $C = F(\wp^{-1}(\Delta))$ with respect to $\mathbb{F}_{2^{ln}}$ if and only if,

either $\mathrm{Tr}_{\mathbb{F}_{2^{ln}}/\mathbb{F}_2}(\gamma_3) = 0$ or $\mathrm{Tr}_{\mathbb{F}_{2^{ln}}/\mathbb{F}_{2^l}}(\gamma_1) \neq 0$ or $\mathrm{Tr}_{\mathbb{F}_{2^{ln}}/\mathbb{F}_{2^l}}(\gamma_2) \neq 0$. 　　(6)

In addition, the genus of C is given by

$$g_C = 2^m - 2^{m-\deg(\mathrm{Ord}_{\gamma_1})} - 2^{m-\deg(\mathrm{Ord}_{\gamma_2})} + 1$$

and there exists a hyperelliptic curve H with genus g_C that can be related to an elliptic curve $E/\mathbb{F}_{2^{ln}} : y^2 + xy = x^3 + ax^2 + b$ with $a = \gamma_3$ and $b = (\gamma_1\gamma_2)^2$.

4.2　Using Isogenies to Extend the Attacks

Let E and E' be two ordinary elliptic curves defined over $\mathbb{F}_{2^{ln}}$ and given by the equation (3). A rational map $\Psi\colon E \to E'$ over $\mathbb{F}_{2^{ln}}$ is an element of the elliptic curve $E'(\mathbb{F}_{2^{ln}}(E))$. An isogeny $\Phi\colon E \to E'$ over $\mathbb{F}_{2^{ln}}$ is a non-constant rational map over $\mathbb{F}_{2^{ln}}$ and is also a group homomorphism from $E(\mathbb{F}_{2^{ln}})$ to $E'(\mathbb{F}_{2^{ln}})$. In that case, we say that E and E' are isogenous. It is known that E and E' are isogenous over $\mathbb{F}_{2^{ln}}$ if and only if $\#E(\mathbb{F}_{2^{ln}}) = \#E'(\mathbb{F}_{2^{ln}})$ [37].

An isogeny $\Phi\colon E \to E'$ induces a map $\Phi^*\colon \mathbb{F}_{2^{ln}}(E') \to \mathbb{F}_{2^{ln}}(E)$, called the pullback of Φ [10], which is necessarily injective,

$$\Phi^* : \mathbb{F}_{2^{ln}}(E') \to \mathbb{F}_{2^{ln}}(E)$$
$$\theta \to \theta \circ \Phi.$$

If $x \in E'(\mathbb{F}_{2^{ln}})$, we can pull back x along Φ, and obtain a divisor $D = \sum_{P \in \Phi^{-1}(x)} \nu_P(\Phi)(P)$. The degree δ of Φ is defined by the integer $[\mathbb{F}_{2^{ln}}(E) \colon \Phi^*(\mathbb{F}_{2^{ln}}(E'))]$ and we say that Φ is a δ-isogeny.

The authors in [11] propose to extend the range of vulnerable curves against the GHS attack (and equivalently the gGHS attack) by finding an explicit representation for an isogeny $\Phi \colon E \to E'$ and determining if there exists at least one elliptic curve E' against which the attack is effective.

5   Analyzing the GLS Elliptic Curves

Let $\mathbb{F}_{2^{2n}}$ be a degree-2 extension of $\mathbb{F}_{2^n}$. Let $E/\mathbb{F}_{2^n}$ be an ordinary elliptic curve given by the equation

$$E/\mathbb{F}_{2^n} \colon y^2 + xy = x^3 + ax^2 + b \qquad a \in \mathbb{F}_{2^n}, \ b \in \mathbb{F}_{2^n}^*, \tag{7}$$

with $\mathrm{Tr}(a) = 1$. We know that $\#E(\mathbb{F}_{2^n}) = q + 1 - t$, where t is the trace of E over $\mathbb{F}_{2^n}$. It follows that $\#E(\mathbb{F}_{2^{2n}}) = (q+1)^2 - t^2$. Let $a' \in \mathbb{F}_{2^{2n}}$ such that $\mathrm{Tr}(a') = 1$. Then we can construct the GLS curve,

$$E'/\mathbb{F}_{2^{2n}} \colon y^2 + xy = x^3 + a'x^2 + b. \tag{8}$$

Which is isomorphic to E over $\mathbb{F}_{2^{4n}}$ under the involutive isomorphism $\tau \colon E \to E'$. The GLS endomorphism can be constructed by applying ϕ with the Frobenius automorphism σ as follows, $\psi = \tau \sigma \tau^{-1}$.

5.1   Applying the gGHS Weil Descent Attack

The theoretical security of a given binary GLS curve $E/\mathbb{F}_{2^{2n}}$ depends basically on the complexity of solving the DLP on its group of points $E(\mathbb{F}_{2^{2n}})$. As discussed in Sect. 1, the usual approach is to apply the Pollard Rho algorithm for elliptic curves, which runs in approximately $\sqrt{\frac{\pi 2^{2n}}{2}}$ operations [32].

However, after the publication of the gGHS reduction, it is also necessary to check whether the complexities of solving the DLP on $J_H(\mathbb{F}_2)$, $J_H(\mathbb{F}_{2^2})$ or $J_H(\mathbb{F}_{2^n})$ are lower than solving it on $E(\mathbb{F}_{2^{2n}})$. If such is the case, the smallest complexity provides us the real security of the curve E.

Let us assume that the number of isogenous curves E' is smaller than the number the of vulnerable isogeny classes, then the following steps describe a method for determining if a given GLS curve is vulnerable against the extended gGHS attack:

1. **Setting the environment.** Let us have a GLS curve $E_{\tilde{a},\tilde{b}}/\mathbb{F}_{2^{2n}}$ given by the equation (8) but defined with the particular parameters $\tilde{a}$ and $\tilde{b}$. In the context of the gGHS attack, the extension field $\mathbb{F}_{2^{2n}}$ can be seen as a degree-n extension of $\mathbb{F}_{2^2}$ or a degree-$2n$ extension of $\mathbb{F}_2$. For the sake of simplicity, we will represent the base field as $\mathbb{F}_{2^2}$. Nonetheless, the steps must be executed on both base representations.

2. **Checking the $\tilde{b}$ parameter.** We know that $(x^n + 1)(\sigma) = x^{2^{2n}} + x = 0 \Leftrightarrow x^{2^{2n}} = x$. In addition, $\mathrm{Ord}_\gamma | (x^n + 1)$. Given that the polynomial $x^n + 1$ factorizes as $(x + 1) \cdot f_i \cdot \ldots \cdot f_s$, let $d = \deg(f_i)$. Then, search a pair of polynomials $s_1 = (x + 1)^{j_1} \cdot f_i^{j_2}$ and $s_2 = (x + 1)^{j_3} \cdot f_i^{j_4}$, with positive integers j_i and find a representation of $\tilde{b}as(\gamma_1\gamma_2)^2$, such that $\mathrm{Ord}_{\gamma_i} = s_i(\sigma)$ and $\mathrm{Ord}_{\gamma_1, \gamma_2, \tilde{a}}$ derive a small associated value g_C.

3. **Solving the DLP on a hyperelliptic curve.** If such minimum pair s_1, s_2 exists (i), apply the Weil descent on E to construct a hyperelliptic curve H. Check if the complexity of solving the DLP on $J_H(\mathbb{F}_{2^2})$ is smaller than solving it on $E(\mathbb{F}_{2^{2n}})$ (ii). If that is the case, the curve E is vulnerable against the gGHS attack. If either (i) or (ii) is false, go to step 4.

4. **The extended gGHS attack.** For each isogenous curve E' to E, perform the check (steps 2 and 3). If there is no vulnerable elliptic curve E' isogenous to E, then the curve E is not vulnerable against the extended gGHS attack.

If the number of isogenous curves E' is greater than the number of vulnerable isogeny classes, a more efficient method to perform the vulnerability check is to list all vulnerable parameters $\hat{b}$ and store all of the related group orders $\#E_{a,\hat{b}}(\mathbb{F}_{2^{2n}})$ in a set L. The check consists in verifying whether $\#E_{a,b}(\mathbb{F}_{2^{2n}}) \in L$ [19].

The extension field $\mathbb{F}_{2^{2n}}$ can also be represented as a degree-2 extension of $\mathbb{F}_{2^n}$. However, as analyzed in [19], in this setting the gGHS attack generates hyperelliptic curves of genus 2 or 3. Solving the DLP on the Jacobian of these curves is not easier than solving it on $E(\mathbb{F}_{2^{2n}})$ with the Pollard Rho method.

The complexity of solving the DLP on $J_H(\mathbb{F}_{2^2})$ (or $J_H(\mathbb{F}_2)$) is determined by the genus of the curve H (see Sect. 3). In the gGHS attack context, the genus of the constructed hyperelliptic curve H is given by the degree of the minimum pair of polynomials (s_1, s_2). For each extension degree n, these values are derived from the factors of the polynomial $x^n + 1$. For this reason, in characteristic two, we have many extensions where the genus of H is large and consequently, the gGHS attack is ineffective for any GLS curve defined over such extension fields.

To illustrate those cases, we present in Table 1 the costs of solving the DLP with the gGHS/Enge-Gaudry approach and the Pollard Rho algorithm on binary GLS curves $E/\mathbb{F}_{2^{2n}}$ with $n \in [5, 257]$. We chose all the hidden constant factors in the Enge-Gaudry algorithm complexity to be one, and suppressed all fields whose genus of the generated curve H is higher than 10^6. In addition, the effort of finding a vulnerable curve against the gGHS attack is not included in the cost for solving the DLP.

5.2   A Mechanism for Finding Vulnerable Curves

In this part, we propose a mechanism for performing the step 3 check on the parameter b of a given GLS curve $E_{a,b}$. This mechanism is useful when the number of gGHS vulnerable isogeny classes is greater than the number of isogenous curves to $E_{a,b}$. Similarly to the previous section, $\mathbb{F}_{2^{2n}}$ is a degree-n extension field of $\mathbb{F}_{2^2}$. However, the method can be easily adapted to any field representation.

Table 1. Different binary GLS curves and their security. The smallest complexity is written in bold type.

Base field of H	Base Field of E	Genus of H	$E(\mathbb{F}_{2^{2n}})$ order ($\approx$) (bits)	Cost for solving the DLP (bits)	
				Pollard Rho algorithm on E	Enge-Gaudry algorithm on H
$\mathbb{F}_{2^{2\cdot5}}$	$\mathbb{F}_2$	32	9.08	**4.87**	16.91
	$\mathbb{F}_{2^2}$	15			16.20
$\mathbb{F}_{2^{2\cdot7}}$	$\mathbb{F}_2$	16	13.02	**6.83**	10.53
	$\mathbb{F}_{2^2}$	7			9.58
$\mathbb{F}_{2^{2\cdot11}}$	$\mathbb{F}_2$	2048	21.00	**10.82**	207.10
	$\mathbb{F}_{2^2}$	1023			206.98
$\mathbb{F}_{2^{2\cdot13}}$	$\mathbb{F}_2$	8192	25.00	**12.82**	452.03
	$\mathbb{F}_{2^2}$	4095			451.96
$\mathbb{F}_{2^{2\cdot17}}$	$\mathbb{F}_2$	512	33.00	**16.82**	93.13
	$\mathbb{F}_{2^2}$	255			92.92
$\mathbb{F}_{2^{2\cdot19}}$	$\mathbb{F}_2$	524288	37.00	**18.82**	4400.90
	$\mathbb{F}_{2^2}$	262143			4400.90
$\mathbb{F}_{2^{2\cdot23}}$	$\mathbb{F}_2$	4096	45.00	**22.82**	306.55
	$\mathbb{F}_{2^2}$	2047			306.46
$\mathbb{F}_{2^{2\cdot31}}$	$\mathbb{F}_2$	64	61.00	30.82	26.46
	$\mathbb{F}_{2^2}$	31			**25.93**
$\mathbb{F}_{2^{2\cdot43}}$	$\mathbb{F}_2$	32768	85.00	**42.82**	973.85
	$\mathbb{F}_{2^2}$	16383			973.82
$\mathbb{F}_{2^{2\cdot73}}$	$\mathbb{F}_2$	1024	145.00	**72.82**	139.27
	$\mathbb{F}_{2^2}$	511			139.12
$\mathbb{F}_{2^{2\cdot89}}$	$\mathbb{F}_2$	4096	177.00	**88.82**	306.55
	$\mathbb{F}_{2^2}$	2047			306.46
$\mathbb{F}_{2^{2\cdot127}}$	$\mathbb{F}_2$	256	253.00	126.83	61.84
	$\mathbb{F}_{2^2}$	127			**61.56**
$\mathbb{F}_{2^{2\cdot151}}$	$\mathbb{F}_2$	65536	301.00	**150.83**	1424.00
	$\mathbb{F}_{2^2}$	32767			1424.00
$\mathbb{F}_{2^{2\cdot257}}$	$\mathbb{F}_2$	131072	513.00	**256.82**	2077.90
	$\mathbb{F}_{2^2}$	65535			2077.90

Let $x^n + 1 = (x+1)f_1 \cdots f_s$, where each irreducible polynomial $f_i \in \mathbb{F}_2[x]$ has degree d and $f_i \neq f_j$ for $i \neq j$. Also, let $S = \{(x+1)^{j_1}f_i^{j_2}, (x+1)^{j_3}f_i^{j_4}\}_{j_i \in \{0,1\}}$ be a nonempty finite set where for each pair $(s_1, s_2) \in S$, $\deg(s_1) \geq \deg(s_2)$. Then let $B = \{b = (\gamma_1\gamma_2)^2 \colon \gamma_1, \gamma_2 \in \mathbb{F}_{2^{2n}}^*, \exists (s_1, s_2) \in S | s_1(\sigma)(\gamma_1) = 0 \wedge s_2(\sigma)(\gamma_2) = 0\}$.

Let $f, g \in \mathbb{F}_2[x]$ be two degree-d polynomials. Then we have the following theorems:

Theorem 1. $(f \cdot g)(\sigma)(x) = (f(\sigma) \circ g(\sigma))(x) = (g(\sigma) \circ f(\sigma))(x).$

Proof. Let $q = 2^2$. The expression $(f \cdot g)(x)$ can be written as $(\sum_{i=0}^{d} f_i x^i)(\sum_{j=0}^{d} g_j x^j) = \sum_{i=0}^{d} \sum_{j=0}^{d} f_i g_i x^{i+j}$. Then,

$$(f \cdot g)(\sigma)(x) = \sum_{i=0}^{d} \sum_{j=0}^{d} f_i g_j x^{q^{i+j}} = \sum_{i=0}^{d} \sum_{j=0}^{d} f_i \left(g_i x^{q^j} \right)^{q^i} = \sum_{i=0}^{d} f_i \sum_{j=0}^{d} \left(g_j x^{q^j} \right)^{q^i}$$

$$= \sum_{i=0}^{d} f_i \left(\sum_{j=0}^{d} g_j x^{q^j} \right)^{q^i} = (f(\sigma) \circ g(\sigma)) (x)$$

The proof for the theorem $(f \cdot g)(\sigma)(x) = (g(\sigma) \circ f(\sigma))(x)$ is similar.

Theorem 2. $f(\sigma)(x) \mid (f \cdot g)(\sigma)(x)$ over $\mathbb{F}_{2^{2n}}(\alpha_1, \ldots, \alpha_{q^d})$ where each α_i is a root of the polynomial f.

Proof. Given that $p(\sigma)(x)$ has a zero at $x = 0$ for all polynomial $p \in \mathbb{F}_2[x]$, let α be a root of $f(\sigma)(x)$ over its splitting field. Then $g(\sigma)(f(\sigma)(\alpha)) = g(\sigma)(0) = 0$, i.e., α is also a root of $(f \cdot g)(\sigma)$. As a result, $f(\sigma) = \prod(x - \alpha_i)$ divides $(f \cdot g)(\sigma)$ over $\mathbb{F}_{2^{2n}}(\alpha_1, \ldots, \alpha_{q^d})$.

Theorem 3. $\forall \gamma \in \mathbb{F}_{2^{2n}}$, we have that $Ord_\gamma(\sigma)(x)$ splits over $\mathbb{F}_2[x]$.

Proof. Following the the previous theorem, we have that for a given polynomial $p(x) \in \mathbb{F}_2[x]$, every factor $\bar{p}(x)$ (not necessarily irreducible) of $p(x)$ satisfies $\bar{p}(\sigma)(x) \mid p(\sigma)(x)$ over $\mathbb{F}_{2^{2n}}(\alpha_1, \ldots, \alpha_{\deg(\bar{p})})$ where each α_i is a root of $\bar{p}$. Then, since $Ord_\gamma(x) \mid x^n + 1$ we have that $Ord_\gamma(\sigma)(x) \mid (x^{q^{2^n}} + x)$ over $\mathbb{F}_{2^{2n}}(\alpha_1, \ldots, \alpha_{\deg(Ord_\gamma)})$ where each α_i is a root of $Ord_\gamma(\sigma)(x)$. Consequently, $\mathbb{F}_{2^{2n}} = \mathbb{F}_{2^{2n}}(\alpha_1, \ldots, \alpha_{\deg(Ord_\gamma)})$ and $Ord_\gamma(\sigma)(x)$ splits over $\mathbb{F}_2[x]$.

For a given $b \in \mathbb{F}_{2^{2n}}^*$, we can determine whether b is in B as follows. For all $(s_1, s_2) = (\sum_j s_{1,j} x^j, \sum_j s_{2,j} x^j) \in S$, let $b_i(x) = s_i(\sigma)(b^{1/2} x) = \sum_j (b^{1/2} s_{i,j}) x^{q^j}$ and let $\bar{s}_i(x) = x^{\deg(s_i)} s_i(\sigma)(\frac{1}{x})$. Then,

$$\gcd (b_1(x), \bar{s}_2(x)) \neq 1 \Leftrightarrow \exists \gamma \in \mathbb{F}_{2^{2n}}^* \text{ such that } s_1(\sigma)(b^{1/2} \gamma) = 0 \text{ and } s_2(\sigma)(\tfrac{1}{\lambda}) = 0$$

$$\Leftrightarrow s_1(\sigma)(x) \text{ has a zero at } \gamma_1 = b^{1/2} \gamma \text{ and}$$

$$s_2(\sigma)(x) \text{ has a zero at } \gamma_2 = \frac{1}{\gamma}$$

$$\Leftrightarrow b = (\gamma_1 \gamma_2)^2 \text{ where } s_1(\sigma)(\gamma_1) = 0 \text{ and}$$

$$s_2(\sigma)(\gamma_2) = 0$$

$$\Leftrightarrow b = (\gamma_1 \gamma_2)^2 \in B.$$

Let us now assume that S contains only the pairs of polynomials (s_1, s_2) that construct parameters b which are vulnerable against the gGHS attack. Then, for an arbitrary GLS curve E we have that,

$$E_{a,b} \text{ is vulnerable} \Leftrightarrow \exists (s_1, s_2) \in S \text{ such that } \gcd(b_1(x), \bar{s}_2(x)) \neq 1. \quad (9)$$

Complexity analysis. Let $s_1 = (x+1)f_i$ be the maximum degree polynomial of all pairs in S and the complexity of computing the greatest common divisor over elements in S be $O((q^{d+1})^2)$. Then the complexity for checking a parameter b with the above mechanism is $O(\#S(q^{d+1})^2)$. This complexity is an upper bound because in practice we see that a $gcd(,)$ in S requires a smaller number of operations.

At last, we summarize the mechanism in Algorithm 1. Note that, for determining whether a given b is a vulnerable parameter, the algorithm must be executed in all base field representations.

Algorithm 1. Mechanism for verifying binary curve parameter b.

Require: The element $b \in \mathbb{F}_{2^{2n}}^*$, the polynomial lists $b_1, \bar{s}_2$ obtained from the set S.
Ensure: *True* if the binary curve defined with a parameter b is vulnerable against the gGHS attack and *False* otherwise.
  aux $\leftarrow 1$, $j \leftarrow 0$
  while aux $= 1$ **and** $j < \#S$ **do**
    $j \leftarrow j + 1$
    aux $\leftarrow \gcd(b_1[j], \bar{s}_2[j])$
  end while
  if aux $\neq 1$ **then**
    return *True*
  else
    return *False*
  end if

---

6   A Concrete Attack on the GLS Curve $E/\mathbb{F}_{2^{62}}$

In order to understand the practical implications of the gGHS Weil descent algorithm over a binary GLS curve, we implemented a complete attack on a curve defined over the field $\mathbb{F}_{2^{31 \cdot 2}}$. Such field was chosen for two reasons: (i) solving the DLP on the Jacobian of a hyperelliptic curve obtained by the gGHS attack is easier then solving it on the elliptic curve (see Table 1); (ii) the small amount of resources required for solving the DLP in this curve allowed us to experiment different approaches to the problem.

6.1   Building a Vulnerable Curve

Let $\mathbb{F}_{2^{2 \cdot 31}}$ be an extension field of $\mathbb{F}_q$ with $q = 2^{2 \cdot 31/n}$ where $n \in \{31, 62\}$. Then we can represent the field $\mathbb{F}_{2^{62}}$ as follows,

- $n = 62, q = 2$, $\mathbb{F}_{2^{62}} \cong \mathbb{F}_2[v]/f(v)$, with $f(v) = v^{62} + v^{29} + 1$.
- $n = 31, q = 2^2$, $\mathbb{F}_{2^2} \cong \mathbb{F}_2[z]/g(z)$, with $g(z) = z^2 + z + 1$.
  $\mathbb{F}_{2^{62}} \cong \mathbb{F}_{2^2}[u]/h(u)$, with $h(u) = u^{31} + u^3 + 1$

Also, let E be a binary GLS curve given by the following equation

$$E_{a,b}/\mathbb{F}_{2^{62}} : y^2 + xy = x^3 + ax^2 + b \qquad a \in \mathbb{F}_{2^{62}},\ b \in \mathbb{F}_{2^{31}}^*$$

Given that the parameter a can be chosen arbitrarily subject to the constraint $\mathrm{Tr}_{\mathbb{F}_{2^{62}}/\mathbb{F}_2}(a) = 1$, we chose $a = z^2$. The next step was to find a vulnerable parameter $b \in \mathbb{F}_{2^{31}}^*$ which constructs a curve $E_{a,b}$ which is vulnerable against the gGHS attack. Moreover, to simulate a cryptographic environment, we must have $\#E_{a,b}(\mathbb{F}_{2^{62}}) = c \cdot r$, with small c and prime r.

Let $x^{31} + 1 = (x+1)f_1 \cdots f_6$. We have $\deg(f_i) = 5$. Then Tables 2 and 3 give us a list of polynomials that generate the vulnerable parameters $b = (\gamma_1 \gamma_2)^2$.

Table 2. Polynomials Ord_{γ_i} which generate low-genus hyperelliptic curves for the case $n = 31$, $q = 2^2$.

Ord_{γ_1}	Ord_{γ_2}	$\deg(\mathrm{Ord}_{\gamma_1})$	$\deg(\mathrm{Ord}_{\gamma_2})$	m	genus	E-G algorithm complexity
$(x+1)f_i$	$x+1$	6	1	6	32	26.46
f_i	$x+1$	5	1	6	31	25.93

Table 3. Polynomials Ord_{γ_i} which generate low-genus hyperelliptic curves for the case $n = 62$, $q = 2$.

Ord_{γ_1}	Ord_{γ_2}	$\deg(\mathrm{Ord}_{\gamma_1})$	$\deg(\mathrm{Ord}_{\gamma_2})$	m	genus	E-G algorithm complexity
$(x+1)^2 f_i$	$x+1$	7	1	7	64	26.46

At first, we looked for vulnerable parameters $b \in \mathbb{F}_{2^{31}}^*$ by obtaining the roots of the polynomials listed in Tables 2 and 3. However, for all those b parameters, $\log_2 |r| < 52$. For that reason, we considered non-GLS vulnerable parameters $b \in \mathbb{F}_{2^{62}}^*$ for which $\log_2 |r| \geq 52$. As a result, 61 isogeny classes were found. Let L be the set of its group orders. Then, in a 20-core Intel Xeon E5-2658 2.40GHz, we executed for 70 hours an extensive search through all $b \in \mathbb{F}_{2^{31}}^*$ checking if $\#E_{a,b}(\mathbb{F}_{2^{62}}) \in L$. However, no isogenous curves were found and the extended gGHS attack could not be carried out.

Next, under the setting $(n = 31, q = 2^2)$, we chose the vulnerable parameter $b = u^{24} + u^{17} + u^{16} + u^{12} + u^5 + u^4 + u^3 + u + 1$, which allowed us to construct a group with order $\#E_{a,b}(\mathbb{F}_{2^{62}}) = 4611686014201959530$. The size of our subgroup of interest is of about 51.38 bits. In theory, solving the DLP on this subgroup through the Pollard Rho method would take about 2^{26}, which is the same cost as solving the DLP with the gGHS/Enge-Gaudry approach.

Finally, we created an order-r generator point $P \in E_{a,b}(\mathbb{F}_{2^{62}})$ with the Magma Random function:

$$
\begin{aligned}
P(x,y) =&(u^{30} + z^2 u^{28} + z u^{27} + u^{26} + z u^{25} + z u^{24} + u^{23} + z^2 u^{20} + u^{18} + z u^{17}\\
&+ z u^{16} + u^{15} + u^{12} + z^2 u^{10} + z u^9 + z^2 u^8 + u^7 + z u^6 + u^4 + u^2\\
&+ z^2 u + z,\\
&z u^{30} + z^2 u^{29} + z^2 u^{26} + z^2 u^{25} + z u^{24} + u^{23} + z^2 u^{22} + z^2 u^{21} + z u^{20}\\
&+ u^{19} + z u^{18} + u^{17} + u^{15} + z u^{14} + z u^{13} + z^2 u^{12} + z^2 u^{10} + z u^9 + u^8\\
&+ z u^7 + u^6 + u^2 + z u).
\end{aligned}
$$

The challenge Q was generated with the same function:

$$
\begin{aligned}
Q(x,y) =&(u^{29} + z^2 u^{28} + u^{27} + u^{26} + z^2 u^{25} + z u^{24} + u^{23} + z u^{22} + z^2 u^{20} + z^2 u^{17}\\
&+ z^2 u^{16} + z u^{12} + u^{11} + z u^{10} + z^2 u^9 + z^2 u^8 + z u^7 + z u^6 + z^2 u^5 + z u^4 +\\
&z^2 u^2 + u + z^2,\\
&u^{30} + u^{29} + z^2 u^{28} + u^{27} + z u^{26} + z^2 u^{24} + z u^{22} + u^{21} + z^2 u^{20} + z^2 u^{19}\\
&+ z u^{18} + z u^{17} + z u^{15} + u^{14} + z u^{12} + z^2 u^{11} + u^{10} + z^2 u^9 + u^6 + u^5\\
&+ z^2 u^3 + z^2 u^2 + z^2 u + z).
\end{aligned}
$$

Then we constructed the following genus-32 hyperelliptic curve with the Weil descent method:

$$
H(\mathbb{F}_{2^2}): \; y^2 + (z^2 x^{32} + x^{16} + z^2 x^8 + z^2 x^2 + x)y =
$$
$$
x^{65} + x^{64} + z^2 x^{33} + z x^{32} + x^{17} + z^2 x^{16} + x^8 + x^5 + x^4 + z^2 x^3 + z x^2 + z x.
$$

The points P, Q were mapped to the $J_H(\mathbb{F}_{2^2})$, which generated the divisors D_P and D_Q.

6.2 Adapting the Enge-Gaudry Algorithm

To solve the DLP on $J_H(\mathbb{F}_q)$, with $q = 2^2$ and genus $g = 32$, we adapted the Enge-Gaudry algorithm by restricting the factor base size in order to balance the relation collection and the linear algebra phases. According to [6], we can balance the two phases by selecting the factor base degree as $m = \lceil \log_q L_{q^g}[\varrho] \rceil$ where $\varrho = \sqrt{\frac{1}{2} + \frac{1}{4\vartheta}} - \sqrt{\frac{1}{4\vartheta}}$ for some positive integer ϑ which complies with (i) $g \geq \vartheta \log q$ and (ii) $q \leq L_{q^g}[\frac{1}{\sqrt{\vartheta}}]$. Similarly to the Sect. 5, we chose the constant factors of the algorithm complexity to be one. For all values of ϑ that satisfy the restrictions (i) and (ii), we have $m = [4, 6]$.

However, in practice, we constructed the factor base dynamically. At first, we initialized our base $\mathcal{F}$ as an empty set and imposed a restriction so that $\mathcal{F}$ can contain polynomials up to degree m. Next, for each valid relation in the Enge-Gaudry algorithm, that is, when the polynomial U of a divisor $D = \text{div}(U, V)$ is d-smooth, we included in $\mathcal{F}$ all irreducible factors of U which were not in $\mathcal{F}$. Finally, when the number of relations were equal to the number of factors in $\mathcal{F}$, we concluded the relations collection phase.

Experimentally, we saw that, at the end of the relations collection phase, just a part of the irreducible polynomials of degree less or equal than m were included in $\mathcal{F}$. For that reason, in order to have approximately the same factor base size as if we had constructed a factor base with all irreducible polynomials of degree up to 6, we chose $m = 7$. The algorithm was executed within the Magma v2.20-2 system, in one core of a Intel Core i7-4700MQ 2.40 GHz machine. The timings of each phase is presented below (Table 4).

Table 4. Timings for the adapted Enge-Gaudry algorithm

Random walk initialization	3.00 s
Relations collection	284.52 s
Linear Algebra (Lanczos)	0.11 s

At the end of the relation collection phase, our factor basis had 1458 elements, which is 44.12 % of the total number of irreducible polynomials of degree 7 and below. Although the algorithm phases were not balanced as expected, solving the linear algebra system was trivial, and we considered our degree selection satisfactory. Finally, the computed discrete logarithm is given as $\lambda = 2344450470259921$.

An Analysis of the Algorithm Balance. In order to verify the theoretical balance of [6] in the context of the dynamic factor base construction, we executed the algorithm with different factor base degree limits. The results are presented in Table 5.

The theoretical costs for the relation collection phase were obtained by multiplying the inverse of the probability of having a d-smooth degree-32 polynomial by the factor base size. The linear algebra step theoretical cost was computed as the square of the factor base size multiplied by the average number of irreducible factors in each d-smooth degree-32 polynomial, which was calculated experimentally.

Here we can see that, regarding the theoretical costs, setting the factor base degree limit to 6 results in the most balanced implementation. However, the practical timings demonstrate against this affirmative. This is because factorizing a degree-32 polynomial in $\mathbb{F}_{2^2}[x]$, which is the relations collection step, is more expensive than the linear algebra step.

On the other hand, if we consider practical timings, the degree-11 setting offers the most balanced version. However, it is clearly more important to have

Table 5. Details of different Enge-Gaudry algorithm setting

	Factor base maximum degree (d)							
	5	6	7	8	9	10	11	12
Relations collection phase								
Number of irreducible poly of degree $\leq d$ (α)	294	964	3304	11464	40584	145338	526638	1924378
Factor base size (β)	152	474	1458	4352	12980	34883	91793	214116
Ratio β/α	0.517	0.492	0.441	0.380	0.320	0.240	0.174	0.111
Theoretical cost (bits) *	23.24	20.88	19.50	19.08	19.16	19.49	20.03	20.57
Average timing per relation (s)	20.250	1.469	0.195	0.050	0.019	0.012	0.009	0.011
Timing (s)	3078.14	646.43	284.52	220.12	252.05	413.15	909.12	2451.80
Original E-G timing estimation (s)	5953.50	1416.12	644.28	573.20	771.10	1744.06	4739.74	21168.16
Linear algebra phase								
Theoretical cost (bits) *	17.49	20.58	23.71	26.75	29.78	32.55	35.29	37.65
Timing (s)	0.01	0.03	0.11	0.87	9.62	169.78	1288.65	6774.26

* The steps in the relations collection and the linear algebra phases have different costs. Since we do not have access to the Magma algebra system code, we could give the exact timings of each step.

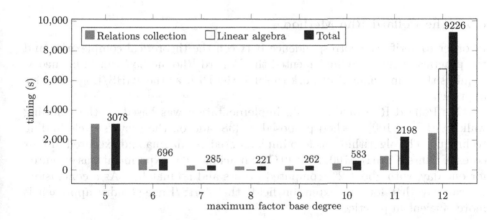

Fig. 1. Timings for the Enge-Gaudry algorithm with dynamic factor base

the lowest overall timings, which is achieved by the degree-8 setting. The results for the degree settings from 5 to 12 are presented in Fig. 1.

The problem of balancing the Enge-Gaudry method with dynamic factor base is slightly different from the traditional algorithm. In the former, the cost of finding a valid relation and the ratio α/β (see Table 5) decreases as we increase the factor base degree limit. However, because of the larger number of irreducible polynomials, the probability of having relations with factors which are not included in our factor base increases. As a consequence, for each valid relation, more factors are added and the cost to achieve a matrix with the same number of columns and rows also increases. This effect is shown in the Fig. 2.

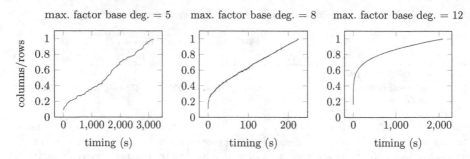

Fig. 2. The ratio of the matrix columns (polynomials in the factor base) and rows (valid relations) per time. The relation collections phase ends when the ratio is equal to one.

One possible solution for achieving a balanced algorithm is to restrict the size of the dynamic factor base. Ultimately, although unbalanced, constructing our factor base dynamically was useful in our context, since it allowed us to conclude the relations collection phase more efficiently when compared with the original Enge-Gaudry algorithm (see Table 5).

Further analysis can be found in the Appendix A, where the algorithm is applied to a Jacobian of a hyperellitpic curve of genus 45.

6.3 The Pollard Rho Method

In order to verify the correspondence between the theoretical complexities and the practical results, we implemented the Pollard Rho method which, in theory, requires the same amount of work to solve the DLP as the gGHS/Enge-Gaudry approach.

Our Pollard Rho random walk implementation was based on the r-adding walk (with $r = 100$) method proposed in [38] and on the Floyd's cycle-finding technique. The algorithm was also implemented on Magma and executed in four cores of a Intel Core i7-4700MQ 2.40 GHz machine. The experiment was executed for one day, with about 2^{31} computed points and no matches. As a conclusion, we can say that for this extension field, the gGHS/Enge-Gaudry approach is more efficient in practice.

7   Conclusion

In this work, we implemented a successful attack against a vulnerable-constructed GLS binary curve defined over $\mathbb{F}_{2^{62}}$ with the Magma algebra system. The discrete logarithm was solved using a dynamic large-base version of the Enge-Gaudry algorithm, which used only 44.12 % of the total base size.

Our future work is to analyze GLS curves defined over other extension fields. We are particularly interested in the field $\mathbb{F}_{2^{254}}$, which is a formidable challenge in terms of CPU and memory resources. We also want to understand the practical implications of different approaches to solve the DLP on elliptic curves, such as the Gaudry and Diem methods [4,16] based on the Semaev work [35].

Acknowledgments. The authors would like to thank Sorina Ionica, Gora Adj and the reviewers for their valuable comments and suggestions.

A Analyzing the Enge-Gaudry Algorithm Balance

In this Section, we analyzed the balance between the relations collection and the linear algebra phases of the dynamic-base Enge-Gaudry algorithm over a Jacobian of a hyperelliptic curve of genus 45 defined over $\mathbb{F}_{2^2}$. The subgroup of interest is of size $r = 2934347646102267239451433$ of approximately 81.27 bits.

After performing the theoretical balancing computations presented in Sect. 6, we saw that our factor base should be composed by irreducible polynomials of degree up to $m = [5, 8]$. For that reason, we used this range as a reference for our factor base limit selection. The results are presented below (Table 6).

Table 6. Details of different Enge-Gaudry algorithm settings

	Factor base maximum degree (d)						
	7	8	9	10	11	12	13
Relations collection phase							
Number of irreducible poly of degree $\leq d$ (α)	3304	11464	40584	145338	526638	1924378	7086598
Factor base size (β)	1626	5227	16808	52366	158226	460240	1268615
Ratio β/α	0.492	0.456	0.414	0.360	0.300	0.239	0.179
Theoretical cost (bits) *	27.46	25.71	24.88	24.62	24.73	25.09	25.60
Average timing per relation (s)	29.450	4.604	1.048	0.290	0.119	0.071	0.085
Timing (s)	47895.19	24067.58	17621.50	15204.72	18909.36	32630.56	107902.72
Original E-G timing estimation (s)	97319.00	52780.25	42532.03	42148.02	62669.92	136630.84	602360.83
Linear algebra phase							
Theoretical cost (bits) *	24.62	27.74	31.07	34.32	37.37	40.45	43.26
Timing (s)	0.62	3.79	39.18	421.33	4803.53	48660.84	417920.24

[a] The steps in the relations collection and the linear algebra phases have different costs. Since we do not have access to the Magma algebra system code, we could give the exact timings of each step.

[b] The value were not computed until the date of publication of this document.

Compared with the example in Sect. 6, we had a large number of factors per relation. As a result, more irreducible polynomials were added to the factor base and consequently, the relations collection phase became more costly. In addition, the ratios α/β were greater than the ones presented in the genus-32 example (see Table 5).

The most efficient configuration ($d = 10$) was unbalanced, the relations collection was about 36 times slower than the linear algebra phase. However, the genus-45 example provided a more balanced Enge-Gaudry algorithm, since the

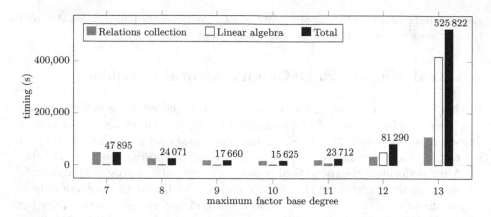

Fig. 3. Timings for the Enge-Gaudry algorithm with dynamic factor base

best setting for the genus-32 curve was unbalanced by a factor of 253. One possible reason is that, here, each linear algebra steps computed over operands of about 81 bits, which are 2^{30} bigger than the operands processed in the genus-32 linear algebra steps.

We expect that, for curves with larger genus, with respectively larger subgroups, a fully balanced configuration can be found. The results for each setting in the 45-genus example is shown in Fig. 3.

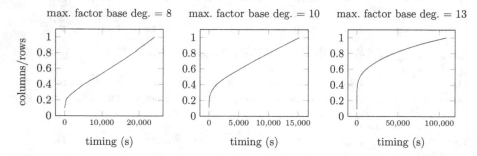

Fig. 4. The ratio of the matrix columns (polynomials in the factor base) and rows (valid relations) per time. The relation collections phase ends when the ratio is equal to one.

In Fig. 4, we show the progression of the ratio $\dfrac{\text{number of valid relations}}{\text{factor base size}}$ during the relations collection phase. Similarly to the genus-32, for bigger d values, the rate of the factor base growth stalled the progress of the relations collection algorithm. Again, one potential solution to this issue is to impose limits on the factor base size.

The challenge for obtaining an optimal relations collection phase is to find a balance between the average timing per relation and the factor base growth rate. The goal is to have a graph which, after the initial vertical rise, directs toward the ratio one as a linear function, such as the $d = 8, 10$ cases.

References

1. Barbulescu, R., Gaudry, P., Joux, A., Thomé, E.: A heuristic quasi-polynomial algorithm for discrete logarithm in finite fields of small characteristic. In: Nguyen, P.Q., Oswald, E. (eds.) EUROCRYPT 2014. LNCS, vol. 8441, pp. 1–16. Springer, Heidelberg (2014)
2. Bos, J.W., Costello, C., Hisil, H., Lauter, K.: High-performance scalar multiplication using 8-dimensional GLV/GLS decomposition. In: Bertoni, G., Coron, J.-S. (eds.) CHES 2013. LNCS, vol. 8086, pp. 331–348. Springer, Heidelberg (2013)
3. Cohen, H., Frey, G., Avanzi, R., Doche, C., Lange, T., Nguyen, K., Vercauteren, F.: Handbook of Elliptic and Hyperelliptic Curve Cryptography, (2nd edn). Chapman & Hall/CRC (2012)
4. Diem, C.: On the discrete logarithm problem in elliptic curves. Compositio Mathematica **147**, 75–104 (2011)
5. Dierks, T., Rescorla, E.: The Transport Layer Security (TLS) Protocol Version 1.2. RFC 5246 (Proposed Standard) (2008). http://www.ietf.org/rfc/rfc5246.txt
6. Enge, A., Gaudry, P.: A general framework for subexponential discrete logarithm algorithms. Acta Arithmetica **102**(1), 83–103 (2002)
7. Faugère, J.-C., Perret, L., Petit, C., Renault, G.: Improving the complexity of index calculus algorithms in elliptic curves over binary fields. In: Pointcheval, D., Johansson, T. (eds.) EUROCRYPT 2012. LNCS, vol. 7237, pp. 27–44. Springer, Heidelberg (2012)
8. Faz-Hernández, A., Longa, P., Sánchez, A.H.: Efficient and secure algorithms for GLV-based scalar multiplication and their implementation on GLV-GLS curves. In: Benaloh, J. (ed.) CT-RSA 2014. LNCS, vol. 8366, pp. 1–27. Springer, Heidelberg (2014)
9. Frey, G.: How to disguise an elliptic curve. In: Talk at ECC 1998 (Workshop on Elliptic Curve Cryptography), Waterloo (1998). http://www.cacr.math.uwaterloo.ca/conferences/1998/ecc98/frey.ps
10. Galbraith, S.D.: Mathematics of Public Key Cryptography, 1st edn. Cambridge University Press, New York, NY, USA (2012)
11. Galbraith, S.D., Hess, F., Smart, N.P.: Extending the GHS Weil Descent Attack. In: Knudsen, L.R. (ed.) EUROCRYPT 2002. LNCS, vol. 2332, pp. 29–44. Springer, Heidelberg (2002)
12. Galbraith, S.D., Lin, X., Scott, M.: Endomorphisms for Faster Elliptic Curve Cryptography on a Large Class of Curves. J. Cryptology **24**(3), 446–469 (2011)
13. Galbraith, S.D., Smart, N.P.: A cryptographic application of weil descent. In: Walker, M. (ed.) Cryptography and Coding 1999. LNCS, vol. 1746, pp. 191–200. Springer, Heidelberg (1999)
14. Gallant, R.P., Lambert, R.J., Vanstone, S.A.: Faster point multiplication on elliptic curves with efficient endomorphisms. In: Kilian, J. (ed.) CRYPTO 2001. LNCS, vol. 2139, pp. 190–200. Springer, Heidelberg (2001)
15. Gaudry, P.: An Algorithm for Solving the Discrete Log Problem on Hyperelliptic Curves. In: Preneel, B. (ed.) EUROCRYPT 2000. LNCS, vol. 1807, pp. 19–34. Springer, Heidelberg (2000)
16. Gaudry, P.: Index calculus for abelian varieties of small dimension and the elliptic curve discrete logarithm problem. J. Symbolic Comput. **44**(12), 1690–1702 (2009)
17. Gaudry, P., Hess, F., Smart, N.P.: Constructive and destructive facets of Weil descent on elliptic curves. J. Cryptology **15**(1), 19–46 (2002)

18. Gaudry, P., Thomé, E., Thériault, N., Diem, C.: A double large prime variation for small genus hyperelliptic index calculus. Math. Comput. **76**, 475–492 (2007)
19. Hankerson, D., Karabina, K., Menezes, A.: Analyzing the Galbraith-Lin-Scott point multiplication method for elliptic curves over binary fields. IEEE Trans. Comput. **58**(10), 1411–1420 (2009)
20. Hess, F.: Generalising the GHS attack on the elliptic curve discrete logarithm problem. LMS J. Comput. Math. **7**, 167–192 (2004)
21. Hu, Z., Longa, P., Xu, M.: Implementing the 4-dimensional GLV method on GLS elliptic curves with j-invariant 0. Des. Codes Crypt. **63**(3), 331–343 (2012)
22. Joux, A., Vitse, V.: Cover and decomposition index calculus on elliptic curves made practical. In: Pointcheval, D., Johansson, T. (eds.) EUROCRYPT 2012. LNCS, vol. 7237, pp. 9–26. Springer, Heidelberg (2012)
23. Koblitz, N.: Elliptic curve cryptosystems. Math. Comput. **48**(177), 203–209 (1987)
24. Longa, P., Sica, F.: Four-dimensional Gallant-Lambert-Vanstone scalar multiplication. J. Cryptology **27**(2), 248–283 (2014)
25. Menezes, A., Qu, M.: Analysis of the Weil descent attack of Gaudry, Hess and Smart. In: Naccache, D. (ed.) CT-RSA 2001. LNCS, vol. 2020, pp. 308–318. Springer, Heidelberg (2001)
26. Menezes, A., Teske, E., Weng, A.: Weak Fields for ECC. In: Okamoto, T. (ed.) CT-RSA 2004. LNCS, vol. 2964, pp. 366–386. Springer, Heidelberg (2004)
27. Menezes, A.J., Okamoto, T., Vanstone, S.A.: Reducing elliptic curve logarithms to logarithms in a finite field. IEEE Trans. Inf. Theor. **39**(5), 1639–1646 (1993)
28. Miller, V.S.: Use of elliptic curves in cryptography. In: Williams, H.C. (ed.) CRYPTO 1985. LNCS, vol. 218, pp. 417–426. Springer, Heidelberg (1986)
29. Nagao, K.: Decomposition attack for the Jacobian of a hyperelliptic curve over an extension field. In: Hanrot, G., Morain, F., Thomé, E. (eds.) ANTS-IX. LNCS, vol. 6197, pp. 285–300. Springer, Heidelberg (2010)
30. National Institute of Standards and Technology: FIPS PUB 186-4. Digital Signature Standard (DSS), Department of Commerce, U.S (2013)
31. Oliveira, T., López, J., Aranha, D.F., Rodríguez-Henríquez, F.: Two is the fastest prime: lambda coordinates for binary elliptic curves. J. Cryptographic Eng. **4**(1), 3–17 (2014)
32. Pollard, J.: Monte Carlo methods for index computation (mod p). Math. Comput. **32**, 918–924 (1978)
33. Sarkar, P., Singh, S.: A New Method for Decomposition in the Jacobian of Small Genus Hyperelliptic Curves. Cryptology ePrint Archive, Report 2014/815 (2014). http://eprint.iacr.org/
34. Sarkar, P., Singh, S.: A simple method for obtaining relations among factor basis elements for special hyperelliptic curves. Cryptology ePrint Archive, Report 2015/179 (2015). http://eprint.iacr.org/
35. Semaev, I.: Summation polynomials and the discrete logarithm problem on elliptic curves. Cryptology ePrint Archive, Report 2004/031 (2004). http://eprint.iacr.org/
36. Stebila, D.: Elliptic Curve Algorithm Integration in the Secure Shell Transport Layer. RFC 5656 (Proposed Standard) (2009). http://www.ietf.org/rfc/rfc5656.txt
37. Tate, J.: Endomorphisms of abelian varieties over finite fields. Inventiones math. **2**(2), 134–144 (1966)
38. Teske, E.: Speeding up Pollard's Rho method for computing discrete logarithms. In: Buhler, J.P. (ed.) ANTS 1998. LNCS, vol. 1423, pp. 541–554. Springer, Heidelberg (1998)

Cryptographic Engineering

Fast Implementation of Curve25519 Using AVX2

Armando Faz-Hernández[✉] and Julio López

Institute of Computing, University of Campinas,
1251 Albert Einstein, Cidade Universitaria, Campinas, Brazil
{armfazh,jlopez}@ic.unicamp.br

Abstract. AVX2 is the newest instruction set on the Intel Haswell processor that provides simultaneous execution of operations over vectors of 256 bits. This work presents the advances on the applicability of AVX2 on the development of an efficient software implementation of the elliptic curve Diffie-Hellman protocol using the Curve25519 elliptic curve. Also, we will discuss some advantages that vector instructions offer as an alternative method to accelerate prime field and elliptic curve arithmetic. The performance of our implementation shows a slight improvement against the fastest state-of-the-art implementations.

Keywords: AVX2 · SIMD · Vector instructions · Elliptic Curve Cryptography · Prime Field Arithmetic · Curve25519 · Diffie-Hellman Protocol

1   Introduction

Nowadays, the use of Elliptic Curve Cryptography (ECC) schemes has been widely spread in secure communication protocols, such as the key agreement Elliptic Curve Diffie-Hellman (ECDH) protocol. In terms of performance, the critical operation in elliptic curve protocols is the computation of point multiplication. This operation can be accelerated by performing efficient computation of the underlying prime field arithmetic.

Recently, proposals of new elliptic curves defined over prime fields that accelerate the finite field arithmetic operations have appeared in [2,3,10]. Such proposals use pseudo-Mersenne primes ($p = 2^m - c$) which enable fast modular reduction. One of these proposals is based on the curve named *Curve25519*, which has been gained a lot of relevance due to their efficiency and secure implementation. The Curve25519 is a Montgomery elliptic curve defined over $\mathbb{F}_{2^{255}-19}$. On this curve, only the x-coordinate of a point P is required to compute the x-coordinate of the point multiplication kP, for any integer k.

For the past few years, processors have benefited from increasing support for vector instructions, which operate each instruction over a vector of data.

Armando Faz-Hernández and Julio López were partially supported by the Intel Labs University Research Office.
Julio López was partially supported by FAPESP, Projeto Temático grant number 2013/25.977-7.

K. Lauter and F. Rodríguez-Henríquez (Eds.): LatinCrypt 2015, LNCS 9230, pp. 329–345, 2015.
DOI: 10.1007/978-3-319-22174-8_18

The AVX2 instruction set extends the capabilities of the processor with 256-bit registers and instructions that are able to compute up to four simultaneous 64-bit operations. Thus, it is relevant to study how to benefit from vector instructions for the acceleration of ECC protocols. In this work, we exhibit implementation techniques using AVX2 instructions to compute the ECDH protocol based on Curve25519.

The rest of the document is presented as follows: in Sect. 2, the features of the AVX2 instruction set are described; in Sect. 3, we detail the prime field arithmetic for pseudo-Mersenne primes; in Sect. 4, Curve25519 is described along with the arithmetic of the Montgomery elliptic curve; in Sect. 5, we present the implementation techniques for the field $\mathbb{F}_{2^{255}-19}$ using AVX2 instructions; in Sect. 6, the performance results of our implementation are summarized; and finally in Sect. 7, we present the conclusions of this work.

2   The AVX2 Instruction Set

An interesting trend of micro-architecture design is the Single Instruction Multiple Data (SIMD) processing; in this setting, processors contain a special bank of vector registers and associated vector instructions, which are able to compute an operation on every element stored in the vector register. Since 1997, SIMD processing has been present on processors; first, starting with the MMX instruction set [12] which contains 64-bit vectors; and then followed by a number of Streaming SIMD Extensions (SSE) instruction sets [13] that extended the size of vector registers to 128 bits.

In 2011, the Advanced Vector eXtensions (AVX) instruction set was released. It extended the size of vector registers to 256 bits. However, most of the AVX instructions were focused on the acceleration of floating point arithmetic targeting applications for graphics and scientific computations, postponing the integer arithmetic instructions to later releases. Therefore, in 2013 Intel released the Haswell micro-architecture with the AVX2 instruction set [14], which contained plenty of new instructions not only to support integer arithmetic, but also to compute other versatile operations. For the purpose of this work, we detail the most relevant AVX2 instructions used, which will be referred by a mnemonic described in Table 4 (in Appendix A):

- Logic. The XOR and AND instructions were extended to operate over every bit in a 256-bit register. The ALIGN instruction concatenates simultaneously the lower (and higher) 128-bit parts from two 256-bit registers into a 256-bit temporary result, then shifts the result right by a multiple of 8 bits, and stores the lower (and higher) 128 bits in a destination register.
- Integer addition/subtraction (ADD/SUB). AVX2 extended the integer addition and subtraction instructions from SSE and AVX to 256-bit vectors, thus enabling the computation of four simultaneous 64-bit operations. On AVX2 both addition and subtraction are unable to handle input carry and borrow.
- Integer multiplication (MUL). AVX2 is able to compute four products of 32×32 bits, storing four 64-bit results on a 256-bit vector register.

- Variable shifts. Former instruction sets were able to compute logical shifts SHL/SHR using the same fixed (resolved at compile time) shift displacement for every word stored in the vector register. Now, AVX2 added the new SHLV/SHRV variable shift instructions; thus, the displacement used for each word can be determined at run time. This feature adds more flexibility for the implementation of asymmetric operations over vector registers.
- Combination. The BLEND instruction fills the content of a vector register with the words from two different register sources chosen through a binary selection mask register; such mask can be defined either at compile or run time. The UNPCK instruction sets a register with the interleaved words of two registers.
- Permutation. The PERM, BCAST and PERM128 instructions move the words stored in a 256-bit vector register using a permutation pattern that can be specified either at compile or at run time.

In terms of performance, it is worth to say that the 4× speedup factor expected for 64-bit operations using AVX2 can be attained only for some instructions; in practice, factors like the execution latency of the instruction, the number of execution units available and the throughput of every instruction reduce the acceleration.

3 Prime Field Arithmetic Using Pseudo-Mersenne Primes

This section describes the techniques used for the efficient computation of the prime field arithmetic using a pseudo-Mersenne prime modulus. First, the representation of elements in a prime field is detailed; we then show how to perform each prime field operation under such a representation.

3.1 Representation of Prime Field Elements

Given an integer n (e.g. the size of machine registers), a commonly used approach to represent a field element $a \in \mathbb{F}_p$ is using a *multiprecision representation*:

$$A(n) = \sum_{i=0}^{s-1} u_i 2^{in}, \tag{1}$$

such that $a \equiv A(n) \bmod p$ for $n \in \mathbb{Z}^+$, $0 \le u_i < 2^n$ and $s = \lceil \frac{m}{n} \rceil$. Using this representation an element is stored using s words of n bits. Multiprecision representation has been widely used in several multiprecision and cryptographic libraries [1,17,23].

However, one of the disadvantages of using a multiprecision representation on an n-bit architecture is that some arithmetic operations impose a sequential evaluation of integer operations; e.g. in the modular addition, the carry bits must be propagated from the least to the most significant coefficient, and this behavior limits the parallelism level of the computations. Since AVX2 has no support for additions with carry, then a representation that minimizes the propagation of carry bits is required.

The *redundant representation* meets the criteria, because it relies on the selection of a real number $n' < n$; thus each word will have enough bits to store the carry bits produced by several modular additions. Thus, a field element $a \in \mathbb{F}_p$ in this representation is denoted by the tuple of coefficients $\mathbf{A} = \{a_{s'-1}, \ldots, a_0\}$ of the following number:

$$A(n') = \sum_{i=0}^{s'-1} a_i 2^{\lceil in' \rceil}, \tag{2}$$

where $a \equiv A(n') \bmod p$ for $n' \in \mathbb{R}$ and $s' = \lceil \frac{m}{n'} \rceil$. The fact that n' is a non-integer number produces that every coefficient a_i has an asymmetric amount of bits $\beta_i = \lceil n'(i+1) \rceil - \lceil n'i \rceil$ for $i \in [0, s')$.

The redundant representation introduces a significant improvement in the parallel execution of operations, a proof of such an acceleration was reported in [3], where the author proposed to use $n' = 25.5$ for speed up the elliptic curve arithmetic over $\mathbb{F}_{2^{255}-19}$.

3.2   Prime Field Operations

In order to compute prime field operations, the operands must be converted from binary to redundant representation and, at the end of the whole computation, the result must be converted back to binary.

Addition/Subtraction. Given two tuples $\mathbf{A}$ and $\mathbf{B}$, the operation $\mathbf{R} = \mathbf{A} \pm \mathbf{B}$ can be computed by performing the addition/subtraction coefficient-wise, e.g. $r_i = a_i \pm b_i$ for $i \in [0, s')$. Notice that these operations are totally independent and admit a parallel processing provided that no overflow occurs.

Multiplication. The computation of a prime field multiplication is usually processed in two parts: the integer multiplication and then the modular reduction; however as a pseudo-Mersenne prime ($p = 2^m - c$) is used to define the finite field then both operations can be computed in the same step. Therefore, given $\mathbf{A}$ and $\mathbf{B}$, the tuple $\mathbf{R} = \mathbf{A} \times \mathbf{B}$ is computed in the following manner:

$$r_i = \sum_{j=0}^{s'-1} (2^{\eta_{j,t}} \delta_{j,t}) a_j b_t \quad \text{for } i \in [0, s') \text{ and } t = i - j \bmod s', \tag{3}$$

where the terms $\delta_{x,y}$ and $\eta_{x,y}$ are constants defined as follows:

$$\delta_{x,y} = \begin{cases} c & \text{if } x + y \geq s', \\ 1 & \text{otherwise.} \end{cases} \tag{4}$$

$$\eta_{x,y} = \lceil xn' \rceil + \lceil yn' \rceil - \lceil (x + y \bmod s')n' \rceil \bmod m. \tag{5}$$

Since $\forall i \in [0, s')$ $\beta_i \leq \lceil n' \rceil$ is true, this implies that the products in r_i on Eq. 3 will not overflow the $2n$-bit boundary for some $n' < n$. Additionally, whenever

$\delta_{x,y} \neq 1$ denotes those products that were moved around to the corresponding power of two because of modular reduction; and $\eta_{x,y} \neq 1$ indicates that some products must be adjusted as a consequence of n' not being an integer.

Squaring. Following the same idea for multiplication; in the squaring some products appear twice and then can be replaced with multiplications by 2, as denoted by term $\nu_{x,y} = 2$ if $x \neq y$, otherwise $\nu_{x,y} = 1$. Given a tuple $\mathbf{A}$, the square $\mathbf{R} = \mathbf{A}^2$ is computed as:

$$r_i = \sum_{j=\lceil \frac{i}{2} \rceil}^{\lfloor \frac{1}{2}(s'+i) \rfloor} (2^{\eta_{j,t}} \delta_{j,t} \nu_{j,t}) a_j a_t \quad \text{for } i \in [0, s'),\ t = i - j \bmod s', \qquad (6)$$

Coefficient Reduction. Every time an addition, a subtraction, a multiplication or a squaring is computed, the result fits on s' words of n bits, so it is possible to continue processing more additions and subtractions. However, if the result of the operation is the input of a multiplication or of a squaring, then a coefficient reduction must be processed.

The coefficient reduction over a tuple $\mathbf{A}$ is an operation that ensures that every coefficient a_i verifies the following condition: $|a_i| \leq \beta_i + 1$, where β_i was defined in the previous section. This operation keeps the size of coefficients under a safe range to process another modular operation without overflowing registers.

Given a tuple $\mathbf{A}$, every coefficient is splitted into three parts, namely $a_i = h_i \parallel m_i \parallel l_i$, where $|l_i| = \beta_i$, $|m_i| = \beta_{i+1 \bmod s'}$, $|h_i| = n - |l_i| - |m_i|$ for $i \in [0, s')$, thus the coefficient reduction is computed as follows: $a_0' = l_0 + t_0$, $a_1' = l_1 + m_0 + t_1$, and $a_i' = l_i + m_{i-1} + h_{i-2}$ for $i \in [2, s')$, where the terms t_0 and t_1 are computed using the following equation: $t_1 2^{\beta_0} + t_0 = c \cdot (h_{s'-1} 2^{\beta_0} + m_{s'-1} + h_{s'-2})$.

Multiplicative Inverse. In order to compute the multiplicative inverse of an element $a \in \mathbb{F}_p^*$, the following identity is used: $a^{-1} \equiv a^{p-2} \pmod{p}$; part of this exponentiation can be calculated using an addition chain as shown by Itoh-Tsujii in [18]. Let $x, y \in \mathbb{Z}^+$ and $x \leq y$, define the term $\alpha_x = a^{2^x - 1}$ and the relation $\alpha_x \to \alpha_y$ as $\alpha_y = (\alpha_x)^{2^{y-x}} \alpha_{y-x}$. In [3] was given an addition chain for $\mathbb{F}_{2^{255}-19}$, starting with $\alpha_5 \to \alpha_{10} \to \alpha_{20} \to \alpha_{40} \to \alpha_{50} \to \alpha_{100} \to \alpha_{200} \to \alpha_{250}$, the multiplicative inverse is obtained as $a^{-1} = a^{2^{255}-21} = (\alpha_{250})^{2^5} a^{11}$ using 11 multiplications, 254 squarings and 265 coefficient reductions.

4 Elliptic Curve Diffie-Hellman on Curve25519

4.1 Safe Elliptic Curves

Around 2000, the National Institute of Standards and Technology (NIST) standardized a set of elliptic curves and associated parameters of finite fields to provide elliptic curve cryptography schemes for different security levels [20]. The selected elliptic curves were defined over both binary and prime fields. For the case of prime fields, prime numbers were selected as Generalized Mersenne

Table 1. Recent proposals of elliptic curves for three different security levels.

Security Level	Elliptic Curve	Prime Number (p)
128	NIST-P256	$2^{256} - 2^{224} + 2^{192} + 2^{96} - 1$
	Curve25519	$2^{255} - 19$
	Curve1174	$2^{251} - 9$
192	NIST-P384	$2^{384} - 2^{128} - 2^{96} + 2^{32} - 1$
	M-383	$2^{383} - 187$
	E-382	$2^{382} - 105$
256	NIST-P521	$2^{521} - 1$
	M-511	$2^{511} - 187$
	E-521	$2^{521} - 1$

primes, defined by Solinas in [22]; these primes have the property of allowing faster modular reduction compared to random selected primes.

Recently, new proposals have appeared for different elliptic curves which associate a different construction of prime modulus, such as [2,3,10]. The proposals have in common the use of pseudo-Mersenne primes ($p = 2^m - c$), with m being close to twice the targeted security level, and c as small as possible for the acceleration of prime field operations.

Nowadays, the study of the prime field implementation not only impacts on the efficiency of the cryptographic protocols but also on its security. An implementation of prime field arithmetic could cause leakage of secret information when is not implemented properly. Recently, Bernstein and Lange started the project called *SafeCurves* [7] with the aim to ensure elliptic curve cryptography security through the design of simple and secure implementations. The SafeCurves project evaluates the fulfillment of some security criteria over several elliptic curves. Table 1 lists some of the recent proposals of elliptic curves, notice that prime numbers selected are simpler than those from NIST's recommendation.

4.2   Arithmetic of Curve25519

Focussing on the 128-bit security level, the elliptic curve named Curve25519 has attracted some attention due to its efficient and secure implementation. For example, it has been proposed for inclusion in the DNS protocol (DNSCurve project [5]); additionally, the OpenSSH library has chosen Diffie-Hellman over Curve25519 as the default key-exchange protocol [21].

Curve25519 was proposed by Bernstein [3] for the acceleration of the elliptic curve Diffie-Hellman protocol targeting 128-bit security level. This curve is defined over the prime field $\mathbb{F}_{2^{255}-19}$ and has the following form:

$$\text{Curve25519:} \quad y^2 = x^3 + \hat{a}_2 x^2 + x, \qquad \hat{a}_2 = 486662. \tag{7}$$

This curve belongs to the family of Montgomery elliptic curves, which were used in [19] to accelerate the elliptic curve method for factoring (ECM).

Algorithm 1. Ladder Step Algorithm Tuned for SIMD Processing

Input: $X_{P-Q}, Z_{P-Q}, X_P, Z_P, X_Q, Z_Q \in \mathbb{F}_p$ and the coefficient $\hat{a}_2$ from Eq. (7).
Output: $X_{2P}, Z_{2P}, X_{P+Q}, Z_{P+Q} \in \mathbb{F}_p$.

1: $A \leftarrow X_P + Z_P$	$C \leftarrow X_Q + Z_Q$	[add]
2: $B \leftarrow X_P - Z_P$	$D \leftarrow X_Q - Z_Q$	[sub]
3: $DA \leftarrow A \times D$	$CB \leftarrow C \times B$	[mul]
4: $t_1 \leftarrow DA + CB$	$t_0 \leftarrow DA - CB$	[add/sub]
5: $t_1 \leftarrow t_1^2$	$t_0 \leftarrow t_0^2$	[sqr]
6: $X_{P+Q} \leftarrow t_1 \times Z_{P-Q}$	$Z_{P+Q} \leftarrow t_0 \times X_{P-Q}$	[mul]
7: $A' \leftarrow A^2$	$B' \leftarrow B^2$	[sqr]
8: $A'x \leftarrow \frac{1}{4}(\hat{a}_2 + 2) \cdot A'$	$B'y \leftarrow \frac{1}{4}(\hat{a}_2 - 2) \cdot B'$	[mul-cst]
9: $E \leftarrow A' - B'$	$F \leftarrow A'x - B'y$	[sub]
10: $X_{2P} \leftarrow A' \times B'$	$Z_{2P} \leftarrow E \times F$	[mul]
11: **return** $X_{2P}, Z_{2P}, X_{P+Q}, Z_{P+Q}$.		

In the same paper, Montgomery devised an algorithm to efficiently compute the x-coordinate of kP using only the x-coordinate of the point P; the technique uses the projective representation of points on the curve.

The computation of point multiplication using Montgomery Ladder algorithm is shown in Algorithm 6 in Appendix B.3. For each bit of the scalar, the procedure updates the values of two points P and Q through the ladder step algorithm (Algorithm 1), which computes a point doubling of P and a differential point addition of P and Q. The results are conditionally stored in temporary registers depending on a bit of the scalar k; the conditional update must be protected to avoid leaking the bits of k using either arithmetic or logic operations. Finally, the affine version of the x-coordinate of Q is recovered.

5   The AVX2 Implementation

This section starts discussing some performance penalties encountered in AVX2, then we describe some ways of getting a better performance through parallel computations in the Montgomery ladder algorithm, and finally we will show the techniques used to implement the prime field $\mathbb{F}_{2^{255}-19}$ with AVX2 instructions.

5.1   Performance Challenges Using AVX2

Before proceeding to the implementation of prime field operations, we detail a relevant issue on the implementation of modular multiplication. Recall that the AVX2 instruction MUL is able to process four integer multiplications of 32×32 bits, so in order to compute some products of the modular multiplication the natural approach is to pack four consecutive products. However, the way that the products are packed is critical in terms of performance; as an illustrative example, we present two cases that compute four products required in the modular multiplication, under the assumption that there are four registers initialized with the following values: $R_0 = [a_3, a_2, a_1, a_0]$, $R_1 = [b_3, b_2, b_1, b_0]$, $R_2 = [b_7, b_6, b_5, b_4]$ and $R_3 = [\Box, \Box, b_9, b_8]$.

Example 1. To calculate the vector $[a_0 b_3, a_0 b_2, a_0 b_1, a_0 b_0]$, only two instructions are required: first, we fill a register with a_0 using the BCAST instruction and then we apply the MUL instruction with the register R_1.

Example 2. Computing $[a_3 b_0, a_3 b_9, a_3 b_8, a_3 b_7]$ requires the following set of operations:

$$
\begin{array}{ll}
X \leftarrow \text{BCAST}(R_0) & [a_3, a_3, a_3, a_3] \\
Y \leftarrow \text{BLEND}(R_1, R_2, 1100) & [b_7, b_6, b_1, b_0] \\
Y \leftarrow \text{PERM}(Y) & [b_0, b_6, b_1, b_7] \\
U \leftarrow \text{PERM}(R_3) & [\Box, b_9, b_8, \Box] \\
Y \leftarrow \text{BLEND}(Y, U, 0110) & [b_0, b_9, b_8, b_7] \\
Z \leftarrow \text{MUL}(X, Y) & [a_3 b_0, a_3 b_9, a_3 b_8, a_3 b_7]
\end{array}
$$

In the second example, the computation takes 5 instructions just to place the operands in the right position to be multiplied, while in the first example it only takes 1 instruction. Products arranged as in the second example appear more frequently in the computation of the modular multiplication; although we could compute them using permutation instructions, the use of these instructions impacts negatively on the performance of the operations.

The high latency of permutation instructions is the result of the architectural design of Haswell micro-architecture. The previous instruction sets (SSE and AVX) operate with an execution network that computes vector instructions on 128-bit registers. On the other hand, Haswell contains an additional network of 128-bit registers to represent the higher part of a 256-bit register, so both networks compute in parallel most of the AVX2 instructions. Consequently, any data transfer between such networks will incur a performance penalty.

5.2 The SIMD Montgomery Ladder

Since an efficient implementation of the prime field will improve the elliptic curve arithmetic; so, we also focus on the flow of operations in the curve level. Analyzing the ladder step algorithm, we noticed that there are several opportunities to compute two prime field operations in parallel without dependency between the elements involved in the operation.

The general idea is simple: it is possible to compute two prime field operations by packing the operands in the lower and higher parts of a 256-bit vector register, thus the arithmetic operations will be computed on both parts at the same time. At this point, some natural questions are raised: why do we not use a 4-way version in the evaluation of the operations? The answer comes from the evaluation of Montgomery ladder step algorithm, which processes a nearly symmetrical computation over two sets of prime field elements; this does not restrict the use of, for example, a 4-way version applied to computations with four independent operations in other scenarios. A second natural question is: using 2-way prime field operations, are the benefits brought by the use of vector instructions lost? Working in this scenario with 256-bit registers, each prime field operation still takes advantage from the use of 128-bit registers.

A parallel computation of the ladder step algorithm was suggested in [11]: one of the parallel units will compute the point doubling while the other unit will produce the differential point addition. Another interesting idea is scheduling operations in a SIMD fashion, which was demonstrated to be efficient on the implementation presented in [9]; such an implementation takes advantage of the use of the NEON instructions to compute two finite field operations independently.

In this work, we go further by exploiting the parallelism at two levels: first at high level, the SIMD execution of prime field operations; and at low level, the computations inside of the prime field operation can also benefit from SIMD execution. The right-hand side of Algorithm 1 lists the operations computed in each row; as one may notice, the same operation is applied to two different data sets exhibiting exactly the spirit behind the SIMD paradigm. The way that the ladder step algorithm was presented in Algorithm 1 gives an insight of the register allocation and of the scheduling of operations.

5.3   Implementation of $\mathbb{F}_{2^{255}-19}$

For the implementation of $\mathbb{F}_{2^{255}-19}$ the most efficient approach is to set $n' = 51$ on a 64-bit architecture and $n' = 25.5$ for a 32-bit architecture. As was mentioned before, Bernstein in [3] encouraged the use of $n' = 25.5$; such a implementation used the floating point registers to emulate integer arithmetic operations because in that scenario a double precision floating point register is able to store a 53-bit number without loss of information. In our scenario, we also choose $n' = 25.5$ as working on a 32-bit architecture, because the wider multiplier available in AVX2 is a 32-bit multiplier, even though other fundamental integer operations support 64-bit operands.

Summarizing the parameters described in Sect. 3.1, our implementation of $\mathbb{F}_{2^{255}-19}$ sets $n' = 25.5$, $s' = 10$, $n = 32$ for integer multiplication and $n = 64$ for integer, shift and logic operations.

Interleaving Tuples. As it was shown in the previous section, the ladder step function computes two field operations at each time. We denote by $\langle \mathbf{A}, \mathbf{B} \rangle$ the interleaved tuples $\mathbf{A}$ and $\mathbf{B}$ which represents five 256-bit registers, such that $\langle \mathbf{A}, \mathbf{B} \rangle_i = [a_{2i+1}, a_{2i}, b_{2i+1}, b_{2i}]$ for $i \in [0, 5)$. Thus, two coefficients from tuple $\mathbf{A}$ will be stored in the higher 128-bit register and two coefficients from tuple $\mathbf{B}$ will be stored in the lower 128-bit register.

Addition/Subtraction. The addition and subtraction operations require only five addition (ADD) or subtraction (SUB) instructions, respectively.

Multiplication. Algorithm 2 shows the computation of two interleaved tuples $\langle \mathbf{A}, \mathbf{C} \rangle$ and $\langle \mathbf{B}, \mathbf{D} \rangle$. Values on the right hand side show the content stored in the destination register. The lines 2 to 5 compute the multiplication by the $\eta_{x,y}$ term using variable shift instructions. In the main loop (lines 6–18), a temporary register U contains the first and third words of $\langle \mathbf{A}, \mathbf{C} \rangle_i$ in the lower and higher 128-bit parts of the register, respectively; this task can be efficiently performed using a SHUF instruction. Once that U was computed, in the inner loop U will be multiplied by $\langle \mathbf{B}, \mathbf{D} \rangle_j$ and the result will be accumulated in Z_{i+j}. Analogously to U, the products resulting from the V register will be accumulated in Z_{i+j+1}. Register W contains some products that must be accumulated in either Z_i or Z_{i+5}, thereby a BLEND instruction masks the appropriate products to be added. Finally, the products for which $\delta_{x,y} = 19$ will be multiplied using shift instructions.

Algorithm 2. Instruction scheduling to compute a modular multiplication in $\mathbb{F}_{2^{255}-19}$ using AVX2 instructions.

Input: Two interleaved tuples $\langle \mathbf{A}, \mathbf{C} \rangle$ and $\langle \mathbf{B}, \mathbf{D} \rangle$.
Output: An interleaved tuple $\langle \mathbf{E}, \mathbf{F} \rangle$ such that $\mathbf{E} = \mathbf{A} \times \mathbf{B}$ and $\mathbf{F} = \mathbf{C} \times \mathbf{D}$.

1: $Z_i \leftarrow 0$ for $i \in [0, 10)$
2: **for** $i \leftarrow 0$ **to** 4 **do**
3:   $\langle \mathbf{B}', \mathbf{D}' \rangle_i \leftarrow \text{ALIGN}(\langle \mathbf{B}, \mathbf{D} \rangle_{i+1 \bmod 5}, \langle \mathbf{B}, \mathbf{D} \rangle_i)$ $[b_{2i+2}, b_{2i+1}, d_{2i+2}, d_{2i+1}]$
4:   $\langle \mathbf{B}', \mathbf{D}' \rangle_i \leftarrow \text{SHLV}(\langle \mathbf{B}', \mathbf{D}' \rangle_i, [0,1,0,1])$ $[b_{2i+2}, 2b_{2i+1}, d_{2i+2}, 2d_{2i+1}]$
5: **end for**
6: **for** $i \leftarrow 0$ **to** 4 **do**
7:   $U \leftarrow \text{SHUF}(\langle \mathbf{A}, \mathbf{C} \rangle_i, 0)$ $[a_{2i}, a_{2i}, c_{2i}, c_{2i}]$
8:   **for** $j \leftarrow 0$ **to** 4 **do**
9:     $Z_{i+j} \leftarrow \text{ADD}(Z_{i+j}, \text{MUL}(U, \langle \mathbf{B}, \mathbf{D} \rangle_j))$ $[a_{2i}b_{j+1}, a_{2i}b_j, c_{2i}d_{j+1}, c_{2i}d_j]$
10:   **end for**
11:   $V \leftarrow \text{SHUF}(\langle \mathbf{A}, \mathbf{C} \rangle_i, 1)$ $[a_{2i+1}, a_{2i+1}, c_{2i+1}, c_{2i+1}]$
12:   **for** $j \leftarrow 0$ **to** 3 **do**
13:     $Z_{i+j+1} \leftarrow \text{ADD}(Z_{i+j+1}, \text{MUL}(V, \langle \mathbf{B}', \mathbf{D}' \rangle_j))$
    $[a_{2i+1}b_{2j+2}, 2a_{2i+1}b_{2j+1}, c_{2i+1}d_{2j+2}, 2c_{2i+1}d_{2j+1}]$
14:   **end for**
15:   $W \leftarrow \text{MUL}(V, \langle \mathbf{B}', \mathbf{D}' \rangle_4)$ $[a_{2i+1}b_0, 2a_{2i+1}b_9, c_{2i+1}d_0, 2c_{2i+1}d_9]$
16:   $Z_i \leftarrow \text{ADD}(Z_i, \text{BLEND}(W, [0,0,0,0], 0101))$ $[a_{2i+1}b_0, 0, c_{2i+1}d_0, 0]$
17:   $Z_{i+5} \leftarrow \text{ADD}(Z_{i+5}, \text{BLEND}(W, [0,0,0,0], 1010))$ $[0, 2a_{2i+1}b_9, 0, 2c_{2i+1}d_9]$
18: **end for**
19: **for** $i \leftarrow 0$ **to** 4 **do**
20:   $19Z_{i+5} \leftarrow \text{ADD}(\text{ADD}(\text{SHL}(Z_{i+5}, 4), \text{SHL}(Z_{i+5}, 1)), Z_{i+5})$
21:   $\langle \mathbf{E}, \mathbf{F} \rangle_i \leftarrow \text{ADD}(Z_i, 19Z_{i+5})$
22: **end for**
23: **return** $\langle \mathbf{E}, \mathbf{F} \rangle$

Squaring/Coefficient Reduction. The description of the instruction scheduling for these operations can be found in Appendices B.1 and B.2, respectively.

Conditional Swapping. The Montgomery point multiplication requires the conditional swapping of register values depending on the bits of the scalar; usually, the scalar represents a secret key, thereby this operation must be computed without branches and must run in constant time. In order to implement these requirements, the conditional swapping is computed using logic operations as shown in Algorithm 3.

Algorithm 3. Conditional Swapping.

Input: $b \in \{0,1\}$ a conditional bit, X and Y two registers to be swapped.
Output: $X \leftarrow Y$ and $Y \leftarrow X$ if $b = 1$, otherwise remain unchanged.
1: $M \leftarrow \text{BCAST}(-b)$
2: $T \leftarrow \text{AND}(\text{XOR}(X,Y),M)$
3: $X' \leftarrow \text{XOR}(X,T)$
4: $Y' \leftarrow \text{XOR}(Y,T)$
5: **return** X',Y'

---

6   Performance Results

Benchmarking was performed on a Haswell processor (Core i7-4770) at 3.4 GHz, where the Intel Turbo Boost and Intel Hyper Threading technologies were disabled. Our code was compiled using the GNU C Compiler v4.9.0 and timings were measured as the average time of 10^6 and 10^4 computations for prime field operations and point multiplication, respectively.

Prime Field Operations. Table 2 shows the performance of the field arithmetic operations using AVX2 instructions. The first row exhibits the clock cycles required to compute one single arithmetic operation over a tuple **A**; the second row represents the clock cycles used to compute two simultaneous arithmetic operations over interleaved tuples $\langle \mathbf{A}, \mathbf{B} \rangle$; and the last row shows the speedup factor obtained by the 2-way against the single implementation.

The acceleration of the 2-way operations was achieved by minimizing the use of permutation instructions and working with interleaved tuples. Recently, an algorithm to compute a modular multiplication on the field $\mathbb{F}_{2^{521}-1}$ using a redundant representation was presented in [16]. This algorithm requires $\frac{1}{2}s'(s' + 1)$ word multiplications and $2(s'^2 - 1)$ word additions. The paper also shows a formulation for the field $\mathbb{F}_{2^{255}-19}$; and following this idea, we implemented a 2-way multiplier whose performance was 117 clock cycles, and this result is 48 % slower than our schoolbook 2-way multiplier that takes 79 clock cycles. The main issue observed was an overhead produced by arranging the vectors to be multiplied, as permuting words between registers is costly.

Elliptic Curve Diffie-Hellman. In order to illustrate the performance of our software implementation of Elliptic Curve Diffie-Hellman protocol using Curve25519, we follow the guidelines presented in [4] to implement the following algorithms:

– **Key Generation.** Let G be the generator point of the Curve25519 where $x(G) = 9$, the key generation algorithm computes a public key $x(kG)$ given a secret key $k \in [0, 2^{256})$.
– **Shared Secret.** Given the x-coordinate of the public key, $x(P)$, and a secret key k, the shared secret algorithm computes the x-coordinate of kP.

Table 3 shows the timings obtained by our implementation and compares the performance against the state-of-the-art implementations. For the key generation

Table 2. Cost in clock cycles to compute one prime field operation using single implementation, and two prime field operations using the 2-way implementation.

	Addition Subtraction	Multiplication	Squaring	Coefficient Reduction	Inversion
Single (1 op)	4	57	47	26	16,500
2-way (2 ops)	8	79	57	33	21,000
Speedup Factor	1×	1.44×	1.64×	1.57×	1.57×

Table 3. Timings obtained for the computation of the Elliptic Curve Diffie-Hellman protocol. The entries represent 10^3 clock cycles. The timings in the rows with (α) were measured on our Core i7-4770 processor and the rest of the entries were taken from the corresponding references.

Implementation	Processor	Key Generation	Shared Secret
NaCl [8]	Core i7-4770 (α)	261.0	261.0
amd64-51 [6]	Core i7-4770	172.6	163.2
amd64-51 [6]	Xeon E3-1275 V3	169.9	161.6
Our work	Core i7-4770 (α)	156.5	156.5

algorithm the implementations listed in Table 3 use the same routine to compute both key generation and shared secret.

As it can be seen from Table 3, the performance achieved in both the key generation and the shared secret computations brings a moderate speedup compared against the implementations reported in the eBACS website [6]. Notice that non-vector implementations found in the literature take advantage of the native 64-bit multiplier that takes 3 clock cycles; whereas, for AVX2, the same computation is performed using a 32-bit multiplier that takes 5 clock cycles. Additionally, the high latency of some instructions guides the optimization to use a more reduced set of instructions.

On a side note, it is to be noted that, in the key generation algorithm, since $x(G) = 9$, the modular multiplication on line 6 of Algorithm 1 can be replaced by only a few shift instructions. In our implementation, this gives a 13.5 % speedup, computing the key generation step in only 135.5×10^3 clock cycles.

7   Conclusions

Applying vector instructions to an implementation requires a careful knowledge of the target architecture, thus selecting the best scheduling of instructions is not a straightforward task because it demands a meticulous study of the instruction set and of the architectural capabilities. On the presence of architectural issues that limit the performance, we found a way to overcome some of them and produced an efficient implementation as fast and secure as the best optimized implementations for 64-bit architectures.

Our main contribution is a fast implementation of the elliptic curve Diffie-Hellman protocol based on Curve25519 with a minor improvement over the state-of-the-art implementations. This performance was mainly achieved due to the efficient implementation of $\mathbb{F}_{2^{255}-19}$ using AVX2 vector instructions.

In this work, we expose the versatility of the AVX2 instructions, where the SIMD processing was applied at two levels: at the higher level, we showed how to schedule arithmetic operations in the Montgomery ladder step algorithm to be computed in parallel; and at the lower level, the computation of prime field operations also benefited from vector instructions.

We remark that the algorithms used to implement prime field arithmetic using vector instructions can also be extended for other prime fields that use pseudo-Mersenne primes. The applicability of these techniques to other elliptic curve models is left as a future work. Also, it would be interesting to know how the upcoming architectures will impact on the performance of AVX2 instructions. In particular, the new Intel Skylake micro-architecture that has support for 512-bit registers should promote a high applicability of the results of this work.

Acknowledgments. The authors would like to thank the anonymous reviewers for their helpful suggestions and comments. Additionally, they would like to show their gratitude to Jérémie Detrey for his valuable comments on an earlier version of the manuscript.

Table 4. Latency and reciprocal throughput of some AVX2 instructions.

Type	Mnemonic	Assembler Instructions	Latency (cycles)	Reciprocal Throughput (cycles/op)
Arithmetic	ADD/SUB	VPADDQ/VPSUBQ	1	0.5
	MUL	VPMULDQ	5	1
Logic	SHL/SHR	VPSLLQ/VPSRLQ	1	1
	SHLV/SHRV	VPSLLVQ/VPSRLVQ	2	2
	ALIGN	VPALIGNR	1	1
	AND/XOR	VPAND/VPXOR	1	0.33
Combination	BLEND	VPBLENDD	1	0.33
		VPBLENDVB	2	2
	SHUF	VSHUFPD	1	1
	UNPCK	VPUNPCKHQDQ	1	1
		VPUNPCKLQDQ	1	1
Permutation	PERM	VPERMQ	3	1
	BCAST	VPBROADCASTQ	5	0.5
	PERM128	VPERM2I128	3	1

A   Relevant AVX2 Instructions

A list of the most relevant instructions used in this work is presented. For clarity, instructions were grouped according to their functionality. Table 4 shows in the second column a mnemonic used in this document; in the third column is described the specific assembler name of the instruction, and the last columns show the latency and the reciprocal throughput of every instruction, the entries were taken from the Agner Fog's measurements published in [15].

Algorithm 4. Instruction scheduling to compute a modular squaring in $\mathbb{F}_{2^{255}-19}$ using AVX2 instructions.

Input: An interleaved tuple $\langle \mathbf{A}, \mathbf{B} \rangle$.
Output: An interleaved tuple $\langle \mathbf{E}, \mathbf{F} \rangle$ such that $\mathbf{E} = \mathbf{A}^2$ and $\mathbf{F} = \mathbf{B}^2$.

1: **for** $i \leftarrow 0$ **to** 4 **do**
2:     $U_{2i} \leftarrow \langle \mathbf{A}, \mathbf{B} \rangle_i$ $[a_{2i+1}, a_{2i}, b_{2i+1}, b_{2i}]$
3:     $U_{2i+1} \leftarrow \text{ALIGN}(\langle \mathbf{A}, \mathbf{B} \rangle_{i+1 \bmod 5}, \langle \mathbf{A}, \mathbf{B} \rangle_i)$ $[a_{2i+2}, a_{2i+1}, b_{2i+2}, b_{2i+1}]$
4:     $U_{2i+1} \leftarrow \text{SHLV}(U_{2i+1}, [0,1,0,1])$ $[a_{2i+2}, 2a_{2i+1}, b_{2i+2}, 2b_{2i+1}]$
5:     $V_{2i} \leftarrow \text{SHUF}(\langle \mathbf{A}, \mathbf{B} \rangle_i, 0)$ $[a_{2i}, a_{2i}, b_{2i}, b_{2i}]$
6:     $V_{2i+1} \leftarrow \text{SHUF}(\langle \mathbf{A}, \mathbf{B} \rangle_i, 1)$ $[a_{2i+1}, a_{2i+1}, b_{2i+1}, b_{2i+1}]$
7: **end for**
8: **for** $i \leftarrow 0$ **to** 4 **do**
9:     $T \leftarrow \text{MUL}(U_i, V_i)$ $[a_{i+1}a_i, a_i a_i, b_{i+1}b_i, b_i b_i]$
10:     $Z_i \leftarrow \text{BLEND}(T, [0,0,0,0], 1010)$ $[0, a_i a_i, 0, b_i b_i]$
11:     $W \leftarrow \text{BLEND}(T, [0,0,0,0], 0101)$ $[a_{i+1}a_i, 0, b_{i+1}b_i, 0]$
12:     **for** $j \leftarrow 1$ **to** i **do**
13:         $t \leftarrow i - j \bmod 10$
14:         $W \leftarrow \text{ADD}(W, \text{MUL}(U_j, V_t))$ $[a_{j+1}a_t, a_j a_t, b_{j+1}b_t, b_j b_t]$
15:     **end for**
16:     $Z_i \leftarrow \text{ADD}(Z_i, \text{SHL}(W, 1))$
17:     $S \leftarrow \text{MUL}(U_{i+5}, V_{i+5})$ $[a_{i+6}a_{i+5}, a_{i+5}a_{i+5}, b_{i+6}b_{i+5}, b_{i+5}b_{i+5}]$
18:     $Z_{i+5} \leftarrow \text{BLEND}(S, [0,0,0,0], 1010)$ $[0, a_{i+5}a_{i+5}, 0, b_{i+5}b_{i+5}]$
19:     $X \leftarrow [0,0,0,0]$
20:     **for** $j \leftarrow i+1$ **to** 4 **do**
21:         $t \leftarrow i - j \bmod 10$
22:         $X \leftarrow \text{ADD}(X, \text{MUL}(U_j, V_t))$ $[a_{j+1}a_t, a_j a_t, b_{j+1}b_t, b_j b_t]$
23:     **end for**
24:     $Z_{i+5} \leftarrow \text{ADD}(Z_{i+5}, \text{SHL}(X, 1))$
25: **end for**
26: **for** $i \leftarrow 0$ **to** 4 **do**
27:     $19Z_{i+5} \leftarrow \text{ADD}(\text{ADD}(\text{SHL}(Z_{i+5}, 4), \text{SHL}(Z_{i+5}, 1)), Z_{i+5})$
28:     $\langle \mathbf{E}, \mathbf{F} \rangle_i \leftarrow \text{ADD}(Z_i, 19Z_{i+5})$
29: **end for**
30: **return** $\langle \mathbf{E}, \mathbf{F} \rangle$

B   Algorithms

B.1   Implementation of Modular Squaring Using AVX2

To compute the modular squaring we follow a similar approach like in the case of modular multiplication. Algorithm 4 shows the scheduling of instructions used to compute the modular squaring of an interleaved tuple $\langle \mathbf{A}, \mathbf{B} \rangle$. The products $a_{x,y}$ such that $\nu_{x,y} = 2$ are computed in the inner loops (lines 12 to 15 and 20 to 23) and once that these products were accumulated, they are multiplied by 2 using shift instructions. At the end, the lines from 26 to 29 compute the modular reduction.

B.2   Implementation of Coefficient Reduction Using AVX2

The coefficient reduction is processed coefficient-wise. We split each coefficient into three parts $a_i = h_i \parallel m_i \parallel l_i$ and compute the process described in Sect. 3.2. Simultaneously, each m_i (medium coefficient) is added to the correspondent l_{i+1} (low coefficient) and to the h_{i-1} (high coefficient). For those coefficients that need to be reduced modulo p, we compute the multiplication by c using just shift instructions. After the coefficient reduction is processed, the size of each coefficient in the updated tuple will have at most $\beta_i + 1$ bits.

Algorithm 5. Instruction scheduling for computing a coefficient reduction in $\mathbb{F}_{2^{255}-19}$ using AVX2 instructions.

Input: An interleaved tuple $\langle \mathbf{A}, \mathbf{B} \rangle$.
Output: An updated interleaved tuple $\langle \mathbf{A}, \mathbf{B} \rangle$ such that $|a_i| \leq \beta_i + 1$ and $|b_i| \leq \beta_i + 1$
for $i \in [0, 10)$.
1: **for** $i \leftarrow 0$ **to** 4 **do**
2:     $L_i \leftarrow \text{AND}(\langle \mathbf{A}, \mathbf{B} \rangle_i, [2^{\beta_{2i+1}} - 1, 2^{\beta_{2i}} - 1, 2^{\beta_{2i+1}} - 1, 2^{\beta_{2i}} - 1])$
3:     $M_i \leftarrow \text{SRLV}(\langle \mathbf{A}, \mathbf{B} \rangle_i, [\beta_{2i+1}, \beta_{2i}, \beta_{2i+1}, \beta_{2i}])$
4:     $M_i \leftarrow \text{AND}(M_i, [2^{\beta_{2i+2}} - 1, 2^{\beta_{2i+1}} - 1, 2^{\beta_{2i+2}} - 1, 2^{\beta_{2i+1}} - 1])$
5:     $H_i \leftarrow \text{SRLV}(\langle \mathbf{A}, \mathbf{B} \rangle_i, [\beta_{2i+1} + \beta_{2i+2}, \beta_{2i} + \beta_{2i+1}, \beta_{2i+1} + \beta_{2i+2}, \beta_{2i} + \beta_{2i+1}])$
6: **end for**
7: **for** $i \leftarrow 0$ **to** 4 **do**
8:     $M_i' \leftarrow \text{ALIGN}(M_i, M_{i-1 \bmod 5})$
9: **end for**
10: $H_4 \leftarrow \text{SHRV}(\langle \mathbf{A}, \mathbf{B} \rangle_8, [\beta_8 + \beta_9, \beta_8, \beta_8 + \beta_9, \beta_8])$
11: $U \leftarrow \text{ADD}(H_4, \text{SHR}(H_4, 64))$
12: $19U \leftarrow \text{ADD}(\text{ADD}(\text{SHR}(U, 4), \text{SHR}(U, 1)), U)$
13: $T \leftarrow \text{AND}(19U, [0, 2^{\beta_0} - 1, 0, 2^{\beta_0} - 1])$
14: $S \leftarrow \text{SHR}(19U, [0, \beta_0, 0, \beta_0])$
15: $H_4 \leftarrow \text{UPCK}(T, S)$
16: **for** $i \leftarrow 0$ **to** 4 **do**
17:     $\langle \mathbf{A}, \mathbf{B} \rangle_i \leftarrow \text{ADD}(\text{ADD}(L_i, M_i'), H_{i-1 \bmod 5})$
18: **end for**
19: **return** $\langle \mathbf{A}, \mathbf{B} \rangle$

Algorithm 6. Point Multiplication using Montgomery Ladder.

Input: $k \in [0, 2^t)$ and $x(P) \in \mathbb{F}_p$, be the x-coordinate of an elliptic curve point P.
Output: $x(Q)$, the x-coordinate of $Q = kP$.
1: Let $k = (0, k_{t-1}, \ldots, k_0)_2$
2: $X_{P-Q} \leftarrow x(P)$
3: $X_P \leftarrow x(P)$, $Z_P \leftarrow 1$
4: $X_Q \leftarrow 1$, $Z_Q \leftarrow 0$
5: **for** $i \leftarrow t - 1$ **to** 0 **do**
6:     $b \leftarrow k_i \oplus k_{i+1}$
7:     $X_Q, X_P \leftarrow \text{CondSwap}(b, X_Q, X_P)$
8:     $Z_Q, Z_P \leftarrow \text{CondSwap}(b, Z_Q, Z_P)$
9:     $X_Q, Z_Q, X_P, Z_P \leftarrow \text{Ladder}(X_{P-Q}, X_Q, Z_Q, X_P, Z_P)$
10: **end for**
11: **return** $x(Q) \leftarrow X_Q(Z_Q)^{-1}$

B.3   Point Multiplication Using Montgomery Ladder

Algorithm 6 shows the computation of the Montgomery point multiplication to calculate the x-coordinate of kP given the x-coordinate of P and an integer scalar k. This algorithm also requires the ladder step presented in Algorithm 1.

For its use in the computation of the elliptic curve Diffie-Hellman protocol using the Curve25519, the document [4] describes an encoding for the secret key when is given as a string of bytes. Then, the description of Algorithm 6 assumes that the secret key was already encoded.

References

1. Aranha, D.F., Gouvêa, C.P.L.: RELIC is an Efficient LIbrary for Cryptography. http://code.google.com/p/relic-toolkit/
2. Aranha, D.F., Barreto, P.S.L.M., Pereira, G.C.C.F., Ricardini, J.E.: A note on high-security general-purpose elliptic curves. Cryptology ePrint Archive, Report 2013/647 (2013). http://eprint.iacr.org/
3. Bernstein, D.J.: Curve25519: new Diffie-Hellman speed records. In: Yung, M., Dodis, Y., Kiayias, A., Malkin, T. (eds.) PKC 2006. LNCS, vol. 3958, pp. 207–228. Springer, Heidelberg (2006)
4. Bernstein, D.J.: Cryptography in NaCl, March 2009. http://cr.yp.to/highspeed/naclcrypto-20090310.pdf
5. Bernstein, D.J.: DNSCurve: usable security for DNS, June 2009. http://dnscurve.org
6. Bernstein, D.J., Lange, T.: eBACS: ECRYPT benchmarking of cryptographic systems, March 2015. Accessed on 20 March 2015 http://bench.cr.yp.to/supercop.html
7. Bernstein, D.J., Lange, T.: SafeCurves: choosing safe curves for elliptic-curve cryptography (2015). Accessed 20 March 2015 http://safecurves.cr.yp.to
8. Bernstein, D.J., Lange, T., Schwabe, P.: NaCl: Networking and Cryptography library, October 2013. http://nacl.cr.yp.to/

9. Bernstein, D.J., Schwabe, P.: NEON Crypto. In: Prouff, E., Schaumont, P. (eds.) CHES 2012. LNCS, vol. 7428, pp. 320–339. Springer, Heidelberg (2012). http://dx.doi.org/10.1007/978-3-642-33027-8_19

10. Bos, J.W., Costello, C., Longa, P., Naehrig, M.: Selecting Elliptic Curves for Cryptography: An Efficiency and Security Analysis. Cryptology ePrint Archive, Report 2014/130 (2014). http://eprint.iacr.org/

11. Cohen, H., Frey, G., Avanzi, R., Doche, C., Lange, T., Nguyen, K., Vercauteren, F.: Handbook of Elliptic and Hyperelliptic Curve Cryptography, (2nd edn). Chapman & Hall/CRC (2012)

12. Corporation, I.: Intel Pentium processor with MMX technology documentation, January 2008. http://www.intel.com/design/archives/Processors/mmx/

13. Corporation, I.: Define SSE2, SSE3 and SSE4, January 2009. http://www.intel.com/support/processors/sb/CS-030123.htm

14. Corporation, I.: Intel Advanced Vector Extensions Programming Reference, June 2011. https://software.intel.com/sites/default/files/m/f/7/c/36945

15. Fog, A.: Instruction tables: Lists of instruction latencies, throughputs and micro-operation breakdowns for Intel, AMD and VIA CPUs, December 2014

16. Granger, R., Scott, M.: Faster ECC over $\mathbb{F}_{2^{521}-1}$. Cryptology ePrint Archive, Report 2014/852 (2014). http://eprint.iacr.org/

17. Granlund, T., the GMP development team: GNU MP: The GNU Multiple Precision Arithmetic Library, (5.0.5 edn) (2012). http://gmplib.org/

18. Itoh, T., Tsujii, S.: A fast algorithm for computing multiplicative inverses in $GF(2^m)$ using normal bases. Inf. Comput. **78**(3), 171–177 (1988). http://dx.doi.org/10.1016/0890-5401(88)90024-7

19. Montgomery, P.L.: Speeding the pollard and elliptic curve methods of factorization. Math. Comput. **48**(177), 243–264 (1987). http://dx.doi.org/10.2307/2007888

20. National Institute of Standards and Technology: Digital Signature Standard (DSS). FIPS Publication 186, may 1994. http://www.bibsonomy.org/bibtex/2a98c67565fa98cc7c90d7d622c1ad252/dret

21. Shell, O.S.: OpenSSH, January 2014. http://www.openssh.com/txt/release-6.5

22. Solinas, J.A.: Generalized Mersenne Numbers. Technical report,Center of Applied Cryptographic Research (CACR) (1999)

23. The OpenSSL Project: OpenSSL: The Open Source toolkit for SSL/TLS, April 2003. www.openssl.org

High-Performance Ideal Lattice-Based Cryptography on 8-Bit ATxmega Microcontrollers

Thomas Pöppelmann$^{(\boxtimes)}$, Tobias Oder, and Tim Güneysu

Horst Görtz Institute for IT-Security, Ruhr-University Bochum,
Bochum, Germany
{thomas.poeppelmann,tobias.oder,tim.gueneysu}@rub.de

Abstract. Over the last years lattice-based cryptography has received much attention due to versatile average-case problems like Ring-LWE or Ring-SIS that appear to be intractable by quantum computers. But despite of promising constructions, only few results have been published on implementation issues on very constrained platforms. In this work we therefore study and compare implementations of Ring-LWE encryption and the Bimodal Lattice Signature Scheme (BLISS) on an 8-bit Atmel ATxmega128 microcontroller. Since the number theoretic transform (NTT) is one of the core components in implementations of lattice-based cryptosystems, we review the application of the NTT in previous implementations and present an improved approach that significantly lowers the runtime for polynomial multiplication. Our implementation of Ring-LWE encryption takes 27 ms for encryption and 6.7 ms for decryption. To compute a BLISS signature, our software takes 329 ms and 88 ms for verification. These results outperform implementations on similar platforms and underline the feasibility of lattice-based cryptography on constrained devices.

Keywords: Ideal lattices · NTT · RLWE · BLISS · ATxmega

1 Introduction

RSA and ECC-based schemes are the most popular asymmetric cryptosystems to date, being deployed in billions of security systems and applications. Despite of their predominance, they are known to be susceptible to attacks using quantum computers [50] on which significant resources are spent to boost their further development [47]. Additionally, RSA and ECC have been shown to be quite inefficient on very small and constrained devices like 8-bit AVR microcontrollers [26,31]. A possible alternative are asymmetric cryptosystems based

This work was partially funded by the European Union H2020 SAFEcrypto project (grant no. 644729), European Union H2020 PQCRYPTO project (grant no. 645622), German Research Foundation (DFG), and DFG Research Training Group GRK 1817/1.

© Springer International Publishing Switzerland 2015
K. Lauter and F. Rodríguez-Henríquez (Eds.): LatinCrypt 2015, LNCS 9230, pp. 346–365, 2015.
DOI: 10.1007/978-3-319-22174-8_19

on hard problems in ideal lattices. The special algebraic structure of ideal lattices [36] defined in $\mathcal{R} = \mathbb{Z}_q[\mathbf{x}]/\langle x^n + 1\rangle$ allows a significant reduction of key and ciphertext sizes and enables efficient arithmetic using the number theoretic transform (NTT)[1] [5,41,52]. To realize lattice-based public key encryption several proposals exists (see [9] for a comparison) like classical NTRU [30] (defined in $\mathbb{Z}_q[\mathbf{x}]/\langle x^n - 1\rangle$), provably secure NTRU [51] (defined in $\mathbb{Z}_q[\mathbf{x}]/\langle x^n + 1\rangle$), or a scheme based on the ring learning with errors (RLWE) problem [32,36] (from now on referred to as RLWEenc). From an implementation perspective, the RLWEenc scheme is currently one of the best-studied lattice-based public key encryption schemes (see [7,10,12,35,44,48]) and is similar to a recently proposed key exchange protocol [8]. Concerning signature schemes, several proposals exist like GLP [24], BG [1], PASSSign [29], a modified NTRU signature scheme [38], or a signature scheme derived from a recently proposed IBE scheme [19]. However, so far the Bimodal Lattice Signature Scheme (BLISS) [17] seems superior in terms of signature size, performance, and security. Despite their popularity, implementation efforts so far mainly led to very efficient hardware designs for RLWEenc [48] and BLISS [43] and fast software on 32-bit microcontrollers [42] and 64-bit microprocessors [17,25] but only few works cover constrained 8-bit architectures [6,7,35]. Additionally, current works usually rely on the straightforward Cooley-Tukey radix-2 decimation-in-time algorithm (e.g., [7,12,43,48]) to implement the NTT and thus to realize polynomial multiplication $\mathbf{c} = \mathbf{a} \cdot \mathbf{b}$ for $\mathbf{a}, \mathbf{b}, \mathbf{c} \in \mathcal{R}$ as $\mathbf{c} = \mathsf{INTT}(\mathsf{NTT}(\mathbf{a}) \circ \mathsf{NTT}(\mathbf{b}))$. However, by taking a closer look at works on the implementation [11,15] of the highly related fast Fourier transform (FFT) it becomes evident that the sole focus on Cooley-Tukey radix-2 decimation-in-time algorithms prevents further optimizations of the NTT, especially given the constraints of an 8-bit architecture.

Contribution. The contribution of this work is twofold. We first review different approaches and varieties of NTT algorithms and then adapt and optimize these algorithms for the polynomial multiplication use-case prevalent in ideal lattice-based cryptography. Improvements compared to previous work are mainly achieved by merging certain operations into the NTT itself (multiplication by n^{-1}, powers of ψ and ψ^{-1}) and by removing the expensive bit-reversal step. In the second part we provide an efficient implementation of these NTT algorithms on the 8-bit AVR/ATxmega architecture. With respect to the constrained resources of our target platform, we further speed up the NTT by applying a modular reduction algorithm that does not require integer division. By using this optimized NTT we achieve high performance for RLWEenc and BLISS. Our work shows that lattice-based cryptography can be used to realize the two most basic asymmetric primitives (public key encryption and signatures) on very constrained devices and with high performance. To the best of our knowledge, we provide the smallest implementation of BLISS on AVR in terms of flash consumption and also the fastest implementation in terms of runtime since one signature computation requires only 329 ms and verification requires 88 ms. By performing

---

[1] The NTT can be regarded as Fast Fourier Transform over $\mathbb{Z}_q$.

encryption in 27 ms and decryption in 6.7 ms, our implementation of RLWEenc also outperforms previous work on AVR. To allow third-party evaluation of our results, source code and documentation is available on our website[2]. For more information regarding the Schooolbok and Karatsuba algorithms we refer to the full version of this work [45].

2   Background

In this section we introduce the NTT and explicitly describe its application in the algorithms of the RLWEenc public key encryption scheme and the BLISS signature scheme.

2.1   The Number Theoretic Transform and Negacyclic Convolutions

The number theoretic transform (NTT) [5,41,52] is similar to the discrete Fourier transform (DFT) but all complex roots of unity are exchanged for integer roots of unity and arithmetic is also carried out modulo an integer q in the field $GF(q)$[3]. Main applications of the NTT, besides ideal lattice-based cryptography, are integer multiplication (e.g., Schönhage and Strassen [49]) and signal processing [5]. The forward transformation $\tilde{\mathbf{a}} = \mathrm{NTT}(\mathbf{a})$ of a length n sequence $(\mathbf{a}[0], \ldots, \mathbf{a}[n-1])$ to $(\tilde{\mathbf{a}}[0], \ldots, \tilde{\mathbf{a}}[n-1])$ with elements in $\mathbb{Z}_q$ is defined as $\tilde{\mathbf{a}}[i] = \sum_{j=0}^{n-1} \mathbf{a}[j]\omega^{ij} \bmod q$, $i = 0, 1, \ldots, n-1$ where ω is an n-th primitive root of unity. The inverse transform $\mathbf{a} = \mathrm{INTT}(\tilde{\mathbf{a}})$ is defined as $\mathbf{a}[i] = n^{-1} \sum_{j=0}^{n-1} \tilde{\mathbf{a}}[j]\omega^{-ij} \bmod q$, $i = 0, 1, \ldots, n-1$ where ω is exchanged by ω^{-1} and the final result scaled by n^{-1}. For an n-th primitive root of unity ω_n it holds that $\omega_n^n = 1 \bmod q$, $\omega_n^{n/2} = -1 \bmod q$, $\omega_{\frac{n}{2}} = \omega_n^2$, and $\omega_n^i \neq 1 \bmod q$ for any $i = 1, \ldots, n-1$.

The main operation in ideal lattice-based cryptography is polynomial multiplication[4]. Schemes are usually defined in $\mathcal{R} = \mathbb{Z}_q[\mathbf{x}]/\langle x^n + 1 \rangle$ with modulus $x^n + 1$ where n is a power of two and one can make use of the *negacyclic convolution* property of the NTT that allows carrying out a polynomial multiplication in $\mathbb{Z}_q[\mathbf{x}]/\langle x^n + 1 \rangle$ using length-n transforms and no zero padding. When $\mathbf{a} = (\mathbf{a}[0], \ldots \mathbf{a}[n-1])$ and $\mathbf{b} = (\mathbf{b}[0], \ldots \mathbf{b}[n-1])$ are polynomials of length n with elements in $\mathbb{Z}_q$, ω be a primitive n-th root of unity in $\mathbb{Z}_q$ and $\psi^2 = \omega$, then we define $\mathbf{d} = (\mathbf{d}[0], \ldots \mathbf{d}[n-1])$ as the negative wrapped convolution of $\mathbf{a}$ and $\mathbf{b}$ so that $\mathbf{d} = \mathbf{a} \cdot \mathbf{b} \bmod x^n + 1$. We then define

---

[2] See http://www.sha.rub.de/research/projects/lattice/.

[3] Actually, this is overly restrictive and the NTT is also defined for certain composite numbers (n has to divide $p - 1$ for every prime factor p of q). However, for the given target parameter sets common in lattice-based cryptography we can restrict ourselves to prime moduli and refer to [41] for further information on composite moduli NTTs.

[4] Similar to exponentiation being the main operation of RSA or point multiplication being the main operation of ECC.

$\bar{\mathbf{a}} = \mathsf{PowMul}_{\psi}(\mathbf{a}) = (\mathbf{a}[0], \psi\mathbf{a}[1], \ldots, \psi^{n-1}\mathbf{a}[n-1])$ as well as the inverse multiplication by powers of ψ^{-1} denoted as $\mathbf{a} = \mathsf{PowMul}_{\psi^{-1}}(\bar{\mathbf{a}})$. Then it holds that $\mathbf{d} = \mathsf{PowMul}_{\psi^{-1}}(\mathsf{INTT}(\mathsf{NTT}(\mathsf{PowMul}_{\psi}(\mathbf{a})\circ\mathsf{NTT}(\mathsf{PowMul}_{\psi}(\mathbf{b})))))$ [14,15,52], where $\circ$ denotes point-wise multiplication. For simplicity, we do not always explicitly apply PowMul_{ψ} or $\mathsf{PowMul}_{\psi^{-1}}$ when it is clear from the context that a negacyclic convolution is computed.

2.2 The RLWEenc Cryptosystem

The semantically secure public key encryption scheme RLWEenc was proposed in [32,36,37] and is also used as a building block in the identity-based encryption scheme (IBE) by Ducas et al. [19]. We provide the key generation procedure $\mathsf{RLWEenc}_{\mathrm{GEN}}$ in Algorithm 1, the encryption procedure $\mathsf{RLWEenc}_{\mathrm{ENC}}$ in Algorithm 2, and the decryption procedure $\mathsf{RLWEenc}_{\mathrm{DEC}}$ in Algorithm 3. All algorithms explicitly use calls to the NTT and function names used later on during the evaluation of our implementation (see Sect. 5). The exact placement of NTT transformations is slightly changed compared to [48], which saved one transformation compared to [44], as $\mathbf{c}_2$ is not transmitted in NTT form and thus removal of least significant bits is still possible (see [19,44]).

Algorithm 1 RLWEenc Key Generation

Precondition: Access to global constant $\tilde{\mathbf{a}} = \mathsf{NTT}(\mathbf{a})$
1: $\mathsf{RLWEenc}_{\mathrm{GEN}}()$
2:     $\tilde{\mathbf{r}}_1 \leftarrow \mathsf{NTT}(\mathsf{SampleGauss}_{\sigma}())$
3:     $\tilde{\mathbf{r}}_2 \leftarrow \mathsf{NTT}(\mathsf{SampleGauss}_{\sigma}())$
4:     $\tilde{\mathbf{p}} = \tilde{\mathbf{r}}_1 - \tilde{\mathbf{a}}\circ\tilde{\mathbf{r}}_2$
5:     **return** $(pk, sk) = (\tilde{\mathbf{p}}, \tilde{\mathbf{r}}_2)$
6: **end**

Algorithm 2 RLWEenc Encryption

Precondition: Access to global constant $\tilde{\mathbf{a}} = \mathsf{NTT}(\mathbf{a})$
1: $\mathsf{RLWEenc}_{\mathrm{ENC}}(\tilde{\mathbf{a}}, \tilde{\mathbf{p}}, \mu \in \{0,1\}^n)$
2:     $\tilde{\mathbf{e}}_1 = \mathsf{NTT}(\mathsf{SampleGauss}_{\sigma}())$
3:     $\tilde{\mathbf{e}}_2 = \mathsf{NTT}(\mathsf{SampleGauss}_{\sigma}())$
4:     $\tilde{\mathbf{c}}_1 = \tilde{\mathbf{a}}\circ\tilde{\mathbf{e}}_1 + \tilde{\mathbf{e}}_2$
5:     $\tilde{\mathbf{h}}_2 = \tilde{\mathbf{p}}\circ\tilde{\mathbf{e}}_1$
6:     $\mathbf{e}_3 \leftarrow \mathsf{SampleGauss}_{\sigma}()$
7:     $\mathbf{c}_2 = \mathsf{INTT}(\tilde{\mathbf{h}}_2) + \mathbf{e}_3 + \mathsf{Encode}(m)$
8:     **return** $(\tilde{\mathbf{c}}_1, \mathbf{c}_2)$
9: **end**

Algorithm 3 RLWEenc Decryption

1: $\mathsf{RLWEenc}_{\mathrm{DEC}}(\mathbf{c} = [\tilde{\mathbf{c}}_1, \mathbf{c}_2], \tilde{\mathbf{r}}_2)$
2:     **return** $\mathsf{Decode}(\mathsf{INTT}(\tilde{\mathbf{c}}_1\circ\tilde{\mathbf{r}}_2)+\mathbf{c}_2)$.
3: **end**

The main idea of the scheme is that during encryption the n-bit encoded message $\bar{\mathbf{m}} = \mathsf{Encode}(m)$ is added to $\mathbf{pe}_1 + \mathbf{e}_3$ (in NTT notation $\mathsf{INTT}(\tilde{\mathbf{h}}_2) + \mathbf{e}_3$) which is uniformly random and thus hides the message. Decryption is only possible with knowledge of $\mathbf{r}_2$ since otherwise the large term $\mathbf{ae}_1\mathbf{r}_2$ cannot be eliminated when computing $\mathbf{c}_1\mathbf{r}_2 + \mathbf{c}_2$. The encoding of the message of length n is necessary as the noise term $e_1r_1 + e_2r_2 + e_3$ is still present after calculating $c_1r_2 + c_2$ and would prohibit the retrieval of the binary message after decryption. With the simple threshold encoding $\texttt{encode}(m) = \frac{q-1}{2}m$ the value $\frac{q-1}{2}$ is assigned only to each binary one of the string m. The corresponding decoding function needs to test whether a received coefficient $z \in [0, q-1]$ is in the interval $\frac{q-1}{4} \leq z < 3\frac{q-1}{4}$ which is interpreted as one and zero otherwise. As a consequence, the maximum error added to each coefficient must not be larger than $\lfloor \frac{q}{4} \rfloor$ in order to decrypt correctly. The probability for decryption errors is mainly determined by the tailcut τ and the standard deviation σ of the polynomials $e_1, e_2, e_3 \leftarrow D_{\mathbb{Z}^n, \sigma}$, which follow a small discrete Gaussian distribution (sampled by $\mathsf{SampleGauss}_\sigma$). To support the NTT, Göttert et al. [23] proposed parameter sets (n, q, s) where $\sigma = s/\sqrt{2\pi}$ denoted as RLWEenc-Ia $(256, 7681, 11.31)$ and RLWEenc-IIa $(512, 12289, 12.18)$. Lindner and Peikert [32] originally proposed the parameter sets RLWEenc-Ib $(192, 4093, 8.87)$, RLWEenc-IIb $(256, 4093, 8.35)$ and RLWEenc-IIIb $(320, 4093, 8.00)$. The security levels of RLWEenc-Ia and RLWEenc-IIb are roughly comparable and RLWEenc-IIb provides 105.5 bits of pre-quantum security[5], according to a refined security analysis by Liu and Nguyen [33] for standard LWE and the original parameter sets. The RLWEenc-IIa parameter set uses a larger dimension n and should thus achieve even higher security than the 156.9 bits obtained by Liu and Nguyen for RLWEenc-IIIb. For the IBE scheme in [19] the parameters $n = 512$, $q \approx 2^{23}$ and a trinary error/noise distribution are used.

2.3   The BLISS Cryptosystem

In this work we only consider the efficient ring-based instantiation of BLISS [17]. We recall the key generation procedure $\text{BLISS}_{\text{GEN}}$ in Algorithm 4, the signing procedure $\text{BLISS}_{\text{SIGN}}$ in Algorithm 5, and the verification procedure $\text{BLISS}_{\text{VER}}$ in Algorithm 6. Key generation requires uniform sampling of sparse and small polynomials $\mathbf{f}, \mathbf{g}$, rejection sampling ($N_\kappa(\mathbf{S})$), and computation of an inverse. To sign a message, two masking polynomials $\mathbf{y}_1, \mathbf{y}_2 \leftarrow D_{\mathbb{Z}^n, \sigma}$ are sampled from a discrete Gaussian distribution using the $\mathsf{SampleGauss}_\sigma$ function. The computation of $\mathbf{ay}_1$ is performed using the NTT and the compressed $\mathbf{u}$ is then hashed together with the message μ by Hash. The binary string c' is used by $\mathsf{GenerateC}$ to generate a sparse polynomial $\mathbf{c}$. Polynomials $\mathbf{y}_1, \mathbf{y}_2$ then hide the secret key which is multiplied with the sparse polynomials using the $\mathsf{SparseMul}$ function. This function exploits that only κ coefficients in $\mathbf{c}$ are set and only $d_1 + d_2$ coefficients in $\mathbf{s}_1$ and $\mathbf{s}_2$. After a rejection sampling and compression step the signature $(\mathbf{z}_1, \mathbf{z}_2^\dagger, \mathbf{c})$ is returned. The verification procedure $\text{BLISS}_{\text{VER}}$ in Algorithm 6 just checks norms of signature components and compares the hash output with $\mathbf{c}$ in the signature.

---

[5] Up to our knowledge, all security evaluations of RLWEenc (and also BLISS) only consider best known attacks executed on a classical computer. The security levels are thus denoted as *pre-quantum*. A security assessment that considers quantum computers is certainly necessary but is not in the scope of this paper.

In this work we focus on the 128-bit pre-quantum secure BLISS-I parameter set which uses $n = 512$ and $q = 12289$ (same base parameters as RLWEenc-IIa). The density of the secret key is $\delta_1 = 0.3$ and $\delta_2 = 0$, the standard deviation of the coefficients of $\mathbf{y}_1$ and $\mathbf{y}_2$ is $\sigma = 215.73$ and the repetition rate is 1.6. The number of dropped bits in $\mathbf{z}_2$ is $d = 10$, $\kappa = 23$, and $p = \lfloor 2q/2^d \rfloor$. The final size of the signature is 5600 kb with Huffman encoding and approx. 7680 kb without Huffman encoding.

Algorithm 4 BLISS Key Generation

1: $\text{BLISS}_{\text{GEN}}()$
2:    Choose $\mathbf{f}, \mathbf{g}$ with $d_1 = \lceil \delta_1 n \rceil$ entries in $\{\pm 1\}$ and $d_2 = \lceil \delta_2 n \rceil$ entries in $\{\pm 2\}$
3:    $\mathbf{S} = (\mathbf{s}_1, \mathbf{s}_2)^t \leftarrow (\mathbf{f}, 2\mathbf{g} + 1)^t$
4:    **if** $N_\kappa(\mathbf{S}) \geq C^2 \cdot 5 \cdot (\lceil \delta_1 n \rceil + 4\lceil \delta_2 n \rceil) \cdot \kappa$ **then restart**
5:    $\mathbf{a}_q = (2\mathbf{g} + 1)/\mathbf{f} \bmod q$ **(restart if $\mathbf{f}$ is not invertible)**
6:    **Return**$(pk = \mathbf{A}, sk = \mathbf{S})$ where $\mathbf{A} = (\tilde{\mathbf{a}}_1 = \text{NTT}(\mathbf{a}_q), q - 2) \bmod 2q$
7: **end**

Algorithm 5 BLISS Signing	**Algorithm 6** BLISS Verification
1: $\text{BLISS}_{\text{SIGN}}(\mu \in \{0,1\}^*, pk=\mathbf{A}, sk=\mathbf{S})$	1: $\text{BLISS}_{\text{VER}}(\mu \in \{0,1\}^*, pk=\mathbf{A}, sk=\mathbf{S})$
2: $\mathbf{y}_1 \leftarrow \text{SampleGauss}_\sigma()$	2: **if** $\|(\mathbf{z}_1 \| 2^d \cdot \mathbf{z}_2^\dagger)\|_2 > B_2$ **then** Reject
3: $\mathbf{y}_2 \leftarrow \text{SampleGauss}_\sigma()$	3: **if** $\|(\mathbf{z}_1 \| 2^d \cdot \mathbf{z}_2^\dagger)\|_\infty > B_\infty$ **then** Reject
4: $\tilde{\mathbf{t}} \leftarrow \tilde{\mathbf{a}}_1 \circ \text{NTT}(\mathbf{y}_1)$	4: $\mathbf{r} \leftarrow \text{INTT}(\tilde{\mathbf{a}}_1 \circ \text{NTT}(\mathbf{z}_1))$
5: $\mathbf{u} \leftarrow 2\zeta \cdot \text{INTT}(\tilde{\mathbf{t}}) + \mathbf{y}_2 \bmod 2q$	5: $\mathbf{t} \leftarrow \lfloor 2\zeta \cdot \mathbf{r} + \zeta \cdot q \cdot \mathbf{c} \rceil_d$
6: $c' \leftarrow \text{Hash}(\lfloor \mathbf{u} \rceil_d \bmod p, \mu)$	6: $c' \leftarrow \text{Hash}(\mathbf{t} + \mathbf{z}_2^\dagger \bmod p, \mu)$
7: $\mathbf{c} \leftarrow \text{GenerateC}(c')$	7: **Accept iff** $\mathbf{c} = \text{GenerateC}(c')$
8: Choose a random bit b	8: **end**
9: $\mathbf{z}_1 \leftarrow \mathbf{y}_1 + (-1)^b \text{SparseMul}(\mathbf{s}_1, \mathbf{c})$	
10: $\mathbf{z}_2 \leftarrow \mathbf{y}_2 + (-1)^b \text{SparseMul}(\mathbf{s}_2, \mathbf{c})$	
11: **Continue** with probability	
12: $1 \Big/ \left(M \exp\left(-\frac{\|\mathbf{Sc}\|^2}{2\sigma^2}\right) \cosh\left(\frac{\langle \mathbf{z}, \mathbf{Sc} \rangle}{\sigma^2}\right) \right)$	
13: otherwise **restart**	
14: $\mathbf{z}_2^\dagger \leftarrow (\lfloor \mathbf{u} \rceil_d - \lfloor \mathbf{u} - \mathbf{z}_2 \rceil_d) \bmod p$	
15: **Return** $(\mathbf{z}_1, \mathbf{z}_2^\dagger, \mathbf{c})$	
16: **end**	

---

3   Faster NTTs for Lattice-Based Cryptography

In this section we examine fast algorithms for the computation of the number theoretic transform (NTT) and show techniques to speed up polynomial multiplication for lattice-based cryptography[6]. The most straightforward implementation

---

[6] Most of the techniques discussed in this section have already been proposed in the context of the fast Fourier transform (FFT). However, they have not yet been considered to speed up ideal lattice-based cryptography (at least not in works like [7,12,43,48]). Moreover, some optimizations and techniques are mutually exclusive and a careful selection and balancing has to be made.

of the NTT is a Cooley-Tukey radix-2 decimation-in-time (DIT) approach [13] that requires a bit-reversal step as the algorithm takes bit-reversed input and produces naturally ordered output (from now on referred to as $\mathrm{NTT}_{bo \to no}^{CT}$). To compute the NTT as defined in Sect. 2.1 the $\mathrm{NTT}_{bo \to no}^{CT}$ algorithm applies the Cooley-Tukey (CT) butterfly, which computes $a' \leftarrow a + \omega b$ and $b' \leftarrow a - \omega b$ for some values of $\omega, a, b \in \mathbb{Z}_q$, overall $\frac{n \log_2(n)}{2}$ times. The biggest disadvantage of relying solely on the $\mathrm{NTT}_{bo \to no}^{CT}$ algorithm is the need for bit-reversal, multiplication by constants, and that it is impossible to merge the final multiplication by powers of ψ^{-1} into the twiddle factors of the inverse NTT (see [48]). With the assumption that twiddle factors (powers of ω) are stored in a table and thus not computed on-the-fly it is possible to further simplify the computation and to remove bit-reversal and to merge certain steps. This assumption makes sense on constrained devices like the ATxmega which have a rather large read-only flash.

3.1 Merging the Inverse NTT and Multiplication by Powers of ψ^{-1}

In [48] Roy et al. use the standard $\mathrm{NTT}_{bo \to no}^{CT}$ algorithm for a hardware implementation and show how to merge the multiplication by powers of ψ (see Sect. 2.1) into the twiddle factors of the forward transformation. However, this approach does not work for the inverse transformation due to the way the computations are performed in the CT butterfly as the multiplication is carried out before the addition. In this section we show that it is possible to merge the multiplication by powers of ψ^{-1} during the inverse transformation using a fast decimation-in-frequency (DIF) algorithm [22]. The DIF NTT algorithm splits the computation into a sub-problem on the even outputs and a sub-problem on the odd outputs of the NTT and has the same complexity as the $\mathrm{NTT}_{bo \to no}^{CT}$ algorithm. It requires usage of the so-called Gentlemen-Sande (GS) butterfly which computes $a' \leftarrow a + b$ and $b' \leftarrow (a - b)\omega$ for some values of $\omega, a, b \in \mathbb{Z}_q$. Following [11, Sect. 3.2], where ω_n is an n-th primitive root of unity and by ignoring the multiplication by the scalar n^{-1}, the inverse NTT and application of PowMul_ψ can be defined as

$$
\mathbf{a}[r] = \psi^{-r} \sum_{\ell=0}^{n-1} \mathbf{A}[\ell]\omega_n^{-r\ell} = \psi^{-r} \left(\sum_{\ell=0}^{\frac{n}{2}-1} \mathbf{A}[\ell]\omega_n^{-r\ell} + \sum_{\ell=0}^{\frac{n}{2}-1} \mathbf{A}[\ell + \frac{n}{2}]\omega_n^{-r(\ell+\frac{n}{2})} \right) \tag{1}
$$

$$
= \psi^{-r} \sum_{\ell=0}^{\frac{n}{2}-1} \left(\mathbf{A}[\ell] + \mathbf{A}[\ell + \frac{n}{2}]\omega_n^{-r\frac{n}{2}} \right) \omega_n^{-r\ell}, r = 0, 1, \dots, n-1. \tag{2}
$$

When r is even this results in

$$
\mathbf{a}[2k] = \sum_{\ell=0}^{\frac{n}{2}-1} \left(\mathbf{A}[\ell] + \mathbf{A}[\ell + \frac{n}{2}] \right) \omega_{\frac{n}{2}}^{-k\ell}(\psi^2)^{-k} \tag{3}
$$

and for odd r in

$$\mathbf{a}[2k+1] = \sum_{\ell=0}^{\frac{n}{2}-1} \left(\mathbf{A}[\ell] - \mathbf{A}[\ell + \frac{n}{2}]\omega_n^{-\ell} \right) \omega_{\frac{n}{2}}^{-k\ell} \psi^{-(2k+1)} \tag{4}$$

$$= \sum_{\ell=0}^{\frac{n}{2}-1} \left(\left(\mathbf{A}[\ell] - \mathbf{A}[\ell + \frac{n}{2}] \right) \psi^{-1}\omega_n^{-\ell} \right) \omega_{\frac{n}{2}}^{-k\ell} (\psi^2)^{-k}, k = 0, 1, \ldots, \frac{n}{2} - 1. \tag{5}$$

The two new half-size sub-problems where ψ is exchanged by ψ^2 can now be again solved using the recursion. As a consequence, when using an in-place radix-2 DIF algorithm it is necessary to multiply all twiddle factors in the first stage by ψ^{-1}, all twiddle factors in the second stage by ψ^{-2} and in general by ψ^{-2^s} for stage $s \in [0, 1, \ldots, \log_2(n) - 1]$ to merge the multiplication by powers of ψ^{-1} into the inverse NTT (see Fig. 1 for an illustration). In case the PowMul$_\psi$ or PowMul$_{\psi^{-1}}$ operation is merged into the NTT computation we denote this by an additional superscript ψ or ψ^{-1}, e.g., as NTT$_{bo \to no}^{CT,\psi}$.

3.2 Removing Bit-Reversal

For memory efficient and in-place computation a reordering or so-called bit-reversal step is usually applied before or after an NTT/FFT transformation due to the required reversed input ordering of the NTT$_{bo \to no}^{CT}$ algorithm used in works like [7,12,43,48]. However, by manipulation of the standard iterative algorithms and independently of the used butterfly (CT or GS) it is possible to derive natural order to bit-reversed order ($no \to bo$) as well as bit-reversed to natural order ($bo \to no$) forward and inverse algorithms. The derivation of FFT algorithms with a desired ordering of inputs and outputs is described in [11] and we followed this description to derive the NTT algorithms NTT$_{bo \to no}^{CT}$, NTT$_{no \to bo}^{CT}$, NTT$_{no \to bo}^{GS}$, and NTT$_{bo \to no}^{GS}$, as well es their respective inverse counterparts. It is also possible to construct self-sorting NTTs ($no \to no$) but in this case the structure becomes irregular and temporary memory is required (see [11]).

3.3 Tuning for Lattice-Based Cryptography

The optimizations discussed in this section so far can be used to generically optimize polynomial multiplication in $\mathbb{Z}_q[\mathbf{x}]/\langle x^n + 1 \rangle$. However, for lattice-based cryptography there are special conditions that hold for most practical algorithms; in the NTT-enabled algorithms of RLWEenc and BLISS every point-wise multiplication (denoted by $\circ$) is performed with a constant and a variable, usually a randomly sampled polynomial. Thus, the most common operation in lattice-based cryptography is *not* simple polynomial multiplication but multiplication of a (usually random) polynomial by a constant polynomial (i.e., global constant, or public key). Thus, the scaling factor n^{-1} can be multiplied into the pre-computed and pre-transformed constant $\tilde{\mathbf{a}} = n^{-1}\text{NTT}_{no \to bo}^{CT}(\mathbf{a})$. Taking into

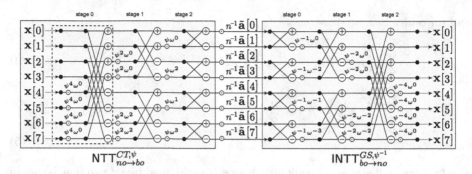

Fig. 1. Signal flow graph for multiplication of a polynomial $\mathbf{x}$ by a pre-transformed polynomial $\tilde{\mathbf{a}} = \mathrm{NTT}_{no \to bo}^{CT,\psi}(\mathbf{a})$, using the $\mathrm{NTT}_{no \to bo}^{CT,\psi}$ and $\mathrm{INTT}_{bo \to no}^{GS,\psi^{-1}}$ algorithms.

account that we also want to remove the need for bit-reversal and want to merge the multiplication by powers of ψ into the forward and inverse transformation (as discussed in Sect. 3.1) we propose to use an $\mathrm{NTT}_{no \to bo}^{CT,\psi}$ for the forward transformation and an $\mathrm{INTT}_{bo \to no}^{GS,\psi^{-1}}$ for the inverse transformation. In this case a polynomial multiplication $\mathbf{c} = \mathbf{a} \cdot \mathbf{e}$ can be implemented without bit-reversal as $\mathbf{c} = \mathrm{INTT}_{bo \to no}^{GS,\psi^{-1}} \left(\mathrm{NTT}_{no \to bo}^{CT,\psi}(\mathbf{a}) \circ \mathrm{NTT}_{no \to bo}^{CT,\psi}(\mathbf{e}) \right)$. An example flow diagram for $n = 8$ is provided in Fig. 1. For more details, pseudo-code of $\mathrm{NTT}_{no \to bo}^{CT,\psi}$ is provided in Algorithm 7 and pseudo-code of $\mathrm{INTT}_{bo \to no}^{GS,\psi^{-1}}$ is given in Algorithm 8.

4   Implementation of Lattice-Based Cryptography on ATxmega128

In this section we provide details on our implementation of the NTT as well as RLWEenc, and BLISS on the ATxmega128. Note that our implementation does not take protection against timing side channels into account which is required for most practical and interactive applications. In this work we solely focused on performance and to a lesser extent on a small memory and code footprint.

4.1   Implementation of the NTT

For the use in RLWEenc, and BLISS we focus on the optimization of the $\mathrm{NTT}_{no \to bo}^{CT,\psi}$ and $\mathrm{INTT}_{bo \to no}^{GS,\psi^{-1}}$ transformations. We implemented both algorithms in C and optimized modular multiplication using assembly language.

Modular Multiplication. To implement the NTT according to Sect. 3 a DIT $\mathrm{NTT}_{no \to bo}^{CT,\psi}$ and a DIF $\mathrm{INTT}_{bo \to no}^{GS,\psi^{-1}}$ transformation are required and the most expensive computation in both algorithms is integer multiplication and reduction modulo q ($\frac{n \log_2(n)}{2}$ times per NTT). In [7] Boorghany, Sarmadi, and Jalili report that most of the runtime of their FFT/NTT is spent on the computation of modulo operations. They review modular reduction algorithms for suitability and propose an approximate variant of Barrett reduction that consumes approx. $\frac{754668}{\frac{256}{2}\log_2(256)} = 736$ cycles for one FFT/NTT butterfly.

We use a method proposed by Liu et al. [35] who introduce an assembly optimized shifting-addition-multiplication-subtraction-subtraction algorithm (SAMS2). They achieve, for $q = 7681$ case, a modular multiplication in 53 clock cycles which beats the subtract-and-shift approach mentioned above. For all reductions modulo $q = 7681$ and $q = 12289$ we thus rely on the SAMS2 method.

Algorithm 7 Optimized CT Forward NTT

Precondition: Store n powers of ψ in bit-reversed order in psi^*

1: $\text{NTT}_{no\to bo}^{CT,\psi}(\mathbf{a})$
2:   $m \leftarrow 1$
3:   $k \leftarrow n/2$
4:   **while** $m < n$ **do**
5:     **for** $i = 0$ to $m - 1$ **do**
6:       $jFirst \leftarrow 2 \cdot i \cdot k$
7:       $jLast \leftarrow jFirst + k - 1$
8:       $\psi_i \leftarrow psi^*[m + i]$
9:       **for** $j = jFirst$ to $jLast$ **do**
10:        $l \leftarrow j + k$
11:        $t \leftarrow \mathbf{a}[j]$
12:        $u \leftarrow \mathbf{a}[l] \cdot \psi_i$
13:        $\mathbf{a}[j] \leftarrow t + u \mod q$
14:        $\mathbf{a}[l] \leftarrow t - u \mod q$
15:      **end for**
16:    **end for**
17:    $m \leftarrow m \cdot 2$
18:    $k \leftarrow k/2$
19: **end while**
20: **Return a**
21: **end**

Algorithm 8 Optimized GS Inverse NTT

Precondition: Store n powers of ψ^{-1} in bit-reversed order in $invpsi^*$

1: $\text{INTT}_{bo\to no}^{GS,\psi^{-1}}(\mathbf{a})$
2:   $m \leftarrow n/2$
3:   $k \leftarrow 1$
4:   **while** $m > 1$ **do**
5:     **for** $i = 0$ to $m - 1$ **do**
6:       $jFirst \leftarrow 2 \cdot i \cdot k$
7:       $jLast \leftarrow jFirst + k - 1$
8:       $\psi_i \leftarrow invpsi^*[m + i]$
9:       **for** $j = jFirst$ to $jLast$ **do**
10:        $l \leftarrow j + k$
11:        $t \leftarrow \mathbf{a}[j]$
12:        $u \leftarrow \mathbf{a}[l]$
13:        $\mathbf{a}[j] \leftarrow t + u \mod q$
14:        $\mathbf{a}[l] \leftarrow (t-u) \cdot \psi_i \mod q$
15:      **end for**
16:    **end for**
17:    $m \leftarrow m/2$
18:    $k \leftarrow k \cdot 2$
19: **end while**
20: **Return a**
21: **end**

Extraction of Stages. As additional optimization we use specific routines for the first and the last stage of each NTT. A common optimization is to recognize that $\omega^0 = 1$ in the first stage of the $\text{NTT}_{no\to bo}^{CT}$ so that only additions are required. As we merge the multiplication by powers of ψ into the NTT this is not the case anymore. However, it is still beneficial to write a specific loop that performs the first stage of the $\text{NTT}_{no\to bo}^{CT,\psi}$ and the last stage of the $\text{INTT}_{bo\to no}^{GS,\psi^{-1}}$ transformation to achieve less loop overhead (simpler address generation) and fewer loads and stores.

Usage of Look-Up Tables for Narrow Input Distributions. As discussed in Sect. 3.3 it is common in lattice-based cryptography to apply forward transformations mostly to values sampled from a narrow Gaussian error/noise distribution (other polynomials are usually constants and pre-computed). In this case only a limited number of possible inputs to the butterfly of the first stage of the $\mathrm{NTT}^{CT,\psi}_{no \to bo}$ transformation exist and it is possible to pre-compute look-up tables to speed-up the modular multiplication. The range of possible input coefficients to the first stage butterfly is rather small, since they are Gaussian distributed and bounded by $[-\tau\sigma, \tau\sigma]$ for standard deviation σ and tail-cut factor τ. Additionally, we only store the result of multiplications of two positive factors. That means that for negative inputs, we invert before the look-up and again after the look-up. The same approach would also work for the binary error distribution used for the IBE scheme in [19] and it would be possible to cover even two or more stages due to the very limited input distribution.

4.2   Implementation of LP-RLWE

Our implementation of $\mathsf{RLWEenc}_{\mathrm{ENC}}$ and $\mathsf{RLWEenc}_{\mathrm{DEC}}$ of the $\mathsf{RLWEenc}$ scheme as described in Sect. 2.2 mainly consists of forward and inverse NTT transformations (NTT and INTT), Gaussian sampling ($\mathsf{SampleGauss}_\sigma$) and point-wise multiplication ($\mathsf{Pointwise}$, also denoted as $\circ$). We assume that secret and public keys are stored in the read-only flash memory, but loading from RAM would also be possible, and probably even faster. As described in Sect. 4.1 we can optimize the $\mathrm{NTT}^{CT,\psi}_{no \to bo}$ transformation for the Gaussian input distribution and with either $\sigma = 4.52$ or $\sigma = 4.85$ it is possible to substitute approx. 99.96% of the multiplications in stage one of $\mathrm{NTT}^{CT,\psi}_{no \to bo}$ by look-ups to a table of 16 entries that requires only 32 bytes of flash memory. For the sampling of the Gaussian distributed polynomials with high precision[7] we use a cumulative distribution table (CDT) [16,21]. We construct the table M with entries $p_z = \Pr(x \le z : x \leftarrow D_\sigma)$ for $z \in [0, \tau\sigma]$ with a precision of $\lambda = 128$ bits. The tail-cut factor τ determines the number of lines $|z_t| = \lceil \tau\sigma \rceil$ of the table and reduces the output to the range $x \in \{-\lceil \tau\sigma \rceil, \ldots, \lceil \tau\sigma \rceil\}$. To sample a value we choose a uniformly random y from the interval $[0, 1)$ and a bit b and return the integer $(-1)^b z \in \mathbb{Z}$ such that $y \in [p_{z-1}, p_z)$. Further we store only the positive half of the tables and then sample a sign bit. For this, the probability of sampling zero has been pre-halved when constructing the table. For efficiency reasons we just work with the binary expansion of the fractional part instead of floating point arithmetic as all numbers used are smaller than 1.0. The constant CDF matrix M is stored in the read-only flash memory with $k = \lceil \sigma\tau \rceil$ rows and $l = \lceil \lambda/8 \rceil$ columns. In order to sample a Gaussian distributed value we perform a linear search in the table

---

[7] It is debatable which precision is really necessary in $\mathsf{RLWEenc}$ and what impact less precision would have on the security of the scheme, e.g., $\lambda = 40$. But as the implementation of the CDT for small standard deviations σ is rather efficient and for better comparison with related work like [6,7,12] we chose to implement high precision sampling and set $\lambda = 128$.

to obtain z. Another option would be binary search, however, for this table size with $x \leftarrow D_\sigma$ being small, the evaluation can already be stopped after only a few comparisons with high probability. The test if the random y is in the range $[p_{z-1}, p_z)$ is performed in a lazy manner on bytes. Laziness means that a comparison is finished when the first bit (or byte on an 8-bit architecture) has been found that differs between two values. Thus we do not need to sample the full λ bits of y and obtain the result of the comparisons early. Random numbers are obtained from a pseudo random number generator (PRNG) using the hardware AES-128 engine running in counter mode. The PRNG is seeded by noise from the LSB of the analog digital converter. For the state (key, plaintext) 32 bytes of statically allocated memory are necessary. The final table size is 624 bytes for $q = 7681$ and 660 bytes for $q = 12289$.

4.3   Implementation of BLISS

Besides polynomial arithmetic the most expensive operation for BLISS is the sampling of $\mathbf{y}_1$ and $\mathbf{y}_2$ from a discrete Gaussian distribution ($\mathsf{SampleGauss}_\sigma$). For the rather large standard deviation of $\sigma = 215.73$ (RLWEenc requires only $\sigma = 4.85$) a straightforward CDT sampling approach, even with binary search, would lead to a large table with roughly $\tau\sigma = 2798$ entries of approx. 30 to 40 kilobytes overall (see [6]). Another option for embedded devices would be the Bernoulli approach from [17] implemented in [7] but the reported performance of 13,151,929 cycles to sample one polynomial would cause a massive performance penalty. As a consequence, we implemented the hardware-optimized sampler from [43] on the ATxmega. It uses the convolution property of Gaussians combined with Kullback-Leibler divergence and mainly exploits that it is possible to sample a Gaussian distributed value with variance σ^2 by sampling x_1, x_2 from a Gaussian distribution with smaller standard deviation σ' such that $\sigma'^2 + k^2\sigma'^2 = \sigma^2$ and combining them as $x_1 + kx_2$ (for BLISS-I $k = 11$ and $\sigma' = 19.53$). Additionally, the performance of the σ'-sampler is improved by the use of short-cut intervals where each possible value of the first byte of the uniformly distributed input is assigned to an interval that specifies the range of the possible sampled values. This approach reduces the number of necessary comparisons and nearly compensates for the additional costs incurred by the requirement to sample two values (x_1, x_2 with $\sigma' = 19.53$) instead of one directly (with $\sigma = 215.73$). The sampling is again performed in a lazy manner and we use the same PRNG based on AES-128 as for RLWEenc.

To implement the $\mathrm{NTT}_{no \rightarrow bo}^{CT}$ we did not use a look-up table for the first stage as the input range $[-\sigma\tau, \sigma\tau]$ of $\mathbf{y}_1$ or $\mathbf{z}_1$ is too large. For the instantiation of the random oracle (Hash) that is required during signing and verification we have chosen the official AVR implementation of Keccak [4]. From the output of the hash function the sparse polynomial $\mathbf{c}$ with κ coefficients equal to one is generated by the GenerateC (see [18, Sect. 4.4]) routine. We store only κ indices where a coefficient of $\mathbf{c}$ is one. This reduces the dynamic RAM consumption and allows a more efficient implementation of the multiplication of $\mathbf{c}$ by $\mathbf{s}_1$ and $\mathbf{s}_2$ using the SparseMul routine. By using column-wise multiplication and by ignoring all zero coefficients, the multiplication can be performed more efficiently than with the NTT.

5   Results and Comparison

All implementations are measured on an ATxmega128A1 8-bit microcontroller running at 32 MHz featuring 128 Kbytes read-only flash, 8 Kbytes RAM and 2 Kbytes EEPROM. Cycle accurate performance measurement were obtained using two coupled 16-bit timer/counters and dynamic RAM consumption is measured using stack canaries. All public and private keys are assumed to be stored in the flash of the microcontroller and we consider the .text + .data + .bootloader sections to determine the flash memory utilization. For our implementation we used the avr-gcc compiler in version 4.7.0. and no calls to the standard library.

LP-RLWE. Detailed cycle counts for the encryption and decryption as well as the most expensive operations are given in Table 1. The costs of the encryption are dominated by the NTT (two calls of $\mathrm{NTT}_{no \to bo}^{CT,\psi}$ and one call of $\mathrm{INTT}_{bo \to no}^{GS,\psi^{-1}}$) which requires approx. 62 % of the overall cycles for RLWEenc-Ia. The Gaussian sampling requires 29 % of the overall cycles which is approx. 328 (RLWEenc-Ia) or 334 (RLWEenc-IIa) cycles per sample. The reason that $\mathrm{NTT}_{no \to bo}^{CT,\psi}$ is slightly faster than $\mathrm{INTT}_{bo \to no}^{GS,\psi^{-1}}$ is that we use table look-ups for the first stage of the forward transformation (see Sect. 4.2). The remaining amount of cycles (9 %) is consumed by additions, point-wise multiplications by a constant/key stored in the flash (PwMulFlash), and message encoding (Encode). In Table 1 we also list cycle counts of operations that are now obsolete, especially BitRev and PowMul. For PowMul we assume an implementation where the powers of ψ are computed on-the-fly to save flash memory, otherwise the costs are the same as PwMulFlash. Decryption is extremely simple, fast, and basically calls $\mathrm{INTT}_{bo \to no}^{GS,\psi^{-1}}$, the decoding and an addition so that roughly 148 decryption operation could be performed per second on the ATxmega128. Note that we also evaluated RLWEenc-Ib, RLWEenc-IIb, RLWEenc-IIIb using classic schoolbook multiplication and Karatsuba multiplication but could not achieve better performance (see the full version [45]). The NTT can be performed in place so that no additional large temporary memory on the stack is needed. But storing the NTT twiddle factors for forward and inverse transforms in flash consumes $2n$ words $= 4n$ bytes which is around 15 % of the allocated flash memory for $q = 7681$ and around 22 % for $q = 12289$.

BLISS. In Table 2, we present detailed cycle counts for signing and verifying as well as for the most expensive operations in BLISS-I. Due to the rejection sampling and the chosen parameter set 1.6 signing attempts are required on average to create one signature. One attempt requires 6,381,428 cycles on average and only a small portion of the computation, i.e. the hashing of the message, does not have to be repeated in case of a rejection. During a signing attempt the

Table 1. Cycle counts and Flash memory consumption in bytes for the implementation of RING-LWEENCRYPT on an 8-bit ATxmega128 microcontroller using the NTT. The stack usage is divided into a fixed amount of memory necessary for plaintext, ciphertext, and additional components (like random number generation) and the dynamic consumption of the encryption and decryption routine. We encrypt a message of n bits.

Operation	$(n = 256, q = 7681)$	$(n = 512, q = 12289)$
	Cycle counts and stack usage	
RLWEenc$_{\text{ENC}}$	874,347 (109 bytes)	2,196,945 (102 bytes)
RLWEenc$_{\text{DEC}}$	215,863 (73 bytes)	600,351 (68 bytes)
NTT$_{no \to bo}^{CT,\psi}$	185,360	502,896
INTT$_{bo \to no}^{GS,\psi^{-1}}$	168,853	427,827
SampleGauss$_\sigma$	84,001	170,861
PwMulFlash	22,012	53,891
AddEncode	16,884	37,475
Decode	4,407	8,759
	Cycle counts of obsolete functions	
NTT$_{bo \to no}^{CT}$	198,491	521,872
BitRev	29,696	75,776
BitrevDual	32,768	79,872
PowMul$_\psi$	35,068	96,603
	Static memory consumption in bytes	
Complete binary	6,668	9,258
RAM	1,088	2,144

most expensive operation is the sampling of two Gaussian distributed polynomials which takes $2 \times 1{,}140{,}600 = 2{,}281{,}200$ cycles (36 % of the overall cycles). The calls of NTT$_{no \to bo}^{CT,\psi}$ and INTT$_{bo \to no}^{GS,\psi^{-1}}$ account for 16 % of the overall cycles of one attempt. In contrast to the RLWEenc implementation we do not use a look-up table for the first stage of NTT$_{no \to bo}^{CT,\psi}$. Additionally, we do not implement a separate modulo reduction after the subtraction in the GS butterfly and reduce the result after the multiplication which explains the slightly better performance of INTT$_{bo \to no}^{GS,\psi^{-1}}$. Hashing the compressed **u** and the message μ is time consuming and accounts for roughly 21 % of the overall cycles during one attempt. Savings would be definitely possible by using a different hash function (see [2] for an evaluation of different functions) but Keccak appears to be a conservative choice that matches the 128-bit security target very well. The sparse multiplication takes only 503,627 cycles for one multiplication. This makes it a favorable approach on the ATxmega and overall the sparse multiplication is 3.6 times faster than an NTT multiplication approach that would require one NTT$_{no \to bo}^{CT,\psi}$, two INTT$_{bo \to no}^{GS,\psi^{-1}}$ and two PwMulFlash calls. The flash memory consumption includes

Table 2. Cycle counts and Flash memory consumption in bytes for the implementation of BLISS on an 8-bit ATxmega128 microcontroller. The stack usage is divided into a fixed amount of memory necessary for message, signature, and additional components (like random number generation) and the dynamic consumption of the signing and verification routine. We sign a message of n bits.

Operation	(n = 512, q = 12289)
	Cycle counts and stack usage
BLISS$_{\text{SIGN}}$	10,537,981 (4,012 bytes)
BLISS$_{\text{VER}}$	2,814,118 (1,103 bytes)
NTT$^{CT,\psi}_{no \to bo}$	521,872
INTT$^{GS,\psi^{-1}}_{bo \to no}$	497,815
SampleGauss$_{\sigma}$	1,140,600
SparseMul	503,627
Hash	1,335,040
GenerateC	4,410
DropBits	11,826
$\cosh\left(\frac{\langle \mathbf{z}, \mathbf{Sc} \rangle}{\sigma^2}\right)$	75,601
$M \exp\left(-\frac{\|\mathbf{Sc}\|^2}{2\sigma^2}\right)$	37,389
	Static memory consumption in bytes
Complete binary	18,674
RAM	2,411

$2n$ words which equals $4n = 2048$ bytes for the NTT twiddle factors and 3374 bytes for look-up tables of the sampler.

Comparison. A detailed comparison of our implementation with related work that also targets the AVR platform[8] and provides 80 to 128-bits of security is given in Table 3. Our implementation of RLWEenc-Ia encryption outperforms the software from [7] by a factor of 3.5 and results from [6] by a factor of 5.7 in terms of cycle counts. Decryption is 6.3 times and 11.4 times faster, respectively. Compared to the work by Liu et al. [35] we achieve similar results. The authors decided to use a Knuth-Yao Sampler that performs better than our CDT and are faster for the RLWEenc-Ia parameter set but slower for RLWEenc-IIa. For the RLWEenc-IIa we compensate our slower sampler by the faster NTT implementation. Since there is no sampling in the decryption, our decryption is faster for both security levels (39 % for RLWEenc-Ia , 17 % for RLWEenc-IIa). A faster implementation of our Gaussian sampler, e.g., by also using hash tables and

---

[8] While the ATxmega128 and ATxmega64 compared to the ATmega64 differ in their operation frequency and some architectural differences cycle counts are mostly comparable.

Table 3. Comparison of our implementations with related work. Cycle counts marked with (*) indicate the runtime of BLISS-I as authentication protocol instead of signature scheme. By AX128 we identify the ATxmega128A1 clocked with 32 MHz, by AX64 the ATxmega64A3 clocked with 32 MHz, by AT64 the ATmega64 clocked with 8 MHz, by AT128 the ATmega128 clocked with 8 MHz, and by AT2560 the ATmega2560 clocked with 16 MHz. Implementations marked with (†) are resistant against timing attacks.

Scheme	Device	Operation	Cycles		OP/s	
RLWEenc-Ia ($n = 256$)	AX128	Enc/Dec	874,347	215,863	36.60	148.24
RLWEenc-IIa ($n = 512$)	AX128	Enc/Dec	2,196,945	600,351	14.57	53.30
RLWEenc-Ia ($n = 256$)[35]	AX128	Enc/Dec	666,671	299,538	48	106.83
RLWEenc-IIa ($n = 512$)[35]	AX128	Enc/Dec	2,721,372	700,999	11.76	45.65
RLWEenc-Ia ($n = 256$)[7]	AT64	Enc/Dec	3,042,675	1,368,969	2.63	5.84
RLWEenc-Ia ($n = 256$)[6]	AX64	Enc/Dec	5,024,000	2,464,000	6.37	12.98
BLISS-I	AX128	Sign/Verify	10,537,981	2,814,118	3.04	11.37
BLISS-I (Bernoulli) [7]	AT64	Sign + Verify	42,069,682*		0.19	
BLISS-I (CDT) [6]	AX64	Sign + Verify	19,328,000*		1.65	
$\text{NTT}^{CT,\psi}_{no \to bo}$ ($n = 256$)	AX128	NTT	198,491		161.21	
$\text{NTT}^{CT}_{bo \to no}$ ($n = 256$) [35]	AX128	NTT	169,423		188.88	
$\text{NTT}^{CT}_{bo \to no}$ ($n = 256$) [7]	AT64	NTT	754,668		10.60	
$\text{NTT}^{CT}_{bo \to no}$ ($n = 256$) [6]	AX64	NTT	1,216,000		26.32	
$\text{NTT}^{CT,\psi}_{no \to bo}$ ($n = 512$)	AX128	NTT	521,872		61.31	
$\text{NTT}^{CT}_{bo \to no}$ ($n = 512$) [35]	AX128	NTT	484,680		66.02	
$\text{NTT}^{CT}_{bo \to no}$ ($n = 512$) [7]	AT64	NTT	2,207,787		3.62	
$\text{NTT}^{CT}_{bo \to no}$ ($n = 512$) [6]	AX64	NTT	2,752,000		11.63	
QC-MDPC [27]	AX128	Enc/Dec	26,767,463	86,874,388	1.20	0.36
Ed25519† [31]	AT2560	Sign/Verify	23,211,611	32,619,197	0.67	0.49
RSA-1024 [26]	AT128	Enc/Dec	3,440,000	87,920,000	2.33	0.09
RSA-1024† [34]	AT128	priv. key	75,680,000		0.11	
Curve25519† [20]	AT2560	Point mul	13,900,397		1.15	
ECC-ecp160r1 [26]	AT128	Point mul	6,480,000		1.23	

binary search could provide a certain speed up but would also result in more flash consumption.

A comparison between our implementation of BLISS-I and the implementations of [7] and [6] is difficult since the authors implemented the signature as authentication protocol. Therefore, they only provide the runtime of a complete protocol run that corresponds to one signing operation and one verification in our results, but without the expensive hashing as the sparse polynomial $\mathbf{c}$ is not obtained from a random oracle but randomly generated by the other protocol party. However, our implementations of BLISS$_{\text{SIGN}}$ and BLISS$_{\text{VER}}$ still require less cycles than the implementation of BLISS-I. The biggest improvement stems from the usage of the KL-convolution sampler from [43] which is superior (1,140,600 cycles per polynomial) compared to the Bernoulli approach (13,151,929 cycles per polynomial [7]) and the straight-forward CDT approach used in [6]. As our implementation of BLISS-I needs 18,674 bytes of flash memory, it is also smaller than the implementation of [6] that requires 66.5 kB of flash memory and the implementation of [7] that needs 25.1 kB of flash memory.

Compared with the 80-bit secure McEliece cryptosystem based on QC-MDPC codes from [27] we get 31 times less cycles for the encryption and even 402 times less cycles for decryption. Our 128-bit secure BLISS-I implementation is 2.2 times faster for signing and 11.6 faster for verification compared to an implementation of the Ed25519 signature scheme for an ATmega2560 [31]. Translating the implementation results for RSA and ECC given in [26] to cycle counts, it turns out that an ECC secp160r1 operation requires 6.5 million cycles. RSA-1024 encryption with public key $e = 2^{16} + 1$ takes 3.4 million cycles [26] and RSA-1024 decryption with Chinese Remainder Theorem (CRT) requires 75.68 million cycles (this number is taken from [34]). A comparison with NTRU implementations is currently not easily possible due to lack of published results for the AVR platform[9].

References

1. Bai, S., Galbraith, S.D.: An improved compression technique for signatures based on learning with errors. In: Benaloh, J. (ed.) CT-RSA 2014. LNCS, vol. 8366, pp. 28–47. Springer, Heidelberg (2014)
2. Balasch, J., Ege, B., Eisenbarth, T., Gérard, B., Gong, Z., Güneysu, T., Heyse, S., Kerckhof, S., Koeune, F., Plos, T., Pöppelmann, T., Regazzoni, F., Standaert, F.-X., Van Assche, G., Van Keer, R., van Oldeneel tot Oldenzeel, L., von Maurich, I.: Compact implementation and performance evaluation of hash functions in ATtiny devices. In: Mangard, S. (ed.) CARDIS 2012. LNCS, vol. 7771, pp. 158–172. Springer, Heidelberg (2013)
3. Batina, L., Robshaw, M. (eds.): CHES 2014. LNCS, vol. 8731. Springer, Heidelberg (2014)
4. Bertoni, G., Daemen, J., Peeters, M., Assche, G.V., Keer, R.V.: Keccak implementation overview, version 3.2 (2012). http://keccak.noekeon.org/Keccak-implementation-3.2.pdf
5. Blahut, R.E.: Fast Algorithms for Signal Processing. Cambridge University Press, Cambridge (2010). http://amazon.com/o/ASIN/0521190495/
6. Boorghany, A., Jalili, R.: Implementation and comparison of lattice-based identification protocols on smart cards and microcontrollers. Eprint 2014, 78 (2014). http://eprint.iacr.org/2014/078, preliminary version of [7]
7. Boorghany, A., Sarmadi, S.B., Jalili, R.: On constrained implementation of lattice-based cryptographic primitives and schemes on smart cards. Eprint 2014, 514 (2014). http://eprint.iacr.org/2014/514, successive version of [6]
8. Bos, J.W., Costello, C., Naehrig, M., Stebila, D.: Post-quantum key exchange for the TLS protocol from the ring learning with errors problem. Eprint 2014, 599 (2014). http://eprint.iacr.org/2014/599

---

[9] One exception is a Master thesis by Monteverde [39], but the implemented NTRU251:3 variant is not secure anymore according to recent recommendations in [28].

9. Cabarcas, D., Weiden, P., Buchmann, J.: On the efficiency of provably secure NTRU. In: Mosca [40], pp. 22–39

10. Chen, D.D., Mentens, N., Vercauteren, F., Roy, S.S., Cheung, R.C.C., Pao, D., Verbauwhede, I.: High-speed polynomial multiplication architecture for ring-LWE and SHE cryptosystems. IEEE Trans. Circuits Syst. **62**–I(1), 157–166 (2015)

11. Chu, E., George, A.: Inside the FFT Black Box Serial and Parallel Fast Fourier Transform Algorithms. CRC Press, Boca Raton (2000)

12. de Clercq, R., Roy, S.S., Vercauteren, F., Verbauwhede, I.: Efficient software implementation of Ring-LWE encryption. Eprint 2014, 725 (2014). http://eprint.iacr.org/2014/725

13. Cooley, J.W., Tukey, J.W.: An algorithm for the machine calculation of complex Fourier series. Math. Comput. **19**, 297–301 (1965)

14. Crandall, R., Fagin, B.: Discrete weighted transforms and large-integer arithmetic. Math. Comput. **62**(205), 305–324 (1994)

15. Crandall, R., Pomerance, C.: Prime Numbers: A Computational Perspective. Springer, Heidelberg (2001)

16. Devroye, L.: Non-uniform Random Variate Generation. Springer, New York (1986). http://luc.devroye.org/rnbookindex.html

17. Ducas, L., Durmus, A., Lepoint, T., Lyubashevsky, V.: Lattice signatures and bimodal Gaussians. In: Canetti, R., Garay, J.A. (eds.) CRYPTO 2013, Part I. LNCS, vol. 8042, pp. 40–56. Springer, Heidelberg (2013)

18. Ducas, L., Durmus, A., Lepoint, T., Lyubashevsky, V.: Lattice signatures and bimodal Gaussians. Eprint 2013, 383 (2013). http://eprint.iacr.org/2013/383, full version of [18]

19. Ducas, L., Lyubashevsky, V., Prest, T.: Efficient identity-based encryption over NTRU lattices. In: Sarkar, P., Iwata, T. (eds.) ASIACRYPT 2014, Part II. LNCS, vol. 8874, pp. 22–41. Springer, Heidelberg (2014)

20. Düll, M., Haase, B., Hinterwälder, G., Hutter, M., Paar, C., Sánchez, A.H., Schwabe, P.: High-speed Curve25519 on 8-bit, 16-bit and 32-bit microcontrollers. Des. Codes Crypt. (to appear)

21. Dwarakanath, N.C., Galbraith, S.D.: Sampling from discrete Gaussians for lattice-based cryptography on a constrained device. Appl. Algebra Eng. Commun. Comput. **25**(3), 159–180 (2014)

22. Gentleman, W.M., Sande, G.: Fast Fourier transforms: for fun and profit. In: AFIPS Conference Proceedings, AFIPS 1966, vol. 29, pp. 563–578. AFIPS/ACM/Spartan Books, Washington D.C. (1966)

23. Göttert, N., Feller, T., Schneider, M., Buchmann, J., Huss, S.A.: On the design of hardware building blocks for modern lattice-based encryption schemes. In: Prouff and Schaumont [46], pp. 512–529

24. Güneysu, T., Lyubashevsky, V., Pöppelmann, T.: Practical lattice-based cryptography: a signature scheme for embedded systems. In: Prouff and Schaumont [46], pp. 530–547

25. Güneysu, T., Oder, T., Pöppelmann, T., Schwabe, P.: Software speed records for lattice-based signatures. In: Gaborit, P. (ed.) PQCrypto 2013. LNCS, vol. 7932, pp. 67–82. Springer, Heidelberg (2013)

26. Gura, N., Patel, A., Wander, A., Eberle, H., Shantz, S.C.: Comparing elliptic curve cryptography and RSA on 8-bit CPUs. In: Joye, M., Quisquater, J.-J. (eds.) CHES 2004. LNCS, vol. 3156, pp. 119–132. Springer, Heidelberg (2004)

27. Heyse, S., von Maurich, I., Güneysu, T.: Smaller keys for code-based cryptography: QC-MDPC McEliece implementations on embedded devices. In: Bertoni, G.,

Coron, J.-S. (eds.) CHES 2013. LNCS, vol. 8086, pp. 273–292. Springer, Heidelberg (2013)
28. Hirschhorn, P.S., Hoffstein, J., Howgrave-Graham, N., Whyte, W.: Choosing NTRUEncrypt parameters in light of combined lattice reduction and MITM approaches. In: Abdalla, M., Pointcheval, D., Fouque, P.-A., Vergnaud, D. (eds.) ACNS 2009. LNCS, vol. 5536, pp. 437–455. Springer, Heidelberg (2009)
29. Hoffstein, J., Pipher, J., Schanck, J.M., Silverman, J.H., Whyte, W.: Practical signatures from the partial Fourier recovery problem. In: Boureanu, I., Owesarski, P., Vaudenay, S. (eds.) ACNS 2014. LNCS, vol. 8479, pp. 476–493. Springer, Heidelberg (2014)
30. Hoffstein, J., Pipher, J., Silverman, J.H.: NTRU: a ring-based public key cryptosystem. In: Buhler, J.P. (ed.) ANTS 1998. LNCS, vol. 1423, pp. 267–288. Springer, Heidelberg (1998)
31. Hutter, M., Schwabe, P.: NaCl on 8-bit AVR microcontrollers. In: Youssef, A., Nitaj, A., Hassanien, A.E. (eds.) AFRICACRYPT 2013. LNCS, vol. 7918, pp. 156–172. Springer, Heidelberg (2013)
32. Lindner, R., Peikert, C.: Better key sizes (and attacks) for LWE-based encryption. In: Kiayias, A. (ed.) CT-RSA 2011. LNCS, vol. 6558, pp. 319–339. Springer, Heidelberg (2011)
33. Liu, M., Nguyen, P.Q.: Solving BDD by enumeration: an update. In: Dawson, E. (ed.) CT-RSA 2013. LNCS, vol. 7779, pp. 293–309. Springer, Heidelberg (2013)
34. Liu, Z., Großschädl, J., Kizhvatov, I.: Efficient and side-channel resistant RSA implementation for 8-bit AVR microcontrollers. In: SECIOT 2010. IEEE Computer Society Press (2010)
35. Liu, Z., Seo, H., Roy, S.S., Großschädl, J., Kim, H., Verbauwhede, I.: Efficient Ring-LWE encryption on 8-bit AVR processors. Eprint 2015, 410 (2015). http://eprint.iacr.org/2015/410, to appear in CHES 2015
36. Lyubashevsky, V., Peikert, C., Regev, O.: On ideal lattices and learning with errors over rings. In: Gilbert, H. (ed.) EUROCRYPT 2010. LNCS, vol. 6110, pp. 1–23. Springer, Heidelberg (2010)
37. Lyubashevsky, V., Peikert, C., Regev, O.: On ideal lattices andlearning with errors over rings (2010), presentation of [36] givenby Chris Peikert at Eurocrypt 2010. http://www.cc.gatech.edu/~cpeikert/pubs/slides-ideal-lwe.pdf
38. Melchor, C.A., Boyen, X., Deneuville, J., Gaborit, P.: Sealing the leak on classical NTRU signatures. In: Mosca [40], pp. 1–21
39. Monteverde, M.: NTRU software implementation for constrained devices. Master's thesis, Katholieke Universiteit Leuven (2008)
40. Mosca, M. (ed.): PQCrypto 2014. LNCS, vol. 8772. Springer, Heidelberg (2014)
41. Nussbaumer, H.J.: Fast Fourier Transform and Convolution Algorithms, Springer Series in Information Sciences, vol. 2. Springer, Heidelberg (1982)
42. Oder, T., Pöppelmann, T., Güneysu, T.: Beyond ECDSA and RSA: lattice-based digital signatures on constrained devices. In: DAC 2014, pp. 1–6. ACM (2014)
43. Pöppelmann, T., Ducas, L., Güneysu, T.: Enhanced lattice-based signatures on reconfigurable hardware. In: Batina and Robshaw [3], pp. 353–370
44. Pöppelmann, T., Güneysu, T.: Towards practical lattice-based public-key encryption on reconfigurable hardware. In: Lange, T., Lauter, K., Lisoněk, P. (eds.) SAC 2013. LNCS, vol. 8282, pp. 68–86. Springer, Heidelberg (2014)
45. Pöppelmann, T., Oder, T., Güneysu, T.: High-performance ideal lattice-based cryptography on ATXmega 8-bit microcontrollers. Eprint 2015, 382 (2015). http://eprint.iacr.org/2015/382

46. Prouff, E., Schaumont, P. (eds.): CHES 2012. LNCS, vol. 7428. Springer, Heidelberg (2012)
47. Rich, S., Gellman, B.: NSA seeks quantum computer that could crack most codes. The Washington Post (2013). http://wapo.st/19DycJT
48. Roy, S.S., Vercauteren, F., Mentens, N., Chen, D.D., Verbauwhede, I.: Compact ring-LWE cryptoprocessor. In: Batina and Robshaw [3], pp. 371–391
49. Schönhage, A., Strassen, V.: Schnelle multiplikation grosser zahlen. Computing 7(3), 281–292 (1971)
50. Shor, P.: Algorithms for quantum computation: discrete logarithms and factoring. In: Proceedings of the 35th Annual Symposium on Foundations of Computer Science, 1994, pp. 124–134. IEEE (1994)
51. Stehlé, D., Steinfeld, R.: Making NTRU as secure as worst-case problems over ideal lattices. In: Paterson, K.G. (ed.) EUROCRYPT 2011. LNCS, vol. 6632, pp. 27–47. Springer (2011)
52. Winkler, F.: Polynomial Algorithms in Computer Algebra. Texts and Monographs in Symbolic Computation, 1st edn. Springer, Heidelberg (1996)

An Efficient Software Implementation of the Hash-Based Signature Scheme MSS and Its Variants

Ana Karina D.S. de Oliveira[1]([✉]) and Julio López[2]

[1] Federal University of Mato Grosso Do Sul (FACOM-UFMS),
Campo Grande, MS, Brazil
anakarina@facom.ufms.br
[2] State University of Campinas (IC-UNICAMP),
Campinas, SP, Brazil
jlopez@ic.unicamp.br

Abstract. In this work, we describe an optimized software implementation of the Merkle digital signature scheme (MSS) and its variants GMSS, XMSS and XMSS^{MT} using the vector instruction set AVX2 on Intel's Haswell processor. Our implementation uses the multi-buffer approach for speeding up key generation, signing and verification on these schemes. We selected a set of parameters to maintain a balance among security level, key sizes and signature size. We aligned these parameters with the ones used in the hash-based signature schemes LDWM and XMSS. We report the performance results of our implementation on a modern Intel Core i7 3.4 GHz. In particular, a signing operation in the XMSS scheme can be computed in 2,001,479 cycles (1,694 signatures per second) at the 128-bit security level (against quantum attacks) using the SHA2-256 hash function, a tree of height 60 and 6 layers. Our results indicate that the post-quantum hash-based signature scheme XMSS^{MT} offers high security and performance for several parameters on modern processors.

Keywords: Digital signature · Scheme xmss · Merkle tree · Post-quantum cryptography

1 Introduction

A digital signature scheme is an important cryptographic tool from the work of public-key cryptography. It is an essential technology for providing authenticity, integrity and non-repudiation of data and is widely used. Therefore, the design and implementation of secure and practical digital signature schemes is crucial for applications that require data integrity assurance and data origin authentication.

J. López—The second author was partially supported by FAPESP Projeto Temático under grant 2013/25.977-7 and research productivity scholarship from CNPq Brazil.

© Springer International Publishing Switzerland 2015
K. Lauter and F. Rodríguez-Henríquez (Eds.): LatinCrypt 2015, LNCS 9230, pp. 366–383, 2015.
DOI: 10.1007/978-3-319-22174-8_20

Nowadays, the most commonly used digital signature schemes are ECDSA [11], RSA [3] and DSA [8]. These schemes have their security based on the difficulty of factoring large integers and computing discrete logarithms. In 1994, Shor [19] introduced a quantum algorithm for factoring integers and computing discrete logarithms in the relevant groups in polynomial time, using simple operations on a quantum computer. Thus, resistant digital signature schemes against quantum computing are an active research field.

The one-time signature scheme [6] is based on a hash function and the security relies on the collision resistance of that function [14]. These one-time schemes are inadequate for the most practical situations since each key pair is only used for a single signature. Merkle [2] proposed a solution to this problem, which transforms the validity of an arbitrary and fixed number of one-time verification keys for one single public key, which is the root of the hash tree. The tree leaves are the one-time key pairs. The Merkle signature scheme (MSS) uses a hash function, and its security is based on the collision resistance of the hash function used.

The extended Merkle signature scheme (XMSS), a hash-based digital signature system, was proposed in [24,25]. This scheme is based on the Merkle signature scheme and uses WOTS$^+$, a modified version of the Winternitz one-time signature scheme [26]. The security of XMSS does not rely on the collision resistance of the hash function used but on weaker properties. The variants GMSS [16] and XMSS^{MT} [18] increase the number of signatures and reduce the memory consumption. They also decrease the runtime of key generation and signing phases, but slightly increase the key sizes.

Our Contributions. We describe an efficient software implementation of MSS and its variants based on the vector instruction set AVX2 on an Intel's Haswell processor. We show how to speed up key generation, signing and verification by taking advantage of the multi-buffer SHA2 algorithm implementation. In our implementation, we compute four hashes in parallel with a implementation of SHA2-384 using 256-bit vector registers, and eight hashes in parallel with a 256-bit version of SHA2-256. To illustrate the performance of our implementation, we select a set of parameters for 128-bit security against quantum computers, similar to the parameters used in LDWM [13] and XMSS [25] signature schemes. We report the performance results of LDWM [13] and XMSS [25] on a modern Intel Core i7 3.4 GHz. We also show several trade-offs among runtime and key sizes.

This paper is organized as follows. Section 2 presents some related works. Section 3 describes the Winternitz one-time signature scheme (WOTS) and the variant WOTS$^+$.The description of the signature scheme MSS and its variants are presented in Sect. 4. Some considerations about the security of MSS and XMSS schemes are given in Sect. 5. Section 6 presents our software implementation and performance results. Section 7 presents the conclusions.

2   Related Works

One-time signature schemes (OTS) first appeared in the work of Lamport [6]. Winternitz [23] proposed an improved Lamport one-time signature scheme, reducing the size of public and private keys. Merkle [23] introduced a *many-time signature scheme* that constructs a binary tree where the leaves are hashes of OTS public keys, and the root is the public key. Moreover, Merkle was the first to present an algorithm for computing the authentication path (tree traversal), which is used to validate the public key. The Fractal Merkle Tree Representation and Traversal [12] algorithms, presented in 2003, generate a sequence of leaves along with their associated authentication. In 2004, Szydlo [15] proposed a technique for Merkle Tree Traversal, which requires only logarithmic space and time. Buchmann et al. [17] showed an efficient algorithm that computes the authentication path based on the Szydlo approach.

Some MSS variants were proposed: CMSS [7], GMSS [16], XMSS [24] and $XMSS^{MT}$ [18]. CMSS [7] builds two chained trees reducing the generation runtime of key pair and signature, and allowing the signature of 2^{40} documents. CMSS reduced the size of the private key using a pseudorandom generator (PRNG). In 2007, GMSS [16] allowed a significant large number of documents to be signed with one key pair and reduced the signature sizes and the signature generation costs. In 2011, XMSS was introduced, a provably forward-secure and practical signature scheme with minimal security requirements. This scheme uses a modified $WOTS^+$ version proposed in [26], which is existentially unforgeable under adaptative chosen message attacks when instantiated with a family of pseudorandom functions. Later, $XMSS^{MT}$ [18] was introduced which allows to sign a large fixed number of documents.

In 2014, McGrew and Curcio [13] proposed an Internet-Draft that describes a digital signature system based on cryptographic hash functions. This system, named LDWM, specifies a one-time signature scheme based on the work of Lamport [6], Diffie [1], Winternitz [23] and Merkle [23], and a general signature scheme, called Merkle tree signature (MTS). Moreover, it relates a modern security analysis of those algorithms. In 2015, Hülsing et al. [25] proposed an Internet-Draft that specifies XMSS: Extended hash-based signatures with the one-time signature scheme $WOTS^+$, a single-tree (XMSS) and a multi-tree ($XMSS^{MT}$).

SPHINCS [22] is a practical stateless hash-based signature scheme and was designed to provide 2^{128} security even against attackers equipped with quantum computers. This scheme introduces a new method to randomize tree-based stateless signatures and uses HORS with trees (HORST) [5] for signing messages.

3   Winternitz One-Time Signature Scheme (WOTS)

The Winternitz one-time signature scheme (WOTS) [23] is a modification of the Lamport one-time signature scheme (LOTS) [6]. WOTS uses a parameter w which is the number of bits to be signed simultaneously. This scheme produces smaller signatures than Lamport, but increases the number of one-way function evaluations from 1 to $2^w - 1$, in each element of the signing key. The larger w, the smaller the signature and longer the time for signing and verification. WOTS uses a one-way function $f : \{0,1\}^n \to \{0,1\}^n$ and a cryptographic hash function

$g : \{0,1\}^* \rightarrow \{0,1\}^m$, where n and m are positive integers. The chaining function f^e computes e iterations of f on input $x \in \{0,1\}^n$ where $e \in \mathbb{N}$. The chaining function is defined as:

$$f^e(x) = \begin{cases} x & \text{if } e = 0; \\ f(f^{e-1}(x)) & \text{if } e > 0. \end{cases}$$

WOTS Key Pair Generation. A parameter $w \in \mathbb{N}$ is given and $t = t_1 + t_2$ is computed, where $t_1 = \lceil m/w \rceil$, $t_2 = \lceil (\lfloor log_2 t_1 \rfloor + 1 + w)/w \rceil$. The private key sk_i, for $i = 0, ..., t - 1$, is chosen uniformly at random. Each sk_i is a string of n bits. The verification key pk is computed as:

$$pk = (pk_0, ..., pk_{t-1}) \in \{0,1\}^{(n,t)} \qquad where \qquad pk_i = f^{2^w-1}(sk_i).$$

WOTS Signature Generation. To generate the signature, first compute the message digest $d = g(M)$. If necessary, add zeros to the left of d, such that d be divisible by w. Then d is split into t_1 binary blocks of size w, resulting in $d = (m_0||...||m_{t_1-1})$, where $||$ denotes concatenation. The m_i blocks are represented as integers in $\{0, 1, ..., 2^w - 1\}$.

The checksum is computed as $c = \sum_{i=0}^{t_1-1}(2^w - m_i)$. If necessary, add zeros to the left of the c until the extended string c is divisible by w. Then the extended string c can be divided into t_2 blocks $c = (c_0||...||c_{t_2-1})$ of length w. Let $b = d||c$ be the concatenation of the extended string d with the extended string c. Thus $b = (b_0||b_1||...||b_{t-1}) = (m_0||...||m_{t_1-1}||c_0||...||c_{t_2-1})$. The signature is:

$$Sig = (sig_0, ..., sig_{t-1}) = (f^{b_0}(sk_0), ..., f^{b_{t-1}}(sk_{t-1})).$$

WOTS Verification. To verify the signature $Sig = (sig_0, ..., sig_{t-1})$ of the message M, we calculate $(b_0, ..., b_{t-1})$ in the same way as it was calculated during signature generation. Then, we compute: $Sig' = (sig'_0, ..., sig'_{t-1}) = (f^{2^w-1-b_0}(sig_0), ..., f^{2^w-1-b_{t-1}}(sig_{t-1}))$. If $sig'_i = pk_i$ for $i = 0, 1, ..., t - 1$, then the signature is valid.

3.1 Modification on Winternitz One-Time Signature Scheme (WOTS$^+$)

Hülsing [26] proposed WOTS$^+$, a modification of WOTS that uses a chaining function f^e starting from random inputs. This modification allows for shorter signatures without lowering the security. WOTS$^+$ uses a family of hash functions $F_n = \{f_K : \{0,1\}^n \rightarrow \{0,1\}^n | K \in \{0,1\}^n\}$.

Given $f_K \in F_n$, the chaining function f^e computes e iterations of f_K on inputs $x \in \{0,1\}^n$ and bitmask $bm = (bm_1, ..., bm_{w-1})$ with $e \in \mathbb{N}$. The chaining function is defined as:

$$f^e(x, bm) = \begin{cases} x & \text{if } e = 0; \\ f_K(f^{e-1}(x, bm) \oplus bm_e) & \text{if } e > 0. \end{cases}$$

WOTS$^+$ is parameterized by a security parameter n, the binary message length m, and the Winternitz parameter $w \in \mathbb{N}$, for $w > 1$. The parameters m and w are used to compute: $l_1 = \lceil m/(log_2(w)) \rceil$, and $l_2 = \lfloor (log_2(l_1(w-1)))/(log_2(w)) \rfloor + 1$, $l = l_1 + l_2$. These parameters are public known.

WOTS$^+$ Key Pair Generation. On input seed $Seed$ and the bitmask $bm \in \{0,1\}^{(n,w-1)}$, the key generation algorithm computes private keys $sk = (sk_0, \ldots, sk_{l-1})$. The public verification key is:

$$pk = (pk_0, \ldots, pk_{l-1}) = (f^{w-1}(sk_0, bm), \ldots, f^{w-1}(sk_{l-1}, bm)).$$

WOTS$^+$ Signature Generation. Given an m-bit message digest $d = g(M)$ and the bitmask bm, the signature algorithm firstly computes a base-w representation of $g(M) = d = (m_0, \ldots, m_{l_1-1})$, $m_i \in \{0, \ldots, w-1\}$. After, the checksum value $C = \sum_{i=0}^{l_1-1}(w-1-m_i)$ in base w representation and length l_2 is appended to d, resulting in $b = (b_0, \ldots, b_{l-1})$. The size of the signing key and private keys are l strings of n bits. The signature is:

$$Sig = (sig_0, \ldots, sig_{l-1}) = (f^{b_0}(sk_0, bm), \ldots, f^{b_{l-1}}(sk_{l-1}, bm)).$$

WOTS$^+$ Verification. To check the signature, the verifier constructs the values $b = (b_0, \ldots, b_{l-1})$ as the signature. If $(f^{w-1-b_0}(sig_0, bm), \ldots, f^{w-1-b_{l-1}}(sig_{l-1}, bm)) = (pk_0, \ldots, pk_{l-1})$, then the signature is valid.

WOTS$ [24] is a modified version of WOTS that allows to eliminate the need for a collision resistant hash function family. For this modified version, the iterated evaluations of the function $f_K^e(x) = f_K(\ldots(f_K(x))$ are a random walk through the function family F_n, as described below:

$$f_K^e(x) = \begin{cases} K & \text{if } e = 0; \\ f_{K'}(x) & \text{if } e > 0 \text{ and } K' = f_K^{e-1}(x), \end{cases}$$

where $K, x \in \{0,1\}^n$, $e \in \mathbb{N}$, and $f_K \in F_n$.

4   Merkle Signature Scheme (MSS)

MSS [23] is a digital signature scheme that consists of three algorithms: key generation, signing and verification. This scheme constructs a binary tree where the one-time signing and verification keys are the tree leaves, and the public key is the root. A tree with height h and 2^h leaves will have 2^h one-time key pairs.

MSS Key Pair Generation. In the Merkle key generation algorithm, the signer must select the tree height $h \in N$, $h \geq 2$. Figure 1 shows this process. The left picture represents the WOTS key generation algorithm described in Sect. 3. This algorithm is executed to generate the WOTS key pairs. Merkle uses the cryptographic hash function $g : \{0,1\}^* \rightarrow \{0,1\}^n$ where n is a positive integer. This key pair will be able to sign/verify documents, and the digest $g(pk_0||\ldots||pk_{t-1})$ will be a leaf of the Merkle tree.

Merkle Public Key Generation. Appendix A describes the treehash algorithm for generating the public key pub. The tree is built using a stack. Each
time a leaf s is calculated, it is stacked. The treehash algorithm checks if the
nodes at the top of the *stack* have the same heights. If they are of the same
height, the two nodes will be unstacked and the hash value of the concatenation
of both nodes will be pushed into the *stack*. The algorithm terminates when the
tree root is found.

On the right of Fig. 1, we observe the order in which the nodes are stacked
on the tree for generating the public key pub. The first node receives Y_0. The
third node is the hash of the concatenation of the first and second nodes. The
gray nodes represent $AUT = \{Aut_0, \ldots, Aut_h\}$, the first authentication path.
This AUT is formed by the sibling right nodes, connecting the leaf up to the
tree root, which are used to validate the public key. The authentication path
AUT is saved during the execution of the treehash algorithm.

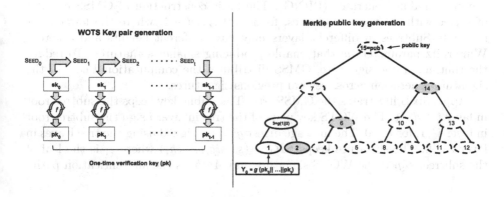

Fig. 1. Merkle key generation.

MSS Signature Generation. The signature generation consists of two steps:
first, the signature of the message digest $g(M)$ is generated using the WOTS
signature algorithm and the corresponding secret key sk_s of the leaf s. Then
the signature $SIG = (s, sig_s, AUT)$ contains the index of the leaf s, the WOTS
signature sig_s and the authentication path AUT. In the second step, the next
authentication path AUT is generated. For generating the next authentication
path, we use the BDS algorithm [17] which is a modification of the classic authentication path algorithm proposed by Merkle [2].

MSS Verification. The signature verification consists of two steps: first, the
signature sig_s is used to recover a leaf of the tree Y_s; second, the public key of
the Merkle tree is validated as the following. The receiver can reconstruct the
path $(p_0, \ldots, p_h)$ from leaf s to root. The index s is used to decide the order
in which the authentication path is reconstructed. Initially, $p_0 = Y_s$. For each

$i = 1, 2, \ldots, h$ is computed p_i using the condition (if $\lfloor s/(2^{i-1}) \rfloor \equiv 1 \bmod 2$) and the recursive formula:

$$p_i = \begin{cases} g(Aut_{i-1} \| p_{i-1}) \text{ if } \lfloor s/(2^{i-1}) \rfloor \equiv 1 \bmod 2; \\ g(p_{i-1} \| Aut_{i-1}) \text{ otherwise.} \end{cases}$$

Finally, if the value p_h is equal to the public key pub, the signature is valid.

4.1 Merkle Signatures with Virtually Unlimited Signature Capacity (GMSS)

The GMSS algorithm was published in 2007 [16]. The authors proposed a change in the Merkle signature scheme, allowing the signature of 2^{80} messages with one key pair. This modification is based on CMSS [7] that reduces the key pair and signature generation runtime and enable the signature of 2^{40} documents with the construction of two chained trees. Moreover, it reduces the private keys using a pseudorandom generator (PRNG). The basic construction of GMSS consists of a tree with d layers of subtrees, for $i = 0, \ldots, d - 1$, where the lower layer is $i = 0$. Subtrees in different layers may have different heights and different Winternitz parameters w that enable producing smaller signatures. To reduce the runtime of one signature, GMSS distributes the computational cost of the signature generation across several previous signatures.

Appendix B illustrates the GMSS tree. The public key keeps the subtree root in layer $i = d - 1$. The WOTS key pairs of the tree in layer i sign the subtree root in layer $i - 1$, for $i > 0$. The message digest $g(M)$ is signed using the tree leaves in layer $i = 0$. The signature keeps $SIG = (s_i, sig_i, Auth_i)$ where s_i is the leaf of the subtree, sig_i is the WOTS signature and $Auth_i$ is the authentication path.

4.2 Extended Merkle Signature Scheme (XMSS)

XMSS [24] is a modification of MSS. This scheme uses a slightly modified version of Winternitz WOTS$^+$ described in Sect. 3.1. To capture the formality of the standard definition of digital signature schemes, the authors of XMSS used the same definition of key evolving signature schemes [4]. XMSS is provably forward-secure and efficient when instantiated with two secure and efficient function families: one second-preimage resistant hash function family G_n and the other a pseudorandom function family F_n, where $G_n = \{g_K : \{0,1\}^{2n} \to \{0,1\}^n | K \in \{0,1\}^n\}$ and $F_n = \{f_K : \{0,1\}^n \to \{0,1\}^n | K \in \{0,1\}^n\}$.

The parameters of XMSS are: $n \in \mathbb{N}$, the security parameter; $w \in \mathbb{N}(w > 1)$, the Winternitz parameter; $m \in \mathbb{N}$, the message digest length; and $h \in \mathbb{N}$, the height tree.

An XMSS binary tree is constructed to generate the public key pub. Appendix C shows the XMSS treehash algorithm and Appendix D shows the XMSS tree construction. The XMSS tree is a modification of the Merkle tree. A tree of height h has $h + 1$ levels. The nodes on level j are $node_{i,j}$, for $0 < j \leq h$ and $0 \leq i < 2^{h-j}$. XMSS uses the hash function g_K and bitmask (bitmask-Tree) $bm \in \{0,1\}^{2n}$, chosen uniformly at random, where bm_{2i+2j} is the left bitmask and $bm_{2i+2j+1}$ is the right bitmask. The bitmasks are the main difference among the others Merkle tree constructions, since they allow to replace the collision resistant hash function family by a second-preimage resistant hash function family. The nodes are computed as: $node_{i,j} = g_K((node_{2i,j-1} \oplus bm_{2i+2j}) \| (node_{2i+1,j-1} \oplus bm_{2i+2j+1}))$.

To generate a leaf Y_s in the XMSS tree, an L-tree is used. The L-tree uses bitmasks (bitmaskLtree) in the same form as in the XMSS tree. The WOTS$^+$ public verification keys $(pk_0, \ldots, pk_{l-1})$ are the first l leaves of an L-tree. If l is not a power of 2, then there are not sufficiently leaves to build a binary tree. Therefore, a node that has a not right sibling is lifted to a higher level of the L-tree until it becomes the right sibling of another node.

4.3 Multi Tree XMSS (XMSS^{MT})

The XMSS^{MT} algorithm proposed in [18] allows a virtually unlimited number of signatures while it preserves its desirable properties. XMSS^{MT} is existentially unforgeable under a chosen message attack (EU-CMA) secure in the standard model when instantiated with a pseudorandom function family and a second preimage resistant hash function family. In addition, XMSS^{MT} is forward-secure.

Many layers of trees are used in XMSS^{MT} as in GMSS [16]. The number of layers is $d \in N$ where $0 \le i < d - 1$. The leaves of the tree in layer $i = 0$ are used to sign the messages. The leaves of the trees in layers i, $i > 0$, are used to sign the roots of the trees in layer $i - 1$. The public key is the tree root in layer $i = d - 1$. XMSS^{MT} uses the WOTS$^+$ algorithm as in XMSS.

5 Security Considerations

In this section, we describe a summary of several formal security proofs of MSS and its variants. McGrew and Curcio [13] specifies that LDWM and MSS rely on the security of the hash function used. The security of MSS, according to [14], relies on the fact that the hash function is collision resistant. Furthermore, MSS is forward-secure if the PRNG used is a cryptographically secure pseudorandom generator.

Buchmann et al. [14] show that the existential unforgeability of MSS under an adaptive chosen message attack can be reduced to the collision resistance of the family G and the existential unforgeability of the underlying one-time signature scheme, where $G = \{g_K : \{0,1\}^* \to \{0,1\}^n | k \in K\}$ is a family of hash functions. The security level of the signature schemes is the base 2 logarithm of the average attack complexity. The authors proved an attack of complexity 2^n for classical computers and $2^{n/2}$ for quantum computers on preimage and second-preimage resistances. For collision resistance is different, the bounds are $2^{n/2}$ for classical computers and $2^{n/3}$ for quantum computers [14].

XMSS with WOTS\$ [24] is forward-secure and existentially unforgeable under chosen message attacks if the function g_K is a second preimage resistant hash function and f_K is a pseudorandom function. If a randomized message hashing is used, then the requirement is that the hash function must be second preimage resistant. If a plain hash is used, then the hash function must be collision resistant. XMSS with WOTS$^+$ [20] needs a function f_K to be second preimage resistant. In addition, a PRNG is required to generate the keys.

In summary, the security requirements for XMSS are minimal, because the existence of the two function families described above is known to be necessary for the existence of digital signature schemes [24]. As demonstrated in [20], if an n-bit hash is used, then n-bit security against classical computers and $n/2$-bit security against quantum computers is obtained.

6 Implementation

In this section, we describe the details of our efficient implementations of the algorithms: MSS, GMSS, XMSS and XMSS^{MT}. For key generation, we used the standard CTR_DRBG [21] as cryptographically secure pseudorandom number generator (PRNG) to generate the secret keys sk. According to [7], PRNG enables the recovery of all signature keys based only on the initial seed and avoids the storage of the full secret key. To implement the one-way function f_K and the cryptographic hash function g_K, we used the SHA2 hash function. To compute the authentication path in the signature generation, we used the BDS algorithm [17]. We will provide the software described in this paper into the public domain.

6.1 Multi-buffer Hash Optimization

A leaf of the XMSS tree involves the following computations: first, the l secret key sk_i are computed using a PRNG. This secret key sk_i is generated in sequence. Then, for each private key, we can compute the corresponding public key pk_i by applying $l(w - 1)$ times the function f_K. Unlike of secret key generation, the WOTS public keys can be generated in parallel. Finally, for each internal node of the L-tree, the function g_K is applied to a binary string computed by the children nodes, and such computation is independent among siblings nodes.

Therefore, the computation of the internal nodes of the XMSS tree is much faster than the leaves. Recovering a leaf from the Merkle tree impacts the performance of the whole algorithm. This is a bottleneck of MSS and XMSS schemes. The computation of hash functions occur over independent messages, and in this scenario the application of vector instructions have an important role in the acceleration of the scheme.

The vector instructions present on the Haswell processor are suitable for the computation of independent hash functions in parallel, where a SIMD processing takes place. The AVX2 instructions process bit operations over 256-bit vector registers, thus each vector register can be seen as either four 64-bit words or eight 32-bit words. In this setting, the operations used to compute the SHA2 function can be applied on vectors with the purpose of computing simultaneous hash values at the same time, which we will further refer as a multi-buffer implementation.

The SHA2-256 algorithm uses operations over words of 32 bits while the SHA2-384 algorithm operates over words of 64 bits. Then, using a multi-buffer approach we can compute either four or eight hash values for the SHA2-384 and SHA2-256, respectively. This parallel hashing technique can be applied to accelerate the operations on WOTS key generation and the generation of the nodes of the L-tree. Thus, to generate the l private key pk_i, we can reduce the number of executions of the function f_K to $\lceil (l(w-1)/8) \rceil$ times with SHA2-256 using the vector instruction set of 256 bits. The number of SHA2-256 evaluations required to computing the L-tree root can also be reduced.

6.2   Optimization of the Function f^e

In this subsection, we highlight the development of an optimized implementation of the function f^e using SHA2 [9]. To reduce the runtime of the function f^e, the core operation of the schemes, we implemented the function f^e by using the multi-buffer technique [27], which allows parallelizing message schedules.

As presented in MSS and XMSS, the generation of the WOTS verification keys involves many applications of the function f^e in each element of the private key. The chaining function used in WOTS is $f^e(sk)$ and in WOTS$^+$ is $f^e(sk, bm)$. Figures 2 and 3 show our optimized implementation of the function f^e using the SHA2 hash function with a fixed pad. The pad values are fixed because the block sizes of sk are the same for SHA2. Since these values have a fixed pad, we reduced the number of operations of the function f^e for getting a better performance. In WOTS, f^e is executed $t(2^w - 1)$ times and in WOTS$^+$ is executed $l(w-1)$ times for the generation of the one-time verification keys.

The multi-buffer C code SHA2 from [10] was modified to use 256-bit AVX2 instructions. We adapted this new code to reduce the runtime of the chaining function f^e. Figure 4 shows our implementation of the function f^e for MSS with SHA2-384 which computes four instances in parallel, generating four public keys pk at the same time. We load four secret keys sk_i into the 256-bit vector registers, perform e iterations of the f^e function and then store four private keys pk_i on

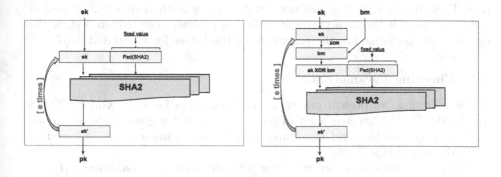

Fig. 2. Function $f^e(sk)$ in WOTS.   **Fig. 3.** Function $f^e(sk, bm)$ in WOTS$^+$.

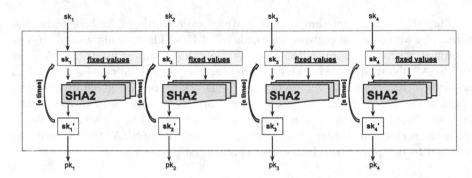

Fig. 4. Implementation of the chaining function $f^e(sk)$ in MSS with SHA2-384.

memory. We can compute the private keys pk_i in parallel because its generation is independent.

The parallel technique is also used in the signature generation to update the authentication path because some tree leaves need to be recovered. The use of SIMD instructions helped to reduce the runtime of the function f^e, which are the computationally most expensive parts for key and signature generation.

6.3   Optimization of the L-Tree Implementation

The L-tree from XMSS [24] is used to generate the leaves of XMSS. In this section, we show an optimization in the generation of the L-tree for improving the performance of generating each leaf of the XMSS tree. This optimization is similar to the one given in [22]. The function g_K is applied to each concatenation of children nodes to generate the parent node. Then, we modified the L-tree algorithm to perform eight evaluations of the function g_K at the same time. We generate eight internal nodes at the same time, from 16 children nodes, which are concatenated 2 by 2. If the amount of remaining nodes is not multiple of 16, we generate the next internal nodes one by one like the traditional way. If l is not a power of 2, then there are not sufficient leaves to build a binary tree. Therefore, a node that has not a right sibling is lifted to a higher level of the L-tree until it becomes the right sibling of another node. Figure 5 show the optimization performed in the generation of the L-tree for w=16 and l=67.

6.4   Choosing Parameters

In the following, we present our choice of parameters for MSS, GMSS, XMSS and XMSS^{MT}. We selected a set of parameters to get a good trade-off among level security, runtime and signature size. These parameters were aligned to LDWM [13] and XMSS [25].

The chosen parameters are: $w \in \mathbb{N}(w > 1)$, the Winternitz parameter; $H \in \mathbb{N}$, the total height of the tree; d, the number of layers; m, the message length in bits, and n the output size of the functions f_K and g_K. To implement f_K and

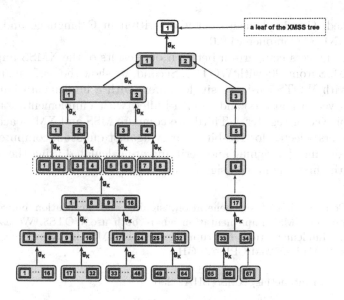

Fig. 5. Construction of the L-tree.

g_K, we use the hash function SHA2-384 in MSS and SHA2-256 in XMSS for 128-bit security against quantum computers.

In MSS with WOTS, each WOTS signature size is tn bits and in XMSS with WOTS$^+$ is ln bits. These parameters are determined according to the chosen parameter w. The value of w influences the runtime and the signature size. Higher w values imply on greater runtime, but smaller signature sizes. The values H and d directly affect the signature size. We set the maximum height $H = 60$ and $d = 6$ to limit the signature size and $w = 64$ to have an acceptable runtime.

The MSS secret key $SK = (s_0, Seed_0, \ldots, s_{d-1}, Seed_{d-1})$ contains one index s per layer and one $Seed$ per layer. The public key is PK=(pub), which is the root of the layer $d - 1$ and the signature $SIG = (s_0, sig_0, Auth_0, \ldots, s_{d-1}, sig_{d-1}, Auth_{d-1})$ contains the index s, one WOTS signature and one authentication path per layer.

In XMSS, the secret key $SK = (Bitmask, s_0, Seed_0, \ldots, s_{d-1}, Seed_{d-1})$ contains the same elements as MSS, also $Bitmask = (bitmaskTree, bitmaskL\ tree, bm)$. The public key is $PK = (pub, Bitmasks)$ and the signature is $SIG = (s, sig_0, Auth_0, \ldots, sig_{d-1}, Auth_{d-1})$ as in MSS.

6.5 Performance Results

This section shows the experimental results for MSS and its variants of our implementation using AVX2 instructions. The results of our implementation were obtained by running benchmarks on a Haswell processor Core i7-4770 at 3.4 GHz, where the Intel Turbo Boost and the Intel Hyper-Threading technologies were disabled. Our implementation was written in C language and compiled using the GNU C Compiler v4.9.0.

First, we make a comparison between our results of the XMSS implementation and XMSS from [20] with WOTS\$. Second, we show the performance results of XMSS with WOTS$^+$ using a single-buffer (with a implementation of SHA2 using 64-bit vector registers) and a multi-buffer (with a implementation of SHA2 using 256-bit vector registers). Third, we compare MSS and XMSS according to the parameters selected for 128-bit security against quantum computers. In our work, the runtimes for signing and verifying are calculated using the arithmetic average of the first one million signatures.

Table 1. Comparasion of key generation, signing and verification between XMSS from [20] and our XMSS implementation where both use WOTS\$. We use a single-buffer (64-bit) implementation. The running times of XMSS [20] were obtained on the Intel(R) Core(TM) i5-2520M CPU 2.5 GHz.

runtime(ms) XMSS SHA2-256							
		[20]			our		
H	w	KeyGen	Sig	Ver	KeyGen	Sig	Ver
20	4	868,647	7.62	0.44	382,903	3.26	0.20
20	16	1,534,748	13.71	0.76	762,132	6.56	0.36
20	64	4,012,157	35.60	1.97	2,038,605	17.51	1.00

We scale the running times of XMSS [20] from 2.5 to 3.4 GHz in order to estimate the running times for $H = 20$ and $w = 4$, so we have 638,711 ms for key generation, 5.60 ms for signing and 0.32 ms for verification. We compared these estimates with our results in Table 1. We have improved the performance achieving an speedup factor of 1.66x, 1.71x and 1.61x for key generation, signing and verification, respectively.

On 256-bit messages, the single-buffer implementation of SHA2 takes 121.25 cycles per byte for eight hashes, while the multi-buffer implementation takes 29.31 cycles per byte for eight hashes, so we obtain an acceleration of 4.13x for hashing. However, this speed factor is reduced for the signature operations of XMSS. In Table 2, we show that the acceleration of the performance due to the multi-buffer optimization, ranges from 2.4x for smaller w to 3.7x for larger w for key generation and signing operations. For verification, the muti-buffer implementation slightly improves the single-buffer implementation, since the optimized implementation of the function f^e is not used in that operation.

Comparing the results of the single-buffer implementations of XMSS using WOTS\$ shown in Table 1 with XMSS using WOTS$^+$ shown in Table 2, we observe that XMSS using WOTS$^+$ is around 44–49 % faster than XMSS using WOTS\$.

Since the requirements for the security of XMSS are minimal [24], one could use a smaller hash function output. Then, for the same security level, the runtime

Table 2. Performance figures of XMSS with WOTS$^+$ for parameters $H = 20$, $K = 2$. The chaining function is implemented using SHA2-256. We compare our results using single-buffer and using a multi-buffer for different values of w.

runtime(ms) XMSS SHA2-256							
		Single buffer			Multi buffer		
H	w	KeyGen	Sig	Ver	KeyGen	Sig	Ver
20	4	191,982	1.81	0.11	80,323	0.71	0.07
20	8	214,820	2.04	0.12	81,130	0.82	0.10
20	16	282,582	2.71	0.14	89,121	0.94	0.12
20	32	438,249	4.24	0.19	123,588	1.35	0.18
20	64	693,187	5.95	0.35	187,192	2.04	0.34

Table 3. Results of our implementation of LDWM [13] with WOTS and XMSS [25] with WOTS$^+$ for 128-bit security against quantum computer. We use a multi-buffer (256-bit) implementation. XMSS^{MT} and GMSS with $d = 1$ correspond to XMSS and MSS, respectively.

gmss_sha2-384_m48_n48							
			runtime(ms)			sizes(bytes)	
H	d	w	KeyGen	Sig	Ver	SK	PK	SIG
20	1	2	183,623	1.64	0.13	100	32	6,980
20	2	2	358	1.06	0.26	300	32	13,320
20	1	4	240,368	2.25	0.26	100	32	3,844
20	2	4	471	1.5	0.52	300	32	7,048
40	4	4	940	1.5	1.02	700	32	14,096
60	6	4	1,411	1.5	1.54	1,100	32	21,144

xmssmt_sha2-256_m32_n32							
			runtime(ms)			sizes(bytes)	
H	d	w	KeyGen	Sig	Ver	SK	PK	SIG
20	1	4	80,323	0.71	0.07	1,956	1,920	4,932
20	2	4	156	0.47	0.14	1,036	1,280	9,228
20	1	16	89,121	0.94	0.12	2,276	2,240	2,820
20	2	16	174	0.59	0.25	1,772	1,600	5,004
40	4	16	347	0.58	0.52	2,044	1,600	10,012
60	6	16	523	0.59	0.75	2,316	1,600	15,020

of key generation, signature generation and verification of XMSS is faster and uses smaller signature size than MSS. However, the private and public keys of XMSS are larger than MSS, because XMSS stores the bitmasks, as it can be seen on Table 3.

Notice that for larger values of w, the size of the public key and signature are smaller, without a significant impact the runtime. On the other hand, by increasing the number of layers the runtime is reduced for key generation and signing; however, the runtime for verification increases because the signatures of all layers must be checked. Additionally, the size of the secret key and signature is larger, because they store information of each layer. Increasing the tree height allows to produce more signatures without impacting the performance of signing and verifying, but increasing the signature size.

7   Conclusion

In this paper, we addressed the efficient software implementation of hash-based signature schemes on an Intels Haswell processor. Our implementation benefits from the vector instruction set AVX2 and a multi-buffer implementation of the SHA2 hash function. In order to implement the GMSS and XMSS^{MT} schemes, we selected a set of parameters to maintain a balance among security level, key sizes, signature size, and running times. We achieved the same security level among the schemes implemented by using a different output hash size. For example, GMSS with SHA2-384 and XMSS^{MT} with SHA2-256 provide 128-bit security against quantum computers. Our performance results show that XMSS^{MT} [25] was about 40-50 % faster than GMSS [13] for key generation, signing and verification. In addition, the signature size of XMSS^{MT} is smaller than GMSS, but the public and private key of GMSS are shorter than XMSS^{MT}.

Finally, with the advent of a hardware implementation of SHA2 on modern processors (Intel Skylake processor [28]), the performance of key generation, signing and verification of hash-based schemes will be even improved.

Acknowledgments. The authors would like to thank the anonymous referees for their valuable comments and suggestions to improve the quality of this paper.

A   Merkle public key generation algorithm (Treehash) [23]

Require: Leaf $leafIni$; tree height h; seed $Seed_{in}$.
Ensure: The tree root pub.
1: Create a *stack*.
2: $Seed_0 = Seed_{in}$
3: **for** $(s = leafIni, s < 2^h, s{+}{+})$ **do**
4:     $(Seed_x, Seed_{s+1}) = PRNG(Seed_s)$;
5:     $node_s.digest = Leafcalc(Seed_x)$
6:     **push** $node_s$ in the stack
7:     **while** The nodes at the top of the stack have the same height **do**
8:         **pop** $node_{right}$
9:         **pop** $node_{left}$
10:        Compute $node_{parent}.digest = g(node_{left}.digest \| node_{right}.digest)$
11:        **if** $node_{parent}.height = h$ **then**
12:            **return** $(node_{parent})$
13:        **else**
14:            **push** $node_{parent}$ in the *stack*
15:        **end if**
16:    **end while**
17: **end for**

B   GMSS Tree Construction

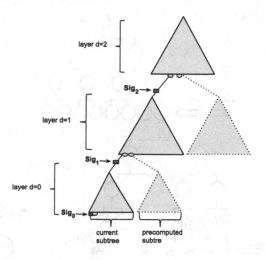

C   The XMSS treehash algorithm [24]

```
 1: procedure treehashXMSS(leafIni, h, Y)
 2:     for (i = leafIni, i < 2^h, i++) do
 3:         j=0;
 4:         node_{i,j} = Y_s;
 5:         push node_{i,j} in the stack;
 6:         while (The nodes at the top of the stack have the same height j) do
 7:             j=j+1;
 8:             pop node_{2i,j-1};
 9:             pop node_{2i+1,j-1};
10:             node_{i,j}=g_K((node_{2i,j-1} ⊕ bm_{2i+2j})||(node_{2i+1,j-1} ⊕ bm_{2i+2j+1}));
11:             if (j=h) then
12:                 return node_{i,j};
13:             else
14:                 push node_{i,j} in the stack;
15:             end if
16:         end while
17:     end for
18: end procedure
```

D   The XMSS Tree Construction

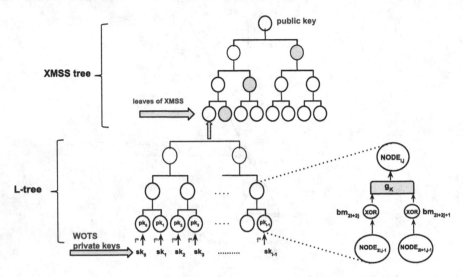

References

1. Diffie, W., Hellman, M.E.: New directions in cryptography. IEEE Trans. Inf. theory **22**, 44–654 (1976)
2. Bremen, L., Kluge, J., Ziefle, M., Modabber, A., Goloborodko, E., Hölzle, F.: "Two faces and a hand scan"- pre- and postoperative insights of patients undergoing an orthognathic surgery. In: Stephanidis, C. (ed.) HCI 2014, Part II. CCIS, vol. 435, pp. 389–394. Springer, Heidelberg (2014)
3. Rivest, R.L., Shamir, A., Adleman, L.: A method for obtaining digital signatures and public-key cryptosystems. Commun. ACM **21**(2), 120–126 (1978)
4. Bellare, M., Miner, S.K.: A forward-secure digital signature scheme. In: Wiener, M. (ed.) CRYPTO 1999. LNCS, vol. 1666, pp. 431–448. Springer, Heidelberg (1999)
5. Reyzin, L., Reyzin, N.: Better than BiBa: short one-time signatures with fast signing and verifying. In: Batten, L.M., Seberry, J. (eds.) ACISP 2002. LNCS, vol. 2384, pp. 144–153. Springer, Heidelberg (2002)
6. Lamport, L.: Constructing Digital Signatures from a One Way Function Technical report SRI-CSL-98, SRI International Computer Science Laboratory (1979)
7. Buchmann, J., García, L.C.C., Dahmen, E., Döring, M., Klintsevich, E.: CMSS – an improved merkle signature scheme. In: Barua, R., Lange, T. (eds.) INDOCRYPT 2006. LNCS, vol. 4329, pp. 349–363. Springer, Heidelberg (2006)
8. NIST.: Digital Signatures Algorithm (DSA). FIPS-186 (1994). http://www.itl.nist.gov/fipspubs/fip186.htm
9. eBACS: ECRYPT Benchmarking of Cryptographic Systems SUPERCOP 20140924 (2014). http://hyperelliptic.org/ebats/supercop-20140924.tar.bz2
10. Gosney, J.: The sse2/xop implementation of sha256 (2013). http://www.openwall.com/lists/john-dev/2013/04/10/6
11. Johnson, D., Menezes, A., Vanstone, S.: Elliptic curve digital signature algorithm ECDSA. Int. J. Inf. Secur. **1**, 36–63 (2001)

12. Jakobsson, M., Leighton, T., Micali, S., Szydlo, M.: Fractal merkle tree representation and traversal. In: Joye, M. (ed.) CT-RSA 2003. LNCS, vol. 2612, pp. 314–326. Springer, Heidelberg (2003)
13. McGrew, D., Curcio, M.: Hash-Based Signatures draft-mcgrew-hash-sigs-02. Crypto Forum Research Group, Internet Draft, Cisco Systems (2014)
14. Buchmann, J., Dahmen, E., Szydlo, M.: Hash-based digital signature schemes. In: Bernstein, D.J., Buchmann, J., Dahmen, E. (eds.) Post-Quantum Cryptography, pp. 35–92. Springer, Heidelberg (2008)
15. Szydlo, M.: Merkle tree traversal in log space and time. In: Cachin, C., Camenisch, J.L. (eds.) EUROCRYPT 2004. LNCS, vol. 3027, pp. 541–554. Springer, Heidelberg (2004)
16. Buchmann, J., Dahmen, E., Klintsevich, E., Okeya, K., Vuillaume, C.: Merkle signatures with virtually unlimited signature capacity. In: Katz, J., Yung, M. (eds.) ACNS 2007. LNCS, vol. 4521, pp. 31–45. Springer, Heidelberg (2007)
17. Buchmann, J., Dahmen, E., Schneider, M.: Merkle tree traversal revisited. In: Buchmann, J., Ding, J. (eds.) PQCrypto 2008. LNCS, vol. 5299, pp. 63–78. Springer, Heidelberg (2008)
18. Hülsing, A., Rausch, L., Buchmann, J.: Optimal parameters for XMSS^{MT}. In: Cuzzocrea, A., Kittl, C., Simos, D.E., Weippl, E., Xu, L. (eds.) CD-ARES Workshops 2013. LNCS, vol. 8128, pp. 194–208. Springer, Heidelberg (2013)
19. Shor, P.: Polynomial-time algorithms for prime factorization and discrete logarithms on a quantum computer, pp. 124–134. IEEE Computer Society Press (1994)
20. Practical Forward Secure Signature using Minimal Security Assumptions. Ph.D. thesis. TU Darmstadt, Darmstadt, August 2013
21. NIST.: Recommendation for Random Number Generation Using Deterministic Random Bit Generators. Computer Security Division - Information Technology Laboratory - NIST Special Publication 800–90A (2012). http://csrc.nist.gov/publications/nistpubs/800-90A/SP800-90A.pdf
22. Bernstein, D., Hopwood, D., Hülsing, A., Lange, T., Niederhagen, R., Papachristodoulou, L., Schwabe, P., O'Hearn, Z.: SPHINCS: practical stateless hash-based signatures. Cryptology ePrint Archive - Report 2014/795 (2014)
23. Merkle, R.C.: Secrecy, Authentication, and Public Key Systems. Stanford Ph.D. thesis (1979)
24. Buchmann, J., Dahmen, E., Hülsing, A.: XMSS - a practical forward secure signature scheme based on minimal security assumptions. In: Yang, B.-Y. (ed.) PQCrypto 2011. LNCS, vol. 7071, pp. 117–129. Springer, Heidelberg (2011). https://huelsing.wordpress.com/publications/
25. Hülsing, A., Butin, D., Gazdag, S.: XMSS: Extended Hash-Based Signatures draft-xmss-00. Crypto Forum Research Group, Internet Draft (2015)
26. Hülsing, A.: W-OTS+ – shorter signatures for hash-based signature schemes. In: Youssef, A., Nitaj, A., Hassanien, A.E. (eds.) AFRICACRYPT 2013. LNCS, vol. 7918, pp. 173–188. Springer, Heidelberg (2013)
27. Guilfor, J., Yap, K., Gopal, V.: Fast SHA-256 Implementations on Intel Architecture Processors. IA Architects Intel Corporation (2012). http://www.intel.com.br/content/dam/www/public/us/en/documents/white-papers/sha-256-implementations-paper.pdf
28. Intel to release first Skylake microprocessors in Q2 2015 (2014). http://www.kitguru.net/components/cpu/anton-shilov/intel-to-release-first-skylake-micropro cessors-in-q2-2015-says-report

Author Index

Printed in the United States
By Bookmasters